Frontiers in Geometry and Topology

Proceedings of Symposia in
PURE MATHEMATICS

Volume 109

Frontiers in Geometry and Topology

Summer School and Research Conference
Frontiers in Geometry and Topology
August 1–12, 2022
The Abdus Salam International Centre for Theoretical Physics
Trieste, Italy

Paul M. N. Feehan
Lenhard L. Ng
Peter S. Ozsváth
Editors

AMS
AMERICAN MATHEMATICAL SOCIETY
Providence, Rhode Island

2020 *Mathematics Subject Classification.* Primary 53D45, 57K10, 57K18, 57K33, 57K41, 57R58.

Library of Congress Cataloging-in-Publication Data

Names: Frontiers in Geometry and Topology (conference) (2022 : Trieste, Italy) | Feehan, Paul M. N., 1961– editor.

Title: Frontiers in geometry and topology : summer school and research conference Frontiers in Geometry and Topology, August 1–12, 2022, The Abdus Salam International Centre for Theoretical Physics, Trieste, Italy / Paul M.N. Feehan, Lenhard L. Ng, Peter S. Ozsváth, editors.

Description: Providence, Rhode Island : American Mathematical Society, [2024] | Series: Proceedings of symposia in pure mathematics, 0082-0717 ; volume 109 | Includes bibliographical references.

Identifiers: LCCN 2024004662 | ISBN 9781470470876 (paperback) | ISBN 9781470477585 (ebook)

Subjects: LCSH: Topology–Congresses. | AMS: Differential geometry – Symplectic geometry, contact geometry – Gromov-Witten invariants, quantum cohomology, Frobenius manifolds. | Manifolds and cell complexes – Differential topology – Floer homology.

Classification: LCC QA611.A1 F76 2024 | DDC 514–dc23/eng/20240223

LC record available at https://lccn.loc.gov/2024004662

∞ The paper used in this book is acid-free and falls within the guidelines
established to ensure permanence and durability.
Visit the AMS home page at `https://www.ams.org/`

10 9 8 7 6 5 4 3 2 1 29 28 27 26 25 24

A Conference in Honor of Tomasz Mrowka
International Centre for Theoretical Physics, Trieste, Italy, August 1–12, 2022

Contents

Preface

The summer school and research conference "Frontiers in Geometry and Topology" was held at the Abdus Salam International Centre for Theoretical Physics (ICTP), August 1–12, 2022. The conference featured twenty-three speakers who spoke on a broad range of subjects centered around research interests of Tom Mrowka, in recognition of his sixtieth birthday. These subjects included knot theory, the topology of three and four-dimensional manifolds, instanton, monopole, and Heegaard Floer homologies, and Khovanov homology. The speakers ranged from early-career mathematicians to senior leaders. The summer school featured ten speakers, with talks on subjects again reflecting Tom's interests. The summer school and the research conference each attracted capacity audiences of 72 participants.

We invited all the speakers to submit contributions to this proceedings volume. We received fifteen excellent articles, covering a broad range of subjects within geometric analysis, differential geometry, and low-dimensional topology. They provide a worthy tribute to Tom on the occasion of his sixtieth birthday.

We are delighted to have this last opportunity to express our gratitude to our co-organizers for their invaluable help: Claudio Arezzo, Yongbin Ruan, and especially Paul Seidel, who tirelessly and cheerfully handled all manner of logistical obstacles and almost single-handedly organized the summer school and conference. We also acknowledge the generous financial support of the National Science Foundation (grant number DMS-2154823), the Clay Mathematical Institute, and ICTP. We express our deep gratitude to the ICTP staff members, especially Monica Petronela Ancuta, Viktoriya Lvova, and Adriana Pinto, who provided superb assistance throughout the summer school and research conference. We thank the speakers for their wonderful talks and thank the participants for helping to make the summer school and research conference a resounding success.

Paul Feehan, Lenny Ng, and Peter Ozsváth

January 2024

Summer School Program

Monday, August 1, 2022

JEN HOM (Georgia Tech, USA)
Heegaard Floer homology and topology in dimensions 3.5 - 1

FRANCESCO LIN (Columbia University, USA)
Families of Dirac operators and applications - 1

JOHN PARDON (Princeton University, USA)
Geometry and algebra of pseudoholomorphic curves - 1

Tuesday, August 2, 2022

JOSHUA GREENE (Boston College, USA)
Lattices in low-dimensional topology - 1

FRANCESCO LIN (Columbia University, USA)
Families of Dirac operators and applications - 2

JEN HOM (Georgia Tech, USA)
Heegaard Floer homology and topology in dimensions 3.5 - 2

Wednesday, August 3, 2022

THOMAS WALPUSKI (Humboldt University, Berlin, Germany)
Generalized Seiberg–Witten equations - 1

JOSHUA GREENE (Boston College, USA)
Lattices in low-dimensional topology - 2

Thursday, August 4, 2022

JOSHUA GREENE (Boston College, USA)
Lattices in low-dimensional topology - 3

GARY GUTH (University of Oregon, USA)
Satellites and stabilizations

JOHN PARDON (Princeton University, USA)
Geometry and algebra of pseudoholomorphic curves - 2

THOMAS WALPUSKI (Humboldt University, Berlin, Germany)
Generalized Seiberg–Witten equations - 2

JEN HOM (Georgia Tech, USA)
Heegaard Floer homology and topology in dimensions 3.5 - 3

Friday, August 5, 2022

THOMAS WALPUSKI (Humboldt University, Berlin, Germany)
Generalized Seiberg–Witten equations - 3

ALFRED HOLMES (University of Oxford, UK)
Witten's proposal for a gauge theoretic description of the Jones polynomial and Khovanov homology

JOHN PARDON (Princeton University, USA)
Geometry and algebra of pseudoholomorphic curves - 3

JUAN MUÑOZ-ECHÁNIZ (Columbia University, USA)
Holomorphic Floer theory

LAURA WAKELIN (Imperial College London, UK)
The Dehn surgery characterisation of Whitehead doubles

DEEPARAJ BHAT (Massachusetts Institute of Technology, USA)
A surgery exact triangle in instanton Floer theory

FRANCESCO LIN (Columbia University, USA)
Families of Dirac operators and applications - 3

Research Conference Program

Monday, August 8, 2022

JEN HOM (Georgia Tech, USA)
Unknotting number and cabling

JOSHUA WANG (Harvard University, USA)
Colored $\mathfrak{sl}(N)$ homology and $SU(N)$ representations of the trefoil

JOHN PARDON (Princeton University, USA)
A representability theorem for moduli spaces of pseudo-holomorphic curves

GORDANA MATIC (University of Georgia, USA)
Spectral order invariant as obstruction to Stein fillability

ANDRÁS STIPSICZ (Alfréd Rényi Institute of Mathematics, Hungary)
Surfaces in four-manifolds

Tuesday, August 9, 2022

CIPRIAN MANOLESCU (Stanford University, USA)
Skein Lasagna modules

DONGHAO WANG (Stony Brook University, USA)
Monopoles and Seidel's spectral sequence

LOTHAR GÖTTSCHE (ICTP, Italy)
Computations of Vafa–Witten invariants in algebraic geometry

JOSHUA GREENE (Boston College, USA)
From ball fillings to cube tilings

FRANCESCO LIN (Columbia University, USA)
Monopole Floer homology and invariant theta-characteristics

Wednesday, August 10, 2022

Robert Lipshitz (University of Oregon, USA)
Khovanov homology and strongly invertible knots

Maggie Miller (Stanford University, USA)
Obstructing isotopy even with an extra dimension of freedom

Paolo Lisca (Università di Pisa, Italy)
Horizontal handlebody decompositions

Thursday, August 11, 2022

David Gabai (Princeton University, USA)
Knotted 3-balls in S^4 and knotted 3-spheres in $S^1 \times S^3$

Juanita Pinzón Caicedo (University of Notre Dame, USA)
Satellite operations that are not homomorphisms

Larry Guth (Massachusetts Institute of Technology, USA)
The degree of a map and its Lipschitz constant

Daniel Ruberman (Brandeis University, USA)
Diffeomorphism groups of 4-manifolds and embedding spaces

Peter Kronheimer (Harvard University, USA)
Algebraic curves defined by singular instanton homology

Friday, August 12, 2022

Lisa Piccirillo (Massachusetts Institute of Technology, USA)
Four-manifolds with boundary and fundamental group $\mathbb{Z}$

Cole Hugelmeyer (Princeton University, USA)
Inscribed squares and linear relation avoiding paths

Claudius Zibrowius (University of Regensburg, Germany)
Classification of link Floer and Khovanov multicurve invariants

Juan Muñoz-Echániz (Columbia University, USA)
Toroidal 3-manifolds, curves, and sutures

Jake Rasmussen (University of Cambridge, UK)
Toroidal 3-manifolds, curves, and sutures

List of Additional Participants

Summer School

(online participants marked by a *)

BAIS, Valentina
Italy

BASCAPE, Giacomo*
Université du Québec à Montréal, Canada

BEKE, Márton
Eötvös Loránd University, Hungary

BERA, Gorapada
Humboldt-Universität zu Berlin, Germany

BHAT, Deeparaj
Massachusetts Institute of Technology, USA

BIANCHI, Filippo*
Università di Pisa, Italy

BINNS, Fraser*
Boston College, USA

BRAND, Jack
Australian National University, Australia

BROĆIĆ, Filip*
Serbia

CANTO COSTA, Vinicius*
Stony Brook University, USA

CAUDELL, Jacob
Boston College, USA

CHATZITHEODORIDIS, Eleftherios*
University of Virginia, USA

CHEN, Daren
Stanford University, USA

COHEN, Asher Jesse*
University of Oregon, USA

CURRIER, Adrien*
Université de Nantes, France

DAI, Xinle
Harvard University, USA

DATTA, Ipsita*
Institute for Advanced Study, USA

GADDAM, Manav*
Institute of Mathematical Sciences, India

GHOSH, Partha Sarathi
Université Libre de Bruxelles, Belgium

GHOSH, Sudipta*
Louisiana State University, USA

GUTH, Matthew Gary
University of Oregon, USA

GUTWEIN, Dominik
Humboldt-Universität zu Berlin, Germany

HOLMES, Alfred
University of Oxford, United Kingdom

JIN, Hyeonhee
Max Planck Institute for Mathematics, Germany

KARADERELI, Şeyma
Boğaziçi University, Türkiye

KEPPLINGER, Willi
University of Vienna, Austria

KHAN, Isabella
Princeton University, USA

KNAVEL, Alexis Sierra*
Georgia Tech, USA

KRUTOWSKI, Romna*
UCLA, USA

KUBASCH, Arnd Alexander*
Eötvös Loránd University, Hungary

LADU, Roberto*
Imperial College London, United Kingdom

LI, Jiakai
Harvard University, USA

LI, Zijun
Duke University, USA

LOKTEVA, Elizaveta
Alfred Renyi Institute, Hungary

MADENGKO, Simo Carole
Université d'Abomey-Calavi, Benin

MALECH, Oliviero
SISSA, Italy

MANIKANDAN, Naageswaran
Humboldt-Universität zu Berlin, Germany

MASSONI, Thomas
Princeton University, USA

MAZUIR, Thibaut*
Sorbonne Université, France

MERZ, Alice
University of Pisa, Italy

MIN, Jie*
Max Planck Institute for Mathematics, Germany

MUÑOZ-ECHÁNIZ, Juan
Columbia University, USA

ORBEGOZO RODRIGUEZ, Miguel*
University of Glasgow, United Kingdom

PARMA, Andrea
Università di Pisa, Italy

PEDROTTI, Riccardo*
University of Texas at Austin, USA

POPPR, Marián
Augsburg University, Germany

QIN, Qianhe
Stanford University, USA

RAMSEY, James Patrick*
Lancaster University, United Kingdom

RAZZAQ, Junaid
University of Bologna, Italy

REINOSO, Braeden
Boston College, USA

RODEWALD, Ryan Thomas*
Georgia Tech, USA

SANO, Taketo*
University of Tokyo, Japan

SANTORO, Diego
Scuola Normale Superiore, Italy

SCHEFLER, Gergő
Alfréd Rényi Institute of Mathematics, Hungary

SERAFINI, Giada*
Heinrich Heine Universität, Germany

SHEN, Weizhe*
Georgia Tech, USA

SORYA, Patricia*
Université du Québec à Montréal, Canada

STELOW, Cecilia Mary
Massachusetts Institute of Technology, USA

STENHEDE, Eric
University of Vienna, Austria

STINE, Ransome Charles
Brandeis University, USA

SWAMINATHAN, Mohan*
Princeton University, USA

TRUÖL, Gill Paula
ETH Zürich, Switzerland

VARVAREZOS, Konstantinos*
Princeton University, USA

WAKELIN, Laura
Imperial College London, United Kingdom

WELLS, Rene Tristan*
Michigan State University, USA

WINDES, Autumn Emily*
USA

WLODEK, Maciej
Princeton University, USA

XU, Alex
Columbia University, USA

YAN, Jiajun*
University of Virginia, USA

ZAMPA, Sarah Juliet
Budapest University of Technology and Economics and Alfréd Rényi Institute of Mathematics, Hungary

ZENDEHBOODI, Roomina*
University of British Columbia, Canada

ZHANG, Yongsheng*
Tongji University, China

Research Conference

(online participants marked by a *)

MROWKA, Tomasz
Massachusetts Institute of Technology, USA

STAFFILANI, Gigliola
Massachusetts Institute of Technology, USA

BHAT, Deeparaj
Massachusetts Institute of Technology, USA

BHUPAL, Lal Mohan*
Middle East Technical University, Türkiye

BIANCHI, Filippo*
Università di Pisa, Italy

BINNS, Fraser*
Boston College, USA

BRAND, Jack
Australian National University, Australia

CAUDELL, Jacob
Boston College, USA

CAZASSUS, Guillem*
University of Oxford, United Kingdom

CHEN, Daren
Stanford University, USA

COHEN, Asher Jesse*
University of Oregon, USA

CONWAY, Anthony
MIT, USA

DAI, Irving
Stanford University, USA

DAI, Xinle
Harvard University, USA

FELLER, Peter
ETH Zürich, Switzerland

GHIGGINI, Paolo
CNRS and Université de Nantes, France

GHOSH, Partha Sarathi*
Université Libre de Bruxelles, Belgium

GOLLA, Marco
CNRS and Université de Nantes, France

GUTH, Matthew Gary
University of Oregon, USA

HAMBLETON, Ian
McMaster University, Canada

HOLMES, Alfred*
University of Oxford, United Kingdom

IIDA, Nobuo*
Japan

JIN, Hyeonhee*
Max Planck Institute for Mathematics, Germany

KARADERELI, Şeyma
Boğaziçi University, Türkiye

KARIMI, Homayun*
McMaster University, Canada

KHAN, Isabella
Princeton University, USA

KIRK, Paul*
Indiana University, USA

KRISHNA, Siddhi
Columbia University, USA

KRUTOWSKI, Romna*
UCLA, USA

LADU, Roberto
Imperial College London, United Kingdom

LENESS, Gibbs Thomas*
Florida International University, USA

LI, Jiakai
Harvard University, USA

LI, Xiaobin*
Southwest Jiaotong University, China

LI, Zijun*
Duke University, USA

MADENGKO, Simo Carole
Université d'Abomey-Calavi, Benin

MALLICK, Abhishek*
Max Planck Institute for Mathematics, Germany

MARENGON, Marco
Alfréd Rényi Institute of Mathematics, Hungary

MARK, Edgar Thomas
University of Virginia, USA

MARTIN, Gage*
Boston College, USA

MASSONI, Thomas
Princeton University, USA

MIN, Hyunki
Massachusetts Institute of Technology, USA

MIN, Jie*
Max Planck Institute for Mathematics, Germany

MONTAGUE, Ian
Brandeis University, USA

MUÑOZ-ECHÁNIZ, Juan
Columbia University, USA

NASSIRI, Meysam
Institute for Research in Fundamental Sciences, Iran

ORBEGOZO RODRIGUEZ, Miguel*
University of Glasgow, United Kingdom

ÖZBAĞCI, Burak
Koc University, Türkiye

PARK, Junghwan
Korea Advanced Institute of Science and Technology, Korea

PARMA, Andrea*
Università di Pisa, Italy

PEDROTTI, Riccardo
University of Texas at Austin, USA

PETKOVA, Tsvetelina Vaneva*
Dartmouth College, USA

PLAMENEVSKAYA, Olga
Stony Brook University, USA

QAZAQZEH, Moham Khaled
Kuwait University, Kuwait

QIN, Qianhe*
Stanford University, USA

RAZZAQ, Junaid
University of Bologna, Italy

REINOSO, Braeden
Boston College, USA

SANO, Taketo*
University of Tokyo, Japan

SANTORO, Diego
Scuola Normale Superiore, Italy

SHEN, Weizhe*
Georgia Tech, USA

STELOW, Cecilia Mary
Massachusetts Institute of Technology, USA

STINE, Ransome Charles
Brandeis University, USA

STRLE, Saso
University of Ljubljana, Slovenia

SUWARA, Piotr
Polish Academy of Sciences, Poland

TANIGUCHI, Masaki*
Institute of Physical and Chemical Research, Japan

TRUÖL, Gill Paula
ETH Zürich, Switzerland

VARVAREZOS, Konstantinos
Princeton University, USA

WAKELIN, Laura
Imperial College London, United Kingdom

WILLIS, Sean Michael
Stanford University, USA

WLODEK, Maciej
Princeton University, USA

XU, Alex
Columbia University, USA

YE, Fan
Cambridge University, United Kingdom

ZENTNER, Raphael
Universität Regensburg, Germany

ZHANG, Melissa*
University of Georgia, USA

ZUDDAS, Daniele*
University of Trieste, Italy

Reminiscences on the Occasion of Tom Mrowka's 60th Birthday

Cliff Taubes

I am writing first to apologize to Tom for not attending his 60th birthday celebration in Italy. I have no excuse except insanity. I hope he forgives me. I am also writing to relate some of my own history with Tom.

My interactions with Tom go back almost 40 years to when he was a graduate student at UC Berkeley and I was starting there as an *Acting* Associate Professor. (By the way, this was the most accurate job title I have had.) Rob Kirby was Tom's official advisor, but Tom was interested in the newfangled mathematics of gauge theory, as pioneered by Simon Donaldson and Karen Uhlenbeck (and Michael Atiyah and Raoul Bott). So Tom came to my office in Evans Hall to ask about a thesis problem in that area. Since we didn't know anything in those days, there were lots of thesis problems to be had. Tom's challenge problem was to figure out the gluing theorems for the self-dual Yang–Mills equations when cutting 4-manifolds along embedded 3-manifolds. When I moved to Harvard a year or so later, Tom came also (but still a graduate student at Berkeley). I recall Tom explaining to John Morgan and me the solution to the gluing problem which made clear to me that Tom knew a lot more about gauge theory than I ever would (which is even more true now). Anyway, this began an ongoing tradition whereby Tom would patiently explain to me his fantastic theorems and me thinking "Wow! That's beautiful."

So Tom got his PhD and a few years later, he was appointed a professor at Caltech. I mention this (you can read his bio on his MIT website) to relate that the salaries at Caltech at the time (and probably still) were way higher than anywhere else. Tom told me what his starting salary was at Caltech (which was way higher than my *present* salary), and I then shamelessly reminded Tom of how much of his success he owed to me as his partial thesis advisor (Rob Kirby being the official advisor). This made Tom feel guilty I suppose, so out of kindness or pity, Tom offered to pay for beers for life whenever we were having a drink somewhere. Thus, I had mixed feelings when Tom left Caltech after only a few a few years for a position at MIT. (I didn't inquire about Tom's MIT salary, but I imagine that it was even higher than the Caltech salary.) Being just down the street, Tom became a good and extremely loyal friend not just to me, but to the members of my family. Indeed, I recall occasions when Tom watched over my two kids, once in an emergency situation and other times for the fun of it. (My kids have fond memories of Tom taking them to movies and treats that I would have never allowed.

When older, both of my kids returned the favor and babysat for Tom and Gigliola's little children.)

I won't attempt the Olympian task of recounting Tom's mathematical contributions. I only say that after the initial work of Simon Donaldson and Andreas Floer, Tom with Peter Kronheimer are ultimately the source of (invented, pioneered, discovered, created) most of the current applications of gauge theory to 3 and 4 dimensional differential topology. It is quite a remarkable track record of creativity that shows no sign of ending. Tom and Peter have also been quite rightly lauded for their writings on these subjects (their book *Monopoles and Three-Manifolds* being a bible of sorts for the Floer homology parts of the subject).

With regards to achievements, I note in particular that Tom with Peter were the recipients of the 2023 Leroy Steele prizes for Seminal Research Contributions, this for their paper *Gauge theory for embedded surfaces I*[1] from 1993. This paper and its immediate sequel led to their even more seminal paper *Embedded surfaces and the structure of Donaldson's polynomial invariants*[2] which put order to what was previously a totally mysterious set of 4-manifold invariants constructed by Simon Donaldson using the SU(2) Yang–Mills equations. Moreover, a case can be made that the formulas in that even more seminal paper presaged Edward Witten's introduction a few years later of the Seiberg–Witten invariants. I mention this paper in particular (any number of papers by Tom and Peter could have been Steele prize papers) because I recall refereeing it (or perhaps it was one of the previous two paper in the series). About that refereeing: the paper invoked a beautiful and fantastical and magical mix of topology, algebraic topology, analysis and geometry. But, a theorem from algebraic geometry was used in one of their proofs with the reference being something like Chapter 6 in C.B. Morrey's book *Multiple Integrals in the Calculus of Variations*. Now I think it fair to say that no one but C.B. Morrey really knows what is in Chapter 6 of his book (or most of the other chapters); this is because the book's notation is (to be charitable) extremely dense. Tom and I have a running joke that you can state any analysis theorem you like and reference some chapter in Morrey's book and no one would be able to dispute the reference. Jokes aside, Morrey's book is a wonderful book if you can translate the notation. Anyway, in my referee report, I wrote that this algebraic geometry theorem could in no way come from Morrey's book because the book has zero algebraic geometry. Some months later, I asked Tom why he referenced Morrey's book for the algebraic geometry theorem; and Tom said "This was to keep the referee on his toes."

This last vignette indicates just a token of Tom's point of view on life, which is (I think) that you should really try to enjoy things. I sometime forget this maxim; and indeed Tom reminded me again this past spring during the PhD thesis defense of Joshua Wang. Tom was on the thesis committee as were I and Peter Kronheimer (who was Joshua's advisor). Joshua's talk at the thesis defense had multiple jokes (some involving ChatGPT) which Tom seemed to enjoy immensely. At one point, Tom was laughing out loud at a page of Greek (it looked like Greek) that Joshua had projected, claiming it was a translation of his thesis. I was in a sour mood

[1] P. B. Kronheimer and T. S. Mrowka, *Gauge theory for embedded surfaces I*, Topology **32** (1993), no. 4, 773–826.

[2] P. B. Kronheimer and T. S. Mrowka, *Embedded surfaces and the structure of Donaldson's polynomial invariants*, J. Differential Geom. **41** (1995), no. 3, 573–634.

that day, but seeing Tom, I thought "Yes, I should remember to laugh more ... like Tom."

Rob Kirby

Let me start by saying that Cliff Taubes is 100% Tom's adviser, and I'm 0%, like a 0% appointment in a department (honorary but not paid). Tom was a Berkeley student when Cliff arrived for a 2-year postdoc at Berkeley in 1981 (Harvard merely loaned him out for a bit of seasoning before recalling him). They got together, especially during the 1982 fall at MSRI when there was a gauge theory program, and Tom went to Harvard to finish his research with Cliff. But he needed a formal advisor from Berkeley, so I got the honor.

Tom was then a postdoc at Stanford and a few years at Caltech, which turned out very well for four of my PhD students. I had tried to learn enough gauge theory to teach it (badly) and encourage my students to enter this very lively subject. But I needed help and Tom generously bailed me out with these students, making himself available, answering questions and giving advice (better than I could).

I never did master Tom's math to the extent I could really talk math with him. But more than once I watched the twinkle in his eye as he, after a bit of thought to get it just right, set me straight with a perspicacious comment.

All these years Tom has been a great guy, and richly deserves the honor of this volume.

András Stipsicz

I met Tom in the early 90's while busy working on my PhD thesis, which was related to his work on understanding the change of moduli spaces of instantons after some topological constructions have been applied to the underlying four-manifold. The first encounter was kind of indirect: we got a few sections of his paper with Bob Gompf on exotic K3's which do not carry any complex structure. Most of the definitions and some of the statements were missing from the file, so it was more like a sudoku or a mad libs assignment than reading a math paper (I should say, this is probably a great way to really learn something...). My impression of Tom in a conference is two-fold: first, we will not be bored, that's for sure. The other thing is that he always asks questions, and the questions are not only right/good/appropriate/smart, but somehow help the speaker to advance in the subject. At least that is what happened to me, and more than once. I discovered later that asking such questions is far from being a routine exercise, it needs focus and to pay attention to the others. And something more: it is a gift not many of us has. Tom, thank you for your questions (and also for the good time at conferences, meetings, wherever).

Peter Ozsváth

I'd like exercise my editorial privilege to indulge in some personal reminiscences of my former postdoctoral mentor, Tom Mrowka. I will constrain my narrative to a brief interval in the mid-1990's, which was a crazy and turbulent time, both from an objective, math-historical perspective, and from my own, personal one.

FIGURE 1. Tom Mrowka and Rob Kirby in Gökova, Turkey at Kirby's 60th birthday conference in 1998.

Simon Donaldson had recently introduced the Yang–Mills equations as a tool for studying the smooth topology of four-dimensional manifolds, and Tom, who had a broad command of both analysis and topology, had established himself as the master of these new techniques. He had just proved, with Bob Gompf, that there are simply-connected, smooth four-manifolds that cannot be written as connected sums of algebraic surfaces; and he had just brought order to the chaos of "Donaldson polynomials", with his structure theorem for Donaldson invariants, proved in collaboration with Peter Kronheimer. Roughly speaking, this latter theorem showed that all of the information from Donaldson's invariants of a given four-manifold X (*a priori* a power series in the exterior algebra over the homology of X) was captured by finitely many "basic classes", which are distinguished two-dimensional cohomology classes on X that constrain the genera of smoothly embedded representatives for any given two-dimensional homology class.

I met Tom in New York when I was a graduate student and I joined him in the Fall of 1994 at Caltech, as an NSF postdoc. Almost immediately afterwards, a new equation appeared magically from the inscrutable world of physics: the Seiberg–Witten equations; and an amazing new picture of four-manifold topology began to emerge, with theorem after theorem being proved at breakneck speed. It was an intense roller-coaster ride for a young mathematician trying to find his place in the world, and I was exposed to a dizzying amount of math in a short period of time. One day, Tom appeared in my office, explaining to me how the Dirac operator works on a Kähler manifold; another day, he dropped by with a complete account of his proof with Kronheimer of the "Thom conjecture" (holomorphic curves in $\mathbb{CP}^2$ minimize genus in their homology class) that they'd come up with on the previous weekend; another day he showed up waving about a preprint by Cliff Taubes relating Seiberg–Witten theory and symplectic geometry. At this time, Tom taught a graduate course on Morse homology for the Chern–Simons–Dirac

FIGURE 2. Tom Mrowka, David Gay, and Tadashi Tokieda on the beach in Lisbon in 1999.

functional, attended by a few graduate students and as many postdocs from the greater LA area as possible.

The mathematical terrain that Tom inhabited had no boundaries: everyone needed to know elliptic operators, Sobolev spaces, four-manifold topology, three-manifold topology, equivariant K-theory, and algebraic geometry. I frantically tried to keep up with all of the well-known prerequisites. One day, it was a paper by Hitchin on harmonic spinors. Another day, a paper by Eliashberg on contact geometry. Next, it was Graeme Segal's *Loop Groups.* And then Atiyah–Patodi–Singer's index theorem for manifolds with boundary. And next, Fulton's *Introduction to Toric Varieties...* Why? because it's cool.

Tom would blow off steam by "going for a drive in the hills", which meant trapping some poor unsuspecting victim in the passenger seat of his Acura while he took his beloved car at top speeds on the serpentine roads in the mountains around Los Angeles. Or he would take some unwitting seminar speaker, along with the local postdocs and graduate students, on an extreme hike in the sun-drenched hills near Pasadena. Or he'd corner various younger colleagues to swim laps in the pool. Particularly memorable was a trip with a fellow colleague to the beach, where we were challenged to classify singularities of smooth maps between Euclidean spaces, etching pictures in the sand with sticks. At the time, Tom was fond of reciting his mantra "Sleep is for the weak".[3] Once I asked him how he could possibly add one more espresso shot to his triple latte at 10 PM, and his response was that this was just the extra help he needed to fall asleep without a headache.

The following academic year, Tom was commuting between Pasadena and Boston. But he was very conscientious about meeting his postdocs: the Starbucks in Pasadena opened at 5:30 AM and we met there there regularly, before his weekly flights out to the East Coast.

[3]After a perfunctory glance at the internet, I'm convinced that the phrase, which is well-known today, was originally coined by Tom.

Tom's love of life and mathematics was infectious, and I have had many great mathematical interactions with him since: espressos in Italy, long sessions at the cafés in Cambridge, mathematical beach parties in Turkey, dinner parties at his place with Gigliola. I couldn't possibly go through the whole list; but I have got to mention one more: a memorable run in 2002 in Princeton on the path along the canal, interrupted by my need to catch my breath and Tom's need to describe the Seiberg–Witten moduli spaces over a two-dimensional family of metrics on $\mathbb{CP}^2$ (with the opposite orientation, of course) minus a circle, which in turn formed a key step in our proof, with Peter Kronheimer and Zoltán Szabó, of the "exact triangle" for Seiberg–Witten monopole Floer homology.

I'm very grateful to have had the fortune of interacting with Tom. In sum: Tom, thanks for all the math; I look forward to much more. And I will try to keep somewhat in shape for our next run!

Paul Feehan

I first met Tom Mrowka around 1992 when he was visiting John Morgan at Columbia University and spoke in John's seminar. I was just about to graduate with my PhD under the supervision of Duong Phong. I had heard of Tom, primarily because of his famous thesis on gluing anti-self-dual Yang–Mills connections over four-manifolds containing long cylinders. I was particularly interested in that topic because I was trying to understand the geometry of the ends of the moduli space of anti-self-dual connections and gluing theory provided the key approach to solving this problem. Tom's reputation even then was legendary, and I and the other graduate students at Columbia working on gauge theory with Robert Friedman and John were all eager to hear his talk.

After spending my final summer in New York in 1992, I moved to Berkeley for a one-year postdoctoral position at MSRI (now SLMath). I had declined a three-year postdoctoral position at McMaster University, Canada, in favor of MSRI, so I was apprehensive about my future, realizing that I would be applying for jobs again during Fall 1992 and that I needed to make my mark in the meantime. My move from New York to California that summer was stressful, for both personal and the usual logistical reasons, and I did not notice until long after the abstract submission deadline (August 3, 1992) that Tom (then at Caltech) and Ron Stern (at the University of California, Irvine) were co-organizing a special session on *Gauge Theory and Four-Manifolds*, November 7–8, 1992 at the University of Southern California, for the Fall AMS Sectional Meeting. However, I contacted Tom and asked if there was any chance that I could still give a talk. There was no way that I could be officially added that late to the program, but he generously offered me an unofficial speaking slot and I spoke about a paper in progress that was an outgrowth of my PhD thesis. His gesture was a kind and thoughtful one that ended up having a major impact on the rest of my year at MSRI, if not my future career. Tom Parker was in the audience and during my talk he pointed out various facts about bubble tree compactifications that I had not clearly appreciated at the time, and those ended up playing a pivotal role in the paper[4] that I wrote during my year at MSRI. I also got to meet other members of the then relatively small community

[4] P. M. N. Feehan, *Geometry of the ends of the moduli space of anti-self-dual connections*, J. Differential Geom. **42** (1995), no. 3, 465–553.

of researchers in gauge theory and four-manifolds: Selman Akbulut, David Austin, Hans Boden, Paul Kirk, Eric Klassen, Ronnie Lee, Johan Råde, and some of Tom Mrowka's early students — including Chris Herald, Yongbin Ruan, Graham Taylor, and Baozhen Yu. Later that year, Johan invited me to visit Stanford for a seminar talk and I got to know him and Rafe Mazzeo, among the few people with whom I had mathematical interests in common near Berkeley for my postdoctoral year there.

During Fall 1992, I applied for an NSF Postdoctoral Fellowship to work with Cliff Taubes at Harvard and, to my immense relief, that application was successful, and I moved to Cambridge in August 1993. One of the projects that I started to pursue at Harvard was to try and show that symplectic four-manifolds had non-vanishing Donaldson invariants, using ideas prompted by my PhD thesis and Donaldson's paper[5] on 'polynomial invariants'. However, I was struggling with that project and in 1994, Edward Witten gave a seminar talk in Cambridge (at Harvard, I think) about his approach to Donaldson invariants, after which the landscape of gauge theory and four-manifolds changed overnight. Much of what I had learned about gauge theory until then now seemed irrelevant, at least anything to do with Yang–Mills gauge theory, energy bubbling, gluing, etc. Cliff gave a seminar talk about Seiberg–Witten gauge theory that year at Harvard, and it soon became clear that many outstanding questions about smooth four-manifolds, such as the irreducibility of symplectic four-manifolds, would quickly be solved by Cliff, Tom, Peter Kronheimer, and others. I was unsure what to work on, but a perfectly timed invitation from Thomas Leness in early 1995 to work with him on the Pidstrigach–Tyurin program to relate the Donaldson and Seiberg–Witten invariants of smooth four-manifolds gave me a uniquely rewarding project that was to occupy me for many years to come.

That project, though suited to my mathematical tastes, was daunting. Fortunately, Tom had just moved to Cambridge for positions as a visiting professor at Harvard for Spring 1995 and then MIT for Fall 1995, before moving permanently to MIT as a full professor in 1996. Tom's encouragement, interest, and suggestions over the remaining time of my four-year postdoc at Harvard and after I left played a critical role in my work to prove Witten's conjecture on the equivalence of Donaldson and Seiberg-Witten invariants. Tom's move to MIT was quickly followed by Peter Kronheimer's move from Oxford to Harvard in Fall 1996. Tom's postdoc at Caltech, Peter Ozsváth, whom I had got to know earlier during my occasional visits to Columbia after my graduation in 1992, also moved with him to MIT. The presence of Cliff and Peter K. at Harvard, Danny Ruberman at Brandeis, and Tom and Peter O. at MIT made Cambridge a uniquely exciting place for researchers in gauge theory. Memories of the various conferences and workshops that many of us in the gauge theory and four-manifolds community attended over the years — Gökova, Oberwolfach, Dubrovnik — are ones that I treasure.

Peter O. and I quickly became good friends, enjoying a shared experience as postdocs in Cambridge. Tom's presence fostered the development of a lively social circle, as well as a lively mathematical one, involving him, Peter O., Lucia Caporaso, and later Peter Kronheimer, with many dinners at our homes or around Cambridge. Some other memories that still stand out include occasional bike rides with Tom

[5]S. K. Donaldson, *Polynomial invariants for smooth four-manifolds*, Topology **29** (1990), no. 3, 257–315.

and Michael Nakamaye and feeling that my life was in danger being driven around the narrow Cambridge streets at top speed in Tom's sports car with Soundgarden blaring at an earsplitting level.

After leaving Harvard to take up a position as a tenure-track assistant professor at Ohio State University, Columbus, in Fall 1997, I would occasionally return to visit Cambridge to give talks and catch up with Tom and Gigliola at their home. They were married in Boston but had a second wedding celebration in Italy, a truly wonderful occasion that I was very happy to attend. I recall visiting them not long after their twins were born and I recognized that while they were thrilled to be parents, they were completely exhausted, not least because as soon as one baby would fall asleep, the other would start to cry.

When I was an assistant professor at Ohio State University, Tom alerted me and Thomas Leness that one likely consequence of our work thus far was a mathematical explanation of the so-called superconformal type phenomenon for the Seiberg–Witten series that had been noticed by Greg Moore and his postdocs, Marcos Mariño and Grigor Peradze, and proved by them using methods of quantum field theory. I have no doubt that Tom and Peter K. could have written up this mathematical explanation without our involvement, but they generously invited us to collaborate with them, leading to our joint paper[6]. Working with them was a great experience and it helped to greatly clarify our approach to proving Witten's conjecture on the equivalence of Donaldson and Seiberg–Witten invariants. Later, Tom's suggestions on how to prove the Kotschick–Morgan Conjecture for the wall-crossing terms for Donaldson invariants of four-manifolds with $b^+ = 1$ and resolve a very similar gluing issue in the Pidstrigach–Tyurin program ultimately led to the development of our monograph[7] with Thomas Leness. While Tom's and my mathematical interests have diverged over years, he has long been a frequent supporter of my work, a source of great ideas, a good friend, a mathematical inspiration, and for all of that I am extremely grateful.

Daniel Ruberman

I met Tom Mrowka early in his career and almost as early in mine. At the urging of Peter Ozsváth, I've written down some recollections of those times. The reader is cautioned that these memories are pretty hazy due to the passage of time, and that dates mentioned should be taken as approximate.

I probably should have met Tom at MSRI during the 1984-85 program in low-dimensional topology, but according to Tom (much later) he was too busy with classes in Berkeley and passing his exams to come around much, and besides it was all over his head. I mention this mainly for younger people who sometimes forget that even their heroes had to start somewhere. We met for real when Tom at some point moved to Harvard to complete his graduate studies with Cliff Taubes. Again, my memories aren't so specific, but my overall impression is of Tom having boundless energy and knowing much more than any student had a right to know.

[6]P. M. N. Feehan, P. B. Kronheimer, T. G. Leness, and T. S. Mrowka, *PU(2) monopoles and a conjecture of Mariño, Moore, and Peradze*, Math. Res. Lett. **6** (1999), no. 2, 169–182.

[7]P. M. N. Feehan and T. G. Leness, *An SO(3)-monopole cobordism formula relating Donaldson and Seiberg–Witten invariants*, Mem. Amer. Math. Soc. **256** (2018), no. 1226.

While it was clear that he had an amazing understanding of the analytical side of gauge theory, Tom could be a little hard to follow. I remember him stopping me in a Harvard hallway to excitedly tell me that he "understood the obstructions from the boundary" in the general gluing problem for 4-manifolds. It all had the ring of truth (and became the heart of his—sadly unpublished—PhD thesis) but the details were hard to follow. At the time, Cliff's *Gauge Theory and Topology* seminar was an informal affair (which grew more formal and prestigious over time) in which we could explain work in progress. Tom gave a few talks to try to explain how to compute the effect of log transforms on Donaldson invariants; my recollection is that these ended with everyone (Tom included) staring at the blackboard. Eventually, Tom worked out how to say all of this coherently, and the result appeared in his wonderful paper with Bob Gompf on homotopy K3 surfaces.

Starting around 1989, I spent parts of a few years at Harvard supported by NSF and Sloan fellowships. Gauge theory was in transition from what seemed to be an astonishing but isolated accident to a powerful theory with broad applications to 4-manifolds. But it was really hard, with hundred-page papers to compute a single Donaldson invariant. John Morgan was around a lot, and following some lectures in which he went through Donaldson's connected sum theorem, I asked him an innocent question: if you split a 4-manifold along a 3-manifold that had no irreducible flat SU(2) connections, does the Donaldson invariant vanish? We quickly worked through some easy examples, but found that the case of circle bundles over the torus posed unique challenges.

The biggest challenge was that the gluing picture we expected to hold just didn't make sense: the moduli spaces that showed up would have to be of larger dimension than the index theorem would suggest. We went up to Cliff's office to ask whether there was a theory of moduli spaces with an L^2 bound on curvature, but with no stronger decay condition. He told us yes (begging the question of why he hadn't told anybody about it) and we had many discussions about what this might look like.

At some point, Cliff suggested to John and me that since Tom had been thinking about related matters, maybe we should join forces with him. So over greasy burgers at Charlie's Kitchen in Harvard Square we talked through the project and decided to work together, eventually producing the book (really an overgrown research paper) known as 'MMR'. The analytical side turned out to be very challenging and it took a while for the right picture to emerge. Trying to keep up with Tom and John, who were incredibly fast, was for me one of the most exhilarating and exhausting experiences of my life.

By this time Tom was in California, and much of the progress happened by emails, phone calls, and occasional visits. I recall an epic session in John's apartment in New York, with Ellen's teenage son Jake staring daggers at us since we were monopolizing the IBM PC that he felt could be put to better use playing computer games. Rather late in the proceedings, we gathered that none of the existing proofs of the 'center manifold theorem' in infinite dimensions (our basic tool for analyzing the moduli space on manifolds with cylindrical ends) actually applied in our situation. This was resolved one day when Tom called, full of excitement, to tell me to go to the library to look at a new paper of Eckmann–Wayne, whose appendix had a proof that would work for us.

At Peter Kronheimer's birthday conference, I told the story of Tom and Peter toasting, over lunch in 1995 in Oxford, to their joint decisions to take up positions at MIT and Harvard. I was just as happy, sensing how much the decisive shift of gauge theory to the Boston area would affect my mathematical life. A nice coda to that lunch was dinner in 2017 with Tom and Peter (this time with spouses) to celebrate all three of us simultaneously finishing terms as chair.

Tom has been incredibly generous over the years, listening to any number of vaguely formulated ideas and as often as not, making a brief comment that pointed me in the right direction. One of these I remember with great clarity–a remark that what I already knew about computing isotopy invariants was enough to finish the calculation I'd been struggling with; I'm still reaping the benefits of that observation 25 years later. Nikolai Saveliev and I were similarly enlightened after a year of struggling with induced maps on Floer homology by Tom pointing out that we weren't really using the definition correctly. Breaking that logjam really got our collaboration off the ground and eventually led to some interesting joint projects that included Tom as well. He was even more generous with several students of mine who had gotten themselves involved with harder analytic problems than they or I had expected.

It's been a really fun ride, and I feel lucky to have Tom as a life-long friend.

Lenny Ng

As one of Tom Mrowka's students from the late nineties and early aughts, I can't relate any firsthand stories about his development of monopole Floer homology or proof of the Thom conjecture. However, I can relate my memories of Tom as a fantastic, if sometimes terrifying, thesis advisor.

Tom was always generous with his time, despite his multitude of students and other obligations. Not surprisingly, my mathematical discussions with him while I was a grad student were crucial formative experiences for me. Back then, Tom already knew more about my eventual specialty (contact and symplectic topology) than I do to this day, even though it's only a tiny part of his portfolio. In our meetings, he invariably made comments that were insightful and essential. Like all good advisors, Tom had an uncanny intuition for what directions were the most interesting for me to pursue, as well as what techniques were likely to get me there; but he also gave me the space to discover (or "discover") all of these on my own. Tom's comments were usually laconic and, to me at least, sometimes inscrutable. I remember often frantically scribbling down his Delphic pronouncements after our meetings—"try a filtration", "look this up in Freed and Uhlenbeck", etc.—and spending the next few days trying to unpack them. That's not to say that I always followed Tom's suggestions once unpacked; there were times when it took me months or years to realize that I should have listened to Tom all along.

However friendly and easygoing Tom may seem to most of the world, I should admit that our meetings sometimes filled me with a healthy sense of unease bordering on fear. Tom would begin most of our meetings by asking me: "So?" I'd then tell him everything that I'd been working on since our previous meeting, usually at some length, and Tom would give me the type of feedback that I've already described. When I was finished, there would be a long pause, followed by Tom asking: "What else?" At this point I'd have to scramble to think of some other interesting thing that I hadn't already told him. This was often pretty tough to do since I'd already exhausted all the news I'd prepared, but I'd think hard and come up with something. Another long pause, and then: "What else?" This would go for several more rounds, until either Tom was satisfied that he'd wrung everything out of me, or it was time for one of his many other students to meet with him.

I'm sure I'm not the only student of Tom's who experienced this sort of process. But even if it was unnerving to me at times, Tom's advising style has clearly worked brilliantly for his (30+ and counting) students, many of whom have become leaders in their own fields. What is especially impressive to me is the sheer breadth of the specialties of Tom's former students: to give one example, when I was a postdoc at Stanford, I was one of three postdocs who were former students of Tom's, and we were all working with different postdoctoral mentors. This breadth of course mirrors the wide range of Tom's interests and areas of expertise, which you can also see in the diversity of subjects covered in this proceedings volume.

Thanks, Tom, for all of your help over the years. I'm still working on emulating you (in all possible ways) in my meetings with my own PhD students.

Gigliola Staffilani

Tom and I met while I was at Stanford in 1999. I learned later that at that time he was already working with "his besty", as our daughter calls Peter Kronheimer, on the book *Monopoles and Three-Manifolds*. They worked on this book for at least six years, meeting every Tuesdays and Thursdays at Harvard, with the exception of a few crazy months when our twins were born. I have always been a little jealous of the mathematical bond that Peter and Tom have, it must be great to have a mathematical soulmate like those two have, but this bond has been wonderful for mathematics as well, since the book I mentioned above definitely is not the only thing they have produced, far from it. In spite of the fact that Tom loves to do math, and certainly it is his favorite activity even above biking and swimming, loving math doesn't even come close to how much he loves his family. He has been my rock and my constant support, he has been a great father for our kids, and more recently their teaching assistant on call, although I am hearing that his explanations and hints for solutions are often starting way too far from the goal. My wish is that Tom can continue doing what he loves most for years to come, we will all benefit enormously from it!

Proceedings of Symposia in Pure Mathematics
Volume **109**, 2024
https://doi.org/10.1090/pspum/109/02000

Lectures on families of Dirac operators and applications

Francesco Lin

Dedicated to Tom Mrowka in the occasion of his 60th birthday.

ABSTRACT. These are the notes for a minicourse taught at the 2022 ICTP summer school *Frontiers in Geometry and Topology.* The goal is to introduce families of Dirac operators and how they can be used to study interactions between geometry and topology. In particular, we discuss the index theorem for families of twisted Dirac operators, and its applications to metrics of positive scalar curvature and the three-dimensional Weinstein conjecture.

Introduction

These are the notes for the three and a half hour long minicourse *Families of Dirac operators and applications* taught at the 2022 ICTP summer school *Frontiers in Geometry and Topology* for an audience of graduate students. The main goal is to show via concrete examples how families of Dirac operators can be used as a bridge between the geometry and the topology of manifolds. These ideas are closely related to current topics of research in gauge theory and low-dimensional topology, and the hope is to provide a friendly first introduction to the subject for those working in the field. In particular, these notes assume the reader to be somehow familiar with the following:

- vector bundles, their classifying spaces, and characteristic classes [**MS74**];
- basic differential geometry, including connections on vector bundles and the various notions of curvature of a Riemannian manifold [**Lee18**];
- basic notions in functional analysis such as separable Hilbert spaces and bounded operators [**Eva10**, Appendix D].

The last part of the notes will also assume the reader to be familiar with the formal properties of either monopole or Heegaard Floer homology (see for example [**Lin16**] and [**Gre21**] respectively for introductions to the topics).

The minicourse consisted of three lectures covering the following:

(1) an informal introduction to spin Dirac operators; the Atiyah-Singer index theorem, the Lichnerowicz formula, and applications to metrics of positive scalar curvature.
(2) K-theory and its relation with Fredholm operators; the index theorem for families of Dirac operators and the Gromov-Lawson theorem on the torus.

Key words and phrases. Dirac operator, index theorem, Weinstein conjecture.
This work was partially supported by the Alfred P. Sloan Foundation and NSF grant DMS-2203498.

(3) the space of self-adjoint Fredholm operators; relations with Taubes' proof of the Weinstein conjecture in dimension three.

This is a very ambitious syllabus - each lecture could have been the topic of a minicourse itself. When preparing the lectures, we have tried to highlight the motivation and key ideas behind the constructions while keeping technicalities at the bare minimum (some important ones are pointed out in the footnotes); we provide precise references for more detailed treatments. These notes follow the content of the lectures very closely, and will have a rather terse style. At the end of each lecture, the interested reader will find some exercises.

Warning. Throughout the text, we will be very imprecise about the analytic setup. For example, we will state most index theorems in terms of bounded operators, but switch to unbounded (differential) ones without further notice when discussing geometric applications.

1. Dirac operators, geometry and topology

1.1. Dirac operators on $\mathbb{R}^n$. The most fundamental differential operator on $\mathbb{R}^n$ is the Laplacian[1]

$$\Delta := -\frac{\partial^2}{\partial x_1^2} - \frac{\partial^2}{\partial x_2^2} - \cdots - \frac{\partial^2}{\partial x_n^2}.$$

Functions satisfying $\Delta u = 0$ are called *harmonic*; these can be used to describe many important objects, such as potentials in electromagnetism. While working on the relativistic quantum mechanics of the electron, Dirac [**Dir28**] was led to look for a first order differential operator D such that $D^2 = \Delta$.

REMARK 1.1. To be precise, Dirac was working with a Lorenzian metric on $\mathbb{R}^4$, so that he considered the d'Alembertian

$$\square = -\frac{\partial^2}{\partial x_1^2} - \frac{\partial^2}{\partial x_2^2} - \frac{\partial^2}{\partial x_3^2} + \frac{\partial^2}{\partial t^2}$$

instead of the Laplacian Δ. We will always work with Riemannian metrics in these notes.

Heuristically, what does such an operator D look like? Let us assume for simplicity that it takes the form

$$D = a_1 \frac{\partial}{\partial x_1} + a_2 \frac{\partial}{\partial x_2} + \cdots + a_n \frac{\partial}{\partial x_n}$$

with the $\{a_i\}$ are constant quantities (belonging to some algebraic structure to be decided). Then

$$D^2 = \sum_i a_i^2 \frac{\partial^2}{\partial x_i^2} + \sum_{i,j} a_i a_j \frac{\partial}{\partial x_i} \frac{\partial}{\partial x_j}.$$

Hence, as derivatives commute, $D^2 = \Delta$ is equivalent to the relations

$$(1) \qquad \begin{cases} a_i^2 & = -1 \\ a_i a_j + a_i a_j & = 0 \text{ if } i \neq j. \end{cases}$$

[1]Our convention is to use all negative signs so that $\langle u, \Delta u \rangle_{L^2} \geq 0$ for all compactly supported u, where we consider the L^2-inner product $\langle u, v \rangle_{L^2} = \int uv\mathrm{dvol}$.

Let us see how to solve such system of equations in the $\{a_i\}$. If $n = 1$ we just need $a_1^2 = -1$; of course, we cannot solve this within real numbers, but if we allow complex valued functions, we can set $a_1 = i$, and

$$D = i\frac{\partial}{\partial x_1}$$

squares to Δ (acting now on complex-valued functions).

When $n = 2$, the anticommutativity relation $a_1a_2 + a_2a_1 = 0$ implies that we cannot work anymore with complex numbers; on the other hand, we can solve (1) if we allow the a_i to be 2×2 matrices by setting

$$a_1 = \begin{bmatrix} 0 & -1 \\ 1 & 0 \end{bmatrix} \quad a_2 = \begin{bmatrix} 0 & i \\ i & 0 \end{bmatrix},$$

so that

$$D = a_1\frac{\partial}{\partial x_1} + a_2\frac{\partial}{\partial x_2} = \begin{bmatrix} 0 & -\frac{\partial}{\partial x_1} + i\frac{\partial}{\partial x_2} \\ \frac{\partial}{\partial x_1} + i\frac{\partial}{\partial x_2} & 0 \end{bmatrix} \tag{2}$$

squares to Δ thought of as an operator on $\mathbb{C}^2$-valued functions. Notice that D is closely related to the holomorphic and anti-holomorphic derivative operators.

In higher dimensions, the situation is similar; when $n = 3$, we can set the a_i's to be the Pauli matrices

$$\sigma_1 = \begin{bmatrix} i & 0 \\ 0 & -i \end{bmatrix} \quad \sigma_2 = \begin{bmatrix} 0 & -1 \\ 1 & 0 \end{bmatrix} \quad \sigma_3 = \begin{bmatrix} 0 & i \\ i & 0 \end{bmatrix},$$

and the corresponding Dirac operator D is again a square root of Δ acting on $\mathbb{C}^2$-valued functions. When $n = 4$, we can set the a_i to be the 4×4 block matrices defined by

$$\begin{bmatrix} 0 & -I_2 \\ I_2 & 0 \end{bmatrix} \text{ and } \begin{bmatrix} 0 & -\sigma_i^* \\ \sigma_i & 0 \end{bmatrix} \text{ for } i = 1, 2, 3$$

where the I_2 is the 2×2 identity matrix and * denotes transpose conjugate. The corresponding Dirac operator D squares to Δ acting on $\mathbb{C}^4$-valued functions, and is the (Riemannian version of the) operator Dirac was looking for. We will leave to the reader to guess the general recipe to define D on $\mathbb{R}^n$ for $n \geq 5$.

REMARK 1.2. Of course, we could have made different choices for the matrices a_i, which would have lead to different Dirac operators; for example, when $n = 1$ we could have taken $D = i\frac{\partial}{\partial x_1}$ to be acting on $\mathbb{C}^m$-valued functions. The specific operators we constructed above are 'minimal' in a specific sense (involving the representation theory of *Clifford algebras*), and are called the *spin Dirac operators*.

1.2. Dirac operators on a manifold. In this lectures we will be interested in how spin Dirac operators can be used as a bridge between the geometry and topology of an oriented Riemannian manifold (M, g); the precise definition of the analogue D is quite involved, and we refer the reader for example to [**Roe98**, Ch. 3,4] for more details. We will only focus on formal properties of such generalization.

First of all, in order to define D, we need to choose a *spin structure* on M; we will not need the precise definition of what this is (see for example [**Kir89**, Ch. IV]), but will only remark that it exists if and only the second Stiefel-Whitney class $w_2(TM)$ is zero. For manifolds of dimension $n \geq 3$, this is equivalent to the restriction of TM to the 2-skeleton of M being trivial. A spin structure affords us an analogue of the target vector spaces $\mathbb{C}^m$ we considered before, which is a hermitian bundle

$$S \to M$$

of rank $2^{\lfloor n/2 \rfloor}$ equipped with a hermitian connection ∇, called the spinor bundle. The spin Dirac operator is then a first order differential operator

$$D : \Gamma(S) \to \Gamma(S)$$

with the following key properties:

- it is *formally self-adjoint in* L^2, i.e. if $\varphi, \psi \in \Gamma(S)$ are compactly supported sections (e.g. if M is compact), then
$$\langle D\varphi, \psi \rangle_{L^2} = \langle \varphi, D\psi \rangle_{L^2}.$$
- it is *elliptic*; we will not define exactly what this means, but this implies for example that *harmonic spinors*, i.e. solutions to the equation $D\varphi = 0$ (which makes sense for φ merely C^1) are automatically smooth.

One should compare the latter with the analogous result for holomorphic and harmonic functions, which are in the kernel of the elliptic operators $\overline{\partial}$ and Δ respectively. We refer the reader to [**War71**, Chapter 6] for a nice introduction to the analysis of elliptic equations on manifolds which only assumes some basic familiarity with Fourier series.

1.3. Dirac operators and topology. Another key consequence of ellipticity is that, when M is compact, the space of harmonic spinors is *finite* dimensional. This is surprisingly related to topological information about M in a way that we now discuss.

Assume that n is even. Looking back at the case of $\mathbb{R}^n$, we see that in this case the matrices corresponding to a_i have a simple structure as block matrices. In the case of a Riemannian manifold, this manifests in a canonical splitting of the spinor bundle

$$S = S^+ \oplus S^-$$

for which we can write the Dirac operator in block form

$$D = \begin{bmatrix} 0 & D^- \\ D^+ & 0 \end{bmatrix} \text{ where } D^\pm : \Gamma(S^\pm) \to \Gamma(S^\mp).$$

We call $D^\pm$ the *chiral* Dirac operators. As D is self-adjoint, $D^\pm$ are adjoint (i.e. transpose) to each other; in particular, the analogue in this situation of the relation $(\operatorname{im} A)^\perp = \ker A^T$ for a matrix A implies that

$$\dim \operatorname{coker} D^+ = \dim \ker D^-,$$

hence is also finite dimensional[2]. In general, an operator T between Hilbert spaces with finite dimensional kernel and cokernel is called *Fredholm*; its *index* is defined to be

$$\operatorname{ind}(T) = \dim \ker T - \dim \operatorname{coker} T \in \mathbb{Z}.$$

The following fundamental result (a special case of the Atiyah-Singer index theorem, one of the cornerstones of 20th century mathematics) relates the index of D^+ (which is an analytical quantity) to the topology of M; this is particularly striking

[2]There is a subtlety here: we use that, after suitably completing the spaces, $\operatorname{im} D^+$ is a closed subspace to identify its orthogonal with $\operatorname{coker} D^+$; closedness also follows from ellipticity.

because dim ker D^+ and dim coker D^+ are not themselves topological invariants (cf. [**Hit74**])!

THEOREM 1.1 ([**AS63**]). *The index of D^+ can be computed in terms of the Pontryagin classes $p_i(TM)$. In particular:*

- *if $n \equiv 2$ modulo 4, $\mathrm{ind}(D^+) = 0$;*
- *if $n = 4$, $\mathrm{ind}(D^+) = -\langle p_1, [M]\rangle/24$;*
- *if $n = 8$, $\mathrm{ind}(D^+) = \langle -4p_2 + 7p_1^2, [M]\rangle/5760$.*

More in general, the index is determined by an expression in the Pontryagin classes of TM called A-hat genus of M, and $\mathrm{ind}(D^+) = \langle \widehat{\mathcal{A}}(M), [M]\rangle$.

REMARK 1.3. When $n = 4$, the Hirzebruch signature formula $p_1 = 3\sigma$ implies that we can also write $\mathrm{ind}(D^+) = -\sigma(M)/8$, where σ denotes the signature.

REMARK 1.4. The index formula shows that in particular the index of D^+ is independent of the choice of spin structure on M.

For a more thorough discussion of the Atiyah-Singer index theorem and its various proofs we refer the reader to [**Sha78**] and [**Roe98**].

1.4. Dirac operators and geometry. Recall that Dirac's goal was to find a differential operator such that $D^2 = \Delta$. The Dirac operator on a spin manifold satisfies an analogous relation. Indeed, recall that our bundle $S \to M$ comes with a hermitian connection

$$\nabla : \Gamma(S) \to \Gamma(T^*M \otimes S).$$

We denote the formal adjoint in L^2 of such operator by

$$\nabla^* : \Gamma(T^*M \otimes S) \to \Gamma(S).$$

This is the differential operator determined by the identity

$$\langle \nabla\varphi, \alpha\rangle_{L^2} = \langle \varphi, \nabla^*\alpha\rangle_{L^2} \text{ for every } \varphi \in \Gamma(S), \alpha \in \Gamma(T^*M \otimes S),$$

and is (after changing sign) the bundle analogue of the classical divergence of a vector field.

The second order operator $\nabla^*\nabla$ is called the *Bochner Laplacian*; it makes sense on any bundle with Hermitian connection, and corresponds in $\mathbb{R}^n$ to the familiar expression

$$\Delta = -\mathrm{div} \circ \mathrm{grad}.$$

The fundamental identity relating the Dirac operator and geometry is then the following.

THEOREM 1.2 (Lichnerowicz '63 [**Lic63**]). *Let s be the scalar curvature of (M, g). Then $D^2 = \nabla^*\nabla + s/4$.*

The magic behind the formula is the following: D^2 and $\nabla^*\nabla$ are second order differential operators which by design have the same second order part; one would expect then their difference to be a first order differential operator, but it turns out to be a zeroth order one! This has surprising consequences for the study of the existence of metrics of positive scalar curvature, which is a quite subtle problem in Riemannian geometry.

PROPOSITION 1.3. *Suppose (M, g) is a compact spin manifold with $s > 0$ everywhere. Then the identity $\mathrm{ind}(D^+) = 0$ (which is topological in nature by Theorem 1.1) holds. In particular, if M is 4-dimensional, then $\sigma(M) = 0$.*

So for example the $K3$ surface, which is a spin 4-manifold with $\sigma = -16$ (see [**GS99**]), does not admit metrics of positive scalar curvature. Notice that the spin assumption is essential, as the Fubini-Study metric on $\mathbb{C}P^2$ has $s > 0$.

To prove the proposition, we just need to show that D (and therefore $D^{\pm}$) has trivial kernel. To see this, assume $D\varphi = 0$ and φ is not identically zero. The Lichnerowicz formula then implies

$$\nabla^*\nabla\varphi + \frac{s}{4}\varphi = 0.$$

Taking the L^2-inner product with φ, we obtain

$$0 = \langle \nabla^*\nabla\varphi, \varphi\rangle_{L^2} + \langle \frac{s}{4}\varphi, \varphi\rangle_{L^2} > \langle \nabla^*\nabla\varphi, \varphi\rangle_{L^2} = \langle \nabla\varphi, \nabla\varphi\rangle_{L^2} \geq 0$$

which is a contradiction.

1.5. Twisted Dirac operators. Consider now the case of the torus T^n, which admits a flat metric, hence a metric with $s \equiv 0$; it is natural to ask whether it admits a metric with $s > 0$. The answer is negative for $n = 2$ by the Gauss-Bonnet theorem. Also, notice that Myers' theorem [**Lee18**, Chapter 12] implies that T^n does not admit metrics with positive Ricci curvature because it has infinite fundamental group.

The tangent bundle of T^n is trivial, so in particular it is spin ($w_2 = 0$) and has vanishing Pontryagin classes; hence $\mathrm{ind}(D^+) = 0$ and the obstruction of Proposition 1.3 does not apply.

The key idea to obtain more subtle obstructions is to study twisted Dirac operators. Going back to the case of a general Riemannian manifold M equipped with a spin structure, consider flat hermitian connections B on the trivial line bundle $\underline{\mathbb{C}} \to M$; up to isomorphism, these are classified by their holonomy around loops, and correspond bijectively to the $b_1(M)$-dimensional torus

$$\mathbb{T}_M = H^1(M;\mathbb{R})/H^1(M;2\pi\mathbb{Z}).$$

To each of these, one can associate a twisted connection ∇_B on S, and correspondingly a twisted Dirac operator D_B on $\Gamma(S)$ which satisfies the analogues of Theorem 1.1 and 1.2[3]. For example, on $M = S^1 = \mathbb{R}/2\pi\mathbb{Z}$, the twisted Dirac operators are of the form

$$i\frac{d}{dt} + c \text{ for } c \in \mathbb{R}.$$

Two such operators are conjugate exactly when $c - c' \in \mathbb{Z}$, so that we get a $\mathbb{T}_{S^1} = S^1$-family of operators. In the next lecture, we will see that the family of Fredholm operators $\{D_B\}$ carries much more information than a single operator.

1.6. Exercises.

(1) The Dirac operator (2) on $\mathbb{R}^2$ is translation invariant, and descends to the Dirac operator on the torus $\mathbb{R}^2/\mathbb{Z}^2$. Determine the space of harmonic spinors, and check that the index formula for D^+ holds in this case.
(2) By looking carefully at the argument in Proposition 1.3, show that if (M, g) is a spin 4-manifold with $s \equiv 0$, then $|\sigma(M)| \leq 16$.
(3) Use Fourier series to compute the spectrum of the twisted Dirac operator $i\frac{d}{dt} + c$ on S^1; show that two such operators are conjugate when $c - c' \in \mathbb{Z}$.

[3] Here it is important that B is flat.

2. K-theory and the index theorem for families

Consider a complex separable Hilbert space H, e.g. $H = L^2(M)$. Recall that a bounded operator $T : H \to H$ is Fredholm if it has finite dimensional kernel and cokernel[4]. Denote by $\mathrm{Fred}(H)$ the space of such operators; one can show that it is an open set of the space of bounded operators, and the map

$$\mathrm{ind} : \mathrm{Fred}(H) \to \mathbb{Z}$$

is locally constant (even though the dimensions of kernel and cokernel are not!). It turns out that the space $\mathrm{Fred}(H)$ has very interesting topology, which we now describe.

2.1. K-theory. We discuss the essential notions in topological K-theory, and refer the reader to [**FF16**, Ch. 38] for more details. Consider a finite CW complex X, which will assume for simplicity to be connected (in our applications X will be the torus $\mathbb{T}_M$). Denote by $\mathrm{Vect}(X)$ the set of isomorphism classes of complex vector bundles on X. The operation of direct sum $\oplus$ and tensor product $\otimes$ make $\mathrm{Vect}(X)$ into a *semiring*, i.e. a structure like a ring but without additive inverses. For example, $\mathbb{N}$ is a semiring, and there is indeed a semiring homomorphism $\mathrm{Vect}(X) \to \mathbb{N}$ given by the rank[5]. Notice that, unlike $\mathbb{N}$, $\mathrm{Vect}(X)$ is not in general *cancellative*, i.e. $a + c = b + c$ does not in general imply $a = b$.

Of course, it is nicer in general to work with rings than semirings. Given any semiring S, there is a natural way to associate a ring called the *Grothendieck ring* $\mathrm{Gr}(S)$. This follows the usual construction of $\mathbb{Z}$ from $\mathbb{N}$; one defines $\mathrm{Gr}(S)$ to be the set of formal differences $a - b$ of elements in S, modulo the relation

$$a - b \sim a' - b' \text{ if } a + b' + e = a' + b + e \text{ for some } e.$$

Here we need the extra e to have a good definition for non-cancellative semirings.

DEFINITION 2.1. For a finite CW complex X, the K-theory of X, denoted by $K(X)$, is the Grothendieck ring of $\mathrm{Vect}(X)$.

Here are some basic properties:

- An element of $K(X)$ is represented by a formal difference of vector bundles $E - F$;
- There is a ring homomorphism $K(X) \to \mathbb{Z}$ given by the rank;
- K gives rise to a contravariant functor from finite CW complexes to rings; homotopic maps induce the same ring homomorphism.

Denote by $\mathrm{Vect}_k(X)$ the isomorphism classes of k-dimensional vector bundles. Recall that we have the identification with the set of homotopy classes

$$\mathrm{Vect}_k(X) = [X, BU(k)]$$

where, concretely, $BU(k)$ is the union of the Grassmannians $G(n,k)$ of k-planes in $\mathbb{C}^n$. Now, for any vector bundle $E \to X$, there is a vector bundle E' such that

[4]Because we are working with Hilbert (hence Banach) spaces, the image of such T is automatically closed; when working with more general topological vector spaces the latter needs to be incorporated in the definition of Fredholm.

[5]If X is not connected, one allows vector bundles with different ranks on each of them, and this map depends on a choice of connected component.

$E \oplus E'$ is trivial. This implies that any element in $K(X)$ has a representative of the form $F - \underline{\mathbb{C}}^m$ for some vector bundle F and $m \in \mathbb{N}$. From this, one shows that

$$K(X) = [X, BU \times \mathbb{Z}]$$

where $BU = \bigcup_{k \geq 1} BU(k)$, and $\mathbb{Z}$ corresponds to the rank.

2.2. Relation with Fredholm operators. It turns out that K-theory is the right algebraic tool to study families of Fredholm operators because of the following.

THEOREM 2.2 (Atiyah-Jänich). *For a finite CW complex X , there is a natural bijection*

$$K(X) = [X, \mathrm{Fred}(H)],$$

i.e. the space of Fredholm operators is also a classifying space for K-theory. In particular, homotopy classes of families of Fredholm operators parametrized by X are in bijection with $K(X)$.

As a consequence, this also tells us that the space $\mathrm{Fred}(H)$ has a very interesting topology.

Following [**Ati67**, Appendix], let us discuss how to associate to a family $\{T_x\}_{x \in X}$ of Fredholm operators $T_x : H \to H$ a class in $K(X)$. Intuitively, this corresponds to the difference between the kernel and cokernel at each point of T_x. Of course, these do not form a vector bundle (the dimensions might not even be constant!) so we need a different approach to formalize this. Because X is compact, one can find a finite dimensional subspace $W \subset H$ such that

$$\mathrm{Im}(T_x) + W = H$$

for all $x \in X$. Then, $V = \bigcup T_x^{-1}(W)$ is a vector bundle over X, and we associate to $\{T_x\}$ the element

$$\mathrm{ind}\{T_x\} := V - \underline{\mathbb{C}}^{\dim W} \in K(X)$$

called the *index* of the family. Notice that the rank map $K(X) \to \mathbb{Z}$ sends the index of the family to the index of a single operator in the family. It turns out that $\mathrm{ind}\{T_x\}$ is independent of choices, and that the assignment is a bijection. A key input in the proof is Kuiper's theorem, which states that the unitary group of a Hilbert space is contractible.[6]

REMARK 2.1. For our purposes, it is important to point out the following: if $\{T_x\}$ is a family of isomorphisms, the corresponding element in $K(X)$ is zero. This is because we can choose $W = \{0\}$.

2.3. The index theorem for families of twisted Dirac operators. Consider now the family of twisted Dirac operators $\{D_B^+\}$ on M parametrized by the torus $\mathbb{T}_M$ as in Section 1.5, and the corresponding element

$$\mathrm{ind}\{D_B^+\} \in K(\mathbb{T}_M).$$

Theorem 1.1 gives us a formula for the image of such element under the dimension map $K(\mathbb{T}_M) \to \mathbb{Z}$; we would like to obtain more detail information. Of course, K-theory is a quite complicated object to deal with, so it would be desirable to work

[6]This also implies that any Hilbert bundle over a CW complex is trivial. In particular, after trivializing, a family of operators on a Hilbert bundle can be considered as a family of operators on fixed Hilbert space; we implicitly did so for the family of twisted Dirac operators on S^1.

with something simpler such as cohomology. Luckily, one can use Chern classes c_i to define a ring homomorphism

$$\mathrm{ch} : K(X) \to H^{\mathrm{even}}(X;\mathbb{Q})$$

called the Chern character; this is given on a vector bundle E by

$$\mathrm{ch}(E) = \dim E + c_1(E) + \frac{c_1^2 - 2c_2}{2} + \dots;$$

in particular, the degree zero part of ch is exactly the dimension. It turns out that ch is a rational isomorphism.

REMARK 2.2. By contrast, the torsion subgroups of $K(X)$ and $H^{\mathrm{even}}(X;\mathbb{Z})$ are in general very different.

The index theorem for families will provide a formula for

$$\mathrm{ch}(\mathrm{ind}\{D_B^+\}) \in H^{\mathrm{even}}(\mathbb{T}_M;\mathbb{Q})$$

in terms of the topology of M. Recall that $\mathbb{T}_M = H^1(M;\mathbb{R})/H^1(M;2\pi\mathbb{Z})$, so that we can identify

$$H_1(\mathbb{T}_M;\mathbb{Z}) = H^1(M;\mathbb{Z}).$$

Via this identification, a basis $x_1,\dots,x_m$ of $H^1(M;\mathbb{Z})$ determines a dual basis $y_1,\dots,y_m$ of $H^1(\mathbb{T}_M;\mathbb{Z}) = H_1(\mathbb{T}_M;\mathbb{Z})^*$. We can then consider the element

$$\Omega = \sum x_i \otimes y_i \in H^1(M;\mathbb{Z}) \otimes H^1(\mathbb{T}_M;\mathbb{Z}) \subset H^2(M \times \mathbb{T}_M;\mathbb{Z}),$$

which is independent of the choice of basis[7]. Recall that the A-hat genus is the combination of Pontryagin classes

$$\widehat{\mathcal{A}}(M) = 1 - \frac{p_1}{24} + \frac{7p_1^2 - 4p_2}{5760} + \dots,$$

which is an element of $H^{\mathrm{even}}(M;\mathbb{Q})$.

THEOREM 2.3 ([**AS71**]). *Consider an even dimensional spin manifold M. Then*

$$\mathrm{ch}(\mathrm{ind}\{D_B^+\}) = \langle e^{\Omega} \cdot \widehat{\mathcal{A}}(M), [M]\rangle \in H^{even}(\mathbb{T}_M.\mathbb{Q}).$$

Here $e^{\Omega} = \sum_{n\geq 0} \Omega^n/n!$ and $e^{\Omega} \cdot \widehat{\mathcal{A}}(M)$ defines an element in $H^{even}(M \times \mathbb{T}_M;\mathbb{Q})$, which can be evaluated on the fundamental class $[M]$ to obtain an element in $H^{even}(\mathbb{T}_M;\mathbb{Q})$.

We refer the reader to [**LM89**] for a more detailed discussion of this result and its generalizations.

2.4. Metrics of positive scalar curvature on the torus. We are now ready to prove the following theorem of Gromov and Lawson.

THEOREM 2.4 ([**GL80**]). *The torus T^n does not admit a metric of positive scalar curvature.*

REMARK 2.3. In fact, they prove the stronger result that a metric on T^n with $s \geq 0$ is necessarily flat.

[7] More intrinsically, Ω is the first Chern class of the so-called *Poincaré line bundle* on $M \times \mathbb{T}_M$.

To see this, we can assume n is even (otherwise we can consider $S^1 \times T^n$). If $s > 0$, then the argument of Proposition 1.3 shows that D_B^+ has trivial kernel and cokernel, hence is an isomorphism, for all B. By Remark 2.1, this implies that $\mathrm{ch}(\mathrm{ind}\{D_B^+\}) = 0$. As $\widehat{\mathcal{A}}(T^n) = 1$, the index formula simplifies to

$$\mathrm{ch}(\mathrm{ind}\{D_B^+\}) = \langle e^{\Omega}, [T^n]\rangle = \langle \frac{\Omega^n}{n!}, [T^n]\rangle =$$
$$= \pm\langle x_1 \cup \cdots \cup x_n \otimes y_1 \cup \cdots \cup y_n, [T^n]\rangle = \pm y_1 \cup \cdots \cup y_n \neq 0$$

because the cup products $x_1 \cup \cdots \cup x_n$ and $y_1 \cup \cdots \cup y_n$ evaluate to ± 1 on the fundamental classes $[T^n]$ and $[\mathbb{T}_{T^n}]$ of the tori respectively, and we obtain a contradiction.

2.5. Exercises.

(1) Recall that BU is simply connected and has $H^2(BU;\mathbb{Z}) = \mathbb{Z}$, generated by the first Chern class c_1. Use this to compute $K(S^2)$, and describe explicitly the generators.

(2) We will discuss a very concrete example of Kuiper's theorem. Consider the family of unitary maps T_ϑ on l^2 parametrized by $\vartheta \in S^1$ given in the standard basis $\{e_i\}$ by

$$T(e_i) = \begin{cases} e^{i\vartheta} e_i \text{ if } i = 0 \\ e_i \text{ if } i > 0. \end{cases}$$

Show that this this family is homotopically trivial by showing that the family

$$\begin{bmatrix} e^{i\vartheta} & 0 \\ 0 & e^{-i\vartheta} \end{bmatrix} \in \mathrm{SU}(2)$$

is trivial, and applying the Mazur swindle

$$1 = 1 + (-1+1) + (-1+1) + \cdots = (1-1) + (1-1) + \cdots = 0.$$

(3) Show that an orientable surface Σ is spin. Use the index theorem for families to show that on any Riemannian surface (M, g) of genus at least one there exists a twisted Dirac operator D_B^+ which has non-trivial kernel.

3. Dirac operators and the Weinstein conjecture in dimension three

3.1. Self-adjoint operators. In our previous examples we focused on the case of even dimensional manifolds, as in this case we have the decomposition of D_B into two chiral operators $D_B^{\pm}$ adjoint to each other. In odd dimensions, there is no such canonical splitting, so we can only work with the Dirac operators D_B themselves. Notice that as these are self-adjoint, $\mathrm{coker}\, D_B = \ker D_B$ so they all have automatically index zero, and hence there is no interesting analogue of Theorem 1.1. On the other hand, it turns out that it is interesting to study families of them. More abstractly, consider again a complex separable Hilbert space H. Recall that an operator T is *self-adjoint* if

$$\langle Tx, y\rangle = \langle x, Ty\rangle \text{ for all } x, y \in H.$$

This is the infinite dimensional analogue of a being symmetric, but one has to be careful as the usual spectral theorem in general does not hold: in fact, T might not even have eigenvalues at all. On the other hand, the differential operators we are interested in (e.g. Δ and D_B) are always diagonalizable on a compact manifold, so for simplicity we will only discuss their case[8]. It turns out that the spectral theory of such operators can have very different qualitative behavior; for example, looking at complex-valued functions on the circle S^1:

- for $c \in \mathbb{R}$ the shifted Laplacian $\Delta + c = -\frac{d^2}{dx^2} + c$ has eigenvalues $n^2 + c$ for $n \in \mathbb{Z}$, hence only finitely many negative eigenvalues and infinitely many positive ones. We call operators like this *essentially positive*;
- on the other hand $-\Delta + c$ has infinitely many negative eigenvalues and finitely many positive ones, and we call it *essentially negative*;
- finally, the twisted Dirac operator $i\frac{d}{dx} + c$ has eigenvalues $n + c$ for $n \in \mathbb{Z}$ hence it has infinitely many negative eigenvalues and infinitely many positive eigenvalues. We call operators like this *indefinite*.

We then have the following.

THEOREM 3.1 ([**AS69**]). *The space* $\mathrm{Fred}^{sa}(H)$ *of Fredholm self-adjoint operators on* H *has three connected components*

$$\mathrm{Fred}^{sa}(H) = \{\textit{essentially positive}\} \amalg \{\textit{essentially negative}\} \amalg \{\textit{indefinite}\}.$$

The first two components are contractible, while the last component has the homotopy type of $U = \bigcup_{n\geq 1} U(n)$.

REMARK 3.1. One should not confuse U with the unitary group $U(H)$ of a Hilbert space H; the latter is contractible by Kuiper's theorem, while for example $\pi_1(U) = \mathbb{Z}$.

As in the case of the circle S^1, a twisted Dirac operator D_B on a compact manifold M is an indefinite operator, hence the family $\{D_B\}$ gives rise to a map

$$\mathbb{T}_M \to U,$$

well-defined up to homotopy.

Now, for a finite connected CW complex X, the set $[X, U]$ can be identified with $K^1(X)$ the *odd* K-theory of X; we also denote $K(X)$ by $K^0(X)$, and call it even K-theory.

REMARK 3.2. More concretely, $K^1(X)$ is usually defined as the the kernel of the rank homomorphism $K^0(\Sigma X) \to \mathbb{Z}$, where ΣX is the reduced suspension of X. This is equivalent to the definition above because

$$\ker\left(K^0(\Sigma X) \to \mathbb{Z}\right) = [\Sigma X, BU] = [X, \Omega BU] = [X, U],$$

where we used that reduced suspension Σ is the adjoint of loopspace Ω, and ΩBG is equivalent to G for any G.

In particular, a family $\{S_x\}$ of indefinite self-adjoint operators parametrized by X determines an index element in $\mathrm{ind}\{S_x\} \in K^1(X)$. As in the even case, there is a Chern character map

$$\mathrm{ch} : K^1(X) \to H^{\mathrm{odd}}(X; \mathbb{Q})$$

which is a rational isomorphism. We are then ready to state the analogue of Theorem 2.3 in the case of odd dimensional manifolds.

---

[8]In general, one needs to rephrase the discussion that follows in terms of the quadratic form $x \mapsto \langle x, Tx \rangle$.

THEOREM 3.2 ([**APS76**]). *Consider an odd dimensional spin manifold M. Then*

$$\mathrm{ch}(\mathrm{ind}\{D_B\}) = \langle e^{\Omega} \cdot \widehat{\mathcal{A}}(M), [M]\rangle \in H^{odd}(\mathbb{T}_M; \mathbb{Q})$$

where again $e^{\Omega} = \sum \Omega^n/n!$.

For example, when $M = S^1$, this shows that the family of twisted Dirac operators $i\frac{d}{dt} + c$ corresponds to a generator of $\pi_1(U) = \mathbb{Z}$.

REMARK 3.3. In fact, $\pi_1(U) = \mathbb{Z}$ is very closely related to the *spectral flow* of a family of operators, a key concept in Floer theory.

3.2. The Weinstein conjecture. We want to discuss how families of Dirac operators are related to a dynamical problem in contact geometry that we now recall (see [**CdS01**, Part IV] for a more detailed discussion). Given an oriented $(2n-1)$-dimensional manifold Y, a *contact form* is a 1-form α such that

$$\alpha \wedge (d\alpha)^{n-1} > 0 \text{ at each point.}$$

The basic example is $\mathbb{R}^{2n-1}$ with coordinates $(x_1, y_1, \dots, x_{n-1}, y_{n-1}, z)$ and $\alpha = dz - \sum y_i dx_i$. Now, at each point $d\alpha$ is a skew-symmetric bilinear form with maximal rank, so there exists a unique vector field R_α, called the *Reeb vector field*, for which

$$d\alpha(R_\alpha, -) = 0 \text{ and } \alpha(R_\alpha) = 1.$$

For example, in the case of $\mathbb{R}^n$ with the contact form above, $R_\alpha = \frac{d}{dz}$. The following conjecture is motivated by questions in classical (Hamiltonian) mechanics.

CONJECTURE 3.3 (Weinstein). *If (Y, α) is a closed contact manifold, then the Reeb vector field R_α admits a closed orbit.*

Of course, the example of $\mathbb{R}^n$ above shows the assumption that M is closed is necessary. The Weinstein conjecture has been proved in many cases, under assumptions of very different nature; we refer for [**Hut10**] for a more thorough discussion of the literature. Our goal here is to discuss how families of Dirac operators enter in the proof of the following celebrated result of Taubes.

THEOREM 3.4 ([**Tau07**]). *The Weinstein conjecture holds in dimension three.*

The rough idea is the following: one can perturb the (curvature part of the) Seiberg-Witten equations on Y [**KM07**, Ch. 4] by a very large multiple of the contact form α; in this limit, solutions to the Seiberg-Witten equations naturally give rise to certain union of orbits of R_α (including the *empty* orbit). Because the count of solutions is independent of the perturbation (in a suitable sense), one can use computations done purely in Seiberg-Witten theory to show that there are closed Reeb orbits.

In fact, the following result (which is really is a consequence of the subsequently proved result that embedded contact homology coincides with monopole Floer homology, see [**Hut10**]) holds.

THEOREM 3.5 (Taubes). *If (Y^3, α) has no closed Reeb orbit, then for the monopole Floer homology*[9] *groups of Y we have*

$$\widehat{HM}_*(Y, \mathfrak{s}) = \begin{cases} \mathbb{Z} \text{ if } \mathfrak{s} \text{ is the spin}^c \text{ structure corresponding to } \ker(\alpha) \\ 0 \text{ for all other spin}^c \text{ structures on } Y. \end{cases}$$

[9]This is phrased in terms of cohomology in [**Hut10**]; the statement here follows from the universal coefficient theorem.

Here, $\mathbb{Z}$ corresponds to the empty Reeb orbit.

The key step is then the following non-triviality result of Kronheimer and Mrowka.

THEOREM 3.6 ([**KM07**], Chapter 35). *For any Y, if $\mathfrak{s}$ is a torsion spin^c structure, $\widehat{HM}_*(Y,\mathfrak{s})$ is non-vanishing (and in fact has positive rank) in infinitely many gradings.*

This concludes the argument because every three-manifold admits a torsion spin^c structure (for example one induced by a spin structure, which exists because TY is trivial [**MS74**, Chapter 12]). Recall that for a torsion spin^c structure $\widehat{HM}_*(Y,\mathfrak{s})$ correponds to the Heegaard Floer homology group $HF^-(Y,\mathfrak{s})$ (see [**CGH11**], [**KLT20**] and subsequent papers).

3.3. Relations with the triple cup product. Theorem 3.6 is proved by showing that the bar version of monopole Floer homology $\overline{HM}_*(Y,\mathfrak{s})$ has positive rank in infinitely many gradings. Recall that using the $\mathbb{Z}[U]$-module structure, this can be thought as the localization

$$\widehat{HM}_*(Y,\mathfrak{s})\otimes_{\mathbb{Z}[U]}\mathbb{Z}[U,U^{-1}],$$

and for a torsion spin^c structure it corresponds to $HF^\infty(Y,\mathfrak{s})$ in Heegaard Floer homology.

The bar version turns out to be a much simpler object to study. Indeed, Kronheimer and Mrowka provide in [**KM07**] a complete computation of $\overline{HM}_*(Y,\mathfrak{s};\mathbb{Q})$. Later, in [**Lid12**] Lidman provided a computation with $\mathbb{Z}/2$-coefficients using surgery techniques in Heegaard Floer homology. Finally, in recent joint work with Miller Eismeier [**LME23**], we built on the work of Kronheimer and Mrowka to provide a full computation with $\mathbb{Z}$-coefficients using techniques from homotopical algebra. We will give the idea of proof of the following statement, which is an important step in the proof of Theorem 3.6.

THEOREM 3.7 ([**KM07**], Ch. 35). *For any Y, if $\mathfrak{s}$ is a torsion spin^c structure (for example a spin^c structure induced by a spin structure), $\overline{HM}_*(Y,\mathfrak{s})$ only depends on the triple cup product map*

$$\begin{aligned}\cup^3_Y:\Lambda^3H^1(Y;\mathbb{Z})&\to\mathbb{Z}\\ \alpha\wedge\beta\wedge\gamma&\mapsto\langle\alpha\cup\beta\cup\gamma,[Y]\rangle.\end{aligned}$$

In particular, $\overline{HM}_*(Y,\mathfrak{s})$ only depends on the algebraic topology of Y, while $\widehat{HM}_*(Y,\mathfrak{s})$ is a much subtler invariant of Y. Intuitively, this is because while $\widehat{HM}_*(Y,\mathfrak{s})$ involves both irreducible and reducible solutions to the Seiberg-Witten equations, $\overline{HM}_*(Y,\mathfrak{s})$ only involves the reducible ones. Because of this, it can be described purely in terms of the family of Dirac operators $\{D_B\}$ parametrized by the torus $\mathbb{T}_Y$.

REMARK 3.4. This is a manifestation in Floer homology of the classical localization theorem in S^1-equivariant cohomology, see for example [**Tu20**].

More specifically, associated with $(Y,\mathfrak{s})$ there is a family of Dirac operators $\{D_B\}$ parametrized by $\mathbb{T}_Y$ which generalizes the family of twisted Dirac operators associated to a spin structure.[10] After choosing a Morse function

$$f : \mathbb{T}_Y \to \mathbb{R},$$

the chain complex computing $\overline{HM}_*(Y, \mathfrak{s})$ is generated by pairs (B_*, E) where B_* is a critical point of f, and E is an eigenspace of D_{B_*}. The differential counts solutions to equations combining flowlines $B(t) \in \mathbb{T}_Y$ of $-\nabla f$ and the family of operators $\{D_{B(t)}\}$.

Then the key point is that this construction can be generalized to any family of indefinite self-adjoint Fredholm operators $\mathbb{T}_Y \to \mathrm{Fred}^{\mathrm{sa}}(H)$ (and not just the family of Dirac operators); the result, called *coupled Morse homology*, only depends on the homotopy class of the family. In our case this is the index

$$\mathrm{ind}\{D_B\} \in [\mathbb{T}_Y, U] = K^1(\mathbb{T}_Y).$$

For dimensional reasons we have $\widehat{A}(Y) = 1$, so the index formula in Theorem 3.2 (which also holds in the slightly more general situation of torsion spin^c structures) gives the computation

$$\mathrm{ch}(\mathrm{ind}\{D_B\}) = \cup_Y^3 \in H^{\mathrm{odd}}(\mathbb{T}_Y)$$

under the identification

$$H^3(\mathbb{T}_Y; \mathbb{Z}) = \Lambda^3 H^1(\mathbb{T}_Y; \mathbb{Z}) = \left(\Lambda^3 H^1(Y; \mathbb{Z})\right)^*.$$

Finally, one can show that $K^1(\mathbb{T}_Y)$ is torsion-free, so that the Chern character (which is a rational isomorphism) is injective and $\mathrm{ind}\{D_B\}$ is determined by its Chern character.

In fact, the same computation can be used to show that the classifying map $\mathbb{T}_Y \to U$ for the family $\{D_B\}$ factors through $\mathbb{T}_Y \to SU(2)$. This is a key ingredient in the proof of Theorem 3.6 because for these simpler families one can provide a concrete description of coupled Morse homology in terms of suitably 'twisted' versions of more familiar theories. In particular, Theorem 3.6 is proved by identifying $\overline{HM}^*(Y, \mathfrak{s})$ over the reals with a version of *twisted de Rham cohomology* for which the underlying vector space is $\Omega^*(\mathbb{T}_Y) \otimes \mathbb{R}[U, U^{-1}]$[11] and the differential is

$$x \mapsto dx + (\zeta \wedge x)U^{-1}$$

for some closed 3-form ζ with $[\zeta] = \cup_Y^3 \in H^3(\mathbb{T}_Y; \mathbb{R})$ [**KM07**, Chapter 34.3]. Notice that this is a differential because $\zeta \wedge \zeta = 0$.

3.4. Exercises.

(1) Let $f : S^1 \to \mathbb{R}$ a non-constant continuous function. Show that the map on $L^2(S^1)$ (complex valued functions) given by multiplication by f is self-adjoint but does not have eigenvalues.

(2) Show that the family of operators T_ϑ in Exercise 2.2 defines a loop in $U = \bigcup U(n)$, and in fact represents a generator of $\pi_1(U)$. Why does the Mazur swindle argument not apply in this situation?

(3) Consider a link $L \subset S^3$. Show that the branched double cover $\Sigma(L)$ has vanishing triple cup product by considering the action of the covering involution on cohomology. Use this to show that the three-torus T^3 is not a branched double cover of S^3.

---

[10] Here one should think of $\mathbb{T}_Y$ as the space of reducible solutions to the Seiberg-Witten equations.

[11] Here we are working with cohomology, so the action of U has degree $+2$.

Acknowledgments

The author thanks the participants of the summer school, whose questions greatly helped to improve these notes. He also thanks the anonymous referee for carefully reading the manuscript.

References

[APS76] M. F. Atiyah, V. K. Patodi, and I. M. Singer, *Spectral asymmetry and Riemannian geometry. III*, Math. Proc. Cambridge Philos. Soc. **79** (1976), no. 1, 71–99, DOI 10.1017/S0305004100052105. MR397799

[AS63] M. F. Atiyah and I. M. Singer, *The index of elliptic operators on compact manifolds*, Bull. Amer. Math. Soc. **69** (1963), 422–433, DOI 10.1090/S0002-9904-1963-10957-X. MR157392

[AS69] M. F. Atiyah and I. M. Singer, *Index theory for skew-adjoint Fredholm operators*, Inst. Hautes Études Sci. Publ. Math. (1969), no. 37, 5–26. MR285033

[AS71] M. F. Atiyah and I. M. Singer, *The index of elliptic operators. IV*, Ann. of Math. (2) **93** (1971), 119–138, DOI 10.2307/1970756. MR279833

[Ati67] M. F. Atiyah, *K-theory*, W. A. Benjamin, Inc., New York-Amsterdam, 1967. Lecture notes by D. W. Anderson. MR224083

[CdS01] Ana Cannas da Silva, *Lectures on symplectic geometry*, Lecture Notes in Mathematics, vol. 1764, Springer-Verlag, Berlin, 2001, DOI 10.1007/978-3-540-45330-7. MR1853077

[CGH11] Vincent Colin, Paolo Ghiggini, and Ko Honda, *Equivalence of Heegaard Floer homology and embedded contact homology via open book decompositions*, Proc. Natl. Acad. Sci. USA **108** (2011), no. 20, 8100–8105, DOI 10.1073/pnas.1018734108. MR2806645

[Dir28] P. A. M. Dirac, *The Quantum Theory of the Electron*, Proceedings of the Royal Society A: Mathematical, Physical and Engineering Sciences **117 (778)** (1928), 610–624.

[Eva10] Lawrence C. Evans, *Partial differential equations*, 2nd ed., Graduate Studies in Mathematics, vol. 19, American Mathematical Society, Providence, RI, 2010, DOI 10.1090/gsm/019. MR2597943

[FF16] Anatoly Fomenko and Dmitry Fuchs, *Homotopical topology*, 2nd ed., Graduate Texts in Mathematics, vol. 273, Springer, [Cham], 2016, DOI 10.1007/978-3-319-23488-5. MR3497000

[GL80] Mikhael Gromov and H. Blaine Lawson Jr., *Spin and scalar curvature in the presence of a fundamental group. I*, Ann. of Math. (2) **111** (1980), no. 2, 209–230, DOI 10.2307/1971198. MR569070

[Gre21] Joshua Evan Greene, *Heegaard Floer homology*, Notices Amer. Math. Soc. **68** (2021), no. 1, 19–33, DOI 10.1090/noti2194. MR4202313

[GS99] Robert E. Gompf and András I. Stipsicz, *4-manifolds and Kirby calculus*, Graduate Studies in Mathematics, vol. 20, American Mathematical Society, Providence, RI, 1999, DOI 10.1090/gsm/020. MR1707327

[Hit74] Nigel Hitchin, *Harmonic spinors*, Advances in Math. **14** (1974), 1–55, DOI 10.1016/0001-8708(74)90021-8. MR358873

[Hut10] Michael Hutchings, *Taubes's proof of the Weinstein conjecture in dimension three*, Bull. Amer. Math. Soc. (N.S.) **47** (2010), no. 1, 73–125, DOI 10.1090/S0273-0979-09-01282-8. MR2566446

[Kir89] Robion C. Kirby, *The topology of 4-manifolds*, Lecture Notes in Mathematics, vol. 1374, Springer-Verlag, Berlin, 1989, DOI 10.1007/BFb0089031. MR1001966

[KLT20] Çağatay Kutluhan, Yi-Jen Lee, and Clifford Henry Taubes, HF = HM*, I: Heegaard Floer homology and Seiberg-Witten Floer homology*, Geom. Topol. **24** (2020), no. 6, 2829–2854, DOI 10.2140/gt.2020.24.2829. MR4194305

[KM07] Peter Kronheimer and Tomasz Mrowka, *Monopoles and three-manifolds*, New Mathematical Monographs, vol. 10, Cambridge University Press, Cambridge, 2007, DOI 10.1017/CBO9780511543111. MR2388043

[Lee18] John M. Lee, *Introduction to Riemannian manifolds*, Graduate Texts in Mathematics, vol. 176, Springer, Cham, 2018. Second edition of [MR1468735]. MR3887684

[Lic63] André Lichnerowicz, *Spineurs harmoniques* (French), C. R. Acad. Sci. Paris **257** (1963), 7–9. MR156292

[Lid12] Tye Lidman, *Triple Cup Products in Heegaard Floer Homology*, ProQuest LLC, Ann Arbor, MI, 2012. Thesis (Ph.D.)–University of California, Los Angeles. MR3034622

[Lin16] Francesco Lin, *Lectures on monopole Floer homology*, Proceedings of the Gökova Geometry-Topology Conference 2015, Gökova Geometry/Topology Conference (GGT), Gökova, 2016, pp. 39–80. MR3526838

[LM89] H. Blaine Lawson Jr. and Marie-Louise Michelsohn, *Spin geometry*, Princeton Mathematical Series, vol. 38, Princeton University Press, Princeton, NJ, 1989. MR1031992

[LME23] Francesco Lin and Mike Miller Eismeier, *Monopoles, twisted integral cohomology, and Hirsch algebras*, to appear in Geometry and Topology (2023).

[MS74] John W. Milnor and James D. Stasheff, *Characteristic classes*, Annals of Mathematics Studies, No. 76, Princeton University Press, Princeton, NJ; University of Tokyo Press, Tokyo, 1974. MR440554

[Roe98] John Roe, *Elliptic operators, topology and asymptotic methods*, 2nd ed., Pitman Research Notes in Mathematics Series, vol. 395, Longman, Harlow, 1998. MR1670907

[Sha78] Patrick Shanahan, *The Atiyah-Singer index theorem*, Lecture Notes in Mathematics, vol. 638, Springer, Berlin, 1978. An introduction. MR487910

[Tau07] Clifford Henry Taubes, *The Seiberg-Witten equations and the Weinstein conjecture*, Geom. Topol. **11** (2007), 2117–2202, DOI 10.2140/gt.2007.11.2117. MR2350473

[Tu20] Loring W. Tu, *Introductory lectures on equivariant cohomology*, Annals of Mathematics Studies, vol. 204, Princeton University Press, Princeton, NJ, 2020. With appendices by Tu and Alberto Arabia. MR4205963

[War71] Frank W. Warner, *Foundations of differentiable manifolds and Lie groups*, Scott, Foresman & Co., Glenview, Ill.-London, 1971. MR295244

Department of Mathematics, Columbia University, New York, New York 10027
Email address: flin@math.columbia.edu

Proceedings of Symposia in Pure Mathematics
Volume **109**, 2024
https://doi.org/10.1090/pspum/109/02003

Lectures on generalised Seiberg–Witten equations

Thomas Walpuski

Dedicated to Tomasz Mrowka on the occasion of his 60th birthday.

These are the notes for a mini-course taught at the 2022 ICTP summer school *Frontiers in Geometry and Topology* consisting of three lectures. The purpose of this course was to remind the audience of the existence of four-dimensional gauge theories beyond anti-self-duality and Seiberg–Witten theory, put these theories into the framework of generalised Seiberg–Witten equations, and provide a glimpse of the new challenges presented by these equations.

1. "New" Four-dimensional gauge theories

This lecture discusses some gauge-theoretic partial differential equations, mostly introduced by physicists, which are known or conjectured to be important. The purpose of this is to illustrate that gauge theory is not just interesting in and of itself, but also has links to other areas of mathematics and physics.

1.1. Anti-self-duality in Yang–Mills theory. A resonable minimalist setup for gauge theory consists of: a connected oriented Riemannian manifold (X, g), a compact connected semi-simple Lie group G, the **structure group**, and a G–principal bundle $(\pi\colon P \to X, P \circlearrowleft G)$. The space of G–principal connections on P is denoted by $\mathscr{A}(P)$. This is an affine space modelled on $\Omega^1(X, \mathrm{Ad}(P))$ with

$$\mathrm{Ad}(P) := P \times_{\mathrm{Ad}} \mathfrak{g} \tag{1.1}$$

denoting the **adjoint bundle**. The **curvature** of $A \in \mathscr{A}(P)$ is denoted by $F_A \in \Omega^2(X, \mathrm{Ad}(P))$. Everything is acted upon by $\mathscr{G}(P)$, the **gauge group**. (Much more detailed treatments of this setup can be found in [**DK90**, §2], [**Bau14**], [**Ham17**, Part I].)

The **Yang–Mills functional** $\mathrm{YM}\colon \mathscr{A}(P) \to [0, \infty]$ is defined by

$$\mathrm{YM}(A) := \frac{1}{2}\int_X |F_A|^2 \,\mathrm{vol}_g. \tag{1.2}$$

Here $|\cdot|$ is induced by minus the Killing form on $\mathfrak{g}$. The Yang–Mills functional is $\mathscr{G}(P)$–invariant. A straight-forward computation reveals that the Euler–Lagrange equation of the Yang–Mills functional is the **Yang–Mills equation**:

$$\mathrm{d}_A^* F_A = 0. \tag{1.3}$$

This work was partially supported by the National Science foundation under grant No. DMS-1928930.

This is a second-order differential equation on the connection A.

It is not trivial to solve (1.3). However, there are (simpler) first-order equations that imply (1.3). Of course, $A \in \mathscr{A}(P)$ satisfies (1.3) if it is **flat**, that is, $F_A = 0$. The monodromy representation induces a bijection

$$\coprod_P \{A \in \mathscr{A}(P) : F_A = 0\}/\mathscr{G}(P) \cong \mathrm{Hom}(\pi_1(X, x_0), G)/G \tag{1.4}$$

with G acting by conjugation. This seems rather satisfactory since it turns a question in geometric analysis into one in algebraic topology. However, $\mathrm{Hom}(\pi_1(X, x_0), G)$ is not as easy to determine as one might expect. In fact, for 3–manifolds, it turns out to be fruitful to use (1.4) in the other direction; cf. Taubes's approach to the Casson invariant [**Tau90**].

An even more intriguing instance of the phenomenon illustrated in the previous paragraph was discovered by Belavin, Polyakov, Schwartz, and Tyupkin [**BPST75**]. They observed that if $\dim X = 4$ and $A \in \mathscr{A}(P)$ is an **anti-self-dual (ASD) connection**,[1] that is,

$$F_A^+ := \tfrac{1}{2}(F_A + *F_A) = 0, \tag{1.5}$$

then the Bianchi identity implies the Yang–Mills equation:

$$\mathrm{d}_A^* F_A = - * \mathrm{d}_A F_A = 0. \tag{1.6}$$

In fact, something even better is true. The Yang–Mills functional satisfies the identity

$$\mathrm{YM}(A) = \int_X |F_A^+|^2 \,\mathrm{vol}_g + 8\pi^2 \check{h}(G) k(A) \tag{1.7}$$

with

$$k(A) := -\frac{1}{8\pi^2 \check{h}(G)} \int_X \langle F_A \wedge F_A \rangle \tag{1.8}$$

denoting the **instanton number**, and $\check{h}(G)$ denoting the dual Coxeter number of G.[2] If X is closed, then by Chern–Weil theory

$$k(A) = k(P) := \frac{1}{2\check{h}(G)} \langle p_1(\mathrm{Ad}(P)), [X] \rangle. \tag{1.9}$$

Therefore, ASD connections are absolute minimisers of the Yang–Mills energy: they saturate a topological lower bound.

Belavin, Polyakov, Schwartz, and Tyupkin also managed to find an explicit solution of (1.5) on $\mathbf{R}^4$: the **BPST instanton**. The latter can conveniently written using the normed division algebra $\mathbf{H} = \mathbf{R}\langle 1, i, j, k \rangle$ of the **quaternions** with relations $i^2 = j^2 = k^2 = ijk = -1$. The unit sphere $\mathrm{Sp}(1) := \{q \in \mathbf{H} : |q| = 1\}$ is a

[1] There is a similar notion of **self-dual (SD) connection**: $F_A^- := \frac{1}{2}(F_A - *F_A) = 0$. The notion of ASD and SD are swapped by flipping the orientation of X. Indeed, the early literature in the subject is concerned with SD connections. In the light of [**Don85**], ASD connections appear preferable. Indeed, on a Kähler surface (X, J, ω), there is a preferred orientation and the (-1)–eigenbundle $\Lambda^+ T^* X$ of the Hodge–$*$–operator agrees with $\Lambda_0^{1,1} T^* X$. Therefore, ASD connections induce holomorphic structures.

[2] [**AHS78**, §8] explains the meaning of the instanton number $k(P)$ and the appearance of $\check{h}(G)$ in (1.8).

Lie group with Lie algebra $\mathfrak{sp}(1) = \operatorname{Im}\mathbf{H}$. The BPST instanton is the connection I on the trivial bundle $\mathbf{H} \times \mathrm{Sp}(1)$ defined by

$$I := \frac{\operatorname{Im}(\bar{q}\,\mathrm{d}\,q)}{|q|^2 + 1} \in \Omega^1(\mathbf{H}, \mathfrak{sp}(1)) \tag{1.10}$$

with $q \in C^\infty(\mathbf{H}, \mathbf{H})$ denoting the identity map. To verify that $F_I^+ = 0$ it is convenient to use

$$\begin{aligned} \mathrm{d}\,q \wedge \mathrm{d}\,\bar{q} = &-2(\mathrm{d}\,q_0 \wedge \mathrm{d}\,q_1 + \mathrm{d}\,q_2 \wedge \mathrm{d}\,q_3) \otimes i \\ &- 2(\mathrm{d}\,q_0 \wedge \mathrm{d}\,q_2 + \mathrm{d}\,q_3 \wedge \mathrm{d}\,q_1) \otimes j \\ &- 2(\mathrm{d}\,q_0 \wedge \mathrm{d}\,q_3 + \mathrm{d}\,q_1 \wedge \mathrm{d}\,q_2) \otimes k \end{aligned} \tag{1.11}$$

and

$$\begin{aligned} \mathrm{d}\,\bar{q} \wedge \mathrm{d}\,q = &\,2(\mathrm{d}\,q_0 \wedge \mathrm{d}\,q_1 - \mathrm{d}\,q_2 \wedge \mathrm{d}\,q_3) \otimes i \\ &2(\mathrm{d}\,q_0 \wedge \mathrm{d}\,q_2 - \mathrm{d}\,q_3 \wedge \mathrm{d}\,q_1) \otimes j \\ &2(\mathrm{d}\,q_0 \wedge \mathrm{d}\,q_3 - \mathrm{d}\,q_1 \wedge \mathrm{d}\,q_2) \otimes k. \end{aligned} \tag{1.12}$$

A brief computation reveals that $k(I) = 1$. The computation involves a simple integral and some bookkeeping of the normalising factors; see [**Bau14**, Satz 7.9].

Although (in hindsight) Belavin, Polyakov, Schwartz, and Tyupkin's observation might appear mundane, the impact that it had on mathematics can hardly be underestimated. Building on foundational work of Atiyah, Nitchin, and Singer [**AHS78**], Uhlenbeck [**Uhl82b, Uhl82a**], Taubes [**Tau82**], Donaldson [**Don83**] discovered that the moduli spaces

$$\mathscr{M}^{\mathrm{ASD}}(P) := \{A \in \mathscr{A}(P) : F_A^+ = 0\}/\mathscr{G}(P) \tag{1.13}$$

and their Uhlenbeck compactifications contain subtle information about the differential topology of X and constructed his **Donaldson polynomials** [**Don90**]. In the hands of Taubes, Donaldson, Floer, Kronheimer, Mrowka, and many other mathematicians moduli spaces of ASD connections have become a crucial tool in the (differential) topology of 3– and 4–manifolds; see [**Don83, Tau87, Flo88, GM93**] and many many more.

1.2. The Seiberg–Witten equation. Witten [**Wit88**] realised that Donaldson theory can be understood physically by considering the ultraviolet behaviour of a topological twist of $N = 2$ supersymmetric Yang–Mills theory in dimension four. Seiberg and Witten [**SW94**] studied the infrared behaviour of the same theory, and Witten [**Wit94**] explains that this should give a way to compute the SU(2) Donaldson invariants in terms of a U(1) gauge theory.[3]

The setup is slightly more elaborate than that of Section 1.1. Here and, indeed, throughout the remainder of these lecture notes, $\dim X = 4$. Instead of a G–principal bundle P, one requires a **spin^c–structure** $\mathfrak{w}$ on (X, g), that is, a reduction of the frame bundle $\mathrm{Fr} = \mathrm{Fr}_{\mathrm{SO}}(TX)$ along

$$\mathrm{Spin}^{\mathrm{U}(1)}(4) := (\mathrm{Spin}(4) \times \mathrm{U}(1))/\{\pm 1\} \to \mathrm{SO}(4).^4 \tag{1.14}$$

[3]The reader is not expected to understand what any of this means; the author makes no pretense to do so either, but has tried to reproduce the physics jargon faithfully.

The latter determines **complex spinor bundles of positive and negative chirality** $W^{\pm}$, both Hermitian vector bundles of rank 2, and a **Clifford multiplication**

$$\gamma\colon T^*X \to \mathrm{Hom}(W^+, W^-) \oplus \mathrm{Hom}(W^-, W^+). \tag{1.15}$$

The latter can be extended to $\gamma\colon \Lambda^2T^*X \to \mathrm{End}(W^+) \oplus \mathrm{End}(W^-)$ such that $\gamma(\alpha \wedge \beta) = \frac{1}{2}(\gamma(\alpha)\gamma(\beta) - \gamma(\beta)\gamma(\alpha))$. In fact, the restriction $\gamma\colon \Lambda^+T^*X \to \mathfrak{su}(W^+)$ is an isomorphism. Instead of arbitrary $A \in \mathscr{A}(\mathfrak{w})$, one restricts to the subset $\mathscr{A}(\mathfrak{w}/\mathrm{LC})$ of those A which induce the Levi-Civita connection LC on Fr, a.k.a., **spin connections**. These are determined by the connection $\det(A)$ which they induce on the Hermitian line bundle $\det(W^+) = \det(W^-)$. The **Seiberg–Witten equation** for $A \in \mathscr{A}(\mathfrak{w}/\mathrm{LC})$ and $\Phi \in \Gamma(W^+)$ is:

$$\begin{aligned} D_A\Phi &= 0 \\ \tfrac{1}{2}F^+_{\det(A)} &= \mu(\Phi). \end{aligned} \tag{1.16}$$

Here

$$D_A := \gamma\nabla_A\colon \Gamma(W^+) \to \Gamma(W^-)$$

is the **Dirac operator** defined by γ and the covariant derivative ∇_A on W^+ induced by A, and $\mu\colon W^+ \to \Lambda^+T^*X \otimes i\mathbf{R}$

$$\mu(\Phi) := \gamma^{-1}\Big(\Phi\Phi^* - \tfrac{1}{2}|\Phi|^2 \cdot \mathbf{1}\Big). \tag{1.17}$$

At first glance, (1.16) might appear to be more complicated than (1.5). However, it is technically much easier to work with because: the structure group U(1) is abelian, bubbling is not an issue, and the spinor field Φ can be bounded a priori. [**Wit94**] already contains a rigorous construction of the **Seiberg–Witten invariants**. Many of the results that were previously proved using the ASD equation with hard work, were reproved with the Seiberg–Witten equation and many new results were obtained; see [**Kot95, Don96**] for reports on the early developments.

Witten [**Wit94**, (2.14)] conjectured that the Donaldson polynomial can be expressed in terms of the Seiberg–Witten invariants. Pidstrigach and Tyurin [**PT95**] put forward a tentative proposal for establishing this conjecture. Feehan and Leness [**FL98a, FL98b, FL01b, FL01a, Fee00, FL18**] have worked on fleshing this out and resolved the conjecture "modulo a gluing theorem".

The ASD equation (1.5) and the Seiberg–Witten equation (1.16) can by now be considered classical and should be familiar to every self-respecting gauge-theorist. The remainder of this lecture discusses two further equations—the Vafa–Witten equation and the Kapustin–Witten equation—which might not be as well-known and seem worthy of more attention.

1.3. The Vafa–Witten equation. $N = 4$ supersymmetric Yang–Mills theory in physics depends on a **structure group** G, an **angle** θ, a **gauge coupling constant** $e > 0$, and a geometric background X. The S–**duality** or **electric-magnetic duality** conjecture originating with Goddard, Nuyts, and Olive [**GNO77**]

[4] If $\lambda\colon H \to G$ is a Lie group homomorphism, and $(\pi\colon P \to X, P \circlearrowleft G)$ is a G–principal bundle, then a reduction of it along ρ is an H–principal bundle $(\rho\colon Q \to X, Q \circlearrowleft H)$ together with an isomorphism of G–principal bundles $Q \times_H G \cong P$.

and Montonen and Olive [**MO77**] predicts that this theory is invariant under simultaneously inverting $4\pi/e^2$ and replacing G with its **GNO dual** or **Langlands dual** ${}^LG$.[5][6][7][8] If θ and e are combined into

$$\tau := \frac{\theta}{2\pi} + \frac{4\pi i}{e^2} \in \mathfrak{H}, \tag{1.18}$$

then the inversion of $4\pi/e^2$ at $\theta = 0$ corresponds to the Möbius transformation

$$S(\tau) := -\frac{1}{\tau} \tag{1.19}$$

of the upper half-plane $\mathfrak{H} := \{\tau \in \mathbf{C} : \operatorname{Im}\tau > 0\}$.

Vafa and Witten [**VW94**] set out to test this prediction with X an oriented closed 4–manifold. $N = 4$ supersymmetric Yang–Mills theory in dimension four admits three topological twists [**Yam88**][**Mar95**][**ESW22**, p. 11]. Vafa and Witten judiciously choose one of them and are led to the **Vafa–Witten equation** for $A \in \mathscr{A}(P)$, $B \in \Omega^+(X, \mathrm{Ad}(P))$, and $C \in \Omega^0(X, \mathrm{Ad}(P))$:

$$\begin{aligned} \mathrm{d}_A C + \mathrm{d}_A^* B &= 0 \\ F_A^+ &= [C, B] + \tfrac{1}{2}[B \times B] \end{aligned} \tag{1.20}$$

with $\cdot \times \cdot \colon \Lambda^+T^*X \otimes \Lambda^+T^*X \to \Lambda^+T^*X$ defined by

$$(\alpha \times \beta)_{ab} := \sum_{c=1}^{4} \alpha_{ac}\beta_{cb} - \beta_{ac}\alpha_{cb}. \tag{1.21}$$

After gauge fixing, (1.20) is an elliptic equation of index zero. (In fact, it has been engineered to be!) Therefore, it does not seem too outlandish to count the moduli spaces

$$\mathscr{M}^{\mathrm{VW}}(P) := \{(A, B, C) \in \mathscr{A}(P) \times \Omega^+(X, \mathrm{Ad}(P)) \times \Omega^0(X, \mathrm{Ad}(P)) : (1.20)\}/\mathscr{G}(P) \tag{1.22}$$

to obtain **Vafa–Witten invariants**

$$\text{“}\#\mathscr{M}^{\mathrm{VW}}(P) \in \mathbf{Z}.\text{”} \tag{1.23}$$

To do this in earnest requires finding a suitable compactification, transversality, etc. Although there has been a lot of work on the Vafa–Witten equation, in particular, by Mares [**Mar10**], Tanaka [**Tan15, Tan17, Tan, Tan19**], and Taubes [**Tau17**], a rigorous mathematical definition of the Vafa–Witten invariants for arbitrary closed oriented 4–manifolds is not available (yet?). The situation in algebraic geometry is much better: Tanaka and Thomas [**TT17a, TT17b**] have defined Vafa–Witten invariants for smooth projective surfaces and $G = \mathrm{U}(n)$.

---

[5] Electric-magnetic duality is not restricted to $N = 4$ supersymmetric Yang–Mills theory. In fact, it plays some role in [**SW94**].

[6] Of course, GNO stands for Goddard, Nuyts, and Olive. The observation that the GNO and Langlands dual agree is due to Atiyah.

[7] This is only true if G is simply-laced and if the Lie algebras of $\mathfrak{g}$ and ${}^L\mathfrak{g}$ are isomorphic. In general, there is an extra factor in front of the inversion; cf. [**KW07**, §2.2], [**Wu08**].

[8] The structure group G (and a maximal torus T) determine the **root datum** $(X^*, X_*, \langle\cdot,\cdot\rangle, R, R^\vee, \vee)$: with $X^* := \mathrm{Hom}(T, \mathrm{U}(1))$ denoting the character lattice, $X_* := \mathrm{Hom}(\mathrm{U}(1), T)$ denoting the cocharacter lattice, $\langle\cdot,\cdot\rangle$ denoting the obvious perfect pairing, $R \subset X^*$ denoting roots, $R^\vee \subset X_*$ denoting the coroots, and $\vee\colon R \to R^\vee$ denoting the bijection between roots and coroots. In fact, the latter determines G up to isomorphism. Exchanging X^* and X_*, and R and $R^\vee$ yields the root datum of ${}^LG$. Evidently, ${}^{L\,L}G = G$. Determining ${}^LG$ from scratch is quite a bit of work. [**KW07**, Table 1] is quite helpful. For example, ${}^L\,\mathrm{SU}(2) = \mathrm{SO}(3)$.

Whatever the precise and rigorous definition of $\#\mathscr{M}^{\mathrm{VW}}(P)$ might be, the **partition function** of Vafa–Witten theory should be

$$Z_X^{\mathrm{VW}}(\tau, G) := \frac{1}{\#Z(G)} \sum_P \#\mathscr{M}^{\mathrm{VW}}(P) \cdot e^{2\pi i \tau (k(P) - s(X))} \tag{1.24}$$

with $Z(G)$ denoting the center of G, the sum being taken over all isomorphism classes of G–principal bundles P over X, $k(P)$ denoting the instanton number (1.8), and the shift in the zero point of the instanton number given by

$$s(X) := \frac{\operatorname{rk} G + 1}{24} \chi(X). \tag{1.25}$$

Finally, according to, the prediction of S–duality is that

$$Z_X^{\mathrm{VW}}(-1/\tau, {}^L G) = \pm \left(\frac{\tau}{i}\right)^{w(X)/2} Z_X^{\mathrm{VW}}(\tau, G) \tag{1.26}$$

with

$$\pm := (-1)^{\frac{\operatorname{rk} G}{4}(\chi(X) + \sigma(X))} \quad \text{and} \quad w(X) := -\chi(X).^9 \tag{1.27}$$

S–duality is a $\mathbf{Z}/2\mathbf{Z}$ symmetry acting by $\tau \mapsto -1/\tau$ and $G \mapsto {}^L G$. This can be enhanced to an (only apparently!) more awe-inspiring symmetry under the **modular group** $\mathbf{PSL}_2(\mathbf{Z}) := \mathrm{SL}_2(\mathbf{Z})/\{\pm \mathbf{1}\}$. The latter acts on $\mathfrak{H}$ by Möbius transformations:

$$\begin{pmatrix} a & b \\ c & d \end{pmatrix} (\tau) := \frac{a\tau + b}{c\tau + d}. \tag{1.28}$$

It is well known that

(1.29)

$$\mathbf{PSL}_2(\mathbf{Z}) = \langle S, T \mid S^2 = (STS)^3 = \mathbf{1} \rangle \quad \text{with} \quad S := \begin{bmatrix} 0 & -1 \\ 1 & 0 \end{bmatrix} \quad \text{and} \quad T := \begin{bmatrix} 1 & 1 \\ 0 & 1 \end{bmatrix}.$$

S acts by (1.19) and sends G to ${}^L G$. Its action on Z_X^{VW} is predicted by (1.26). T acts by $T(\tau) = \tau + 1$ and trivially on G. Its effect on $Z^{\mathrm{VW}}(\tau, G)$ is easy to read off from (1.24); in fact, it is trivial if $k(P) - s(X) \in \mathbf{Z}$. In any case, there is a minimal $N(X, G) \in \mathbf{N}$ such that

$$k(P) - s(X) \in \frac{1}{N(X, G)} \mathbf{Z} \tag{1.30}$$

and $T^{N(X,G)}$ leaves Z_X^{VW} invariant.

The Langlands dual ${}^L G$ can be removed from the picture. This is most straightforward if $N(X, G) = 1$. If $N := N(X, {}^L G)$, then $ST^N S$ and T act trivially on G. Therefore, Z_X^{VW} transforms under the **Hecke congruence subgroup of level** N

$$\Gamma_0(N) := \langle ST^N S, T \rangle = \left\{ \begin{bmatrix} a & b \\ c & d \end{bmatrix} \in \mathbf{PSL}_2(\mathbf{Z}) : c = 0 \bmod N \right\} \tag{1.31}$$

with factor of automorphy

$$j\left(\begin{bmatrix} a & b \\ c & d \end{bmatrix}, \tau \right) := \pm \left(\frac{c\tau + d}{i} \right)^{w(X)/2}. \tag{1.32}$$

[9] The formulae for $s(X)$, $\pm$, and $w(x)$ are from Wu [**Wu08**, (3.9) and (3.10)]. Vafa and Witten are rather vague about this.

This is almost the definition of modular form of level N and weight $w(X)/2$—except that the factor of automorphy is slightly off and no condition on holomorphicity (at ∞) is imposed.

After some leaps of faith, Vafa and Witten arrive at

$$Z_{K3}^{\mathrm{VW}}(\tau, \mathrm{SU}(2)) = \frac{1}{8}G(q^2) + \frac{1}{4}G(q^{1/2}) + \frac{1}{4}G(-q^{1/2}) \tag{1.33}$$

and

$$Z_{K3}^{\mathrm{VW}}(\tau, \mathrm{SO}(3)) = \frac{1}{4}G(q^2) + 2^{21}G(q^{1/2}) + 2^{10}G(-q^{1/2}) \tag{1.34}$$

with $q = e^{2\pi i\tau}$, and $G := \eta^{-24}$ and $\eta(q) := q^{\frac{1}{24}}\prod_{n=1}^{\infty}(1-q^n)$ denoting the **Dedekind eta function**, a modular form of level 1 and weight $\frac{1}{2}$. This can be considered as a "successful test" of S–duality.

It is worthwhile to give some idea for how Vafa and Witten did this. Evidently,

$$\mathscr{M}^{\mathrm{ASD}}(P) \subset \mathscr{M}^{\mathrm{VW}}(P). \tag{1.35}$$

The Vafa–Witten equation has been engineered so that if $\mathscr{M}^{\mathrm{ASD}}(P)$ agrees with $\mathscr{M}^{\mathrm{VW}}(P)$, is compact, cut-out transversely, and contains no reducible solutions, then

$$\#\mathscr{M}^{\mathrm{VW}}(P) = \chi(\mathscr{M}^{\mathrm{ASD}}(P)). \tag{1.36}$$

Of course, that is not the case for $X = K3$ and $G = \mathrm{SU}(2)$ or $\mathrm{SO}(3)$, but it is not too far off. In particular, $\mathscr{M}^{\mathrm{ASD}}(P)$ has an interpretation in complex geometry and a compactification. The Euler characterstic of the latter can be computed and this lead Vafa and Witten to the above formulae.

In general, the Vafa–Witten invariant—if it exists—cannot be an Euler characteristic of a compactified moduli space of ASD connections because the Euler characteristic is only bordism invariant mod 2.[10]

1.4. The Kapustin–Witten equation. The geometric Langlands conjecture predicts a fantastically complicated relation between the moduli spaces of $G^{\mathbf{C}}$ local systems over a Riemann surface Σ and the moduli space of holomorphic ${}^{L}G^{\mathbf{C}}$–principal bundles over Σ. The appearance of ${}^{L}G$ in S–duality suggests a connection between geometric Langlands conjecture and $N = 4$ super-symmetric Yang–Mills theory. Indeed, deep relations have been uncovered—starting with the work of Kapustin and Witten [**KW07**]. The foundation of their work is a topological twist which leads to the following family equations for $A \in \mathscr{A}(P)$ and $\phi \in \Omega^1(X, \mathrm{Ad}(P))$:

$$\begin{aligned} u \cdot \left(F_A - \tfrac{1}{2}[\phi\wedge\phi]\right)^+ - v \cdot \mathrm{d}_A^+\phi &= 0 \\ v \cdot \left(F_A - \tfrac{1}{2}[\phi\wedge\phi]\right)^- + u \cdot \mathrm{d}_A^-\phi &= 0 \\ \mathrm{d}_A^*\phi &= 0 \end{aligned} \tag{1.37}$$

parametrised by $[u : v] \in \mathbf{C}P^1$. It would go far beyond the scope of these lecture notes to even start discussing Kapustin and Witten's work. It is, however, worthwhile to discuss some instances of (1.37).

[10]However, the Euler characteristic is a complex bordism invariant.

Kapustin and Witten's perspective on the geometric Langlands conjecture only requires the **Kapustin–Witten equation**, the specialisation of (1.37) to $[1:1]$:

$$\begin{aligned} F_A - \tfrac{1}{2}[\phi \wedge \phi] - * \,\mathrm{d}_A \phi &= 0 \\ \mathrm{d}_A^* \phi &= 0 \end{aligned} \tag{1.38}$$

The other equations (1.37) play a role in the quantum geometric Langlands conjecture.

For $[1 : \pm i]$ the first two equations (1.37) combine to

$$F_{A \pm i\phi} = 0; \tag{1.39}$$

that is, $A + i\phi$ is a flat $G^{\mathbf{C}}$–principal connection on $P^{\mathbf{C}} := P \times_G G^{\mathbf{C}}$. Corlette [**Cor88**] proved that if $A + i\phi$ is **stable**, that is, its monodromy representation does not factor through a proper parabolic subgroup of $G^{\mathbf{C}}$, then its $\mathscr{G}(P^{\mathbf{C}})$–orbit contains a solution of (1.37) for $[1 : \pm i]$.

The specialisation of (1.37) at $[1 : 0]$ is the following **complex anti-self-duality** condition:

$$\begin{aligned} \left(F_A - \tfrac{1}{2}[\phi \wedge \phi]\right)^+ &= 0 \\ \mathrm{d}_A^- \phi &= 0 \\ \mathrm{d}_A^* \phi &= 0. \end{aligned} \tag{1.40}$$

The specialisation of (1.37) at $[0 : 1]$ is a complex self-duality condition.

Theorem 1.41 (Kapustin and Witten [**KW07**, §3.3] and Gagliardo and Uhlenbeck [**GU12**]). *Suppose that X is closed.*

1. *If $\langle p_1(\mathrm{Ad}(P)), [X]\rangle \neq 0$, then (1.37) admits solutions only for $[u : v] \in \{[1:0],[0,1]\}$.*
2. *If $\langle p_1(\mathrm{Ad}(P)), [X]\rangle = 0$ and (A, ϕ) satisfies (1.37) for any $[u,v] \in \mathbf{C}P^1$, then it also satisfies (1.37) for $[1, \pm 1] \in \mathbf{C}P^1$.*

The Kapustin–Witten equation (1.38) is therefore most relevant if X is not closed; e.g.: if $X = I \times Y$ with I an interval and Y a closed 3–manifold. A brief computation reveals that (1.38) for $\mathbf{A} = A + \xi \,\mathrm{d}t$ and $\boldsymbol{\phi} = \phi + \eta \,\mathrm{d}t$ becomes

$$\begin{aligned} \partial_t \phi - [\xi, \phi] - *_Y F_A + \tfrac{1}{2} *_Y [\phi \wedge \phi] - *_Y \mathrm{d}_A \eta &= 0 \\ \partial_t A - \mathrm{d}_A \xi + [\phi, \eta] - * \,\mathrm{d}_A \phi &= 0 \\ \partial_t \eta + [\xi, \eta] - \mathrm{d}_A^* \phi &= 0. \end{aligned} \tag{1.42}$$

This looks more friendly if $\xi = 0$ (temporal gauge) and $\eta = 0$. It can be interpreted as the gradient flow of a complexified Chern–Simons functional.

The Kapustin–Witten equation does not just play a role in the geometric Langlands conjecture, (1.42) with $I = [0, \infty)$ is also at the heart of Witten's new approach to the Jones polynomial: the coefficients are supposed to arise as "counts" of the moduli space of solutions to (1.42) with a certain charge and the knot encoded as a boundary condition on $\{0\} \times Y$; cf. [**Wit17**]. More propaganda for the Kapustin–Witten equation can be found in [**Uhl12**].

2. The framework of generalised Seiberg–Witten equations

At a very schematic level, the ASD equation (1.5), the Seiberg–Witten equation (1.16), the Vafa–Witten equation (1.20), and the complex ASD equation (1.40) look somewhat similar. Indeed, they can be understood in the unifying framework of generalised Seiberg–Witten equations. The idea of systematically generalising the Seiberg–Witten equation goes back to the work of Taubes [**Tau99**] and Pidstygach [**Pid04**]. Several versions have been discussed in [**Hay06**][**Hay15b**][**Sal13**, §6][**Nak16**, §6(i)][**DW20**, §2.2][**WZ21**, §1.1]. The following construction might, initially, seem a little intimidating, but it is really not much more complicated than what is required for the Seiberg–Witten equation (1.16).

2.1. Spinors in dimension four. As a warm-up it is helpful to recall a few facts about Spin(4). The construction relies on the fact that

$$\mathrm{Spin}(4) = \mathrm{Sp}(1) \times \mathrm{Sp}(1). \tag{2.1}$$

Define the **spinor representations** $\sigma_\pm\colon \mathrm{Spin}(4) \to \mathrm{Sp}(1) < \mathrm{SO}(\mathbf{H})$ and the **vector representation** $\nu\colon \mathrm{Spin}(4) \to \mathrm{SO}(4) = \mathrm{SO}(\mathbf{H})$ by

$$\sigma_\pm(p_+,p_-)\Phi := p_\pm\Phi \quad \text{and} \quad \nu(p_+,p_-)v := p_+ v\bar{p}_-. \tag{2.2}$$

Define the **Clifford multiplication** $\gamma\colon \mathbf{H} \to \mathrm{End}(\mathbf{H}\oplus\mathbf{H})$ by

$$\gamma(v)(\Phi_+,\Phi_-) := (v\Phi_-, -\bar{v}\Phi_+). \tag{2.3}$$

Evidently, γ is equivariant with respect to ν and $\sigma_\pm$; that is, for every $g \in \mathrm{Spin}(4)$, $v, \Phi_\pm \in \mathbf{H}$

$$\gamma(\nu(g)v)(\sigma_+(g)\Phi_+, \sigma_-(g)\Phi_-) = \big[\sigma_+(g)\oplus\sigma_-(g)\big]\gamma(v)(\Phi_+,\Phi_-). \tag{2.4}$$

There are unique $\delta_\pm\colon \mathrm{Spin}(4) \to \mathrm{SO}(3) = \mathrm{SO}(\mathrm{Im}\,\mathbf{H})$ such that the isomorphisms

$$\mathrm{Im}\,\mathbf{H} \to \Lambda^+\mathbf{H}^*, \quad \xi \mapsto \langle \mathrm{d}\,q\wedge\mathrm{d}\,\bar{q}, \xi\rangle \quad \text{and} \quad \mathrm{Im}\,\mathbf{H} \to \Lambda^-\mathbf{H}^*, \quad \xi \mapsto \langle \mathrm{d}\,\bar{q}\wedge\mathrm{d}\,q, \xi\rangle \tag{2.5}$$

are equivariant with respect to the actions of $\delta_\pm$ and ν. Here $\mathrm{d}\,q\wedge\mathrm{d}\,\bar{q} \in \Lambda^+\mathbf{H}^*\otimes\mathrm{Im}\,\mathbf{H}$ and $\mathrm{d}\,\bar{q}\wedge\mathrm{d}\,q \in \Lambda^-\mathbf{H}^*\otimes\mathrm{Im}\,\mathbf{H}$ are as in (1.11) and (1.12) and $\langle\cdot,\cdot\rangle$ is the inner product on $\mathrm{Im}\,\mathbf{H}$.

If $\mathfrak{s}$ is a spin structure on (X,g), then

$$\mathfrak{s}\times_\nu \mathbf{H} \cong TX \cong T^*X \quad \text{and} \quad \mathfrak{s}\times_{\delta_\pm}\mathrm{Im}\,\mathbf{H} \cong \Lambda^\pm T^*X; \tag{2.6}$$

moreover,

$$S^\pm := \mathfrak{s}\times_{\sigma^\pm}\mathbf{H} \tag{2.7}$$

are the **spinor bundles of positive and negative chirality**. Because of (2.4) γ induces a Clifford multiplication $\gamma\colon T^*X \to \mathrm{Hom}(S^+,S^-)\oplus\mathrm{Hom}(S^-,S^+)$. As in Section 1.2, γ extends to $\gamma\colon \Lambda^+T^*X\oplus\Lambda^-T^*X \to \mathrm{End}(S^+)\oplus\mathrm{End}(S^-)$.[11]

[11] The volume element acts as $\gamma(1)\gamma(i)\gamma(j)\gamma(k)(\Phi_+,\Phi_-) = (-\Phi_+,\Phi_-)$. This is contrary to the customary sign conventions for the chirality of *real* spinor bundles, but necessary to guarantee that Λ^-T^*X acts trivially on S^+.

2.2. Generalised Seiberg–Witten equations. The setup for a generalised Seiberg–Witten equation involves choices of algebraic and geometric data.

DEFINITION 2.8.

(1) A **quaternionic Hermitian vector space** is a left $\mathbf{H}$–module S together with an Euclidean inner product $\langle \cdot, \cdot \rangle$ such that i, j, k act by isometries. The **unitary symplectic group** $\mathrm{Sp}(S)$ is the subgroup of $\mathrm{GL}_{\mathbf{H}}(S)$ preserving $\langle \cdot, \cdot \rangle$.
(2) A **quaternionic representation** of a Lie group H is a Lie group homomorphism $\rho\colon H \to \mathrm{Sp}(S)$ for some quaternionic Hermitian vector space S.
(3) A set of **algebraic data** consists of:
  (a) A compact connected Lie group G, the **structure group**.
  (b) A compact connected Lie group K, the **auxiliary group**, and an extension $G \hookrightarrow H \twoheadrightarrow K$.[12]
  (c) A quaternionic representation $\rho\colon H \to \mathrm{Sp}(S)$ together with an $\varepsilon \in Z(H)$ satisfying $\varepsilon^2 = \mathbf{1}$ and $\rho(\varepsilon) = -\mathbf{1}$. •

A *choice of a set of algebraic data* induces further data analogous to Section 2.1. Define $\gamma\colon \mathbf{H} \to \mathrm{End}(S \oplus S)$ and $\underline{\gamma}\colon \mathbf{H} \otimes \mathfrak{g} \to \mathrm{End}(S \oplus S)$ by

(2.9)
$$\gamma(v)(\Phi_+, \Phi_-) := (v\Phi_-, -\bar{v}\Phi_+) \quad \text{and} \quad \underline{\gamma}(v \otimes \xi)(\Phi_+, \Phi_-) := \gamma(v)\,\mathrm{Lie}(\rho)(\xi)(\Phi_+, \Phi_-).$$

The quaternionic representation $\rho|_G\colon G \to \mathrm{Sp}(S)$ has a **distinguished hyperkähler moment map** $\mu\colon S \to (\mathrm{Im}\,\mathbf{H} \otimes \mathfrak{g})^*$ defined by

$$\mu(\Phi) := \tfrac{1}{2}\underline{\gamma}^*(\Phi\Phi^*); \tag{2.10}$$

that is:

$$\langle \mu(\Phi), v \otimes \xi \rangle = \tfrac{1}{2}\langle \underline{\gamma}(v \otimes \xi)\Phi, \Phi \rangle. \tag{2.11}$$

Set

(2.12)
$$\mathrm{Spin}^H(4) := (\mathrm{Spin}(4) \times H)/\langle(-1, \varepsilon)\rangle \quad \text{and} \quad \mathrm{Spin}^K(4) := (\mathrm{Spin}(4) \times K)/\langle(-1, \delta)\rangle$$

with $\delta \in Z(K)$ denoting the image of $\varepsilon \in Z(H)$. Denote by

$$\pi\colon \mathrm{Spin}^H(4) \to \mathrm{Spin}^K(4) \quad \text{and} \quad \nu\colon \mathrm{Spin}^K(4) \to \mathrm{SO}(4) \tag{2.13}$$

the obvious projection maps. Define $\sigma_\pm\colon \mathrm{Spin}^H(4) \to \mathrm{O}(S)$ by

$$\sigma_\pm([p_+, p_-, h])\Phi := p_\pm \rho(h)\Phi. \tag{2.14}$$

Evidently, γ and $\underline{\gamma}$ are equivariant with respect to $\nu \circ \pi$ and $\sigma_\pm$; Moreover, μ is equivariant with respect to the actions of $\sigma_\pm$ on S and $\delta_\pm$ and $\rho|_G$ on $(\mathrm{Im}\,\mathbf{H} \otimes \mathfrak{g})^*$.

As in Section 2.1, in order to globalise these structures to X, further data is required.

DEFINITION 2.15. A set of **geometric data** consists of:

(1) A reduction $\mathrm{Fr}^\star$ of Fr along ν together with a connection $\mathrm{LC}^\star$ which induces LC on Fr, the frame bundle of TX.
(2) A reduction $\mathfrak{s}$ of $\mathrm{Fr}^\star$ along π. •

[12] Often, $H = G \times K$.

Choose a set of geometric data. Define the **spinor bundles of positive and negative chirality** and the **relative adjoint bundle** by

$$S^{\pm} := \mathfrak{s} \times_{\sigma_{\pm}} S \quad \text{and} \quad \mathrm{Ad}(\mathfrak{s}/\mathrm{Fr}^{\star}) := \mathfrak{s} \times_{\mathrm{Spin}^{H}(4)} \mathfrak{g}. \tag{2.16}$$

For $A \in \mathscr{A}(\mathfrak{s})$ denote by

$$F_{A/\mathrm{Fr}^{\star}} \in \Omega^{2}(X, \mathrm{Ad}(\mathfrak{s}/\mathrm{Fr}^{\star})) \tag{2.17}$$

the projection of $F_A \in \Omega^2(X, \mathrm{Ad}(\mathfrak{s}))$ to $\Omega^2(X, \mathrm{Ad}(\mathfrak{s}/\mathrm{Fr}^{\star}))$. The maps γ, $\underline{\gamma}$, and μ induce

$$\gamma\colon T^*X \to \mathrm{End}(S^+, S^-), \quad \underline{\gamma}\colon T^*X \otimes \mathrm{Ad}(\mathfrak{s}/\mathrm{Fr}^{\star}) \to \mathrm{End}(S^+, S^-), \quad \text{and} \tag{2.18}$$

$$\mu\colon S^+ \to \Lambda^+T^*X \otimes \mathrm{Ad}(\mathfrak{s}/\mathrm{Fr}^{\star}). \tag{2.19}$$

Denote by $\mathscr{A}(\mathfrak{s}/\mathrm{LC}^{\star})$ the space of $\mathrm{Spin}^H(4)$–principal connections on $\mathfrak{s}$ which induce $\mathrm{LC}^{\star}$ on $\mathrm{Fr}^{\star}$. This is an affine space modelled on $\Omega^1(X, \mathrm{Ad}(\mathfrak{s}/\mathrm{Fr}^{\star}))$. Every $A \in \mathscr{A}(\mathfrak{s}/\mathrm{LC}^{\star})$ defines a Dirac operator

$$D_A := \gamma\nabla_A\colon \Gamma(S^+) \to \Gamma(S^-). \tag{2.20}$$

DEFINITION 2.21. The **generalised Seiberg–Witten equation** associated with the above is the following equation for $A \in \mathscr{A}(\mathfrak{s}/\mathrm{LC}^{\star})$ and $\Phi \in \Gamma(S^+)$:

$$\begin{aligned} D_A\Phi &= 0 \\ F^{+}_{A/\mathrm{LC}^{\star}} &= \mu(\Phi). \end{aligned} \tag{2.22}$$

Of course, (2.22) is not preserved by the gauge group $\mathscr{G}(\mathfrak{s})$ of $\mathfrak{s}$, but it is preserved by $\mathscr{G}(\mathfrak{s}/\mathrm{Fr}^{\star})$, the subgroup acting trivially on $\mathrm{Fr}^{\star}$.

REMARK 2.23. It is possible to replace S by a hyperkähler manifold together with a trihamiltonian action and a permuting action. In fact, this is done in [**Hay06**][**Hay15b**][**Nak16**, §6(i)][**Sal13**, §6][**Pid04**][**Tau99**].

2.3. Examples of generalised Seiberg–Witten equations. To help digest the above construction and appreciate its scope, the remainder of this lecture discusses a number of examples.

EXAMPLE 2.24 (harmonic spinors). If $G = \{\mathbf{1}\}$, $H = K = \{\pm\mathbf{1}\}$, $\varepsilon = -\mathbf{1}$, and $S = \mathbf{H}$, then $\mathrm{Fr}^{\star}$ is a spin structure, $\mathrm{LC}^{\star}$ is unique, and solutions to (2.22) are harmonic spinors of negative chirality.

EXAMPLE 2.25 (ASD instantons). If $G = H$, $\varepsilon = \mathbf{1}$, and $S = 0$, then the generalised Seiberg–Witten equation (2.22) reduces to the ASD equation (1.5).

EXAMPLE 2.26 (The Seiberg–Witten equation). Let $G = H = \mathrm{U}(1)$, $\varepsilon = (-1, \mathbf{1})$. Define the quaternionic representation $\rho\colon \mathrm{U}(1) \to \mathrm{Sp}(\mathbf{H})$ by

$$\rho(e^{i\alpha})q := qe^{i\alpha}. \tag{2.27}$$

Identifying $(i\mathbf{R} \otimes \mathrm{Im}\,\mathbf{H})^* = i\mathbf{R} \otimes \mathrm{Im}\,\mathbf{H}$, the hyperkähler moment map $\mu\colon \mathbf{H} \to (i\mathbf{R} \otimes \mathrm{Im}\,\mathbf{H})^*$ is

$$\mu(q) = -\frac{i}{2} \otimes qiq^*. \tag{2.28}$$

Splitting $\mathbf{H} = \mathbf{C} \oplus j\mathbf{C}$, we see that $\underline{\gamma}(\mu(q)) \in \mathrm{End}(\mathbf{C}^{\oplus 2})$ for $q = z + jw$ is

$$\tfrac{1}{2}\begin{pmatrix} |z|^2 - |w|^2 & 2z\bar{w} \\ 2\bar{z}w & |w|^2 - |z|^2 \end{pmatrix} = q\langle q, \cdot\rangle_{\mathbf{C}} - \tfrac{1}{2}|q|^2_{\mathbf{C}}\,\mathrm{id}_{\mathbf{C}^{\oplus 2}}. \tag{2.29}$$

In this situation, $\mathrm{Fr}^\star = \mathrm{Fr}$, $\mathfrak{s}$ is a spin^c structure, A is a choice of spin connection on $\mathfrak{s}$, $S^+ = W^+$ is the complex spinor bundle of positive chirality, $F_{A/\mathrm{Fr}^\star} = \frac{1}{2}F_{\det(A)}$. Therefore, the generalized Seiberg–Witten equation (2.22) is the (classical) Seiberg–Witten equation (1.16).

EXAMPLE 2.30 (The Seiberg–Witten equation with n spinors). Let $n \in \mathbf{N}$. Let $G = \mathrm{U}(1)$, $K = \mathrm{SU}(n)$, $H := G \times K$, and $\varepsilon = (-1, 1)$. H acts on $S = \mathbf{H} \otimes_{\mathbf{C}} \mathbf{C}^n = \mathbf{H}^n$ with $\mathrm{U}(1)$ acting on the first factor as in Example 2.26 and $\mathrm{SU}(n)$ acting on the second factor by the defining representation. The moment map is $\mu(q_1, \dots, q_n) = \sum_{a=1}^n \mu(q_a)$ with $\mu(q_a)$ as above. $\mathrm{Fr}^\star$ is equivalent to the choice of a rank n Hermitian vector bundle E (with trivial determinant) and $\mathrm{LC}^\star$ is equivalent to the choice of a unitary connection on E (with trivial determinant), $\mathfrak{s}$ is the choice of a spin^c–structure, and $S^+ = W^+ \otimes E$.

EXAMPLE 2.31 (Vafa–Witten). Let G be a compact connected semi-simple Lie group. Choosing a G–invariant inner product on $\mathfrak{g}$ turns $S := \mathfrak{g} \otimes_{\mathbf{R}} \mathbf{H}$ into a quaternionic Hermitian vector space. The adjoint representation induces a quaternionic representation $\rho\colon G \to \mathrm{Sp}(S)$. The distinguished moment map $\mu\colon S \to \mathrm{Im}\,\mathbf{H} \otimes \mathfrak{g}$ is given by

(2.32)
$$\begin{aligned}\mu(\boldsymbol{\xi}) &= \tfrac{1}{2}[\boldsymbol{\xi}, \boldsymbol{\xi}] \\ &= ([\xi_2, \xi_3] + [\xi_0, \xi_1]) \otimes i + ([\xi_3, \xi_1] + [\xi_0, \xi_2]) \otimes j + ([\xi_1, \xi_2] + [\xi_0, \xi_3]) \otimes k\end{aligned}$$

with $\boldsymbol{\xi} = \xi_0 \otimes 1 + \xi_1 \otimes i + \xi_2 \otimes j + \xi_3 \otimes k \in \mathbf{H} \otimes \mathfrak{g}$. The action of $q \in K = \mathrm{Sp}(1)$ given by right-multiplication with $\bar{q}$ commutes with ρ. Set $\varepsilon := (-\mathbf{1}, \mathbf{1}_G) \in H := G \times K$. The embedding $\mathrm{SO}(4) = (\mathrm{Sp}(1) \times \mathrm{Sp}(1))/\{\pm 1\} \to \mathrm{Spin}^{\mathrm{Sp}(1)}(4)$ defined by

$$[p, q] \mapsto [p, q, p] \tag{2.33}$$

determines the choice of $\mathrm{Fr}^\star$ and $\mathrm{LC}^\star$. The choice of $\mathfrak{s}$ is equivalent to a choice of G–principal bundle P; moreover, $\mathrm{Ad}(\mathfrak{s}/\mathrm{Fr}^\star) = \mathrm{Ad}(P)$ and $\mathscr{A}(\mathfrak{s}/\mathrm{LC}^\star) = \mathscr{A}(P)$. Therefore,

$$S^+ = (\underline{\mathbf{R}} \oplus \Lambda^+ T^*X) \otimes \mathrm{Ad}(P) \quad \text{and} \quad S^- = T^*X \otimes \mathrm{Ad}(P) \tag{2.34}$$

with

$$\gamma(\alpha)(f, \omega) = \alpha f - i_{\alpha^\sharp}\omega. \tag{2.35}$$

In this situation, the generalised Seiberg–Witten equation (2.22) for A and $\Phi = (C, B)$ is the Vafa–Witten equation (1.20) for (A, B, C).

EXAMPLE 2.36 (complex ASD). The algebraic data of Example 2.31 with the geometric data determined by the embedding $\mathrm{SO}(4) = (\mathrm{Sp}(1) \times \mathrm{Sp}(1))/\{\pm 1\} \to \mathrm{Spin}^{\mathrm{Sp}(1)}(4)$ defined by

$$[p, q] \mapsto [p, q, q] \tag{2.37}$$

leads to

$$S^+ = T^*X \otimes \mathrm{Ad}(P) \quad \text{and} \quad S^- = (\underline{\mathbf{R}} \oplus \Lambda^- T^*X) \otimes \mathrm{Ad}(P). \tag{2.38}$$

In this situation, the generalised Seiberg–Witten equation (2.22) is the complex ASD equation (1.40).[13]

[13] The other equations in the family (1.37) do not seem to neatly fit in to the generalised Seiberg–Witten framework; however, see Example 2.39.

It can be insightful to use the above construction, somewhat informally, in infinite dimensions. However, the reader should be warned that, of course, none of the results in Section 3 apply (at least not directly).

EXAMPLE 2.39 (Haydys–Witten equation). Let $I \subset \mathbf{R}$ be an interval. Denote by $\mathrm{pr}_X \colon I \times X \to X$ the projection. Let L be a connected compact semi-simple Lie group. Set $S := C^\infty(I, \mathfrak{l} \otimes_{\mathbf{R}} \mathbf{H})$ and $G := C^\infty(I, L)$. G acts on S via

$$(g\boldsymbol{\xi})(t) := g(t)\boldsymbol{\xi}(t)g(t)^{-1} - \dot{g}(t)g(t)^{-1}. \tag{2.40}$$

A computation reveals that

$$\begin{aligned}\mu(\boldsymbol{\xi}) &= (\dot{\xi}_1 + [\xi_2, \xi_3] + [\xi_0, \xi_1]) \otimes i \\ &\quad + (\dot{\xi}_2 + [\xi_3, \xi_1] + [\xi_0, \xi_2]) \otimes j \\ &\quad + (\dot{\xi}_3 + [\xi_1, \xi_2] + [\xi_0, \xi_3]) \otimes k.\end{aligned} \tag{2.41}$$

Choose $K = \mathrm{Sp}(1)$, H, and $\mathrm{Fr}^\star$ as in Example 2.31. A choice of $\mathfrak{s}$ is induced by a choice of L–principal bundle P on X. In this situation, $\mathscr{A}(\mathfrak{s}/\mathrm{LC}^\star) = \mathscr{A}(\mathrm{pr}_X^* P)$, $\Gamma(\mathrm{Ad}(\mathfrak{s}/\mathrm{Fr}^\star)) = \Gamma(\mathrm{pr}_X^* \mathrm{Ad}(P))$,

$$\begin{aligned}\Gamma(S^+) &= \Gamma(I \times X, \mathrm{pr}_X^*[(\underline{\mathbf{R}} \oplus \Lambda^+ T^*X) \otimes \mathrm{Ad}(P)]) \quad \text{and} \\ \Gamma(S^-) &= \Gamma(I \times X, \mathrm{pr}_x^*(T^*X \otimes \mathrm{Ad}(P))).\end{aligned} \tag{2.42}$$

In this situation, the generalised Seiberg–Witten equation (2.22) for $A \in \mathscr{A}(\mathrm{pr}_X^* P)$, $B \in \Gamma(I \times X, \mathrm{pr}^*[\Lambda^+ T^*X \otimes \mathrm{Ad}(P)])$, and $C \in \Gamma(I \times X, \mathrm{pr}_X^* \mathrm{Ad}(P))$ becomes the **Haydys–Witten equation**:

$$\begin{aligned}\mathrm{d}_A C + \mathrm{d}_A^* B - \partial_t A &= 0 \\ F_{A(t)}^+ - \partial_t B &= [C, B] + \tfrac{1}{2}[B \times B].\end{aligned} \tag{2.43}$$

This is equation was introduced by [**Hay15**, §3] and Witten [**Wit12**]. The Haydys–Witten equation (2.43) is a gauge theory in dimension 5, but it does not make sense on arbitrary oriented 5–manifolds (at least without a further choice of structure).

Evidently, the dimensonal reduction of (2.43) along I is the Vafa–Witten equation (1.20). The dimensonal reduction of (2.43) with $X = \mathbf{R} \times Y$ along $\mathbf{R}$ gives the Kapustin–Witten equation (1.42) on $I \times Y$. In this sense the Kapustin–Witten equation (1.42) is a generalised Seiberg–Witten equation in dimension three. This is explained further in [**Hay15**, §4] and Witten [**Wit12**, §5.3.1].

The ASD equation (1.5), the Seiberg–Witten equation (1.16), the Vafa–Witten equation (1.20), and the complex ASD equation (1.40) cannot just be considered as generalised Seiberg–Witten equations. They can also be understood as arising from the Spin(7)–**instanton equation** by dimensional reduction. This is no coincidence, since (in the relevant cases) the latter itself can be seen as generalised Seiberg–Witten equation. This was observed by Haydys [**Hay08, Hay12**] and Donaldson and Segal [**DS11**, §6].

EXAMPLE 2.44 (Spin(7)–instantons). For simplicity restrict to $X = \mathbf{H}$. Let L be a connected compact semi-simple Lie group. Consider $S := \Omega^1(\mathbf{H}, \mathfrak{l})$ and $G = C^\infty(\mathbf{H}, L)$. S can be thought of as the space of connections on $P := \mathbf{H} \times L$ and G is the gauge group $\mathscr{G}(P)$. It was observed by Atiyah that

$$\mu(I) = F_I^+ \in \Omega^+(\mathbf{H}, \mathfrak{l}). \tag{2.45}$$

Since $X = \mathbf{H}$, the choices of K, H, $\mathrm{Fr}^\star$, and $\mathfrak{s}$ are almost immaterial. A section $I \in \Gamma(S^+)$ is simply a family of connections on P parametrised by $t \in \mathbf{H}$. A connection $A \in \mathscr{A}(\mathfrak{s}, \mathrm{LC}^\star)$ can be written as

$$A = \sum_{a=0}^{3} \xi_a \, \mathrm{d}\, t_a$$

with $\xi_a \in C^\infty(\mathbf{H}, \mathfrak{l})$. A computation shows that (2.22) becomes

$$\begin{aligned} \partial_{t_0} I - \mathrm{d}_I \xi_0 - i(\partial_{t_1} I - \mathrm{d}_I \xi_1) - j(\partial_{t_2} I - \mathrm{d}_I \xi_2) - k(\partial_{t_3} I - \mathrm{d}_I \xi_3) &= 0 \\ F_A^+ &= F_I^+. \end{aligned} \tag{2.46}$$

Another computation shows that this is equivalent to $\mathbf{A} := I + \sum_{a=0}^{3} \xi_a \, \mathrm{d}\, t_a$, considered as a connection on $\mathbf{H}^2 \times L$, satisfying the $\mathrm{Spin}(7)$–**instanton equation**

$$*(F_A \wedge \Phi) = -F_A \wedge \Phi \tag{2.47}$$

with $\Phi \in \Omega^4(\mathbf{H}^2)$ defined by

$$\Phi := \mathrm{vol}_1 + \mathrm{vol}_2 - \frac{1}{4}\langle \mathrm{d}\, q_1 \wedge \mathrm{d}\, \bar{q}_1, \mathrm{d}\, q_2 \wedge \mathrm{d}\, \bar{q}_2 \rangle \tag{2.48}$$

and vol_a denoting the usual volume form on the $\mathbf{H}$ factors.

In general, a variation of the above can be formulated for every oriented Euclidean rank 4 bundle V over an oriented 4–manifold (X, g) together with an isometry $\Lambda^+ V \cong \Lambda^+ T^* X$.

In the previous example, *formally*, the **hyperkähler quotient**

$$S /\!\!/\!\!/ G := \mu^{-1}(0)/G \tag{2.49}$$

is the moduli space of ASD instantons on the trivial bundle. If $L = \mathrm{SU}(r)$ and one defines S and G a little more carefully, then $S /\!\!/\!\!/ G$ turns into the framed moduli space $\mathscr{M}_{k,r}$ of $\mathrm{SU}(r)$ ASD instantons over $\mathbf{R}^4$ with instanton number k. The ADHM construction provides an alternative description of $\mathscr{M}_{k,r}$ and this leads to the following.

EXAMPLE 2.50. For $r, k \in \mathbf{N}$, consider the quaternionic Hermitian vector space

$$S_{r,k} := \mathrm{Hom}_{\mathbf{C}}(\mathbf{C}^r, \mathbf{H} \otimes_{\mathbf{C}} \mathbf{C}^k) \oplus \mathbf{H}^* \otimes_{\mathbf{R}} \mathfrak{u}(k)$$

with

$$G = \mathrm{U}(k) \triangleleft H = \mathrm{SU}(r) \times \mathrm{Sp}(1) \times \mathrm{U}(k) \quad \text{and} \quad \varepsilon := (\mathbf{1}, -\mathbf{1}, -\mathbf{1}).$$

If $r \geqslant 2$, then $S_{r,k} /\!\!/\!\!/ G := \mu^{-1}(0)/G$ is the Uhlenbeck compactification of the moduli space of framed $\mathrm{SU}(r)$ ASD instantons of charge k on $\mathbf{R}^4$ [**AHDM78**]. If $r = 1$, then

$$S_{1,k} /\!\!/\!\!/ G = \mathrm{Sym}^k \mathbf{H} := \mathbf{H}^k / S_k;$$

see [**Nak99**, Proposition 2.9], [**DW19**, Theorem D.2].

The generalized Seiberg–Witten equation associated with the above data is called the **ADHM$_{r,k}$ Seiberg–Witten equation**. It was introduced in [**DW20**, Example A.3][**DW19**, Section 5.1] and is expected to play an important role in gauge theory on G_2–manifolds [**DS11, Wal17, Hay19**]. For $k = 1$, this is essentially the Seiberg–Witten equation with r spinors.

3. The analysis of generalised Seiberg–Witten equations

This lecture discusses some of the analytic difficulties that arise for generalised Seiberg–Witten equations. It will provide a glimpse of what is known and what is not known. Finally, an (internal) application of the currently known results is sketched.

3.1. What seems to be the problem? The key difficulty in working with the generalised Seiberg–Witten equation (2.22) is a potential failure of compactness. If Φ stays bounded, then (2.22) can be treated as mild perturbation of the ASD equation (1.5) and the familiar analysis developed by Uhlenbeck [**Uhl82a, Uhl82b**] and Taubes [**Tau82, Tau84**] largely carries over. Unfortunately, Φ need not be bounded a priori—unless $\mu^{-1}(0) = \{0\}$ (which usually fails).

It might appear naive to expect that if (A_n, Φ_n) is a sequence of solutions to (2.22) with $\|\Phi_n\|_{L^2} \to \infty$, then the renormalised sequence $(A_n, \Phi_n/\|\Phi_n\|_{L^2})$ converges (after passing to a subsequence, modifying by gauge transformations, etc.) to a solution of

$$\begin{aligned} D_A\Phi &= 0 \\ \mu(\Phi) &= 0. \end{aligned} \tag{3.1}$$

Furthermore, this appears to be a horrible equation: it is not elliptic, even after gauge fixing!

However, there is a way to think about (3.1) slightly differently. Consider the hyperkähler quotient

$$Q := S /\!\!/\!\!/ G := \mu^{-1}(0)/G. \tag{3.2}$$

This is usually a singular space but decomposes into hyperkähler manifolds. Let us pretend that Q actually is a hyperkähler manifold. The representation

$$\sigma_+ \colon \operatorname{Spin}^{G\times K}(4) \to \mathrm{O}(S)$$

descends to an action $\operatorname{Spin}^K(4) \circlearrowright Q$ and one can form a bundle of hyperkähler manifolds

$$\underline{Q} := \mathrm{Fr}^\star \times_{\operatorname{Spin}^K(4)} Q. \tag{3.3}$$

This bundle has a non-linear Dirac operator—induced by a Clifford multiplication and a connection. The **Haydys correspondence** relates solutions of (3.1) up to the action of $\mathscr{G}(\mathfrak{s}/\mathrm{Fr}^\star)$ with non-linear harmonic spinors taking values in $\underline{Q}$. The latter is an elliptic equation and, therefore, morally, (3.1) is elliptic after all!

There are many caveats and the actual story is a rather more complicated. It is discussed to some extend in [**DW19**, Appendix C] (in dimension three).

EXAMPLE 3.4. In the situation of Example 2.30 with $n = 2$, $S = \mathbf{H}\otimes_{\mathbf{C}} \mathbf{C}^2$ and

$$(\mathbf{H}\otimes_{\mathbf{C}} \mathbf{C}^2)/\!\!/\!\!/\, \mathrm{U}(1) = \mathrm{Re}(\mathbf{H}\otimes_{\mathbf{C}} \mathbf{C}^2)/\{\pm 1\}.$$

In this case, $\underline{Q} = \mathrm{Re}(S^+ \otimes_{\mathbf{C}} E)/\{\pm 1\}$ with S^+ denoting a positive chirality spinor bundle for some spin structure $\mathfrak{s}$ that induces $\mathfrak{w}$. Therefore, the issue of unbounded Φ is tied up with harmonic spinors defined up to a sign.

EXAMPLE 3.5. Consider the situation of Example 2.31 with $S = \mathbf{H} \otimes \mathfrak{g}$. A computation shows that

$$|\mu(\boldsymbol{\xi})|^2 = \frac{1}{2}\sum_{a,b=0}^{3} |[\xi_a, \xi_b]|^2. \tag{3.6}$$

Therefore, $\mu(\boldsymbol{\xi}) = 0$ implies that the components ξ_a are contained in a maximal torus $\mathfrak{t} \subset \mathfrak{g}$. Consequently

$$S /\!\!/\!\!/ G = (\mathbf{H} \otimes \mathfrak{g})/W_G(T)$$

with $W_G(T) := N_G(T)/T$ denoting the Weyl group. For $G = \mathrm{SO}(3)$

$$S /\!\!/\!\!/ G = \mathbf{H}/\{\pm 1\}$$

and $\underline{Q}^+ = (\underline{\mathbf{R}} \oplus \Lambda^+ T^*X)/\{\pm 1\}$. Therefore, the issue of unbounded Φ is tied up with self-dual harmonic 2–forms defined up to a sign.

3.2. Harmonic $\mathbf{Z}/2\mathbf{Z}$ spinors. The above heuristic indicates that for the Seiberg–Witten equation with two spinors, for the Vafa–Witten equation with $G = \mathrm{SO}(3)$ or $\mathrm{SU}(2)$, and for complex ASD equation with $G = \mathrm{SO}(3)$ or $\mathrm{SU}(2)$ the following objects should play an important role.

DEFINITION 3.7. Let (X, g) be a closed Riemannian manifold and $\mathbf{S}$ a Dirac bundle over X.[14] Denote by $D\colon \Gamma(\mathbf{S}) \to \Gamma(\mathbf{S})$ the associated Dirac operator.

1. A $\mathbf{Z}/2\mathbf{Z}$ **spinor** with values in $\mathbf{S}$ is a triple $(Z, \mathfrak{l}, \Psi)$ consisting of:
   (a) a nowhere-dense closed subset $Z \subset X$,
   (b) a Euclidean line bundle $\mathfrak{l} \to X\backslash Z$, and
   (c) a section $\Psi \in \Gamma(X\backslash Z, \mathbf{S} \otimes \mathfrak{l})$
   such that $|\Psi|$ extends to a Hölder continuous function on X with $|\Psi|^{-1}(0) = Z$ and $|\nabla\Psi| \in L^2(X\backslash Z)$.
2. A $\mathbf{Z}/2\mathbf{Z}$ spinor $(Z, \mathfrak{l}, \Psi)$ is **singular** if $\mathfrak{l}$ does not extend to a Euclidean line bundle on X.
3. A $\mathbf{Z}/2\mathbf{Z}$ spinor $(Z, \mathfrak{l}, \Psi)$ is **harmonic** if

$$D\Psi = 0$$

holds on $X\backslash Z$. •

Embarrassingly little is know about harmonic $\mathbf{Z}/2\mathbf{Z}$ spinors. Taubes [**Tau14**] proved some initial results about the structure of Z. These are superceded by the following.

THEOREM 3.8 (Zhang [**Zha17**, Theorem 1.4]). *Let (X, g) is a Riemannian 4–manifold. Let $(Z, \mathfrak{l}, \Psi)$ be a harmonic $\mathbf{Z}/2\mathbf{Z}$ spinor over (X, g). If there are $c, \varepsilon > 0$ such that for every $x \in X$*

$$\int_{B_x(r)} |\Psi|^2 < cr^{4+\varepsilon},$$

then Z is $\mathscr{H}^2$–rectifiable and has locally finite Minkowski content (that is: it is almost a 2–manifold).

[14] A Dirac bundle is a bundle of Clifford modules together with a metric and a compatible connection; see, [**LM89**, Chapter II, Definition 5.2].

An important *open problem* is to develop a Fredholm deformation theory of harmonic $\mathbf{Z}/2\mathbf{Z}$ spinors. A completely satisfactory theory is not yet available. There has been initial work by Takahashi [**Tak15, Tak17**]. Parker [**Par23b**] has developed a deformation theory for harmonic $\mathbf{Z}/2\mathbf{Z}$ spinors when $\dim X = 3$ and Z is smooth. There is work in progress by Haydys–Mazzeo–Takahashi towards relaxing the smoothness assumption on Z. For harmonic $\mathbf{Z}/2\mathbf{Z}$ 1–forms Donaldson [**Don21**] has developed a substantial part of the desired theory.[15]

3.3. Taubes' compactness theorems. It turns out that the naive expectation discussed above actually is not that far from the truth. There are a number of compactness theorems for concrete generalised Seiberg–Witten equations, mostly due to Taubes. [**Tau13b**] is concerned with flat $\mathbf{PSL}_2(\mathbf{C})$–connection, a generalised Seiberg–Witten equation in dimension three, [**Tau13a**] is concerned with the Kapustin–Witten equation, [**Tau17**] is concerned with the Vafa–Witten equation; see also [**BW96, Doa18, Hay15a, WZ21, He20**]. Here is a simplified version of Taubes's result for U(1) Seiberg–Witten systems.

THEOREM 3.9 (Taubes [**Tau16**]). *Let (A_i, Φ_i) be a sequence of solutions of the Seiberg–Witten equation with n spinors with $\|\Phi_i\|_{L^2} \to \infty$. After passing to a subsequence the following holds:*

(1) *There is a closed nowhere dense subset Z, and a solution (A, Φ) of (3.1) over $X \backslash Z$. Moreover, $|\Phi|$ is Hölder continuous and $|\Phi|^{-1}(0) = Z$.*
(2) *For every compact subset $K \subset X\backslash Z$, there is a sequence (u_i) of gauge transformations defined on $X\backslash Z$ such that, on K, $(u_i^*\Phi_i)$ converges to Φ in the weak $W^{2,2}$–topology and $(u_i^*A_i)$ converges to A in the weak $W^{1,2}$–topology. Moreover, $|\Phi_i|$ converges to $|\Phi|$ in $C^{0,\alpha}$ for some $\alpha > 0$.*

The proof of this theorem is about 100 pages long and arguably among the simplest of Taubes' compactness proofs. A sketch of the proof would go beyond the scope of this lecture; however: it shall be mentioned that a gauge-theoretic version of Almgren's frequency function and the Higgs mechanism play a crucial role.

Although, Theorem 3.9 is a major achievement, it leaves some questions open:

(1) The weak convergence in $W^{2,2}$ and $W^{1,2}$ is quite weak. One does expect C^∞ convergence, but it is not a simple question of elliptic bootstrapping because (3.1) is not elliptic. The recent work of Parker [**Par23a**] makes important progress towards resolving this issue.
(2) The regularity of Z is very weak. If (3.1) induces a harmonic $\mathbf{Z}/2\mathbf{Z}$ spinor, then Theorem 3.8 applies; cf. [**Hay15a**, Appendix A]. In light of the work of Taubes and Wu [**TW20, TW22**], one cannot expect that Z is always smooth, but one might hope (very optimistically) that this is generically true.
(3) Theorem 3.9 makes no assertion whatsoever about the behaviour of (A_i, Φ_i) along Z. Of course, this is tied up closely with the problem of how to explicitly exhibit degenerating (A_i, Φ_i) sequences by a gluing construction. [**DW20**] deals with the case $Z = \varnothing$. See also [**Par22**].

---

[15] In this case $\mathbf{S} = \Lambda^* T^*X$ and $\gamma(\alpha)\Phi := \alpha \wedge \Phi - i_{\alpha^\sharp}\Phi$, and one restricts attention to $\Phi \subset \Gamma(X\backslash Z, T^*X \otimes \mathfrak{l}) \subset \Gamma(X\backslash Z, \mathbf{S} \otimes \mathfrak{l})$.

3.4. An application: existence of harmonic $\mathbf{Z}/2\mathbf{Z}$ spinors. The remainder of this lecture sketches an argument which is spelled out in detail in [**DW21**].

Let (Y, g) be a closed oriented Riemannian 3–manifold with $b_1(Y) \geqslant 2$. Let E be a rank 2 Hermitian vector bundle with a trivialisaton $\det E \cong \underline{\mathbf{C}}$. Let

$$\mathscr{P} := \mathscr{A}(E)$$

be the space of unitary connection on E which induce the product connection on $\underline{\mathbf{C}}$. Fix a spin^c–structure $\mathfrak{w}$ on Y denote the complex spinor bundle by W. For every $B \in \mathscr{P}$ one can consider the Seiberg–Witten equation with two spinors for $A \in \mathscr{A}(\det W)$ and $\Psi \in \Gamma(\mathrm{Hom}(E, W))$:

$$\begin{aligned} D_{A,B}\Psi &= 0 \\ \tfrac{1}{2}F_A &= \mu(\Psi). \end{aligned} \tag{3.10}$$

Consider the moduli space

$$\mathscr{M}_{\mathfrak{w}}(B) := \{(A, B) : (3.10)\}/C^\infty(Y, \mathrm{U}(1)). \tag{3.11}$$

It is natural to ask: can one define an invariant

$$n := \#\mathscr{M}_{\mathfrak{w}}(B) \in \mathbf{Z}$$

by counting $\mathscr{M}_{\mathfrak{w}}(B)$? (Of course, to actually do this, one needs to perturb (3.10). This can be done as in classical Seiberg–Witten theory and shall be swept under the rug here.) The key point is to understand the dependence of the count on B.

Comparing with classical Seiberg–Witten theory in dimension three, what goes wrong is the failure of the usual proof of an a priori bound on Ψ. Instead one has Theorem 3.9 or, rather, its 3–dimensional analogue proved in [**Hay15a**] and generalised in [**WZ21**]. It turns out that there is an explicit mechanism that triggers the failure of invariance and this has an interesting consequence. The failure of compactness is caused by a solution of

$$\begin{aligned} D_{A,B}\Psi &= 0 \\ \mu(\Psi) &= 0 \end{aligned} \tag{3.12}$$

defined away from a nowhere-dense closed subset Z. The Haydys correspondence in this case is not terribly difficult to understand. If a solution of (3.12) on Y exists and Ψ is nowhere-vanishing, then one can prove the following:

(1) $\det W$ is trivial; in particular, $\mathfrak{w}$ is induced by a spin structure. (The set of all spin structures $\mathfrak{s}$ that induce $\mathfrak{w}$ is a torsor over $\ker(\beta\colon \mathrm{H}^1(Y, \mathbf{Z}/2\mathbf{Z}) \to \mathrm{H}^2(Y, \mathbf{Z}))$.)
(2) After applying a gauge transformation, we can assume that A induces the product connection and there is a unique spin structure $\mathfrak{s}$ inducing $\mathfrak{w}$ such that Ψ takes values in $\mathrm{Re}(E \otimes S) \subset \mathrm{Hom}(E, W)$.
(3) $\Psi \in \ker(D_B\colon \Gamma(\mathrm{Re}(S \otimes E)) \to \Gamma(\mathrm{Re}(S \otimes E)))$.

In other words, the failure of compactness is detected by a harmonic $\mathbf{Z}/2\mathbf{Z}$ spinor $(Z, \mathfrak{l}, \Psi)$ with values in $\mathrm{Re}(S \otimes E)$.

It is not difficult to prove that for some $B \in \mathscr{P}$ harmonic spinors $\Psi \in \Gamma(\mathrm{Re}(S \otimes E))$ must exists (using spectral flow arguments). Of course, these define harmonic $\mathbf{Z}/2\mathbf{Z}$ spinors; however, they are non-singular: $\mathfrak{l}$ extends to Y. Therefore, we always have to contend with these and the wall-crossing phenomena they cause.

What is not clear is that singular harmonic $\mathbf{Z}/2\mathbf{Z}$ spinors exist. If they did not for any $B \in \mathscr{P}$, then all of the technology required to understand n would be in place. In light of this, assume that:

There are no *singular* harmonic $\mathbf{Z}/2\mathbf{Z}$ spinors for any $B \in \mathscr{P}$.

This will lead to a contradiction. Therefore, singular harmonic $\mathbf{Z}/2\mathbf{Z}$ spinors must exist for some $B \in \mathscr{P}$. The argument gives more detailed information about what happens and, morally, shows that singular harmonic $\mathbf{Z}/2\mathbf{Z}$ spinors must exist for B in a "codimension one" subset of $\mathscr{P}$.

It is well known that the subset $\mathscr{W} \subset \mathscr{P}$ of those $B \in \mathscr{P}$ for which there is a harmonic spinor has codimension 1. In fact,

$$\mathscr{W}_1 := \{B \in \mathscr{P} : \dim \ker D_B = 1\}$$

is a codimension 1 submanifold and $\mathscr{W}\backslash\mathscr{W}_1$ has codimension 3. One can then define a locally constant function

$$n \colon \mathscr{P}\backslash\mathscr{W} \to \mathbf{Z}. \tag{3.13}$$

$\mathscr{P}$ should be thought of as separated into chambers by the walls $\mathscr{W}$. The question is then how n changes as one crosses these walls. The deformation theory developed in [**DW20**] together with a refinement of Theorem 3.9 shows that

$$n \text{ jumps by } \pm 1 \text{ as one crosses } \mathscr{W}_{1,\varnothing}.$$

Here

$$\mathscr{W}_{1,\varnothing} := \{B \in \mathscr{W}_1 : \text{every non-zero } \Psi \in \ker D_B \text{ is nowhere-vanishing}\}.$$

(It is important that Ψ is nowhere-vanishing, because this amounts to $Z = \varnothing$.)

The actual signs correspond to a coorientation of $\mathscr{W}_{1,\varnothing}$. A naive guess might be that they are simply the coorientation induced by spectral flow. If that were the case, then the coorientation if $\mathscr{W}_{1,\varnothing}$ would extend to a coorientation of $\mathscr{W}_1$. This is not the case: indeed, the coorientation flips along the codimension 1 submanifold

$$\mathscr{W}_{1,\star} = \{B \in \mathscr{W}_1 : \text{every non-zero } \Psi \in \ker D_B \text{ vanishes transversely at precisely one point}\}.$$

The situation is illustrated as follows:

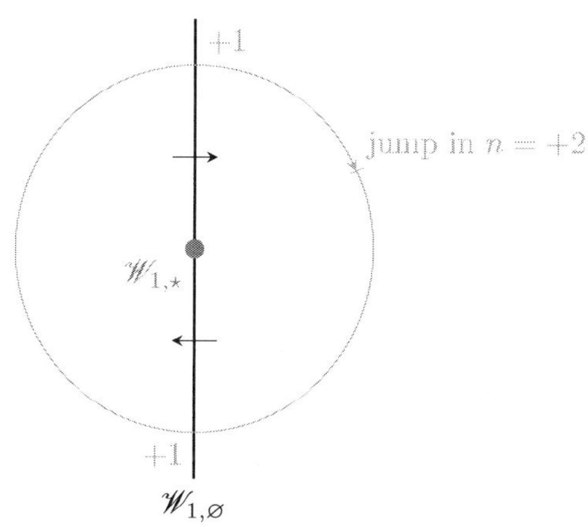

Of course, this leads to a contradiction since n is locally constant on the chambers of $\mathscr{P}\backslash\mathscr{W}$. Therefore, *singular* harmonic $\mathbf{Z}/2\mathbf{Z}$ spinors must exists for some $B \in \mathscr{P}$. □

Acknowledgements

This material is based upon work supported by the National Science Foundation under Grant No. DMS-1928930, while the author was in residence at the Simons Laufer Mathematical Science Institute (previously known as MSRI) Berkeley, California, during the Fall 2022 semester.

References

[AHDM78] M. F. Atiyah, N. J. Hitchin, V. G. Drinfel′d, and Yu. I. Manin, *Construction of instantons*, Phys. Lett. A **65** (1978), no. 3, 185–187, DOI 10.1016/0375-9601(78)90141-X. MR598562

[AHS78] M. F. Atiyah, N. J. Hitchin, and I. M. Singer, *Self-duality in four-dimensional Riemannian geometry*, Proc. Roy. Soc. London Ser. A **362** (1978), no. 1711, 425–461, DOI 10.1098/rspa.1978.0143. MR506229

[Bau14] Helga Baum, *Eichfeldtheorie. Eine Einführung in die Differentialgeometrie auf Faserbündeln* (German), 2nd revised ed., Springer-Lehrb. Mastercl., Heidelberg: Springer Spektrum, 2014.

[BPST75] A. A. Belavin, A. M. Polyakov, A. S. Schwartz, and Yu. S. Tyupkin, *Pseudoparticle solutions of the Yang-Mills equations*, Phys. Lett. B **59** (1975), no. 1, 85–87, DOI 10.1016/0370-2693(75)90163-X. MR434183

[BW96] James A. Bryan and Richard Wentworth, *The multi-monopole equations for Kähler surfaces*, Turkish J. Math. **20** (1996), no. 1, 119–128. MR1392667

[Cor88] Kevin Corlette, *Flat G-bundles with canonical metrics*, J. Differential Geom. **28** (1988), no. 3, 361–382. `http://projecteuclid.org/euclid.jdg/1214442469`. MR0965220

[Doa18] Aleksander Doan, *Adiabatic limits and Kazdan-Warner equations*, Calc. Var. Partial Differential Equations **57** (2018), no. 5, Paper No. 124, 25, DOI 10.1007/s00526-018-1371-5. MR3836210

[DW19] Aleksander Doan and Thomas Walpuski, *On counting associative submanifolds and Seiberg-Witten monopoles*, Pure Appl. Math. Q. **15** (2019), no. 4, 1047–1133, DOI 10.4310/PAMQ.2019.v15.n4.a4. MR4085667

[DW20] Aleksander Doan and Thomas Walpuski, *Deformation theory of the blown-up Seiberg-Witten equation in dimension three*, Selecta Math. (N.S.) **26** (2020), no. 3, Paper No. 48, 48, DOI 10.1007/s00029-020-00574-6. MR4122245

[DW21] Aleksander Doan and Thomas Walpuski, *On the existence of harmonic* Z_2 *spinors*, J. Differential Geom. **117** (2021), no. 3, 395–449, DOI 10.4310/jdg/1615487003. MR4255067

[Don83] S. K. Donaldson, *An application of gauge theory to four-dimensional topology*, J. Differential Geom. **18** (1983), no. 2, 279–315. MR0710056

[Don85] S. K. Donaldson, *Anti self-dual Yang-Mills connections over complex algebraic surfaces and stable vector bundles*, Proc. London Math. Soc. (3) **50** (1985), no. 1, 1–26, DOI 10.1112/plms/s3-50.1.1. MR765366

[Don90] S. K. Donaldson, *Polynomial invariants for smooth four-manifolds*, Topology **29** (1990), no. 3, 257–315, DOI 10.1016/0040-9383(90)90001-Z. MR1066174

[Don96] S. K. Donaldson, *The Seiberg-Witten equations and 4-manifold topology*, Bull. Amer. Math. Soc. (N.S.) **33** (1996), no. 1, 45–70, DOI 10.1090/S0273-0979-96-00625-8. MR1339810

[Don21] Simon Donaldson, *Deformations of multivalued harmonic functions*, Q. J. Math. **72** (2021), no. 1-2, 199–235, DOI 10.1093/qmath/haab018. MR4271385

[DK90] S. K. Donaldson and P. B. Kronheimer, *The geometry of four-manifolds*, Oxford Mathematical Monographs, The Clarendon Press, Oxford University Press, New York, 1990. Oxford Science Publications. MR1079726

[DS11] Simon Donaldson and Ed Segal, *Gauge theory in higher dimensions, II*, Surveys in differential geometry. Volume XVI. Geometry of special holonomy and related topics, Surv. Differ. Geom., vol. 16, Int. Press, Somerville, MA, 2011, pp. 1–41, DOI 10.4310/SDG.2011.v16.n1.a1. MR2893675

[ESW22] Chris Elliott, Pavel Safronov, and Brian R. Williams, *A taxonomy of twists of supersymmetric Yang-Mills theory*, Selecta Math. (N.S.) **28** (2022), no. 4, Paper No. 73, 124, DOI 10.1007/s00029-022-00786-y. MR4468561

[Fee00] Paul M. N. Feehan, *Generic metrics, irreducible rank-one PU(2) monopoles, and transversality*, Comm. Anal. Geom. **8** (2000), no. 5, 905–967, DOI 10.4310/CAG.2000.v8.n5.a1. MR1846123

[FL98a] Paul M. N. Feehan and Thomas G. Leness, PU(2) *monopoles and relations between four-manifold invariants*, Topology Appl. **88** (1998), no. 1-2, 111–145, DOI 10.1016/S0166-8641(97)00201-0. Symplectic, contact and low-dimensional topology (Athens, GA, 1996). MR1634566

[FL98b] Paul M. N. Feehan and Thomas G. Leness, PU(2) *monopoles. I. Regularity, Uhlenbeck compactness, and transversality*, J. Differential Geom. **49** (1998), no. 2, 265–410. MR1664908

[FL01a] Paul M. N. Feehan and Thomas G. Leness, PU(2) *monopoles and links of top-level Seiberg-Witten moduli spaces*, J. Reine Angew. Math. **538** (2001), 57–133, DOI 10.1515/crll.2001.069. MR1855754

[FL01b] Paul M. N. Feehan and Thomas G. Leness, PU(2) *monopoles. II. Top-level Seiberg-Witten moduli spaces and Witten's conjecture in low degrees*, J. Reine Angew. Math. **538** (2001), 135–212, DOI 10.1515/crll.2001.064. MR1855755

[FL18] Paul M. N. Feehan and Thomas G. Leness, *An* SO(3)*-monopole cobordism formula relating Donaldson and Seiberg-Witten invariants*, Mem. Amer. Math. Soc. **256** (2018), no. 1226, xiv+234, DOI 10.1090/memo/1226. MR3897982

[Flo88] Andreas Floer, *An instanton-invariant for* 3*-manifolds*, Comm. Math. Phys. **118** (1988), no. 2, 215–240. MR0956166

[GU12] Michael Gagliardo and Karen Uhlenbeck, *Geometric aspects of the Kapustin-Witten equations*, J. Fixed Point Theory Appl. **11** (2012), no. 2, 185–198, DOI 10.1007/s11784-012-0082-3. MR3000667

[GNO77] P. Goddard, J. Nuyts, and D. Olive, *Gauge theories and magnetic charge*, Nuclear Phys. B **125** (1977), no. 1, 1–28, DOI 10.1016/0550-3213(77)90221-8. MR443696

[GM93] Robert E. Gompf and Tomasz S. Mrowka, *Irreducible* 4*-manifolds need not be complex*, Ann. of Math. (2) **138** (1993), no. 1, 61–111, DOI 10.2307/2946635. MR1230927

[Ham17] Mark J. D. Hamilton, *Mathematical gauge theory: With applications to the standard model of particle physics*, Universitext, Springer, Cham, 2017, DOI 10.1007/978-3-319-68439-0. MR3837560

[Hay19] Andriy Haydys, G_2 *instantons and the Seiberg-Witten monopoles*, Gromov-Witten theory, gauge theory and dualities, Proc. Centre Math. Appl. Austral. Nat. Univ., vol. 48, Austral. Nat. Univ., Canberra, 2019, pp. 12. MR3951401

[Hay06] Andriy Haydys, *Generalized Seiberg–Witten equations and hyperKähler geometry*, 2006, `https://ediss.uni-goettingen.de/handle/11858/00-1735-0000-0006-B381-C`.

[Hay08] Andriy Haydys, *Nonlinear Dirac operator and quaternionic analysis*, Comm. Math. Phys. **281** (2008), no. 1, 251–261, DOI 10.1007/s00220-008-0445-1. MR2403610

[Hay12] Andriy Haydys, *Gauge theory, calibrated geometry and harmonic spinors*, J. Lond. Math. Soc. (2) **86** (2012), no. 2, 482–498, DOI 10.1112/jlms/jds008. MR2980921

[Hay15] Andriy Haydys, *Fukaya-Seidel category and gauge theory*, J. Symplectic Geom. **13** (2015), no. 1, 151–207, DOI 10.4310/JSG.2015.v13.n1.a5. MR3338833

[Hay15a] Andriy Haydys and Thomas Walpuski, *A compactness theorem for the Seiberg-Witten equation with multiple spinors in dimension three*, Geom. Funct. Anal. **25** (2015), no. 6, 1799–1821, DOI 10.1007/s00039-015-0346-3. MR3432158

[Hay15b] Andriy Haydys, *Dirac operators in gauge theory*, New ideas in low dimensional topology, Hackensack, NJ: World Scientific, 2015, pp. 161–188.

[He20] Siqi He, *The behavior of sequences of solutions to the Hitchin–Simpson equations*, Preprint, `arXiv:2002.08109`, 2020.

[KW07] Anton Kapustin and Edward Witten, *Electric-magnetic duality and the geometric Langlands program*, Commun. Number Theory Phys. **1** (2007), no. 1, 1–236, DOI 10.4310/CNTP.2007.v1.n1.a1. MR2306566

[Kot95] D. Kotschick, *Gauge theory is dead!—Long live gauge theory!*, Notices Amer. Math. Soc. **42** (1995), no. 3, 335–338. MR1316132

[LM89] H. Blaine Lawson Jr. and Marie-Louise Michelsohn, *Spin geometry*, Princeton Mathematical Series, vol. 38, Princeton University Press, Princeton, NJ, 1989. MR1031992

[Mar95] Neil Marcus, *The Other topological twisting of N=4 Yang-Mills*, Nucl. Phys. B 452 95, pp. 331-345 arXiv:hep-th/9506002, 1995

[Mar10] Ben A. Mares, *Some Analytic Aspects of Vafa–Witten Twisted* $\mathcal{N} = 4$ *Supersymmetric Yang–Mills Theory*, Massachusetts Institute of Technology, 2010, https://dspace.mit.edu/handle/1721.1/64488.

[MO77] Claus Montonen and David Olive, *Magnetic monopoles as gauge particles?*, Physics Letters B **72** (1977), no. 1, 117–120.

[Nak99] Hiraku Nakajima, *Lectures on Hilbert schemes of points on surfaces*, University Lecture Series, vol. 18, American Mathematical Society, Providence, RI, 1999, DOI 10.1090/ulect/018. MR1711344

[Nak16] Hiraku Nakajima, *Towards a mathematical definition of Coulomb branches of 3-dimensional* $\mathcal{N} = 4$ *gauge theories, I*, Adv. Theor. Math. Phys. **20** (2016), no. 3, 595–669, DOI 10.4310/ATMP.2016.v20.n3.a4. MR3565863

[Par22] Gregory J. Parker, *Concentrating Local Solutions of the Two-Spinor Seiberg-Witten Equations on 3–Manifolds*, Preprint, arXiv:2210.08148, 2022.

[Par23a] Gregory J. Parker, *Concentrating Dirac Operators and Generalized Seiberg-Witten Equations*, Preprint, arXiv:2307.00694, 2023.

[Par23b] Gregory J. Parker, *Deformations of* $\mathbf{Z}_2$*–Harmonic Spinors on 3–Manifolds*, Preprint, arXiv:2301.06245, 2023.

[PT95] Victor Ya. Pidstrigach and Andrei N. Tyurin, *Localisation of Donaldson invariants along the Seiberg–Witten classes*, Preprint, arXiv:dg-ga/9507004, 1995.

[Pid04] V. Ya. Pidstrigach, *Hyper-Kähler manifolds and the Seiberg-Witten equations* (Russian, with Russian summary), Tr. Mat. Inst. Steklova **246** (2004), no. Algebr. Geom. Metody, Svyazi i Prilozh., 263–276; English transl., Proc. Steklov Inst. Math. **3(246)** (2004), 249–262. MR2101297

[Sal13] Dietmar Salamon, *The three-dimensional Fueter equation and divergence-free frames*, Abh. Math. Semin. Univ. Hambg. **83** (2013), no. 1, 1–28, DOI 10.1007/s12188-013-0075-1. MR3055820

[SW94] N. Seiberg and E. Witten, *Electric-magnetic duality, monopole condensation, and confinement in* $N = 2$ *supersymmetric Yang-Mills theory*, Nuclear Phys. B **426** (1994), no. 1, 19–52, DOI 10.1016/0550-3213(94)90124-4. MR1293681

[Tak15] Ryosuke Takahashi, *The moduli space of S1-type zero loci for Z/2-harmonic spinors in dimension 3*, ProQuest LLC, Ann Arbor, MI, 2015. Thesis (Ph.D.)–Harvard University. MR3450477

[Tak17] Ryosuke Takahashi, *Index theorem for the zero loci of* $\mathbf{Z}/2$*–harmonic spinors*, 2017.

[Tan] Yuuji Tanaka, *On the singular sets of solutions to the Kapustin–Witten equations and the Vafa–Witten ones on compact kähler surfaces.*

[Tan15] Yuuji Tanaka, *Stable sheaves with twisted sections and the Vafa-Witten equations on smooth projective surfaces*, Manuscripta Math. **146** (2015), no. 3-4, 351–363, DOI 10.1007/s00229-014-0706-6. MR3312450

[Tan17] Yuuji Tanaka, *Some boundedness properties of solutions to the Vafa-Witten equations on closed 4-manifolds*, Q. J. Math. **68** (2017), no. 4, 1203–1225, DOI 10.1093/qmath/hax015. MR3783982

[Tan19] Yuuji Tanaka, *A perturbation and generic smoothness of the Vafa-Witten moduli spaces on closed symplectic four-manifolds*, Glasg. Math. J. **61** (2019), no. 2, 471–486, DOI 10.1017/S0017089518000307. MR3928649

[TT17a] Yuuji Tanaka and Richard P. Thomas, *Vafa-Witten invariants for projective surfaces II: semistable case*, Pure Appl. Math. Q. **13** (2017), no. 3, 517–562, DOI 10.4310/pamq.2017.v13.n3.a6. MR3882207

[TT17b] Yuuji Tanaka and Richard P. Thomas, *Vafa-Witten invariants for projective surfaces II: semistable case*, Pure Appl. Math. Q. **13** (2017), no. 3, 517–562, DOI 10.4310/pamq.2017.v13.n3.a6. MR3882207

[Tau82] Clifford Henry Taubes, *Self-dual Yang-Mills connections on non-self-dual* 4*-manifolds*, J. Differential Geometry **17** (1982), no. 1, 139–170. MR658473

[Tau84] Clifford Henry Taubes, *Self-dual connections on* 4*-manifolds with indefinite intersection matrix*, J. Differential Geom. **19** (1984), no. 2, 517–560. MR755237

[Tau87] Clifford Henry Taubes, *Gauge theory on asymptotically periodic* 4*-manifolds*, J. Differential Geom. **25** (1987), no. 3, 363–430. MR882829

[Tau90] Clifford Henry Taubes, *Casson's invariant and gauge theory*, J. Differential Geom. **31** (1990), no. 2, 547–599. MR1037415

[Tau99] Clifford Henry Taubes, *Nonlinear generalizations of a* 3*-manifold's Dirac operator*, Trends in mathematical physics (Knoxville, TN, 1998), AMS/IP Stud. Adv. Math., vol. 13, Amer. Math. Soc., Providence, RI, 1999, pp. 475–486, DOI 10.1090/amsip/013/37. MR1708781

[Tau13a] Clifford H. Taubes, *Compactness theorems for* $\mathrm{SL}(2;\mathbf{C})$ *generalizations of the 4-dimensional anti-self dual equations*, Preprint, arXiv:1307.6447, 2013.

[Tau13b] Clifford Henry Taubes, $\mathrm{PSL}(2;\mathbb{C})$ *connections on 3-manifolds with* L^2 *bounds on curvature*, Camb. J. Math. **1** (2013), no. 2, 239–397, DOI 10.4310/CJM.2013.v1.n2.a2. MR3272050

[Tau14] Clifford H. Taubes, *The zero loci of* $\mathbf{Z}/2$ *harmonic spinors in dimension* 2, 3 *and* 4, Preprint, arXiv:1407.6206, 2014.

[Tau16] Clifford H. Taubes, *On the behavior of sequences of solutions to* $\mathrm{U}(1)$ *Seiberg–Witten systems in dimension* 4, Preprint, arXiv:1610.07163, 2016.

[Tau17] Clifford H. Taubes, *The behavior of sequences of solutions to the Vafa–Witten equations*, Preprint, arXiv:1702.04610, 2017.

[TW20] C. H. Taubes and Y. Wu, *Examples of singularity models for* $\mathbb{Z}/2$ *harmonic 1-forms and spinors in dimensional three*, Proceedings of the Gökova Geometry-Topology Conferences 2018/2019, Int. Press, Somerville, MA, 2020, pp. 37–66. MR4251085

[TW22] Clifford H. Taubes and Yingying Wu, *Topological aspects of* $\mathbf{Z}/2\mathbf{Z}$ *eigenfunctions for the Laplacian on* S^2, Preprint, arXiv:2108.05017, 2022.

[Uhl82a] Karen K. Uhlenbeck, *Connections with Lspp bounds on curvature*, Comm. Math. Phys. **83** (1982), no. 1, 31–42. https://projecteuclid.org/euclid.cmp/1103920743. MR0648356

[Uhl82b] Karen K. Uhlenbeck, *Removable singularities in Yang-Mills fields*, Comm. Math. Phys. **83** (1982), no. 1, 11–29. MR0648355

[Uhl12] Karen K. Uhlenbeck, *Equations of gauge theory*, 2012. notes by L. Fredrickson, https://ncatlab.org/nlab/files/UhlenbeckGaugeTheory.pdf.

[VW94] Cumrun Vafa and Edward Witten, *A strong coupling test of S-duality*, Nuclear Phys. B **431** (1994), no. 1-2, 3–77, DOI 10.1016/0550-3213(94)90097-3. MR1305096

[Wal17] Thomas Walpuski, G_2*-instantons, associative submanifolds and Fueter sections*, Comm. Anal. Geom. **25** (2017), no. 4, 847–893, DOI 10.4310/CAG.2017.v25.n4.a4. MR3731643

[WZ21] Thomas Walpuski and Boyu Zhang, *On the compactness problem for a family of generalized Seiberg-Witten equations in dimension 3*, Duke Math. J. **170** (2021), no. 17, 3891–3934, DOI 10.1215/00127094-2021-0005. MR4340726

[Wit88] Edward Witten, *Topological quantum field theory*, Communications in Mathematical Physics **117** (1988), no. 3, 353–386, http://projecteuclid.org/euclid.cmp/1104161738. MR953828

[Wit94] Edward Witten, *Monopoles and four-manifolds*, Math. Res. Lett. **1** (1994), no. 6, 769–796, DOI 10.4310/MRL.1994.v1.n6.a13. MR1306021

[Wit12] Edward Witten, *Fivebranes and knots*, Quantum Topol. **3** (2012), no. 1, 1–137, DOI 10.4171/QT/26. MR2852941

[Wit17] Edward Witten, *Two lectures on the Jones polynomial and Khovanov homology*, Lectures on geometry, Clay Lect. Notes, Oxford Univ. Press, Oxford, 2017, pp. 1–27. MR3676591

[Wu08] Siye Wu, *S-duality in Vafa-Witten theory for non-simply laced gauge groups*, J. High Energy Phys. **5** (2008), 009, 17, DOI 10.1088/1126-6708/2008/05/009. MR2411350

[Yam88] Jonathan P. Yamron, *Topological actions from twisted supersymmetric theories*, Phys. Lett. B **213** (1988), no. 3, 325–330, DOI 10.1016/0370-2693(88)91769-8. MR965719

[Zha17] Boyu Zhang, *Rectifiability and minkowski bounds for the zero loci of* $\mathbf{Z}/2$ *harmonic spinors in dimension 4*, Preprint, arXiv:1712.06254, 2017.

Institut für Mathematik, Humboldt-Universität zu Berlin, Unter den Linden 6, 10099 Berlin, Germany

Email address: thomas@walpu.ski

Proceedings of Symposia in Pure Mathematics
Volume **109**, 2024
https://doi.org/10.1090/pspum/109/02001

A splitting formula in instanton Floer homology

Nima Anvari

ABSTRACT. In a recent paper, Lin, Ruberman and Saveliev proved a splitting formula expressing the Seiberg-Witten invariant $\lambda_{\mathrm{SW}}(X)$ of a smooth 4-manifold with rational homology of $S^1 \times S^3$ in terms of the Frøyshov invariant $h(X)$ and a Lefschetz number in reduced monopole Floer homology. In this note we observe that a similar splitting formula holds in reduced instanton Floer homology.

1. Introduction

Let X be a smooth, oriented spin 4-manifold that has integral homology of $S^1 \times S^3$. A smooth invariant of X, denoted by $\lambda_{\mathrm{SW}}(X)$, was defined by Mrowka, Ruberman and Saveliev [**MRS11**] as a signed count of Seiberg–Witten monopoles plus an index-theoretic correction term. Recently, Lin, Ruberman and Saveliev [**LRS18**] extended the definition of $\lambda_{\mathrm{SW}}(X)$ to all smooth, oriented spin 4-manifolds with rational homology of $S^1 \times S^3$ and proved a splitting formula relating $\lambda_{\mathrm{SW}}(X)$ to the monopole Frøyshov invariant $h(Y, \mathfrak{s})$, where Y is an embedded rational homology 3-sphere generating $H_3(X; \mathbb{Z})$ (assuming such exists) and $\mathfrak{s}$ is the induced spin structure. It is proved in [**Frø10**] that $h(Y, \mathfrak{s})$ is an invariant of X, denoted by $h(X)$. The splitting formula relating these two invariants reads [**LRS18**, Theorem A]

$$\lambda_{SW}(X) + h(X) = -\operatorname{Lef}(W_* : HM^{\mathrm{red}}(Y, \mathfrak{s}) \to HM^{\mathrm{red}}(Y, \mathfrak{s})),$$

where W is the spin cobordism from Y to itself obtained by cutting X open along Y and $HM^{\mathrm{red}}(Y, \mathfrak{s})$ is the reduced monopole Floer homology.

In this note we show that a similar splitting formula holds when reduced monopole Floer homology is replaced by reduced instanton Floer homology, $\lambda_{\mathrm{SW}}(X)$ is replaced by the Furuta–Ohta invariant $\lambda_{\mathrm{FO}}(X)$, as defined in [**FO93**], and the monopole Frøyshov invariant $h(Y)$ is replaced by the instanton Frøshov invariant $h(Y)$ (that the same notation is used for both Frøyshov invariants may be confusing, although the two invariants are known to coincide in all calculated examples).

More precisely, let X be a $\mathbb{Z}[\mathbb{Z}]$-homology $S^1 \times S^3$, that is, a smooth 4-manifold such $H_*(X; \mathbb{Z}) = H_*(S^1 \times S^3; \mathbb{Z})$ and $H_*(\widetilde{X}; \mathbb{Z}) = H_*(S^3; \mathbb{Z})$, where $\widetilde{X}$ is the universal abelian cover of X. The moduli space $\mathcal{M}^*(X)$ of irreducible (perturbed) anti-self dual connections on a trivial $SU(2)$ bundle over X is compact and has formal dimension zero; in fact, it coincides with the moduli space of flat $SU(2)$ connections on X. With an appropriate choice of orientations and admissible perturbations, the Furuta–Ohta invariant is defined as a signed count

$$\lambda_{\mathrm{FO}}(X) = \frac{1}{4}\,\#\mathcal{M}^*(X),$$

see Ruberman–Saveliev [**RS04b**] for details.

THEOREM A. *Let X be a $\mathbb{Z}[\mathbb{Z}]$-homology $S^1 \times S^3$ and suppose that there exists an embedded integral homology sphere Y which generates $H_3(X;\mathbb{Z})$. Cut X open along Y to obtain a homology cobordism W from Y to itself, and denote by W_* and $\widehat{W}_*$ the induced homomorphisms on, respectively, the instanton Floer homology and the reduced instanton Floer homology of Y. Then the quantity*

$$h(X) = \frac{1}{2}\,(\mathrm{Lef}(\widehat{W}_*) - \mathrm{Lef}(W_*)) \tag{1}$$

defined in terms of the Lefschetz numbers is independent of W and coincides with the instanton Frøyshov invariant $h(Y)$. Moreover, the following splitting formula holds

$$\lambda_{\mathrm{FO}}(X) + h(X) = \frac{1}{2}\,\mathrm{Lef}(\widehat{W}_* : \widehat{HF^*}(Y) \longrightarrow \widehat{HF^*}(Y)). \tag{2}$$

We refer the reader to Frøyshov [**Frø02**] for the definitions of the reduced instanton Floer homology and the instanton Frøyshov invariant; see also Section 2 and Section 3. Note that the splitting formula (2) follows easily from (1) and the formula $\lambda_{\mathrm{FO}}(X) = 1/2\ \mathrm{Lef}(W_*)$ of [**RS04b**]. To prove that $h(X)$ is a well-defined invariant, we will show that it coincides with the instanton Frøyshov invariant $h(Y)$ for any choice of Y. When W is a product cobordism $Y \times [0,1]$, both homomorphisms W_* and $\widehat{W}_*$ are identity maps, and (1) reduces to the Frøyshov formula

$$h(X) = h(Y) = \frac{1}{2}(\chi(\widehat{HF_*}(Y)) - \chi(HF_*(Y))).$$

We will prove the general case of (1) in this paper using properties of the special boundary maps of Frøyshov [**Frø02**].

EXAMPLE 1. Let Y be the Brieskorn homology 3-sphere $\Sigma(2,7,13)$ oriented as the link of a complex surface singularity, and τ the involution on $\Sigma(2,7,13)$ induced by complex conjugation. The mapping torus of τ is a smooth 4-manifold $X_\tau = [0,1] \times Y/(0,x) \sim (1,\tau(x))$ and a $\mathbb{Z}[\mathbb{Z}]$-homology $S^1 \times S^3$. The mod 8 graded instanton Floer homology of Y, from grading 0 (mod 8) to 7 (mod 8) , is given by (see [**FS90**])

$$HF_*(Y) = (0, \mathbb{Z}^4, 0, \mathbb{Z}^2, 0, \mathbb{Z}^4, 0, \mathbb{Z}^2)$$

and the reduced instanton Floer homology

$$\widehat{HF}_*(Y) = (0, \mathbb{Z}^2, 0, \mathbb{Z}^2, 0, \mathbb{Z}^2, 0, \mathbb{Z}^2)$$

is completely determined by $HF_*(Y)$ and the invariant $h(Y)$ which was computed by Frøyshov [**Frø04**] to be $h(Y) = 2$. In this case the Furuta–Ohta invariant $\lambda_{\mathrm{FO}}(X_\tau)$ equals the Neumann–Siebenmann $\overline{\mu}$–invariant; see [**RS04a**], and a closed form expression is given by Saveliev [**Sav00**, Theorem 6.28, p.147] as $-b_1 + b_3$, where b_i is the rank of $HF_i(Y)$. Therefore, $\lambda_{\mathrm{FO}}(X_\tau) = -2$. On the other hand, let W be the homology cobordism obtained by cutting X_τ open along Y. According to [**RS04a**, Proposition 9.2], the induced map $W_* = \tau_* : HF_k(Y) \to HF_k(Y)$ is the identity for $k \equiv 1 \pmod 4$ and minus the identity for $k \equiv -1 \pmod 4$. It follows that $\operatorname{Lef}(W_*) = -4$ and $\operatorname{Lef}(\widehat{W}_*) = 0$ hence

$$h(X_\tau) = \frac{1}{2}\,(\operatorname{Lef}(\widehat{W}_*) - \operatorname{Lef}(W_*)) = 2,$$

which matches $h(Y)$ and hence confirms the splitting formula $\lambda_{\mathrm{FO}}(X_\tau) + h(X_\tau) = 0 = 1/2\ \operatorname{Lef}(\widehat{W}_*)$.

EXAMPLE 2. Let W be the Akbulut cork as in [**RS04b**, Section 9.3]. That is, W is a smooth contractible manifold with boundary an integral homology sphere Σ that can be embedded into a blown up elliptic surface $E(n)\,\#\,\overline{\mathbb{CP}}^2$ in such a way that cutting it out and re-gluing by an involution $\tau : \Sigma \to \Sigma$ changes the smooth structure on $E(n)\,\#\,\overline{\mathbb{CP}}^2$ but preserves its homeomorphism type. The instanton homology groups are

$$HF_*(\Sigma) = (0, \mathbb{Z}, 0, \mathbb{Z}, 0, \mathbb{Z}, 0, \mathbb{Z})$$

and, since Σ is homology cobordant to zero, $h(\Sigma) = 0$ and $\widehat{HF}_*(\Sigma) = HF_*(\Sigma)$. Let X_τ be the mapping torus of τ then it follows from [**RS04b**] that $W_* = -\operatorname{Id}$ and $\lambda_{\mathrm{FO}}(X_\tau) = 2$, which again confirms the splitting formula.

2. Reduced instanton Floer (co)homology

In this section, we recall the definition of the reduced instanton Floer homology (cohomology) groups; the reader is referred to Frøyshov [**Frø02**] for all the details. We work throughout with real coefficients and the orientation conventions of [**Frø02**] and use the canonical identification $HF_q(Y) = HF^{5-q}(\overline{Y})$.

The instanton Floer cohomology $HF^*(Y)$ is defined as homology of the mod 8 graded instanton cochain complex $CF^*(Y)$ generated by the irreducible (perturbed) flat $SU(2)$ connections on Y, with respect to the boundary map $d : CF^*(Y) \to CF^{*+1}(Y)$ roughly given by a signed count of anti-self dual connections on $\mathbb{R}\times Y$ of finite energy. The definition of the reduced instanton cohomology further employs special boundary maps δ_0 and δ_0', which are defined as follows. Let θ be the trivial flat connection on a trivialized $SU(2)$ bundle over Y. Define

$$\delta :\ CF^4(Y) \longrightarrow \mathbb{R} \quad \text{by} \quad \delta(\alpha) = \#\check{\mathcal{M}}(\theta, \alpha),$$

where $\mathcal{M}(\theta, \alpha)$ is the moduli space of anti-self dual connections on $\mathbb{R} \times Y$ limiting to θ at $+\infty$ and α at $-\infty$, and $\check{\mathcal{M}}(\theta, \alpha)$ its quotient by translations. Define

$$\delta' : \mathbb{R} \longrightarrow CF^1(Y) \quad \text{by} \quad \delta'(1) = \sum_\beta \#\check{\mathcal{M}}(\theta, \beta)\cdot\beta,$$

with the summation extending over the generators $\beta \in CF^1(Y)$. The maps δ and δ' satisfy equations $d\delta = 0$ and $d\delta' = 0$, thereby inducing maps in cohomology,

$$\delta_0 : HF^4(Y) \longrightarrow \mathbb{R} \quad \text{and} \quad \delta_0' : \mathbb{R} \longrightarrow HF^1(Y).$$

The special boundary maps δ_0 and δ_0' are further included into the sequence of maps

$$\delta_n = [\delta v^n] : HF^{4-4n}(Y) \longrightarrow \mathbb{R} \quad \text{and} \quad \delta_n' = [v^n\delta'] : \mathbb{R} \longrightarrow HF^{1+4n}(Y)$$

using the homomorphism $v : CF^*(Y) \longrightarrow CF^{*+4}(Y)$ defined in [**Frø02**]. It follows from the cochain homotopy formula [**Frø02**, Theorem 4(ii)] that either δ_0 or δ_0' must vanish; moreover, if $\delta_0 = 0$ then $\delta_n = 0$ for all n, and similarly for δ_n'.

We also need to recall the relation between the special boundary maps and the homomorphisms in Floer homology induced by negative definite cobordisms. Let W be a connected Riemannian 4-manifold with two cylindrical ends, $\mathbb{R}_- \times Y_1$ and $\mathbb{R}_+ \times Y_2$, and assume that W has negative definite intersection form, $H_1(W;\mathbb{Z}) = 0$, and both Y_1 and Y_2 are integral homology spheres. Then W induces a degree preserving cochain homomorphism

$$W^* : CF^*(Y_1) \to CF^*(Y_2) \quad \text{by} \quad W^*(\alpha) \;=\; \sum_{\beta} \#\mathcal{M}(W;\beta,\alpha)\cdot\beta,$$

where the summation extends over the generators $\beta \in CF^*(Y_2)$ with the same index as α, and $\mathcal{M}(W;\beta,\alpha)$ is the zero-dimensional part of the moduli space of anti-self dual connections on W limiting to α at $-\infty$ and to β at $+\infty$. Frøyshov [**Frø02**, Theorem 7] shows that $\delta_0 W^* = \delta_0$ and $W^*\delta_0' = \delta_0'$ and, moreover, that there exist integers a_{ij} and b_{ij} such that

$$\delta_n W^* = \delta_n + \sum_{i=0}^{n-1} a_{in}\delta_i, \tag{3}$$

$$W^*\delta_n' = \delta_n' + \sum_{i=0}^{n-1} b_{in}\delta_i'. \tag{4}$$

A straightforward grading count shows that $a_{in} = 0$ and $b_{in} = 0$ whenever i and n have opposite parity.

We are now ready to define the reduced instanton cohomology groups. Let $B^* \subseteq HF^*(Y)$ denote the linear span of the vectors $\delta_n'(1)$ for all n so that $B^1 = \operatorname{span}\{\delta_{2k}'(1)\}$ and $B^5 = \operatorname{span}\{\delta_{2k+1}'(1)\}$ with $k \geq 0$, while $B^q = 0$ for $q \neq 1,5 \pmod 8$. In addition, let

$$Z^* = \bigcap_n \ker(\delta_n) \;\subseteq\; HF^*(Y)$$

so that

$$Z^0 = \bigcap_{k\geq 0} \ker(\delta_{2k+1}) \subseteq HF^0(Y) \quad \text{and} \quad Z^4 = \bigcap_{k\geq 0} \ker(\delta_{2k}) \subseteq HF^4(Y),$$

while $Z^q = HF^q(Y)$ for $q \neq 0,4 \pmod 8$. The reduced Floer cohomology groups are then defined as

$$\widehat{HF}^q(Y) = Z^q/B^q.$$

For any negative definite cobordism W as above, the map W^* leaves the subspaces Z^q and B^q invariant and induces a map [**Frø02**, Theorem 8]

$$W^* : \widehat{HF}^*(Y_1) \longrightarrow \widehat{HF}^*(Y_2).$$

Finally, we mention that both instanton Floer cohomology $HF^*(Y)$ and reduced instanton Floer homology $\widehat{HF}^*(Y)$, which *a priori* have a mod 8 grading, are 4-periodic; see [**Frø02**, Theorem 2 and Corollary 3].

3. Proof of Theorem A

Recall that the (instanton) Frøyshov invariant $h(Y)$ is defined by the formula

$$h(Y) = \frac{1}{2}\,(\chi(HF^*(Y)) - \chi(\widehat{HF}^*(Y)).$$

The theorem will be proved as soon as we show that $h(X) = h(Y)$ for any choice of Y. Using the aforementioned 4-periodicity in Floer homology, we only have two cases to consider, depending on which one of the special boundary maps, δ_0 or δ_0', vanishes.

CASE 1. Suppose that $\delta_0' = 0$ and hence $\delta_n' = 0$ for all n. This implies that $B^1 = 0$ and $\widehat{HF}^1(Y) = Z^1 = HF^1(Y)$. Thus the reduced theory differs from the unreduced one only in degree 0 mod 4, and

$$\begin{aligned} h(X) &= \frac{1}{2}\,(\operatorname{Lef}(W^*) - \operatorname{Lef}(\widehat{W}^*)) \\ &= \operatorname{Tr}(W^*\colon HF^0(Y) \to HF^0(Y)) - \operatorname{Tr}(\widehat{W}^*\colon \widehat{HF}^0(Y) \to \widehat{HF}^0(Y)). \end{aligned}$$

Note that, since $B^0 = 0$, we have

$$\widehat{HF}^0(Y) = Z^0 = \bigcap_{k\geq 0} \ker(\delta_{2k+1}).$$

We claim that $h(X) = \dim(HF^0(Y)/Z^0)$; this will imply that $h(X)$ equals the Frøyshov invariant $h(Y) = \dim HF^0(Y) - \dim \widehat{HF}^0(Y)$.

To prove the claim, we first observe that, since $\delta_1 W^* = \delta_1$, we have a commutative diagram of exact sequences

$$\begin{array}{ccccccc} 0 & \longrightarrow & Z_1^0 & \longrightarrow & HF^0(Y) & \xrightarrow{\delta_1} & \mathbb{R} \\ & & \big\downarrow{\scriptstyle \widehat{W}_1^*} & & \big\downarrow{\scriptstyle W^*} & & \big\downarrow{\scriptstyle \mathrm{Id}} \\ 0 & \longrightarrow & Z_1^0 & \longrightarrow & HF^0(Y) & \xrightarrow{\delta_1} & \mathbb{R} \end{array}$$

where $Z_1^0 = \ker(\delta_1)$ and $\widehat{W}_1^*$ is the restriction of W^* to Z_1^0. It follows that $\operatorname{Tr}(W^*) - \operatorname{Tr}(\widehat{W}_1^*) = 1$ if $\delta_1 \neq 0$ and $\operatorname{Tr}(W^*) - \operatorname{Tr}(\widehat{W}_1^*) = 0$ if $\delta_1 = 0$. Next we use relation (3) to define a sequence of W^*–invariant subspaces

$$Z_k^0 \;=\; \ker(\delta_1) \cap \ldots \cap \ker(\delta_k) \;\subseteq\; HF^0(Y)$$

for k odd, and construct the following commutative diagrams with exact rows,

$$\begin{array}{ccccccc} 0 & \longrightarrow & Z_k^0 & \longrightarrow & Z_{k-2}^0 & \xrightarrow{\delta_k} & \mathbb{R} \\ & & \big\downarrow{\scriptstyle \widehat{W}_k^*} & & \big\downarrow{\scriptstyle \widehat{W}_{k-2}^*} & & \big\downarrow{\scriptstyle \mathrm{Id}} \\ 0 & \longrightarrow & Z_k^0 & \longrightarrow & Z_{k-2}^0 & \xrightarrow{\delta_k} & \mathbb{R} \end{array}$$

where $\widehat{W}_k^*$ denotes the restriction of W^* to subspace Z_k^0. It follows that $\operatorname{Tr}(\widehat{W}_{k-2}^*) - \operatorname{Tr}(\widehat{W}_k^*) = 1$ if $\delta_k \neq 0$ and $\operatorname{Tr}(\widehat{W}_{k-2}^*) - \operatorname{Tr}(\widehat{W}_k^*) = 0$ if $\delta_k = 0$. Apply induction to the sequence of subspaces

$$Z_m^0 \;\subseteq \ldots \subseteq\; Z_3^0 \;\subseteq\; Z_1^0 \;\subseteq\; HF^0(Y)$$

to conclude that

$$\operatorname{Tr}(W^*) \;=\; \dim(HF^0(Y)/Z_m^0) \;+\; \operatorname{Tr}(\widehat{W}_m^*)$$

for all odd m. Since $Z^0 = Z_m^0$ and $\widehat{W}^* = \widehat{W}_m^*$ for all sufficiently large odd m, the claim follows.

CASE 2. Suppose that $\delta_0 = 0$ and hence $\delta_n = 0$ for all n. In this case, $\widehat{HF^0}(Y) = HF^0(Y)$ and $\widehat{HF^1}(Y) = HF^1(Y)/B^1$. The reduced Floer cohomology differs from Floer cohomology only in degree 1 mod 4, and

$$\begin{aligned} h(X) &= \frac{1}{2}\left(\operatorname{Lef}(W^*) - \operatorname{Lef}(\widehat{W^*})\right) \\ &= \operatorname{Tr}(\widehat{W}^*\colon \widehat{HF}^1(Y) \to \widehat{HF}^1(Y)) - \operatorname{Tr}(W^*\colon HF^1(Y) \to HF^1(Y)) \end{aligned}$$

It follows from the commutative diagram of exact sequences

$$\begin{array}{ccccccccc} 0 & \longrightarrow & B^1 & \longrightarrow & HF^1(Y) & \longrightarrow & \widehat{HF^1}(Y) & \longrightarrow & 0 \\ & & \downarrow{\scriptstyle W_1^*} & & \downarrow{\scriptstyle W^*} & & \downarrow{\scriptstyle \widehat{W}^*} & & \\ 0 & \longrightarrow & B^1 & \longrightarrow & HF^1(Y) & \longrightarrow & \widehat{HF}^1(Y) & \longrightarrow & 0 \end{array}$$

that $\operatorname{Tr}(W^*) - \operatorname{Tr}(\widehat{W}^*) = \operatorname{Tr}(W_1^*)$, where W_1^* is the restriction of W^* to B^1. We claim that

$$\operatorname{Tr}(W_1^*) = \dim(B^1) = \dim HF^1(Y) - \dim \widehat{HF}^1(Y).$$

This will imply that $h(X)$ equals $h(Y) = \dim \widehat{HF}^1(Y) - \dim HF^1(Y)$ so the conclusion will follow.

To prove the claim, use relation (4) to define W^*–invariant subspaces $B_k^1 = \operatorname{span}\{\delta_0'(1), \dots, \delta_k'(1)\}$ for k even. We have an increasing sequence of quotient spaces,

$$\widehat{HF}^1(Y) = HF^1(Y)/B_m^1 \subseteq HF^1(Y)/B_{m-2}^1 \subseteq \dots \subseteq HF^1(Y)/B_0^1$$

corresponding to the sequence $B_0^1 \subseteq \dots \subseteq B_m^1 \subset HF^1(Y)$, where $B^1 = B_m^1$ for some sufficiently large even m. We also have the commutative diagrams with exact rows

$$\begin{array}{ccccccc} \mathbb{R} & \xrightarrow{\delta_k'} & HF^1(Y)/B_{k-2}^1 & \longrightarrow & HF^1(Y)/B_k^1 & \longrightarrow & 0 \\ \downarrow{\scriptstyle \mathrm{Id}} & & \downarrow{\scriptstyle \overline{W}^*_{k-2}} & & \downarrow{\scriptstyle \overline{W}^*_k} & & \\ \mathbb{R} & \xrightarrow{\delta_k'} & HF^1(Y)/B_{k-2}^1 & \longrightarrow & HF^1(Y)/B_k^1 & \longrightarrow & 0 \end{array}$$

and the relations $\operatorname{Tr}(\overline{W}^*_{k-2}) - \operatorname{Tr}(\overline{W}^*_k) = 1$ if $\delta_k' \neq 0$ and $\operatorname{Tr}(\overline{W}^*_{k-2}) - \operatorname{Tr}(\overline{W}^*_k) = 0$ if $\delta_k' = 0$. By induction we obtain

$$\operatorname{Tr}(\overline{W}^*_0) = \dim(B^1/B_0^1) + \operatorname{Tr}(\widehat{W}^*).$$

Finally, it follows from the relation $W^*\delta_0' = \delta_0'$ and the exact sequence

$$0 \longrightarrow B_0^1 \longrightarrow HF^1(Y) \longrightarrow HF^1(Y)/B_0^1 \longrightarrow 0$$

that $\operatorname{Tr}(\overline{W}^*_0) = \operatorname{Tr}(W^*) - \dim(B_0^1)$. Therefore, $\operatorname{Tr}(W^*) - \operatorname{Tr}(\widehat{W}^*) = \dim(B^1)$, which completes the proof of the claim.

Acknowledgments

The author thanks Nikolai Saveliev for many helpful discussions, valuable suggestions and for encouragement in writing this paper. The author also thanks the referee for helpful comments and suggestions.

References

[FO93] Mikio Furuta and Hiroshi Ohta, *Differentiable structures on punctured 4-manifolds*, Topology Appl. **51** (1993), no. 3, 291–301, DOI 10.1016/0166-8641(93)90083-P. MR1237394

[Frø02] Kim A. Frøyshov, *Equivariant aspects of Yang-Mills Floer theory*, Topology **41** (2002), no. 3, 525–552, DOI 10.1016/S0040-9383(01)00018-0. MR1910040

[Frø04] Kim A. Frøyshov, *An inequality for the h-invariant in instanton Floer theory*, Topology **43** (2004), no. 2, 407–432, DOI 10.1016/S0040-9383(03)00049-1. MR2052970

[Frø10] Kim A. Frøyshov, *Monopole Floer homology for rational homology 3-spheres*, Duke Mathematical Journal, **155**(3):519–576, 2010.

[FS90] Ronald Fintushel and Ronald J. Stern, *Instanton homology of Seifert fibred homology three spheres*, Proc. London Math. Soc. (3) **61** (1990), no. 1, 109–137, DOI 10.1112/plms/s3-61.1.109. MR1051101

[LRS18] Jianfeng Lin, Daniel Ruberman, and Nikolai Saveliev, *A splitting theorem for the Seiberg-Witten invariant of a homology $S^1 \times S^3$*, Geom. Topol. **22** (2018), no. 5, 2865–2942, DOI 10.2140/gt.2018.22.2865. MR3811774

[MRS11] Tomasz Mrowka, Daniel Ruberman, and Nikolai Saveliev, *Seiberg-Witten equations, end-periodic Dirac operators, and a lift of Rohlin's invariant*, J. Differential Geom. **88** (2011), no. 2, 333–377. MR2838269

[RS04a] Daniel Ruberman and Nikolai Saveliev, *Rohlin's invariant and gauge theory. I. Homology 3-tori*, Comment. Math. Helv. **79** (2004), no. 3, 618–646, DOI 10.1007/s00014-004-0816-y. MR2081729

[RS04b] Daniel Ruberman and Nikolai Saveliev, *Rohlin's invariant and gauge theory. II. Mapping tori*, Geom. Topol. **8** (2004), 35–76, DOI 10.2140/gt.2004.8.35. MR2033479

[Sav00] Nikolai Saveliev, *Invariants for homology 3-spheres*, Encyclopaedia of Mathematical Sciences, vol. 140, Springer-Verlag, Berlin, 2002. Low-Dimensional Topology, I, DOI 10.1007/978-3-662-04705-7. MR1941324

Department of Mathematics, University of Miami, Coral Gables, Florida 33146

Current address: Department of Mathematics & Statistics, McMaster University L8S 4K1 Hamilton, Ontario, Canada

Email address: anvarin@math.mcmaster.ca

Proceedings of Symposia in Pure Mathematics
Volume **109**, 2024
https://doi.org/10.1090/pspum/109/01989

Finite group actions on 4-manifolds and equivariant bundles

Nima Anvari and Ian Hambleton

Abstract. Given a 4-manifold with a homologically trivial and locally-linear cyclic group action, we obtain necessary and sufficient conditions for the existence of equivariant bundles. The conditions are derived from the twisted signature formula and are in the form of congruence relations between the fixed point data and the isotropy representations.

1. Introduction

Finite group actions on 4-manifolds can be studied in various settings. We are mainly interested in comparing smooth actions with those which are topological and locally linear, but important examples arise for symplectic 4-manifolds and complex surfaces. Here is a sampling of survey articles and recent work on aspects of this general theme: [**1**–**3**, **7**–**9**, **12**, **19**, **25**–**28**, **32**, **38**, **45**]. We will focus on the existence and classification of equivariant bundles, and their applications in Yang-Mills gauge theory to the study of finite group actions.

We begin by recalling some standard definitions. Let (X, π) denotean oriented, simply-connected, closed 4-manifold X together with a locally linear and homologically trivial action of a cyclic group $\pi = \mathbb{Z}/p$ of prime order. The fixed point set X^π has Euler characteristic $\chi(X^\pi) = b_2(X) + 2$, by the Lefschetz fixed point formula. In general X^π will consist of isolated fixed points and a disjoint union of fixed 2-spheres (see [**11**, §2]).

Definition 1.1. At each isolated fixed point $x \in X^\pi$, the tangent space admits an equivariant decomposition $(T_{x_i}X, \pi) = \mathbb{C}(a_i) \oplus \mathbb{C}(b_i)$ of complex representation spaces. Let

$$t \cdot (z_1, z_2) = (\zeta^{a_i} z_1, \zeta^{b_i} z_2)$$

denote the action for t a fixed generator in the cyclic group π, $\zeta = e^{2\pi i/p}$, and with integers (a_i, b_i), both non-zero modulo p.

(1) The integers (a_i, b_i) are the local tangential *rotation data*, and are well-defined up to order and simultaneous change in sign.
(2) Similarly, for each point x on a π-fixed 2-sphere F_j, there is a representation $\mathbb{C} \oplus \mathbb{C}(c_j)$ corresponding to the equivariant splitting

$$TX\,|_{F_j} = TF_j \oplus N(F_j)$$

2020 *Mathematics Subject Classification.* Primary 57M60, 58J20.

This research was partially supported by NSERC Discovery Grant A4000.

where $N(F_j)$ is the normal bundle with rotation ζ^{c_j}, and $c_j \not\equiv 0 \pmod p$.

(3) The total *fixed point rotation data* is the collection

$$\mathcal{F} = \{(a_i, b_i), (c_j, \alpha_j) \mid i \in I, j \in J\}.$$

where I and J index the isolated fixed points and 2-spheres respectively and $\alpha_j = [F_j] \cdot [F_j]$ is the self-intersection number of the fixed spheres $[F_j] \in H_2(X; \mathbb{Z})$.

(4) By an equivariant line bundle $(L, \pi) \to (X, \pi)$ we mean a principal $U(1)$-bundle L over X together with a lift of the π-action to the total space. Given such a lift, there exists a set of *isotropy representations* of the π-action on each fiber over the fixed point set which we denote by $L\,|_{x_i} = t^{\lambda_i}$ over isolated fixed points and $L\,|_{F_j} = t^{\lambda_j}$ over a fixed 2-sphere. Denote the collection of these isotropy representations by $\mathcal{I} = \{\lambda_i, \lambda_j \mid i \in I, j \in J\}$.

With this notation we have our main result.

THEOREM A. *Let (X, π) denote an oriented, simply-connected, closed 4-manifold with a locally linear, homologically trivial action of a cyclic group $\pi = \mathbb{Z}/p$ of odd prime order p, with fixed-point rotation data $\mathcal{F} = \{(a_i, b_i), (c_j, \alpha_j) \mid i \in I, j \in J\}$.*

A collection $\mathcal{I}$ of integers $\{\lambda_i, \lambda_j \mid i \in I, j \in J\}$ can be realized (modulo p) as the isotropy representations of an equivariant line bundle $(L, \pi) \to (X, \pi)$ if and only if there is a collection of integers $\{m_j \mid j \in J\}$ such that

$$\sum_{i\in I} \frac{\lambda_i}{a_i b_i} + \sum_{j \in J} \frac{c_j m_j - \lambda_j \alpha_j}{c_j^2} \equiv 0 \pmod p. \tag{1.1}$$

When such a line bundle exists, the integers $m_j = c_1(i^* L)[F_j]$ satisfy equation (1.1).

REMARK 1.2. A special case of this result can be found in [**27**, Proposition 1.4], for a cyclic group π of odd order acting locally linearly and semi-freely on the complex projective plane $\mathbb{C}P^2$ which has three isolated fixed points. The necessary condition (1.1) in Theorem A is established by extending the methods of [**27**, §2] to more general actions.

REMARK 1.3. The definitions above generalize directly to actions (X, π) where π is any finite group, and the structural group G of the principal bundle is any compact Lie group. The details will be left to the reader (see also [**23, 24**]).

For applications in gauge theory, $G = SU(2)$ is an important example. In Theorem 7.1 we obtain the analogous necessary conditions for the existence of equivariant $SU(2)$-bundles, and discuss the standard examples on $X = S^4$ and $X = \overline{\mathbb{C}P}^2$. In Section 8 we apply our methods to compute the formal dimensions of fixed strata equivariant $SU(2)$-moduli spaces (see [**25**]), and work out one special case.

2. Some motivating questions

Here are some questions related to the general theme (all actions will be assumed to preserve orientation). More information about some of these directions can be found in the references.

1. Does there exist a smooth $\mathbb{Z}/p$-action on a (homotopy) K3 surface, which induces the identity on integral homology?

This is a well-known question of Allan Edmonds. Note that results of Edmonds and Ewing [**14**] imply that topological locally linear examples exist for odd primes.

2. Does there exist a smooth $\mathbb{Z}/p$-action on a homotopy K3 surface, which contains an invariant embedded Brieskorn homology 3-sphere $\Sigma(2,3,7)$?

Fintushel and Stern [**16**] showed that many homotopy K3 surfaces admit embeddings of $\Sigma(2,3,7)$. The question concerns the possible existence of an equivariant splitting of the K3 surface along the Brieskorn sphere.

3. A Brieskorn homology 3-sphere $\Sigma(a,b,c)$ admits a free $\mathbb{Z}/p$-action if $p \nmid abc$. Does there exist a smooth, homologically trivial extension of this action with isolated fixed points to any smooth simply connected negative definite 4-manifold X with boundary $\Sigma(a,b,c)$?

Anvari [**1**] proved that the free $\mathbb{Z}/7$ on $\Sigma(2,3,5)$ does not extend in this way over the minimal negative definite 4-manifold obtained by resolving the link singularity. The method of proof involves studying equivariant Yang-Mills gauge theory on the non-compact manifold with cylindrical end obtained from X by attaching the end $\Sigma(2,3,5) \times [0,\infty)$. Information about the Floer homology of Brieskorn spheres is an essential ingredient in tackling this problem (see Saveliev [**39–41**]).

The corresponding extension problem from $\mathbb{Z}/p$ actions on Brieskorn spheres to smooth contractible 4-manifolds was studied in [**2**], [**3**] where the orientation in the instanton moduli space played a key role (see also [**32**]).

4. What sets of rotation numbers can be realized by a smooth, pseudo-free $\mathbb{Z}/n$-action on $X = S^2 \times S^2$?

A pseudo-free action is one with isolated singular points. If the action is semi-free, there are "standard models" with rotation data $\{(a,b),(c,d),(a,-b),(c,-d)\}$ at the four fixed points. This question for $X = \mathbb{C}P^2$ was answered in [**13, 25**], where it was shown that the rotation data is always the same as for the linear actions arising from $PGL_3(\mathbb{C})$.

5. Let (X,π) denote a smooth action of a finite group π on a closed, simply connected smooth 4-manifold. Under what conditions does there exist a π-equivariant principal G-bundle over X with prescribed Chern classes, for $G = U(1)$ or $G = SU(2)$?

Some information about this question was provided in [**25, 27**] for $X = \mathbb{C}P^2$ and π finite cyclic, or more generally for X negative definite (see also [**24**] for the connection between Chern classes and the isotropy representations). For $X = S^4$, the existence and classification of such bundles was applied by Austin [**6**] and Furuta [**19–21**] to study group actions via instanton gauge theory. The compactification of an equivariant version of Donaldson's Yang-Mills moduli space [**10**], [**25**] involves "bubbling" convergence to equivariant instantons over the 4-sphere. Further applications of equivariant bundles arise in studying the equivariant compactification of moduli spaces over cylindrical end 4-manifolds.

3. The G-signature formula

In this section we review the derivation of the G-signature formula as the index of the signature operator, see [**42**, §18], [**5**, pp. 585-586]. The G-signature of a closed, smooth 4-manifold X with an orientation preserving action of a finite group G acting as isometries on X is defined as the virtual representation

$$\operatorname{Sign}(X,G) = [H^2_+(X;\mathbb{C})] - [H^2_-(X;\mathbb{C})] \tag{3.1}$$

where $H^2_\pm(X;\mathbb{C})$ are the maximal positive/negative definite G-invariant subspaces of $H^2(X;\mathbb{C})$. Taking characters gives the g-signatures

$$\operatorname{Sign}(g,X) = \operatorname{tr}_g H^2_+(X) - \operatorname{tr}_g H^2_-(X) \tag{3.2}$$

which by the G-signature formula can be computed in terms of certain Lefschetz numbers from the fixed point set.

Let $D : C^\infty(\Lambda^+) \to C^\infty(\Lambda^-)$ denote the signature operator. The Lefschetz numbers are computed as follows. Let X^g denote the fixed set of g, $n = \dim X^g$ and N^g denotes the normal bundle of X^g in X. Then the Lefschetz number is given by

$$L(g,D) = (-1)^{n(n+1)/2} \frac{\operatorname{ch}_g(\Lambda^+ - \Lambda^-)(TX\,|_{X^g} \otimes \mathbb{C}) Td(TX^g \otimes \mathbb{C})}{e(TX^g)\operatorname{ch}_g(\Lambda_{-1} N^g \otimes \mathbb{C})}[X^g] \tag{3.3}$$

where

$$\begin{aligned}
Td(TX^g \otimes \mathbb{C}) &= \frac{-x^2}{(1-e^{-x})(1-e^x)} \\
\operatorname{ch}(\Lambda^+ - \Lambda^-)(T^g \otimes \mathbb{C}) &= e^{-x} - e^x \\
\operatorname{ch}_g(\Lambda^+ - \Lambda^-)(N^g \otimes \mathbb{C}) &= e^{-y-i\theta} - e^{y+i\theta} \\
\operatorname{ch}_g(\Lambda_{-1}(N^g \otimes \mathbb{C})) &= (1-e^{y+i\theta})(1-e^{-y-i\theta}).
\end{aligned}$$

Here $x = e(TX^g)$ denotes the Euler class of the tangent bundle to the 2-dimentional fixed point set, ch_g is the equivariant Chern character, and θ is the rotation on the normal bundle N^g with Euler class y. Note that $TX\,|_{X^g} = TX^g \oplus N^g$ and

$$\operatorname{ch}_g(\Lambda^+ - \Lambda^-)(TX\,|_{X^g} \otimes \mathbb{C}) = \operatorname{ch}_g(\Lambda^+ - \Lambda^-)(TX^g \otimes \mathbb{C}) \operatorname{ch}_g(\Lambda^+ - \Lambda^-)(N^g \otimes \mathbb{C}). \tag{3.4}$$

Substituting these expressions into the Lefschetz number and letting F denote a 2-dimensional fixed set component, then the contribution to the g-signature is given by

$$\begin{aligned}
L(g,D)\,|_F &= (-1)\frac{(e^{-x}-e^x)(e^{-y-i\theta}-e^{y+i\theta})}{x(1-e^{y+i\theta})(1-e^{-y-i\theta})} \frac{x(-x)}{(1-e^{-x})(1-e^x)}[F] \\
&= \coth\left(\frac{y+i\theta}{2}\right) x \coth\left(\frac{x}{2}\right)[F]
\end{aligned}$$

where the following trigonometric identity is used

$$\coth\left(\frac{x}{2}\right) = \frac{e^{-x}-e^x}{(1-e^{-x})(1-e^x)}. \tag{3.5}$$

To evaluate on $[F]$, we use the Taylor expansions

$$\begin{aligned}
x\coth(x/2) &= 2 + \frac{x^2}{6} + \cdots \\
\coth\left(\frac{y+i\theta}{2}\right) &= \coth(i\theta/2) - \frac{1}{2}\operatorname{csch}^2\left(\frac{i\theta}{2}\right) y
\end{aligned}$$

Thus the contribution to the Lefschetz number is given by

$$\begin{aligned}L(g,D)\,|\,_F &= \{2\coth(i\theta/2) - \operatorname{csch}^2(i\theta/2)y\}[F]\\ &= -\operatorname{csch}^2(i\theta/2)[F]^2 = \csc^2(\theta/2)[F]^2\\ &= \frac{-4t^{c_F}}{(t^{c_F}-1)^2}[F]^2\end{aligned}$$

where $\theta = \frac{2\pi c_F}{p}$ and c_F is the rotation number on the normal fiber of F and $t = e^{2\pi i/p}$ is a primitive pth root of unity. Similarly we can compute the contribution from isolated fixed points:

$$\begin{aligned}L(g,D)\,|\,_{pt} &= \frac{(e^{-i\theta_1} - e^{i\theta_1})(e^{-i\theta_2} - e^{i\theta_2})}{(1-e^{i\theta_1})(1-e^{-i\theta_1})(1-e^{i\theta_2})(1-e^{-i\theta_2})}\\ &= \coth(i\theta_1/2)\coth(i\theta_2/2)\\ &= -\cot(\theta_1/2)\cot(\theta_2/2)\\ &= \frac{(t^a+1)(t^b+1)}{(t^a-1)(t^b-1)}\end{aligned}$$

where $\theta_1 = \frac{2\pi a}{p}$ and $\theta_2 = \frac{2\pi b}{p}$ are the rotation numbers at the fixed point. By summing over the fixed point set components, the G-signature formula is given by

$$\operatorname{Sign}(g,X) = \sum_i \frac{(t^{a_i}+1)}{(t^{a_i}-1)}\frac{(t^{b_i}+1)}{(t^{b_i}-1)} + \sum_j \frac{-4\alpha_j t^{c_j}}{(t^{c_j}-1)^2} \tag{3.6}$$

where α_j denotes the self-intersection $[F_j]\cdot[F_j]$ of the fixed 2-spheres $\{F_j\}$. Note the equivariant signature on the left-hand side of this formula is equal to the usual signature in this case because the action is homologically trivial. We also remark here that this formula can be viewed as an equation in the cyclotomic field $\mathbb{Q}[\zeta] = \mathbb{Q}[t]/\Phi_p(t)$ where $\Phi_p(t)$ is the cyclotomic polynomial $1+t+t^2+\cdots+t^{p-1}$.

REMARK 3.1. The G-signature formula (and its generalizations to the twisted G-signature formlas used in Section 5 and Section 7 also hold in the topological locally linear category. The argument uses bordism theory (see Wall [**43**, Chapter 14B], [**22**]) and the foundational work of Freedman [**17,18**] (see also [**31**, Theorem 1.1]). For the twisted G-signature formula [**4**, Theorem 6.7], the analytic index and the topological index coincide as homomorphisms $K_G(TX) \to R(G)$. If G is a finite group of odd order, Madsen [**36**, Theorem 3.3 and Proposition 3.6] relates equivariant K_G-theory to equivariant bordism after localizing away from 2.

4. Congruence relations for the G-signature formula

In this section we derive congruence relations satisfied by the rotation data $\mathcal{F}$ using the G-signature formula. We first note that since $(t^a-1)/(t-1)$ is a unit in ring of cyclotomic integers $\mathbb{Z}[\zeta]$ when $(a,p)=1$, multiplying both sides of the G-signature formula by $(t-1)^2$ induces an equation in the ring $R = \mathbb{Z}[\zeta]$. The I-adic expansion of the resulting right-hand side leads to congruence relations relating the rotation data, where I denote the ideal generated by $(t-1)$ in R.

Following the method of [**27**] we lift the equation to $\mathbb{Z}[t]$, compute the Taylor expansions about $t=1$ and reduce the coefficients modulo p. Since the indeterminacy of the coefficients are determined from the expansion of the cyclotomic polynomial $\Phi_p(t)$ (for which p divides the coefficients of its Taylor expansion about

$t = 1$ up to order $p-1$) we obtain valid congruence relations by equating coefficients modulo p up to order $p-1$.

The expansion arising from contributions from isolated fixed points are given by

$$\frac{(t^a+1)}{(t^a-1)}\frac{(t^b+1)}{(t^b-1)}(t-1)^2 = \frac{4}{ab} + \frac{4}{ab}(t-1) + \frac{1}{3}\left(\frac{a^2+b^2+1}{ab}\right)(t-1)^2 - \frac{1}{180}\left(\frac{a^4+b^4-5a^2b^2+3}{ab}\right)(t-1)^4 + \cdots$$

Similarly the expansion of the second term is given by expressions of the form

$$\frac{-4\alpha t^c}{(t^c-1)^2}(t-1)^2 = \frac{-4\alpha}{c^2} + \frac{-4\alpha}{c^2}(t-1) + \frac{1}{3}\frac{\alpha(c^2-1)}{c^2}(t-1)^2 - \frac{1}{60}\frac{\alpha(c-1)(1+c+c^2+c^3)}{c^2}(t-1)^4 + \cdots.$$

Equating both sides of the expansion $\pmod p$ from the resulting equation in R we thus obtain the following congruence relations:

Theorem 4.1 ([**27**, p. 625]). *Let (X,π) denote a simply connected, closed 4-manifold with a homologically trivial, locally-linear group action of a finite cyclic group $\pi = \mathbb{Z}/p$ of odd order. Then the following congruence relations hold*

$$(1) \sum_i \frac{1}{a_i b_i} - \sum_j \frac{\alpha_j}{c_j^2} \equiv 0 \pmod p$$

$$(2) \sum_i \frac{a_i^2+b_i^2}{a_i b_i} + \sum_j \alpha_j \equiv 3\,\mathrm{Sign}(X) \pmod p$$

$$(3) \sum_i \frac{a_i^4+b_i^4-5a_i^2b_i^2}{a_i b_i} + 3\sum_j \alpha_j c_j^2 \equiv 0 \pmod p$$

$$(4) \sum_i \frac{2a_i^6-7a_i^4b_i^2-7a_i^2b_i^4+2b_i^6}{a_i b_i} + 10\sum_j \alpha_j c_j^4 \equiv 0 \pmod p$$

Higher-order relations are valid up to and including terms of order $p-1$.

Example 4.2 (Linear models on $\mathbb{C}P^2$). Let $G = \mathbb{Z}/p$ with odd prime p act linearly on $X = \mathbb{C}P^2$ by $t\cdot[z_1 : z_2 : z_3] = [\zeta^a z_1 : z_2 : z_3]$ for $0 < a < p$. The fixed point set consists of one isolated fixed point $[1:0:0]$ with tangential rotation number (a,a) and a fixed projective line $F = \{[z_1 : z_2 : z_3] \mid z_1 = 0\}$ with self-intersection $+1$ and a rotation of $c_F \equiv a \pmod p$ on the normal bundle. Then it is easy to check that the congruence relations are satisfied. Similarly, in the case when the action is given by $t\cdot[z_1 : z_2 : z_3] = [\zeta^a z_1 : \zeta^b z_2 : z_3]$ for $0 < a < b < p$ the action consists of three isolated fixed points $[0:0:1], [1:0:0], [0:1:0]$ with rotation numbers $(a,b), (b-a,-a), (a-b,-b)$ and with some algebra it can be checked that the congruence relations hold. Additional examples can be obtained for $\#_n\mathbb{C}P^2$ by equivariant connected sums along fixed point sets using the linear models.

5. Equivariant line bundles

Let (X, π) denote a simply connected, closed 4-manifold with a homologically trivial action of a finite cyclic group $\pi = \mathbb{Z}/p$ of odd prime p and $L \to X$ an equivariant line bundle. The index of the signature operator twisted by an equivariant line bundle is the same as in the previous section, except that one has to consider the contribution of the equivariant Chern character of the line bundle restricted to the fixed point set. Let F denote a fixed surface as before, then the contribution to the Lefschetz number is given by:

$$\begin{aligned} L(g,D)\,|_F &= \{2\coth(i\theta/2) - \operatorname{csch}^2(i\theta/2)y\}\{\operatorname{ch}_g(i^*L)\}[F] \\ &= \{2\coth(i\theta/2) - \operatorname{csch}^2(i\theta/2)y\}\{e^{z+i\phi}\}[F] \\ &= \{2\coth(i\theta/2) - \operatorname{csch}^2(i\theta/2)y\}\{e^{i\phi} + e^{i\phi}z\}[F] \\ &= \{2\coth(i\theta/2)e^{i\phi}z - \operatorname{csch}^2(i\theta/2)ye^{i\phi} \\ &= 2c_1(i^*L)[F]\frac{(t^{c_F}+1)}{(t^{c_F}-1)}t^\lambda + \frac{-4[F]^2t^{c_F}}{(t^{c_F}-1)^2}t^\lambda \end{aligned}$$

where $\phi = \frac{2\pi\lambda}{p}$ is the rotation on the fiber in L over a fixed point on F and $z = c_1(i^*L)$ denotes the restriction of the first Chern class to the fixed set component. Similarly for the contribution to the isolated fixed points. To summarize, given an equivariant line bundle $(L, \pi) \to (X, \pi)$, the index of the twisted G-signature operator gives a virtual character given by

$$\chi(t) = \sum_i \frac{(t^{a_i}+1)}{(t^{a_i}-1)}\frac{(t^{b_i}+1)}{(t^{b_i}-1)}t^{\lambda_i} + \sum_j \frac{-4\alpha_j t^{c_j}}{(t^{c_j}-1)^2}t^{\lambda_j} + \sum_j 2c_1(i^*L)[F_j]\frac{(t^{c_j}+1)}{(t^{c_j}-1)}t^{\lambda_j}$$

where $\{F_j\}$ are fixed 2-spheres of the action on X with α_j denoting the self-intersection $[F_j]\cdot[F_j]$. Note that

$$\chi(1) = \operatorname{ch}(L)\mathcal{L}(X)[X] = \left(1 + c_1(L) + \frac{1}{2}c_1(L)^2\right)\left(4 + \frac{p_1}{3}\right)[X] \tag{5.1}$$

$$= \left(\frac{p_1}{3} + 2c_1(L)^2\right)[X] = \operatorname{Sign}(X) + 2c_1(L)^2[X]. \tag{5.2}$$

Since $\chi(t)$ is a virtual character for $G = \mathbb{Z}/p$ we may write $\chi(t) = \sum_{i=0}^{p-1} a_i t^i$ for some $a_i \in \mathbb{Z}$ and

$$\chi(t)(t-1)^2 = \chi(1)(t-1)^2 + \text{higher order terms} \pmod p$$

We can then take the Taylor expansion of the right-hand side of the twisted G-signature formula after multiplying by $(t-1)^2$ and equate the first and second order terms to obtain two additional congruence relations. The first order term vanishes while the second order term is congruent to $\chi(1) \pmod p$. Taking Taylor expansions for these terms in the G-signature formula give:

$$\begin{aligned} \frac{(t^a+1)}{(t^a-1)}\frac{(t^b+1)}{(t^b-1)}(t-1)^2t^\lambda &= \frac{4}{ab} + \frac{4(\lambda+1)}{ab}(t-1) + \\ &\quad \frac{1}{3}\frac{(a^2+b^2+1+6\lambda^2+6\lambda)}{ab}(t-1)^2 + \cdots . \end{aligned}$$

Similarly for the second term:

$$\frac{-4\alpha t^c}{(t^c-1)^2}(t-1)^2t^\lambda = \frac{-4\alpha}{c^2} + \frac{-4\alpha(1+\lambda)}{c^2}(t-1)+ \frac{1}{3}\frac{\alpha(c^2-1-6\lambda-6\lambda^2)}{c^2}(t-1)^2+\cdots .$$

and for the third term:

$$2m\frac{(t^c+1)}{(t^c-1)}(t-1)^2t^\lambda = \frac{4m}{c}(t-1)+\frac{2m(2\lambda+1)}{c}(t-1)^2+\cdots$$

where $m = c_1(i^*L)[F]$. Combining these expressions we obtain the following theorem:

THEOREM 5.1. *Let $(L,\pi)\to(X,\pi)$ denote an equivariant line bundle over a simply connected, closed 4-manifold with a homologically trivial group action of a finite cyclic group $\pi=\mathbb{Z}/p$ of odd order. Then the following congruence relation holds*

$$(i)\ \sum_i \frac{\lambda_i}{a_ib_i} - \sum_j \frac{\lambda_j\alpha_j}{c_j^2} + \sum_j \frac{c_1(i^*L)[F_j]}{c_j} \equiv 0 \pmod p$$

$$(ii)\ \sum_i \frac{\lambda_i^2}{a_ib_i} - \sum_j \frac{\lambda_j^2\alpha_j}{c_j^2} + 2\sum_j \frac{\lambda_jc_1(i^*L)[F_j]}{c_j} \equiv c_1(L)^2[X] \pmod p.$$

EXAMPLE 5.2 (Linear models on $\mathbb{C}P^2$). Let p denote an odd prime and consider $X=\mathbb{C}P^2$ and a linear action $t\cdot[z_1:z_2:z_3]=[\zeta^az_1:\zeta^bz_2:z_3]$ for $0<a<b<p$. We give an explicit construction of equivariant line bundles over X. Consider a finite dimensional complex representation space $V=\mathbb{C}(\lambda_1)\oplus\mathbb{C}(\lambda_2)\oplus\mathbb{C}(\lambda_3)$ with action given by $\rho\in GL_3(\mathbb{C})$:

$$\rho = \begin{pmatrix} t^{\lambda_1} & & \\ & t^{\lambda_2} & \\ & & t^{\lambda_3}\end{pmatrix} : \mathbb{C}^3\setminus\{0\}\longrightarrow\mathbb{C}^3\setminus\{0\} \tag{5.3}$$

with the λ_i positive integer weights. Let $S(V)$ denote the unit sphere in V then ρ commutes with the free S^1-action on $S(V)$ and

$$\mathbb{C}P^2 = S(V)/S^1 = S^5/S^1.$$

The ρ-action on $S(V)$ is a lift of the linear action on X if the following system of linear congruences

$$\begin{aligned} &\lambda_1-\lambda_3\equiv a, \quad \lambda_2-\lambda_3\equiv b\\ &\lambda_2-\lambda_1\equiv b-a, \quad \lambda_3-\lambda_1\equiv -a\\ &\lambda_1-\lambda_2\equiv a-b, \quad \lambda_3-\lambda_2\equiv -b.\end{aligned}$$

has a solution. This system has one degree of freedom; let $\lambda_3=\lambda$ be a fixed parameter, then the isotropy representations over the three isolated fixed points are given by:

fixed point $p_1=[0:0:1]$, rotation number (a,b), with isotropy $\lambda_3\equiv\lambda$

fixed point $p_2=[1:0:0]$, rotation number $(b-a,-a)$, with isotropy $\lambda_1\equiv\lambda+a$

fixed point $p_3=[0:1:0]$, rotation number $(-b,a-b)$ with isotropy $\lambda_2\equiv\lambda+b$.

The equivariant line bundle

$$L = S(V) \times_{S^1} \mathbb{C}$$

is the canonical bundle over $\mathbb{C}P^2$ and the congruence relations of the theorem are satisfied:

$$\sum_i \frac{\lambda_i}{a_i b_i} \equiv 0$$

$$\sum_i \frac{\lambda_i^2}{a_i b_i} \equiv 1.$$

In the case when $b = 0$ the action is given by $t \cdot [z_1 : z_2 : z_3] = [\zeta^a z_1 : z_2 : z_3]$ and has a fixed projective line $F = \{z_1 = 0\}$ with normal rotation number $c_F \equiv a \pmod p$ and self-intersection $+1$, while the isolated fixed point $[1 : 0 : 0]$ has rotation number (a, a). The compatibility for a lift of the linear action is the congruence relation:

$$\lambda - \lambda_F \equiv a \pmod p.$$

It is easily seen that the congruence relation of the theorem are satisfied:

$$\frac{\lambda}{a^2} - \frac{\lambda_F}{a^2} + \frac{c_1(i^*L)[F]}{a} \equiv 0.$$

where we used $c_1(i^*L)[F] = -1$ since the first Chern class of the canonical line bundle over $\mathbb{C}P^2$ is negative of the preferred generator $[F] \in H^2(\mathbb{C}P^2; \mathbb{Z})$ [**37**, Theorem 14.10, p. 169].

Similarly, the second relation is

$$\begin{aligned} &\frac{\lambda^2}{a^2} - \frac{\lambda_F^2}{a^2} + \frac{2\lambda_F c_1(i^*L)[F]}{a} \\ &\equiv \frac{(a + \lambda_F)^2 - \lambda_F^2}{a^2} + \frac{2\lambda_F c_1(i^*L)[F]}{a} \\ &\equiv 1 \equiv c_1(L)^2[X]. \end{aligned}$$

EXAMPLE 5.3. Let $\pi = \mathbb{Z}/p$ act on $X = \mathbb{C}P^2$ preserving an almost complex structure. Then the complexified second exterior power of the cotangent bundle $K_X = \Lambda^2_{\mathbb{C}} T^*X$ is an equivariant line bundle see [**27**, Proposition 1.8]). We check the congruence relations in the case when the action has a fixed projective line and leave the remaining case of isolated fixed points to the reader.

The action from the linear model is given by $t \cdot [z_1 : z_2 : z_3] = [\zeta^a z_1 : z_2 : z_3]$ for some $0 < a < p$ and the fixed point set consists of an isolated fixed point with tangential rotation number $(-a, -a)$ and a fixed projective line

$$F = \{[z_1 : z_2 : z_3] \mid z_1 = 0\}$$

with self-intersection $+1$ and a rotation of $c_F \equiv a \pmod p$ on the normal bundle. The isotropy representations over the fixed point set are $K_X \mid_{pt} = t^{2a}$ and $K_X \mid_F = t^{-a}$. With these values substituted it is easy to verify the congruence relations in Theorem 5.1 for $L = K_X$, noting that K_X is the canonical class of X with dual -3 times the standard generator in $H_2(\mathbb{C}P^2; \mathbb{Z})$.

DEFINITION 5.4. Let (X, π) be a homologically trivial of $\pi = \mathbb{Z}/p$ in the setting of Theorem A, with rotation data $\mathcal{F} = \{(a_i, b_i), (c_j, \alpha_j) \mid i \in I, j \in J\}$. We say that (X, π) *satisfies the condition of Theorem A* if there exists a set of isotropy data

$\mathcal{I} = \{\lambda_i, \lambda_j \,|\, i \in I, j \in J\}$, and a set of integers $\{m_j \,|\, j \in J\}$ so that the equation given in Theorem A holds.

In the next statement, we apply the equivariant connected sum operation to line bundles.

LEMMA 5.5. *Suppose that (X, π) satisfies the condition of Theorem A. There there is an equivariant connected sum $(X \sharp \mathbb{C}P^2, \pi)$ with a linear π-action on $\mathbb{C}P^2$ which also satisfies the condition of Theorem A.*

PROOF. We first suppose that (X, π) contains a fixed 2-sphere F_j with data $\{\alpha_j, c_j\}$. Let $(\mathbb{C}P^2, \pi)$ be the linear action given by $t \cdot [z_1 : z_2 : z_3] = [\zeta^{-c_j} z_1 : z_2 : z_3]$ as in Example 4.2. We do the equivariant connected sum (preserving the orientations) along a point in F_j and a point in the fixed 2-sphere of $\mathbb{C}P^2$. The new data is obtained by (i) adding the data $\{(-c_j, -c_j); \lambda_j\}$ for the newly created isolated fixed point (on $\mathbb{C}P^2$), and (ii) the data $\{(c_j, \alpha_j + 1); \lambda_j\}$ for the new fixed 2-sphere. With these choices, it follows that the action $(X \sharp \mathbb{C}P^2, \pi)$ satisfies the condition of Theorem A. The proof in case the action (X, π) has only isolated fixed points is easier, and will be left to the reader. □

REMARK 5.6. Suppose that (X, π) has data satisfying the condition of Theorem A, and contains a fixed 2-sphere F. Let $X_0 \subset X$ denote the complement of a linear π-invariant 4-ball neighbourhood of a point $x \in F$. If L is an equivariant line bundle over $(X \sharp \mathbb{C}P^2, \pi)$, then the restriction of L to X_0 extends to an equivariant line bundle over (X, π) realizing the given data.

6. The proof of Theorem A

The first relation in Theorem 5.1 proves the necessary conditions of Theorem A. To prove sufficiency we will need the following lemmas.

Note that in a standard lens space $Y = L(n; a, b)$, a generator $\mu \in H_1(Y; \mathbb{Z})$ is represented by a circle fibre in the fibration $S^1 \to L(n; a, b) \to S^2$ given by the quotient of a free $\mathbb{Z}/n$ action on S^3.

LEMMA 6.1. *The linking paring $lk\colon H_1(Y) \times H_1(Y) \to \mathbb{Q}/\mathbb{Z}$ in the lens space $Y = L(n; a, b)$ is given by $lk(\mu, \mu) = \dfrac{ab}{n}$ where μ is a generator of $H_1(Y; \mathbb{Z}) = \mathbb{Z}/n$.*

PROOF. In the usual representation of lens spaces $L(n; q)$ as the quotient of the free $\mathbb{Z}/n$ action $t \cdot (z_1, z_2) = (\zeta z_1, \zeta^q z_2)$ on S^3, the linking pairing is given by $lk(\mu, \mu) = \dfrac{q}{n}$. The diffeomorphism $L(n; a^* b) \to L(n; a, b)$ where a^* is the multiplicative inverse mod n arising from changing the generator in $\mathbb{Z}/n$ induces a map on first homology given by multiplication by a. It follows that the linking pairing on Y is given by

$$lk(a\mu, a\mu) = a^2 \cdot \left(\frac{a^* b}{n}\right) = \frac{ab}{n} \in \mathbb{Q}/\mathbb{Z}.$$

□

LEMMA 6.2 ((See [**27**, Proposition 1.4, p. 621], [**30**, Lemma 2.11, p. 95])). *Let $\phi\colon \pi_1(Y) \longrightarrow U(1)$ be the holonomy representation of a flat $U(1)$-bundle over the lens space $Y = L(n; a, b)$ that sends a generator μ to $\exp(2\pi i \lambda / n)$. Then the Poincaré dual of the first Chern class $PD(c_1(L))$ is given by $\dfrac{\lambda}{ab}[\mu]$ in $H_1(Y; \mathbb{Z})$.*

PROOF. The adjoint to the linking form $\Phi\colon H_1(Y;\mathbb{Z}) \to \mathrm{Hom}(H_1(Y;\mathbb{Z}), \mathbb{Q}/\mathbb{Z})$ sends $m[\mu]$ to $lk(m\mu, -)$ which can be identified with the holonomy representation of the flat bundle via :

$$e^{2\pi i \cdot lk(m\mu,-)}\colon H_1(Y;\mathbb{Z}) \to U(1)$$

and this maps the generator to $\exp(2\pi i m\cdot(ab/n))$. It follows that $m \equiv \dfrac{\lambda}{ab} \pmod{n}$. □

If Y is a lens space, we let $\hat{\mu} \in H^2(Y;\mathbb{Z})$ denote a standard cohomology generator: the Poincaré dual to the circle fibre class $\mu \in H_1(Y;\mathbb{Z})$.

LEMMA 6.3. *Let $u\colon Y' \to Y$ be a d-fold regular covering of lens space, where $H_1(Y;\mathbb{Z}) \cong \mathbb{Z}/dn$ and $H_1(Y';\mathbb{Z}) \cong \mathbb{Z}/n$, with $\gcd(d,n) = 1$. Then $u^*(\hat{\mu}) = \hat{\mu}' \in H^2(Y';\mathbb{Z})$.*

PROOF. For a d-fold regular covering $u\colon Y' \to Y$ of lens spaces, we have $u_*[\mu'] = d[\mu] \in H_1(Y,\mathbb{Z})$ and $u_*[Y'] = d[Y] \in H_3(Y;\mathbb{Z})$. The cohomology generator $\hat{\mu} = [Y] \cap \mu \in H^2(Y;\mathbb{Z})$ is the Poincaré dual of μ, and similarly for $\hat{\mu}' \in H^2(Y';\mathbb{Z})$. We have the formula

$$u_*([Y'] \cap u^*(\hat{\mu})) = u_*(\mu') = d\mu.$$

If $H_1(Y;\mathbb{Z}) = \mathbb{Z}/dn$ and $H_1(Y';\mathbb{Z}) = \mathbb{Z}/n$, where $\gcd(d,n) = 1$, then $u^*(\hat{\mu}) = k\hat{\mu}'$ implies that $k \equiv 1 \pmod{n}$. Hence $u^*(\hat{\mu}) = \hat{\mu}' \in H^2(Y';\mathbb{Z})$. □

Suppose that (X,π) satisfies the assumptions of Theorem 5.1. Let $\Sigma \subset X$ denote the singular set of the action, and let $X_0 := X - \Sigma$. Write $\alpha_j = p^{a_j}\overline{\alpha}_j$, where $\overline{\alpha_j}$ is prime to p (for each fixed 2-sphere F_j).

LEMMA 6.4. *If the singular set $\Sigma \subset X$ contains an isolated point, then $H_1(X_0;\mathbb{Z})$ is a quotient of $\bigoplus \mathbb{Z}/\overline{\alpha}_j$ and has order prime to p.*

PROOF. Let $F \subset X$ denote the disjoint union of the fixed 2-spheres, so $F = \bigcup_j F_j$. First note that $H_1(X_0) \cong H^3(X,F)$ and we have an exact sequence

$$\cdots \to H^2(X) \to H^2(F) \to H^3(X,F) \to H^3(X) \to \dots$$

Since $H^3(X) = 0$, and the homology classes of the fixed 2-spheres are linearly independent mod p (by [**11**, Corollary 2.6]), it follows that $H^3(X,F) \cong H_1(X_0)$ is a torsion group of order prime to p.

Moreover, the exact Mayer-Vietoris sequence

$$0 \to H_2(X_0) \oplus H_2(F) \to H_2(X) \to H_1(\partial X_0) \to H_1(X_0) \to 0$$

and the equality $H_1(\partial X_0) = H_1(\partial\nu(F)) \cong \bigoplus \mathbb{Z}/\alpha_j$ completes the proof. □

THE PROOF OF THEOREM A. By Theorem 5.1(i), the indicated formulas hold if (X,π) admits an equivariant line bundle. It remains to prove that a solution $\{\lambda_i, \lambda_j, m_j \,|\, i \in I, j \in J\}$ to the congruence relation

$$\sum_i \frac{\lambda_i}{a_i b_i} - \sum_j \frac{\lambda_j \alpha_j}{c_j^2} + \sum_j \frac{m_j}{c_j} \equiv 0 \pmod{p} \tag{6.1}$$

is sufficient for the existence of an equivariant line bundle with $\{\lambda_i, \lambda_j\}$ isotropy representations over the isolated fixed points and 2-spheres respectively, and $m_j = c_1(i^*L\,|\,_{F_j})$.

For simplicity, we will assume that the action (X, π) contains at least one isolated fixed point. This may always be arranged by taking the equivariant connected sum of (X, π) along a fixed 2-sphere with a suitable linear π-action on $\mathbb{C}P^2$ (see Lemma 5.5).

Let $X_0 = X - N$, where $N = \nu(\Sigma)$ is a π-invariant tubular neighbourhood of the singular set $\Sigma \subset X$. More explicitly, X_0 is the compact 4-manifold with boundary obtained by removing π-invariant 4-balls around each isolated fixed point and π-invariant tubular neighbourhoods $D^2 \to \nu(F_j) \to F_j$ around each π-fixed 2-sphere F_j with rotation t^{c_j} on D^2-fibers. Then $\nu(F_j)$ is a 2-disk bundle over S^2 with Euler class $\alpha_j[F] \in H^2(F; \mathbb{Z}) \cong \mathbb{Z}$, and the lens space $\partial\nu(F_j) = L(\alpha_j, 1)$ inherits a free $\mathbb{Z}/p$ action with rotation number c_j on the circle fibre.

If $W_0 := X_0/\pi$ denotes the quotient manifold with (regular) covering map $q\colon X_0 \to W_0$ classified by $u\colon W_0 \to B\pi$, then the boundary ∂W_0 consists of lens spaces $Y_i = L(p; a_i, b_i)$ and $Y_j = L(p\alpha_j; c_j, c_j)$. Note that

$$H_1(W_0; \mathbb{Z}) \cong \mathbb{Z}/p \oplus H_1(X_0, \mathbb{Z})$$

by the spectral sequence of the covering. By Lemma 6.4, $H_1(X_0; \mathbb{Z})$ is a quotient of $\bigoplus \mathbb{Z}/\overline{\alpha}_j$ and has order prime to p.

Recall that π-equivariant line bundles L over (X, π) are classified by an element

$$\theta(L) \in H^2_\pi(X; \mathbb{Z}) = H^2(X \times_\pi E\pi; \mathbb{Z})$$

in the Borel equivariant cohomology of X (see [**33**]). Since the π-action on X_0 is free, for the restriction $L_0 \searrow X_0$ we have $\theta(L_0) \in H^2_\pi(X_0; \mathbb{Z}) \cong H^2(W_0; \mathbb{Z})$, and $\theta(L_0) = c_1(\bar{L}_0)$, where $\bar{L}_0$ is the line bundle over W_0 obtained by dividing out the free π-action on the total space of L_0. Moreover, in the short exact sequence

$$0 \to H^2(\mathbb{Z}/p; \mathbb{Z}) \xrightarrow{c^*} H^2(W_0; \mathbb{Z}) \xrightarrow{q^*} H^2(X_0; \mathbb{Z}) \to 0$$

the pullback $q^*(\theta(L_0)) = c_1(q^*(\bar{L}_0)) = c_1(L_0) \in H^2(X_0; \mathbb{Z})$.

The strategy will be to find a suitable element $\theta(L) \in H^2_\pi(X; \mathbb{Z})$ by studying the Mayer-Vietoris sequence

$$\cdots \to H^2_\pi(X) \to H^2_\pi(X_0) \oplus H^2_\pi(N) \to H^2_\pi(\partial X_0) \xrightarrow{\delta} H^3_\pi(X) \to H^3_\pi(X_0) \oplus H^3_\pi(N) \to \cdots$$

in Borel cohomology associated to the π-equivariant decomposition $X = X_0 \cup N$.

Obervations:

(1) The Mayer-Vietoris coboundary map $H^2_\pi(\partial X_0) \xrightarrow{\delta} H^3_\pi(X)$ factors as

$$H^2_\pi(\partial X_0) \xrightarrow{\delta} H^3_\pi(X_0, \partial X_0) \cong H^3_\pi(X, N) \to H^3_\pi(X).$$

(2) The cokernel of the map $H^2_\pi(N) \to H^2_\pi(\partial X_0)$ has exponent p. This follows from the commutative diagram of restriction maps

$$\begin{array}{ccccc} H^2_\pi(N) & \longrightarrow & H^2_\pi(\partial X_0) & \longrightarrow & \bigoplus \mathbb{Z}/p\alpha_j \\ \downarrow & & \downarrow & & \downarrow \\ H^2(N) & \longrightarrow & H^2(\partial X_0) & \xrightarrow{\cong} & \bigoplus \mathbb{Z}/\alpha_j \end{array}$$

since the map $H^2_\pi(N) \to H^2(N)$ is surjective (by the Borel spectral sequence) and the map $H^2(N, \partial N) \to H^2(N)$ is adjoint to the (diagonal) intersection form on N, with cokernel $H^2(\partial N) = H^2(\partial X_0)$, hence determined by the self-intersection numbers $\{\alpha_j\}$.

(3) We have $H^3_\pi(X_0, \partial X_0) \cong H^3(W_0, \partial W_0) \cong H_1(W_0) \cong \mathbb{Z}/p \oplus H_1(X_0)$, where $H_1(X_0)$ is a quotient of $\bigoplus \mathbb{Z}/\overline{\alpha}_j$ and has order prime to p (by Lemma 6.4).

(4) The Mayer-Vietoris coboundary map $H^2_\pi(\partial X_0) \xrightarrow{\delta} H^3_\pi(X)$ also factors as

$$H^2_\pi(\partial X_0) \xrightarrow{\delta} H^3_\pi(N, \partial X_0) \cong H^3_\pi(X, X_0) \to H^3_\pi(X).$$

From the first three points above, it follows that $\mathrm{Im}(\delta) = \mathbb{Z}/p \subseteq H^3_\pi(X)$ may be identified with the first summand of the identification $H^3_\pi(X_0, \partial X_0) \cong \mathbb{Z}/p \oplus H_1(X_0)$.

To complete the proof of Theorem A, it is now enough to produce a class

$$\theta_0 \in H^2_\pi(X_0) \cong H^2(W_0),$$

which added together with the classes already found in $H^2_\pi(N)$ will have image zero in $H^2_\pi(\partial X_0)$. By the observations above, this amounts to finding a $U(1)$-bundle $\bar{L}_0$ on ∂W_0 whose first Chern class $\theta_0 = c_1(\bar{L}_0)$ has image of order prime to p under the coboundary map

$$H^2(\partial W_0) \to H^3(W_0, \partial W_0).$$

In other words, we need to find suitable $U(1)$-bundles over each of boundary components of ∂W_0, so that the sum of their first Chern classes is zero (mod p). This required relation is exactly the condition (1.1) given in the statement of Theorem A.

Let Y denote one of the lens spaces in ∂W_0, For convenience, we will identify $\mathbb{Z}/n \cong H^2(Y;\mathbb{Z})$ by $a \mapsto a \cdot \hat{\mu}$, for $a \in \mathbb{Z}/n$, where $\hat{\mu} \in H^2(Y;\mathbb{Z})$ denotes a standard generator, Poincaré dual to the circle fibre class $\mu \in H_1(Y;\mathbb{Z})$ (introduced in Lemma 6.1).

If $Y_i = L(p; a_i, b_i) \subset \partial W_0$ arises from one of the isolated fixed points in X, then choose the holonomy representation that sends a generator in $\pi_1(Y)$ to $\exp(2\pi i \lambda_i / p)$. The first Chern class of the associated flat $U(1)$-bundle is

$$\frac{\lambda_i}{a_i b_i}[\hat{\mu}_i] \in H^2(Y_i;\mathbb{Z}) \cong \mathbb{Z}/p.$$

In the case when the π-action on X has only isolated fixed points, the condition for an extension is that these elements lie in the kernel of δ in $H^2(\partial W_0;\mathbb{Z}) = \bigoplus_i H^2(Y_i;\mathbb{Z})$ which is equivalent to the condition $\sum_i \dfrac{\lambda_i}{a_i b_i} \equiv 0 \pmod{p}$.

In the general case when components of the fixed set contain 2-spheres, we need to consider contributions from the lens spaces $Y_j = L(p\alpha_j; c_j, c_j)$. These lens spaces arise from the free t^{c_j}-action on $\widetilde{Y}_j := \partial\nu(F_j) \approx L(\alpha_j; 1)$. Consider the induced covering spaces of Y_j by the lens spaces $L(p^{a_j+1}, 1)$ and $L(\overline{\alpha}_j, 1)$, where $\alpha_j = p^{a_j}\overline{\alpha}_j$, and note that the covering maps induce an isomorphism:

$$\begin{array}{ccc} H^2(Y_j;\mathbb{Z}) & \xrightarrow{\cong} & H^2(L(p^{a_j+1},1)) \oplus H^2(L(\overline{\alpha}_j,1)) \\ \downarrow\cong & & \downarrow\cong \\ \mathbb{Z}/p\alpha_j & \xrightarrow{\cong} & \mathbb{Z}/p^{a_j+1} \oplus \mathbb{Z}/\overline{\alpha}_j \end{array} \tag{6.2}$$

Under the two covering maps, the standard cohomology generator $\hat{\mu}_j \in H^2(Y_j;\mathbb{Z})$ is sent to the standard generators in $H^2(L(p^{a_j+1},1))$ and $H^2(L(\overline{\alpha}_j,1))$, respectively, by Lemma 6.3. The maps in the lower sequence are the reductions mod p^{a_j+1} and $\overline{\alpha}_j$, after using the identifications provided by the cohomology generators.

The required Chern class $c_1(\bar{L}_j) = \frac{\ell_j}{c_j^2}[\hat{\mu}_j] \in H^2(Y_j;\mathbb{Z})$ for each component Y_j can now be determined uniquely by solving the congruences:

$$\begin{aligned} \ell_j &\equiv -\lambda_j\alpha_j \pmod{p^{a_j+1}}, \\ \ell_j &\equiv c_j m_j \pmod{\overline{\alpha}_j}. \end{aligned} \tag{6.3}$$

and hence we have a $U(1)$-bundle $\bar{L}_j \searrow Y_j$. The minus sign is chosen in the first congruence because the induced orientation on Y_j from ∂W_0 is opposite to its orientation as the disk bundle over S^2 with Euler class α_j.

By diagram (6.2) and Lemma 6.3, the first Chern class has image

$$c_1(\bar{L}_j) = \frac{\ell_j}{c_j^2}[\hat{\mu}_j] = \frac{-\lambda_j\alpha_j + c_j m_j}{c_j^2}[\hat{\mu}_j] \in H^2(Y_j;\mathbb{Z})$$

with respect to the decomposition $H^2(Y_j;\mathbb{Z}) \cong \mathbb{Z}/p^{a_j+1} \oplus \mathbb{Z}/\overline{\alpha}_j$. After substituting these expressions into the congruence relation (6.1), we see that the sum vanishes mod p (from Observation 4 above). Hence we have

$$\delta\Big(\sum_i \frac{\lambda_i}{a_i b_i}[\hat{\mu}_i] + \sum_j \frac{\ell_j}{c_j^2}[\hat{\mu}_j]\Big) = 0$$

under the coboundary map $\delta\colon H^2_\pi(\partial X_0) \to H^3_\pi(X)$, and the required line bundle $\bar{L}_0$ over W_0 exists. □

7. Equivariant $SU(2)$ bundles

In this section we compute a (necessary) congruence relation similar to the previous section, but for equivariant $SU(2)$-bundles. As above, we work over a closed, simply connected, oriented 4-manifold with a finite homologically trivial cyclic group action. We again use the twisted G-signature formula (with the previously established notation). In particular, let D denote the signature operator twisted by an equivariant $SU(2)$-bundle $E \longrightarrow X$, then the contribution to the Lefschetz numbers from isolated fixed points is given by

$$L(g,D)\,|_{\,pt} = \frac{(t^a+1)\,(t^b+1)}{(t^a-1)\,(t^b-1)}(t^\lambda + t^{-\lambda}).$$

We need to compute the contribution from isolated fixed 2-spheres F. Since $E\,|_{\,F} = L \oplus L^{-1}$, we have $\operatorname{ch}_g(L \oplus L^{-1}\,|_{\,F}) = \{e^{\lambda+z} + e^{-\lambda-z}\}[F]$ and

$$\begin{aligned} L(g,D)\,|_{\,F} &= \{2\cot(i\theta/2) - \operatorname{csch}^2(i\theta/2)y\}\operatorname{ch}_g(L\oplus L^{-1}\,|_{\,F})[F] \\ &= \{2\frac{(t^c+1)}{(t^c-1)} - \frac{4t^c y}{(t^c-1)^2}\}\{e^\lambda(1+z) + e^{-\lambda}(1-z)\}[F] \\ &= \{2\frac{(t^c+1)}{(t^c-1)} - \frac{4t^c y}{(t^c-1)^2}\}\{t^\lambda + t^{-\lambda} + z(t^\lambda - t^{-\lambda})\}[F] \\ &= -\frac{4t^c[F]^2}{(t^c-1)^2}(t^\lambda + t^{-\lambda}) + 2c_1(L)[F]\frac{(t^c+1)}{(t^c-1)}(t^\lambda - t^{-\lambda}). \end{aligned}$$

Also note

$$\begin{aligned} \chi(1) = \operatorname{ch}(E)\mathcal{L}(X)[X] &= (2 - c_2(E))(4 + \frac{1}{3}p_1) \\ &= 2\operatorname{Sign}(X) - 4c_2(E). \end{aligned}$$

We now again multiply both sides of the G-signature formula by $(t-1)$, take Taylor expansions about $t = 1$ and reduce coefficients modulo p:

$$\frac{(t^a+1)}{(t^a-1)}\frac{(t^b+1)}{(t^b-1)}(t-1)^2(t^\lambda+t^{-\lambda}) = \frac{8}{ab} + \frac{8}{ab}(t-1) + \frac{2}{3}\frac{(a^2+b^2+1+6\lambda^2)}{ab}(t-1)^2 + \cdots .$$

and for the second term, where we let m denote $c_1(L)[F]$:

$$(t-1)^2\{\frac{-4\alpha t^c}{(t^c-1)^2}(t^\lambda+t^{-\lambda}) + 2m\frac{(t^c+1)}{(t^c-1)}(t^\lambda+t^{-\lambda})\}$$
$$= \frac{-8\alpha}{c^2} + \frac{-8\alpha}{c^2}(t-1) + \frac{2}{3}\frac{(\alpha c^2 - \alpha - 6\alpha\lambda^2 + 12mc\lambda)}{c^2}(t-1)^2 + \cdots$$

Summing over all the fixed sets and simplifying the coefficient of second order term $(t-1)^2$, we obtain:

$$2\,\mathrm{Sign}(X) + \sum_i \frac{4\lambda_i^2}{a_ib_i} - \sum_j \frac{4\alpha_j\lambda_j^2}{c_j^2} + \sum_j \frac{8m_j\lambda_j}{c_j}.$$

Equating this with $\chi(1) = 2\,\mathrm{Sign}(X) - 4c_2(E)$ and reducing coefficients modulo p gives the following congruence relation:

THEOREM 7.1. *Let $(E,\pi) \to (X,\pi)$ denote an equivariant $SU(2)$-bundle over a simply connected, closed 4-manifold with a homologically trivial group action of a finite cyclic group $\pi = \mathbb{Z}/p$ of odd prime order. Then the following congruence relation holds*

$$\sum_i \frac{\lambda_i^2}{a_ib_i} - \sum_j \frac{\alpha_j\lambda_j^2}{c_j^2} + \sum_j \frac{2\lambda_j}{c_j}c_1(i^*L_j)[F_j] \equiv -c_2(E)[X] \pmod{p},$$

where L_j is a local reduction $E\,|_{F_j} = L_j \oplus L_j^{-1}$.

Instanton gauge theory on $c_2(E) = 1$ bundles in the equivariant setting provide many examples of equivariant $SU(2)$-bundles on smooth definite 4-manifolds. We next check these relations for some examples of smooth cyclic group actions in the well-known linear models (see [**6, 8, 19–21, 26, 29**]).

EXAMPLE 7.2 (Linear Models on S^4). Let $X = S^4$ with a linear $\mathbb{Z}/p$-action which gives rotation numbers (a,b) and $(a,-b)$. Let E denote the instanton one equivariant $SU(2)$-bundle, i.e. with $c_2(E) = 1$. Then the congruence relation is given by

$$-c_2(E) = \frac{\lambda_1^2}{ab} - \frac{\lambda_2^2}{ab} \pmod{p}$$

It is elementary to check that this congruence relation is satisfied with the following isotropy representations

$$\lambda_1 = \frac{b-a}{2}$$
$$\lambda_2 = \frac{a+b}{2}.$$

over the fibres of the fixed points.

EXAMPLE 7.3 (Linear Models on $\overline{\mathbb{C}P}^2$). Let $X = \overline{\mathbb{C}P}^2$ with a linear $\mathbb{Z}/p$-action with one isolated fixed point with rotation number $(a, -a)$ for some $a \pmod p$ and a fixed projective line F with rotation number $a \pmod p$ on the normal bundle. Let E again denote the instanton one equivariant $SU(2)$-bundle. The congruence relation gives

$$-1 \equiv -\frac{\lambda^2}{a^2} + \frac{\lambda_F^2}{a^2} + \frac{2m\lambda_F}{a}$$

where $m = c_1(i^*L)[F]$. There exists two distinct lifts giving rise to equivariant bundles which admit G-invariant ASD connections. In the case when the equivariant lift comes from "bubbling" on the isolated fixed point then $m = 0$ and

$$\lambda \equiv a \pmod p$$
$$\lambda_F \equiv 0 \pmod p.$$

Thus the congruence is satisfied. On the other hand, if we choose the equivariant lift associated to the fixed 2-sphere (from 3-dimensional fixed connected component in the moduli space of equivariant ASD connections with $c_2(E) = 1$) then $m = -1$ and

$$\lambda \equiv a/2 \pmod p$$
$$\lambda_F \equiv a/2 \pmod p,$$

again the congruence relation is satisfied.

REMARK 7.4. At present we do not have general sufficient conditions for the existence of equivariant principal $SU(2)$-bundles. However, in the special cases where X is negative definite and $c_2(E) = 1$, for any choice of reduction $E = L \oplus L^{-1}$ from a cohomology class $c_1(L) = \alpha \in H^2(X;\mathbb{Z})$ with $\langle \alpha^2, [X]\rangle = -1$, we have sufficient conditions by applying Theorem A.

8. Equivariant index computation

In this section we compute the dimension of the moduli space of invariant anti-self dual connections for a given equivariant $SU(2)$-bundle over smooth 4-manifolds with a given homologically trivial cyclic group action.

Let X be a simply connected, closed, smooth negative definite 4-manifold, with a homologically trivial action of a finite group G. If $E \searrow X$ is an $SU(2)$-bundle with $c_2(E) = k$, the moduli space $\mathcal{M}_1^*(X)$ of irreducible ASD connections (on an $SU(2)$-bundle E with $c_2(E) = 1$) inherits a G-action, and the connected components of the fixed point set $\mathcal{M}_1^G(X)$ correspond to G-invariant ASD connections for certain equivariant lifts of the G-action on X to E (see [**19**], [**8**], [**25**, Theorem A], [**29**, §2]).

We want to compute the dimension of the moduli space $\mathcal{M}_k^G(X)$ of irreducible G-invariant ASD connections. This is motivated by Example 8.1, for which the formal dimension $\dim \mathcal{M}_1^*(X) = 5$. In this case, we expect a dimension formula that gives 1 and 3-dimensional strata depending on contributions from isolated fixed points or isolated fixed 2-spheres in X and on the isotropy representations from the equivariant lift (see [**8**] and [**26**] for details). There are similar index calculations in the literature in various gauge-theoretic settings (for example, see [**15**, §3], [**6**]), [**34, 35**], [**44**], [**1**]).

We first very briefly review the dimension calculation in the non-equivariant setting to set some notation. Let $D_A^+ = d_A^* + d_A^+ \colon \Omega^1(\operatorname{ad} E) \to \Omega^0(\operatorname{ad} E) \oplus \Omega_+^2(\operatorname{ad} E)$

denote the anti-self duality operator, and let $\mathcal{M}_k$ denote the ASD moduli space with $c_2(E) = k$. Note that the formal dimension is given by $\dim \mathcal{M}_k = -\operatorname{Ind}(D_A^+)$ and this is given by

$$\operatorname{Ind}(D_A^+) = \hat{A}(X)\operatorname{ch}(S^+)\operatorname{ch}(\operatorname{ad}_{\mathbb{C}} E)[X] \tag{8.1}$$

where $S = S^+ \oplus S^-$ and $\hat{A}(X) = \prod \frac{x_i/2}{\sinh(x_i/2)}$ with $\operatorname{ch}(S) = \prod(e^{x_i/2} + e^{-x_i/2})$ and $\operatorname{ch}(S^+) - \operatorname{ch}(S^-) = \prod(e^{x_i/2} - e^{-x_i/2})$. Using this we compute

$$\begin{aligned} 2\hat{A}(X)\operatorname{ch}(S^+)\operatorname{ch}(\operatorname{ad}_{\mathbb{C}} E)[X] &= (4 + \frac{1}{3}p_1 + \chi)(3 - 4c_2(E))[X] \\ &= -16c_2(E) + 3(\frac{p_1}{3} + \chi). \end{aligned}$$

Thus the index $\operatorname{Ind}(D_A^+) = -8c_2(E) + \frac{3}{2}(\operatorname{Sign} + \chi)(X)$ and we get the usual expression $\dim \mathcal{M}_k = 8k - 3/2(\chi + \operatorname{Sign})(X)$ for the dimension of the moduli space. Also note the following alternative expression for the index:

$$\begin{aligned} \operatorname{Ind}(D_A^+) &= \frac{\operatorname{ch}(S^+ - S^-)\operatorname{ch}(S^+)\operatorname{ch}(\operatorname{ad}_{\mathbb{C}} E)Td(TX \otimes \mathbb{C})}{e(X)}[X] \\ &= \hat{A}(X)\operatorname{ch}(S^+ \otimes \operatorname{ad}_{\mathbb{C}} E)[X]. \end{aligned}$$

For the equivariant setting E is an equivariant $SU(2)$-bundle and let $D = D_A^+$ denote the anti-self duality operator $d_A^* + d_A^+ : \Omega^1(\operatorname{ad} E)^G \to \Omega^0(\operatorname{ad} E)^G \oplus \Omega^2_+(\operatorname{ad} E)^G$. We compute the equivariant index by averaging the Lefschetz numbers as in [**15**]:

$$\operatorname{Ind}(D) = \frac{1}{p}\sum_{g\in G} L(g, D)$$

$$\begin{aligned} \operatorname{Ind}(D) &= \frac{1}{p}\{L(1, D) + \sum_{g\neq 1} L(g, D)\} \\ &= \frac{1}{p}\{-8c_2(E) + \frac{3}{2}(\chi + \operatorname{Sign})(X) + \sum_{g\neq 1} L(g, D)\} \\ &= \frac{1}{p}\{-8c_2(E) + \frac{3p}{2}(\chi + \operatorname{Sign})(X/G) - \frac{3}{2}(d_\chi + d_\sigma)(X^G) + \sum_{g\neq 1} L(g, D)\} \end{aligned}$$

where $p\chi(X/G) = \chi(X) + d_\chi$ with $d_\chi = \sum_{g\neq 1}\chi(X^g)$ is the Euler characteristic defect terms and similarly for the signature defect term:

$$\begin{aligned} -\frac{3}{2}(d_\chi + d_\sigma)[pt] &= -\frac{3}{2}(1 - \cot(\theta_1/2)\cot(\theta_2/2)) \\ -\frac{3}{2}(d_\chi + d_\sigma)[F] &= -\frac{3}{2}(2 + [F]^2\csc^2(\theta/2)), \end{aligned}$$

where (θ_1, θ_2) are the rotation numbers at an isolated fixed point and $\theta = c_F$ is the rotation number on the normal bundle to F. Decomposing the contributions from isolated fixed points and 2-spheres:

$$\sum_{g\neq 1} L(g, D)(X^G) = \sum_{g\neq 1}\{\sum_i L(g, D)\,|_{(a_i, b_i)} + \sum_j L(g, D)\,|_{F_j}\}.$$

Now $\operatorname{ch}_g(\operatorname{ad}_{\mathbb{C}} E)(pt) = 3 - 4\sin^2(\frac{\pi k\ell}{p})$, with ℓ the isotropy representation on the fiber of E over the fixed point. The Lefshetz numbers from the fixed sets can be computed directly from the index formula and are given by:

$$L(g,D)\,|\,_{pt} = \frac{-1}{2}[\cot(\theta_1/2)\cot(\theta_2/2) - 1]\,\mathrm{ch}_g(\mathrm{ad}_{\mathbb{C}} E)[pt] \tag{8.2}$$

$$L(g,D)\,|\,_{F} = [-i\cot(\theta/2) + \frac{1}{2}(\chi + \csc^2(\theta/2)y)]\,\mathrm{ch}_g(\mathrm{ad}_{\mathbb{C}} E)[F], \tag{8.3}$$

with χ the Euler class of the tangent bundle to F and y is the Euler class of the normal bundle to F. We first compute the contribution from isolated fixed points.

$$\begin{aligned} L(g,D)\,|\,_{pt} &= \frac{-1}{2}[\cot(\theta_1/2)\cot(\theta_2/2) - 1][3 - 4\sin^2(\frac{\pi k\ell}{p})][pt] \\ &= -\frac{3}{2}[\cot(\theta_1/2)\cot(\theta_2/2) - 1] - 2\sin^2(\frac{\pi k\ell}{p}) \\ &+ 2\cot(\theta_1/2)\cot(\theta_2/2)\sin^2(\frac{\pi k\ell}{p}). \end{aligned}$$

Summing over all isolated fixed points gives

$$\begin{aligned} \frac{1}{p}\sum_{g\neq 1}\sum_{i} L(g,D)\,|\,_{(a_i,b_i)} &= \frac{3}{2p}\sum_i (d_\chi + d_\sigma)(a_i,b_i) - \frac{2}{p}\sum_i\sum_{k=1}^{p-1}\sin(\frac{\pi k\ell_i}{p}) \\ &+ \frac{2}{p}\sum_i\sum_{k=1}^{p-1}\cot(\frac{a_i\pi k}{p})\cot(\frac{b_i\pi k}{p})\sin^2(\frac{\pi k\ell_i}{p}) \\ &= \frac{3}{2p}\sum_i (d_\chi + d_\sigma)(a_i,b_i) + m + \sum_i \rho L(p,a_i,b_i,\ell_i) \end{aligned}$$

where m is the number isolated fixed points with non-trivial representation on the fiber and $\rho L(p,a,b,\ell)$ is the rho invariant of lens spaces.

We need to compute $\mathrm{ch}_g(\mathrm{ad}_{\mathbb{C}} E\,|\,_F)$. Since an $SU(2)$ bundle restricted over a fixed 2-submanifold has a local abelian reduction $E\,|\,_F = L \oplus L^{-1}$ for some L, we have $\mathrm{ad}\, E\,|\,_F = L^2 \oplus \underline{\mathbb{R}}$. We need to compute $\mathrm{ch}_g(\mathrm{ad}_{\mathbb{C}} E\,|\,_F) = \mathrm{ch}_g(L^2) + \mathrm{ch}_g(\overline{L^2}) + 1$ and this contributes

$$\begin{aligned} \mathrm{ch}_g(\mathrm{ad}_{\mathbb{C}} E\,|\,_F) &= (g + gc_1(L^2)) + (g^{-1} + g^{-1}c_1(\overline{L^2})) + 1 \\ &= (g + g^{-1} + 1) + c_1(L^2)(g - g^{-1}) \\ &= (3 - 4\sin^2(\frac{\pi k\ell}{p})) + 2ic_1(L^2)\sin(\frac{2\pi k\ell}{p}), \end{aligned}$$

where now ℓ is the isotropy representation on the fibre over the fixed 2-sphere F. Substituting these terms, the Lesfchetz number $L(g,D)\,|\,_F$ evaluated on fixed 2-spheres gives:

$$\begin{aligned} &L(g,D)\,|\,_F \\ &= [-i\cot(\theta/2) + \frac{1}{2}(\chi + \csc^2(\theta/2)y)][(3 - 4\sin^2(\frac{\pi k\ell}{p})) + 2ic_1(L^2)\sin(\frac{2\pi k\ell}{p})][F] \\ &= \frac{1}{2}[\chi + \csc^2(\theta/2)y][3 - 4\sin^2(\frac{\pi k\ell}{p})] + 2c_1(L^2)\sin(\frac{2\pi k\ell}{p})\cot(\frac{\theta}{2}). \end{aligned}$$

Let us introduce a kind of rho invariant term for fixed surfaces:

$$\rho_F(\ell) = \frac{2}{p}\sum_{k=1}^{p-1}\csc^2(\frac{\pi c_F k}{p})\sin^2(\frac{\pi k\ell}{p})[F]^2 - \frac{4c_1(L)[F]}{p}\sum_{k=1}^{p-1}\sin(\frac{2\pi k\ell}{p})\cot(\frac{\pi k c_F}{p}),$$

with this notation we have

$$\frac{1}{p}\sum_{g\neq 1}\sum_{j} L(g,D)\,|_{F_j} = \frac{3}{2p}\sum_{j}(d_\chi + d_\sigma)[F_j] - \frac{2}{p}\sum_{j}\chi(F_j)\sum_{k=1}^{p-1}\sin^2(\frac{\pi k \ell_j}{p}) - \sum_{j}\rho_{F_j}(\ell_j).$$

Now combining all the terms we obtain:

$$\operatorname{Ind}(D_A) = \frac{-8}{p}c_2(E) + \frac{3}{2}(\chi + \operatorname{Sign})(X/G)) - m + \sum_i \rho L(p, a_i, b_i, \ell_i) - \sum_{j \text{ with } \ell_j\neq 0} \chi(F_j) - \sum_j \rho_{F_j}(\ell_j).$$

Since $\dim \mathcal{M}_k^G(X) = -\operatorname{Ind}(D_A)$, the dimension formula is

$$\dim \mathcal{M}_k^G(X) = \frac{8}{p}c_2(E) - \frac{3}{2}(\chi + \operatorname{Sign})(X/G) + m - \sum_i \rho L(p, a_i, b_i, \ell_i) + \sum_{j \text{ with } \ell_j\neq 0} \chi(F_j) + \sum_j \rho_{F_j}(\ell_j).$$

Before giving an example we note a few special cases. When the action on X only has isolated fixed points, let (a_i, b_i) denote the rotation numbers and ℓ_i the isotropy representation over the points, the formula reduces to the following:

$$\dim \mathcal{M}_k^G(X) = \frac{8c_2(E)}{p} - \frac{3}{2}(\chi + \operatorname{Sign})(X/G) + m - \sum_i \rho L(p, a_i, b_i, \ell_i).$$

For invariant ASD connections on the four-sphere this formula reduces to that of [**6**, p. 394]. In the case of $SO(3)$-bundles in the orbifold setting, see Fintushel and Stern [**15**]. When the action on X is a smooth involution with fixed 2-sphere and non-trivial action on fibre $c_F \equiv \ell \equiv 1 \bmod 2$ the formula above reduces to:

$$\dim \mathcal{M}_k^G(X) = 4c_2(E) - \frac{3}{2}(\chi + \operatorname{Sign})(X/G) + \chi(F) + [F]^2.$$

which matches with Wang [**44**, Theorem 18, p. 130]. We finish this section with an example.

EXAMPLE 8.1. Let $X = \#_3\overline{\mathbb{C}P}^2$ with a linear $\mathbb{Z}/p$-action with $p = 5$ that arises from equivariant connected sums of linear actions in the following way. Take the equivariant connected sum of two copies of $\overline{\mathbb{C}P}^2$ along the two dimensional fixed sets which fixes a projective line and a rotation number of $(1, -1)$ at the isolated fixed points in each copy. Now at one of the isolated fixed points take the equivariant connected sum with $\overline{\mathbb{C}P}^2$ that has a linear action with 3 isolated fixed points with rotation numbers $(1, 1), (2, -1), (2, -1)$.

The result is a smooth, homologically trivial $\mathbb{Z}/5$-action on X that has 3 isolated fixed points with rotation data $\{(1, -1), (2, -1), (2, -1)\}$ and a single fixed 2-sphere F with rotation number $c_F \equiv 1 \pmod p$ on the normal bundle and has self intersection -2.

The compactified, equivariant ASD instanton one moduli space $\mathcal{M}_1(X)$ has dimension 5 with fixed components that are 1 and 3-dimensional which correspond to invariant ASD connections for a lifted action to the $SU(2)$-bundle (see [**26**]).

The boundary of the moduli space is the "bubbling" of highly concentrated ASD connections which can be identified with a copy of X. The isolated fixed points propagate 1-fixed dimensional strata into the moduli space. We will compute the dimension of these strata using the dimension formula from this section and from the fixed point data.

For example, at the isolated fixed point $(2,-1)$ the highly concentrated instantons correspond to ASD connections on the 4-sphere, with equivariant lifts matching the linear models which then pull back to X using the degree 1-map in the formation of the Taubes boundary. This determines the equivariant lift on X and has isotropy representation t^{λ_1} over the fixed point $(2,-1)$ with $\lambda_1 \equiv -3 \pmod{p}$ and t^{λ_2} over all the other fixed point sets with $\lambda_2 \equiv 1 \pmod{p}$. The dimension formula gives:

$$\frac{8}{p} - \rho L(p,2,-1,-3) - \rho L(p,2,-1,1) - \rho L(p,1,-1,1) + \chi(F) + \rho_F(1) = 1.$$

On the other hand, at a point on the fixed 2-sphere F following the same procedure with the degree one Taubes map, we can pull-back an equivariant bundle from the linear model on S^4 with a fixed embedded 2-sphere. This time we get an equivariant $SU(2)$-bundle on X with $c_1(L)[F] = -1$ in the local reduction $E\,|\,_F = L \oplus L^{-1}$ (in this case actually a global reduction as it corresponds to reducible). The isotropy representation is t^{λ} over all the fixed point sets with $\lambda \equiv 1 \pmod{p}$. We then have:

$$\frac{8}{p} - 2\rho L(p,2,-1,1) - \rho L(p,1,-1,1) + \chi(F) + \rho_F(1) = 3.$$

after substituting the data into the dimension formula.

References

[1] Nima Anvari, *Extending smooth cyclic group actions on the Poincaré homology sphere*, Pacific J. Math. **282** (2016), no. 1, 9–25, DOI 10.2140/pjm.2016.282.9. MR3463422

[2] Nima Anvari and Ian Hambleton, *Cyclic group actions on contractible 4-manifolds*, Geom. Topol. **20** (2016), no. 2, 1127–1155, DOI 10.2140/gt.2016.20.1127. MR3493101

[3] Nima Anvari and Ian Hambleton, *Cyclic branched coverings of Brieskorn spheres bounding acyclic 4-manifolds*, Glasg. Math. J. **63** (2021), no. 2, 400–413, DOI 10.1017/S0017089520000269. MR4244205

[4] M. F. Atiyah and I. M. Singer, *The index of elliptic operators. I*, Ann. of Math. (2) **87** (1968), 484–530, DOI 10.2307/1970715. MR236950

[5] M. F. Atiyah and I. M. Singer, *The index of elliptic operators. III*, Ann. of Math. (2) **87** (1968), 546–604, DOI 10.2307/1970717. MR236952

[6] David M. Austin, *SO(3)-instantons on $L(p,q) \times \mathbf{R}$*, J. Differential Geom. **32** (1990), no. 2, 383–413. MR1072911

[7] David Baraglia and Pedram Hekmati, *Brieskorn spheres, cyclic group actions and the milnor conjecture*, 2023.

[8] Peter J. Braam and Gordana Matić, *The Smith conjecture in dimension four and equivariant gauge theory*, Forum Math. **5** (1993), no. 3, 299–311, DOI 10.1515/form.1993.5.299. MR1216037

[9] Weimin Chen, *Group actions on 4-manifolds: some recent results and open questions*, Proceedings of the Gökova Geometry-Topology Conference 2009, Int. Press, Somerville, MA, 2010, pp. 1–21. MR2655301

[10] S. K. Donaldson and P. B. Kronheimer, *The geometry of four-manifolds*, Oxford Mathematical Monographs, The Clarendon Press, Oxford University Press, New York, 1990. Oxford Science Publications. MR1079726

[11] Allan L. Edmonds, *Aspects of group actions on four-manifolds*, Topology Appl. **31** (1989), no. 2, 109–124, DOI 10.1016/0166-8641(89)90075-8. MR994404

[12] Allan L. Edmonds, *A survey of group actions on 4-manifolds*, Handbook of group actions. Vol. III, Adv. Lect. Math. (ALM), vol. 40, Int. Press, Somerville, MA, 2018, pp. 421–460. MR3888625

[13] Allan L. Edmonds and John H. Ewing, *Locally linear group actions on the complex projective plane*, Topology **28** (1989), no. 2, 211–223, DOI 10.1016/0040-9383(89)90021-9. MR1003583

[14] Allan L. Edmonds and John H. Ewing, *Realizing forms and fixed point data in dimension four*, Amer. J. Math. **114** (1992), no. 5, 1103–1126, DOI 10.2307/2374891. MR1183533

[15] Ronald Fintushel and Ronald J. Stern, *Pseudofree orbifolds*, Ann. of Math. (2) **122** (1985), no. 2, 335–364, DOI 10.2307/1971306. MR808222

[16] Ronald Fintushel and Ronald J. Stern, *Homotopy $K3$ surfaces containing $\Sigma(2,3,7)$*, J. Differential Geom. **34** (1991), no. 1, 255–265. MR1114462

[17] Michael H. Freedman, *The disk theorem for four-dimensional manifolds*, Proceedings of the International Congress of Mathematicians, Vol. 1, 2 (Warsaw, 1983), PWN, Warsaw, 1984, pp. 647–663. MR804721

[18] Michael Hartley Freedman, *The topology of four-dimensional manifolds*, J. Differential Geometry **17** (1982), no. 3, 357–453. MR679066

[19] Mikio Furuta, *A remark on a fixed point of finite group action on S^4*, Topology **28** (1989), no. 1, 35–38, DOI 10.1016/0040-9383(89)90030-X. MR991097

[20] Mikio Furuta, *Z_a-invariant* SU(2) *instantons over the four sphere*, Geometry of low-dimensional manifolds, 1 (Durham, 1989), London Math. Soc. Lecture Note Ser., vol. 150, Cambridge Univ. Press, Cambridge, 1990, pp. 161–174. MR1171897

[21] Mikio Furuta and Yoshitake Hashimoto, *Invariant instantons on S^4*, J. Fac. Sci. Univ. Tokyo Sect. IA Math. **37** (1990), no. 3, 585–600. MR1080872

[22] C. McA. Gordon, *On the G-signature theorem in dimension four*, À la recherche de la topologie perdue, Progr. Math., vol. 62, Birkhäuser Boston, Boston, MA, 1986, pp. 159–180. MR900251

[23] Ian Hambleton and Jean-Claude Hausmann, *Equivariant principal bundles over spheres and cohomogeneity one manifolds*, Proc. London Math. Soc. (3) **86** (2003), no. 1, 250–272, DOI 10.1112/S0024611502013722. MR1971468

[24] Ian Hambleton and Jean-Claude Hausmann, *Equivariant bundles and isotropy representations*, Groups Geom. Dyn. **4** (2010), no. 1, 127–162, DOI 10.4171/GGD/77. MR2566303

[25] Ian Hambleton and Ronnie Lee, *Perturbation of equivariant moduli spaces*, Math. Ann. **293** (1992), no. 1, 17–37, DOI 10.1007/BF01444700. MR1162670

[26] Ian Hambleton and Ronnie Lee, *Smooth group actions on definite 4-manifolds and moduli spaces*, Duke Math. J. **78** (1995), no. 3, 715–732, DOI 10.1215/S0012-7094-95-07826-0. MR1334207

[27] Ian Hambleton, Ronnie Lee, and Ib Madsen, *Rigidity of certain finite group actions on the complex projective plane*, Comment. Math. Helv. **64** (1989), no. 4, 618–638, DOI 10.1007/BF02564697. MR1022999

[28] Ian Hambleton and Semra Pamuk, *Rank conditions for finite group actions on 4-manifolds*, Canad. J. Math. **74** (2022), no. 2, 550–572, DOI 10.4153/S0008414X21000018. MR4411001

[29] Ian Hambleton and Mihail Tanase, *Permutations, isotropy and smooth cyclic group actions on definite 4-manifolds*, Geom. Topol. **8** (2004), 475–509, DOI 10.2140/gt.2004.8.475. MR2033485

[30] Matthew Hedden and Paul Kirk, *Chern-Simons invariants,* SO(3) *instantons, and $\mathbb{Z}/2$ homology cobordism*, Chern-Simons gauge theory: 20 years after, AMS/IP Stud. Adv. Math., vol. 50, Amer. Math. Soc., Providence, RI, 2011, pp. 83–114, DOI 10.1090/amsip/050/05. MR2809448

[31] Sławomir Kwasik and Reinhard Schultz, *Branched coverings and equivariant smoothings of 4-manifolds*, Enseign. Math. **63** (2017), no. 1-2, 201–232, DOI 10.4171/LEM/63-1/2-7. MR3777136

[32] Sławomir Kwasik and Terry Lawson, *Nonsmoothable Z_p actions on contractible 4-manifolds*, J. Reine Angew. Math. **437** (1993), 29–54, DOI 10.1515/crll.1993.437.29. MR1212252

[33] R. K. Lashof, J. P. May, and G. B. Segal, *Equivariant bundles with abelian structural group*, Proceedings of the Northwestern Homotopy Theory Conference (Evanston, Ill., 1982), Contemp. Math., vol. 19, Amer. Math. Soc., Providence, RI, 1983, pp. 167–176, DOI 10.1090/conm/019/711050. MR711050

[34] Terry Lawson, *Compactness results for orbifold instantons*, Math. Z. **200** (1988), no. 1, 123–140, DOI 10.1007/BF01161749. MR972399

[35] Terry Lawson, *A note on trigonometric sums arising in gauge theory*, Manuscripta Math. **80** (1993), no. 3, 265–272, DOI 10.1007/BF03026552. MR1240649

[36] Ib Madsen, *Geometric equivariant bordism and K-theory*, Topology **25** (1986), no. 2, 217–227, DOI 10.1016/0040-9383(86)90040-6. MR837622

[37] John W. Milnor and James D. Stasheff, *Characteristic classes*, Annals of Mathematics Studies, No. 76, Princeton University Press, Princeton, NJ; University of Tokyo Press, Tokyo, 1974. MR440554

[38] Daniel Ruberman, *Involutions on spin 4-manifolds*, Proc. Amer. Math. Soc. **123** (1995), no. 2, 593–596, DOI 10.2307/2160919. MR1231042

[39] Nikolai Saveliev, *Floer homology of Brieskorn homology spheres*, J. Differential Geom. **53** (1999), no. 1, 15–87. MR1776091

[40] Nikolai Saveliev, *Invariants for homology 3-spheres*, Encyclopaedia of Mathematical Sciences, vol. 140, Springer-Verlag, Berlin, 2002. Low-Dimensional Topology, I, DOI 10.1007/978-3-662-04705-7. MR1941324

[41] Nikolai Saveliev, *Fukumoto-Furuta invariants of plumbed homology 3-spheres*, Pacific J. Math. **205** (2002), no. 2, 465–490, DOI 10.2140/pjm.2002.205.465. MR1922741

[42] Patrick Shanahan, *The Atiyah-Singer index theorem*, Lecture Notes in Mathematics, vol. 638, Springer, Berlin, 1978. An introduction. MR487910

[43] C. T. C. Wall, *Surgery on compact manifolds*, 2nd ed., Mathematical Surveys and Monographs, vol. 69, American Mathematical Society, Providence, RI, 1999. Edited and with a foreword by A. A. Ranicki, DOI 10.1090/surv/069. MR1687388

[44] Shuguang Wang, *Moduli spaces over manifolds with involutions*, Math. Ann. **296** (1993), no. 1, 119–138, DOI 10.1007/BF01445098. MR1213375

[45] Dariusz M. Wilczyński, *Group actions on the complex projective plane*, Trans. Amer. Math. Soc. **303** (1987), no. 2, 707–731, DOI 10.2307/2000692. MR902793

Department of Mathematics & Statistics, McMaster University L8S 4K1 Hamilton, Ontario, Canada

Email address: anvarin@math.mcmaster.ca

Department of Mathematics & Statistics, McMaster University L8S 4K1 Hamilton, Ontario, Canada

Email address: hambleton@mcmaster.ca

Proceedings of Symposia in Pure Mathematics
Volume **109**, 2024
https://doi.org/10.1090/pspum/109/01990

Exotic families of embeddings

Dave Auckly and Daniel Ruberman

Dedicated to Tom Mrowka on the occasion of his 61st *birthday*

ABSTRACT. We construct a number of topologically trivial but smoothly non-trivial families of embeddings of 3-manifolds in 4-manifolds. These include embeddings of homology spheres in S^4 that are not isotopic but have diffeomorphic complements, and families (parameterized by high-dimensional spheres) of embeddings of any 3-manifold that embeds in a blown-up $K3$ surface. In each case, the families are constructed so as to be topologically trivial in an appropriate sense. We also illustrate a general technique for converting a non-trivial family of embeddings into a non-trivial family of submanifolds.

1. Introduction

Starting with the work of Donaldson [**20**] and Freedman [**24**], many simply connected 4-manifolds have been shown to admit exotic smooth structures. It is natural to look for other exotic phenomena related to 4-manifolds, including exotic embeddings of surfaces, exotic embeddings of 3-manifolds, exotic diffeomorphisms, and exotic families of such embeddings and diffeomorphisms. In this paper, we establish two theorems about exotic embeddings of 3-manifolds in 4-manifolds. The first gives infinitely many exotic embeddings of certain homology 3-spheres in the 4-sphere. The second gives exotic families, parameterized by higher dimensional spheres, of many 3-manifolds in larger 4-manifolds.

THEOREM 1.1. *There exist an infinite collection of integer homology* 3*-spheres,* Y_n*, each with an infinite number of smooth embeddings* $j_{p,n}\colon Y_n \to S^4$ *so that*

(1) $S^4 \backslash j_{p,n}(Y_n) \cong \operatorname{int}(X_+(n)) \amalg \operatorname{int}(X_-(n))$,
(2) $X_+(n) \cong -X_-(n)$,
(3) $j_{p,n}$ *is topologically ambient isotopic to* $j_{q,n}$,
(4) *If* $\psi\colon S^4 \to S^4$ *is a diffeomorphism satisfying* $\psi \circ j_{p,n} = j_{q,n}$*, then* $p = q$.

REMARK 1.2. A recent MathOverflow [**38**] posting by Michael Klug asked if there were non-isotopic embeddings of 3-manifolds into S^4 with diffeomorphic complements. A number of people noticed that the existence of corks with doubles and

2020 *Mathematics Subject Classification.* Primary 57R40; Secondary 57K41.

Both authors were partially supported by NSF Grant DMS-1928930 while they were in residence at the Simons Laufer Mathematical Sciences Institute (formerly known as MSRI). We thank the institute for its hospitality. The first author was partially supported by Simons Foundation grant 585139 and NSF grant DMS-1952755. The second author was partially supported by NSF Grant DMS-1811111 and NSF grant DMS-1952790.

twisted doubles diffeomorphic to S^4 allows one to conclude that the answer to the question is yes. Ian Agol posted this as a reply on MathOverflow. The above theorem is our version, based on the infinite-order corks discovered by Gompf [**28**] to give infinitely many non-isotopic embeddings.

We then turn to the construction and detection of exotic families of embeddings. Here we illustrate a general principle by establishing the following result about smoothly knotted families of codimension one embeddings. Our convention is to denote by $\mathrm{Emb}(Y, Z)$ the smooth embeddings of Y in Z, and by $\mathrm{Emb}^{\mathrm{Top}}(Y, Z)$ the space of locally flat topological embeddings. Note that these are spaces of *maps*, rather than submanifolds; the space $\mathrm{Sub}(Y, Z)$ of submanifolds is the quotient of $\mathrm{Emb}(Y, Z)$ by the diffeomorphism group of Y (with similar notation for locally flat topological submanifolds.) With some additional hypotheses, the same results hold for families of submanifolds. We address the case of submanifolds in the final section of the paper.

To state the main theorem about families of embeddings, we use the notation $E(2)$ for the $K3$ surface, and let T be a smooth fiber in a particular elliptic fibration.

THEOREM 1.3. *Let Y be a 3-manifold that embeds smoothly in $E(2) \# \overline{\mathbb{CP}}^2$ disjointly from T. For any $r > 0$, $k \geqslant 0$ there are smooth simply connected 4-manifolds Z_r^k, together with spherical classes $J_p^{j+1} \in H_{j+1}(\mathrm{Emb}(Y, Z_r^k))$ for $j \leqslant k$ and $j \equiv k$, (mod 2) and $p = 1, \cdots, r$ that generate a rank r summand of*

$$\ker\left[H_{j+1}(\mathrm{Emb}(Y, Z_r^k)) \to H_{j+1}(\mathrm{Emb}^{\mathrm{Top}}(Y, Z_r^k))\right].$$

Furthermore, when Y is a homology sphere, the classes J_p^{j+1} are trivial in $\pi_{j+1}(\mathrm{Emb}^{\mathrm{Top}}(Y, Z_r^k))$, and trivial in $\pi_{j+1}(\mathrm{Emb}(Y, Z_r^k \# S^2 \times S^2))$.

As the reader will see, the constructions that go into the proof are very flexible. Hence the argument would work for many other 3-manifolds, including those that embed disjoint from a nucleus in an arbitrary blowup of the elliptic surface $E(n)$. The specific choices embodied in Theorem 1.3 are described in Section 5.

To prove this theorem, we need a method to construct examples, an invariant to distinguish the smooth families, and a method to compute the invariant. Defining and computing invariants builds on the second author's earlier papers [**58**–**60**] on exotic diffeomorphisms. These defined invariants using both Donaldson theory and Seiberg-Witten theory and used them to show the existence of non-isotopic diffeomorphisms and non-isotopic metrics of positive scalar curvature. The story for families of diffeomorphisms and embeddings is much the same. It was easier to see how to define and compute invariants for families using Donaldson theory. These invariants depend upon a U(2) lift of an SO(3) bundle with suitable first Pontryagin class, and a homology orientation. There are only a finite number of possibilities of this data. This leads [**7**] to results demonstrating large rank summands in the kernel for spaces of diffeomorphisms and embeddings of S^3. Many related works have appeared in the last few years, including [**9, 10, 14, 16, 33, 40, 41, 43**–**46, 54, 55, 62, 68, 69**].

After this paper was written, we have worked out how to define and compute invariants based on the Seiberg-Witten equations [**8**]. Here the data includes the choice of a spin^c structure, of which there are an infinite number of suitable choices. The principle of the definition of the invariants is the same as in [**40, 58**] but the calculation for higher dimensional families is new. This calculation leads to a strengthening of the result to conclude that the kernels of the maps on homology or homotopy contain infinite rank summands. It will gives similar statements about families of positive scalar curvature metrics. These stronger results derived via the using the Seiberg-Witten equations are not stated here, but will appear in a paper [**8**] in preparation. The paper [**7**] also includes analogous results about families of embedded spheres of self-intersection one. Drouin has extended these results to analogous results about families of spheres of arbitrary self-intersection [**22**]. The first author [**4**] has given many new examples of exotically knotted surfaces, even ones that remain exotic after stabilization an arbitrary number of times. Work in progress by the authors along with Konno, Mukherjee, and Taniguchi will further extend results about families of embedded surfaces to all genus and self-intersection numbers.

REMARK 1.4. In discussions about our work with various colleagues, it has been suggested that there should be a sort of 'cork theorem' for diffeomorphisms and embeddings, possibly even a parameterized version. Recall that the cork theorem [**19, 48**] says that two homotopy equivalent closed simply connected manifolds are related by removing a contractible piece (the cork) and regluing by an automorphism of the boundary (the cork twist). One might doing this for exotic pairs of embeddings, or showing that exotic pairs of diffeomorphisms also localize on some simple pieces of the underlying manifold. With this in mind, we point out that all of the exotic behavior in this paper and in [**7**] does stem from a single cork twist, in fact Akbulut's original cork twist from [**2**].

In contrast, the work of Konno-Mukherjee-Taniguchi [**41**] shows that some exotic (up to isotopy) codimension-one embedding behavior persists after an unlimited number of stabilizations. Since any cork twist extends over a bounding manifold after some number of stabilizations, this suggests that there may not be any such a cork theorem for embeddings. This seems very intriguing as a topic for future research in this direction.

Conventions: We will assume that all 4-manifolds are smooth, and oriented and that all 3-manifolds are oriented. In addition, when considering gauge theoretic invariants of a manifold X, we will assume a specific choice of homology orientation, i.e., an orientation of $H^2_+(X;\mathbb{R})$. We take homology and cohomology with integer coefficients unless stated otherwise. We will generally assume that 4-manifolds are simply connected, although many of our results will hold for more general classes of manifolds. A pair of manifolds X, Y is *exotic* if they are homeomorphic but not diffeomorphic; a collection $\{X_p\}$ is exotic if each pair X_i, X_j is exotic (for $i \neq j$).

2. Embeddings into S^4

Theorem 1.1 follows from the existence of chiral infinite order corks with trivial twisted doubles. The existence of these corks follows in turn from the work of Gompf and Tange [**28, 29, 66**]; see also [**3**]. This is the content of the following proposition.

PROPOSITION 2.1. *There are oriented, contractible, compact, smooth manifolds $X_+(n)$ together with orientation-preserving self-diffeomorphisms of the boundary, $f(n)\colon \partial X_+(n) \to \partial X_+(n)$, and with embeddings $i(n)\colon X_+(n) \to E(2) \# 2\overline{\mathbb{C}P}^2$ so that*

(1) *If $(E(2)\#2\overline{\mathbb{C}P}^2 \setminus \mathrm{int}(X_+(n))) \cup_{f(n)^p} X_+(n) \cong (E(2)\#2\overline{\mathbb{C}P}^2 \setminus \mathrm{int}(X_+(n))) \cup_{f(n)^q} X_+(n)$, then $p = q$.*

(2) *There are diffeomorphisms* $\Phi_{p,n}\colon X_+(n) \cup_{f(n)^p} -X_+(n) \to S^4$,
(3) *There is a homomorphism* $\delta\colon \pi_0(\mathrm{Diff}(Y_n)) \to \{\pm 1\}$ *so that*

$$\alpha f(n)\alpha^{-1} = f(n)^{\delta(\alpha)}.$$

PROOF OF PROPOSITION 2.1. We consider a subfamily of the infinite order corks defined by Gompf [**28, 29**] set $X_+(n) = C(2, 1; -2n)$ for $n > 0$. Figure 1 is a special case of [**29**, Figure 3] with the addition of the result after the zero-framed 2-handle is canceled with one of the 1-handles.

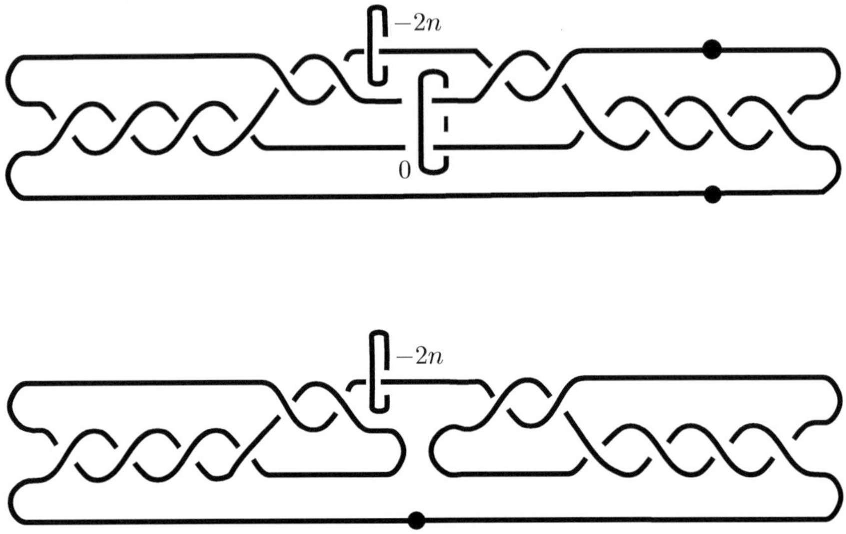

FIGURE 1. Infinite order corks $X_+(n)$

Figure 2 displays the boundary Y_n of $X_+(n)$. It was obtained by a dot-zero swap followed by a Rolfsen twist. We see that this is just $1/2n$ Dehn surgery on a knot K, where K is the sum of the stevedore knot and its reversed mirror, or more briefly, $Y_n = S^3_{1/2n}(K)$. This picture also includes a pair of incompressible tori in Y_n. The torus on the left of this figure separates the manifold Y_n into two components.

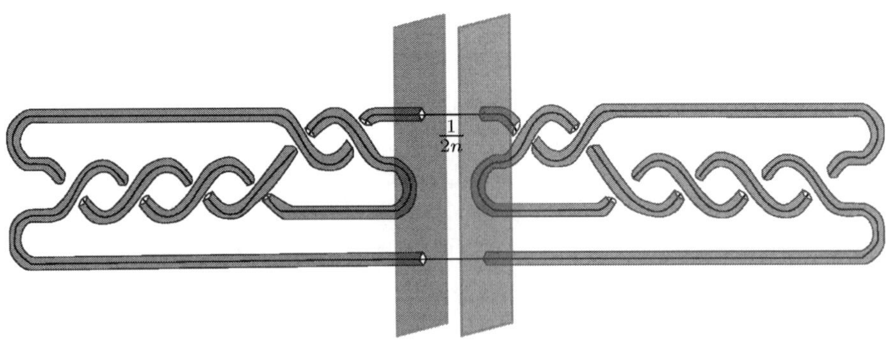

FIGURE 2. The homology spheres Y_n

The component on the left is the (mirror) of the exterior of the stevedore knot. Put coordinates on this torus as $(x, y) \in \mathbb{R}^2/\mathbb{Z}^2$ where $\mathbb{R}/\mathbb{Z} \times$ point is a longitude and point $\times\, \mathbb{R}/\mathbb{Z}$ is a meridian. Coordinates for a tubular neighborhood of the torus will then be (x, y, t) and the diffeomorphism $f(n)$ is given by $f(n)(x, y, t) = (x + t, y, t)$ in the neighborhood and the identity elsewhere. This map is called a torus twist; it is the product of a Dehn twist with the identity on S^1.

We can now prove the items in the proposition. The first two are just quoted results.

(1) This is [**29**, Theorem 3.5],
(2) This is [**66**, Theorem 2],
(3) This follows quickly from the JSJ decomposition [**34**, **35**], as we now describe in more detail.

The manifold Y_n may be described via knot splicing, leading to its complete JSJ decomposition as displayed in Figure 2. In the knot splice construction one identifies the neighborhood of a knot in one manifold with a neighborhood of a knot in a second manifold. One method to see a surgery description of knot splice is to take the product of an interval with the union of the original manifolds. Attaching one end of a 1-handle to a neighborhood of a point of the knot in the first manifold and the other end to a neighborhood of a point produces a manifold with a boundary component that is obtained by just identifying these neighborhoods. Attaching a 2-handle to the remainder of the knots and a pair of arcs along the 1-handle completes the splice. Figure 3 displays Y_n constructed in this way. See also Budney's paper [**15**] for other geometric calculations of this sort.

The knot complements arising in the splicing, going left to right in Figure 3, are: the exterior of the mirror of the stevedore, $T^2 \times I$, the Seifert fiber space with base orbifold having underlying surface a cylinder and one singular point of type $(-2n, 1)$, another copy of $T^2 \times I$, and the exterior of the stevedore. The copies of $T^2 \times I$ are not needed in the JSJ decomposition, but they are included here as a convenient way to describe the gluing via handle calculus. Since the exterior of the stevedore knot admits a complete hyperbolic structure there are no more incompressible tori that are not boundary parallel.

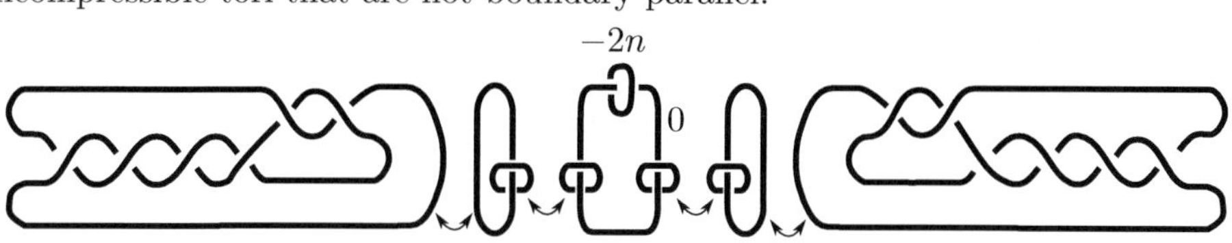

FIGURE 3. The homology sphere Y_n as a splice

The Alexander-Conway polynomial of the stevedore knot is $\nabla(z) = 2z^2 - 1$. It follows that the Alexander-Conway polynomial of K is $\nabla_K(z) = (2z^2-1)^2$. Clearly, $S^3_{1/0}(K) = S^3$, so the Casson invariant satisfies $\lambda(S^3_{1/0}(K)) = 0$. Casson's surgery formula [**1**] reads:

$$\lambda(S^3_{1/(m+1)}(K)) - \lambda(S^3_{1/m}(K)) = \frac{1}{2}\nabla''_K(0),$$

so an induction argument gives $\lambda(Y_n) = \lambda(S^3_{1/2n}(K)) = -8n$. Because Casson's invariant changes sign under orientation reversal [**1**], the manifolds Y_n with $n \neq 0$ do not admit any orientation-reversing diffeomorphisms. Furthermore, the stevedore knot is chiral so there is no diffeomorphism taking the complement of the mirror of the stevedore to the complement of the stevedore. If follows that up to isotopy, any diffeomorphism $\alpha\colon Y_n \to Y_n$ preserves the left torus in the JSJ decomposition as well as the complement of the mirror of the stevedore.

It is known [**39**] that the mapping class group of the exterior of the stevedore knot is $\mathbb{Z}_2^2$ and up to isotopy any diffeomorphism acts as multiplication by ± 1 in the $\mathbb{R}^2/\mathbb{Z}^2$ model of the boundary torus and as $(x, y, t) \mapsto (\pm x, \pm y, t)$ in a tubular neighborhood. Note that the generator of the $\mathbb{Z}_2^2$ that sends $(x, y, t) \to (x, -y, t)$ commutes with $f(n)$, while the generator that sends $(x, y, t) \to (-x, y, t)$ conjugates $f(n)$ to $f(n)^{-1}$. The other generators of $\mathrm{Diff}(Y_n)$ are either twists along one of the tori or diffeomorphisms of the complement of the mirror of the stevedore. It's well-known that twists along a given torus yield an isomorphism $\pi_0(\mathrm{Diff}(T^2 \times I, T^2 \times \partial I)) \cong \mathbb{Z} \oplus \mathbb{Z}$. The main point is to isotope a given diffeomorphism of $T^2 \times I$ to make it level preserving; a detailed argument is written out in [**53**]. Moreover, diffeomorphisms with disjoint support commute.

It follows that for any $\alpha \in \pi_0(\mathrm{Diff}(Y_n))$, either α commutes with $f(n)$ or conjugates it to its inverse. This distinction is encoded in the homomorphism δ. $\square$

We can now prove the first theorem.

Proof of Theorem 1.1. Recall that the goal is to find homology 3-spheres Y_n with topologically isotopic embeddings $j_p : Y_n \to S^4$ where the complements of $j_p(Y_n)$ are all diffeomorphic, but the embeddings are not smoothly isotopic. To this end, let $X_+(n)$ be the manifolds in Proposition 2.1, and set $Y_n = \partial X_+(n)$. By the proposition, we may write S^4 as

$$X_+(n) \cup_{f(n)^p} -X_+(n) = X_+(n) \times \{\pm 1\}/\sim, \text{ where } (y, 1) \sim (f(n)^p(y), -1),$$

oriented so that the orientation on $X_+(n) \times \{1\}$ agrees with the orientation on $X_+(n)$. Define $i_{p,n}\colon Y_n \to X_+(n) \cup_{f(n)^p} -X_+(n)$ by $i_{p,n}(y) = [y, 1]$, and $j_{p,n} = \Phi_{p,n} \circ i_{p,n}$, where $\Phi_{p,n}$ is the diffeomorphism specified in Proposition 2.1. By construction, $S^4 \backslash j_{p,n}(Y_n) \cong \mathrm{int}(X_+(n)) \amalg \mathrm{int}(X_-(n))$ and $X_-(n) \cong -X_+(n)$. The maps $j_{p,n}$ are defined for all p, but we will restrict to $p > 0$ to get the collection required by the theorem.

Let $\psi\colon S^4 \to S^4$ be a diffeomorphism satisfying $\psi \circ j_{p,n} = j_{q,n}$. First assume that ψ is orientation-preserving. Set $\alpha = \Phi_{q,n}^{-1} \circ \psi \circ \Phi_{p,n}$. As $\psi \circ j_{p,n} = j_{q,n}$ we see that α must take $X_+(n) \times \{1\}$ to $X_+(n) \times \{1\}$ or $X_+(n) \times \{-1\}$, and as α is orientation preserving, it must go to $X_+(n) \times \{1\}$. Let $\widehat{\alpha}$ be the diffeomorphism so that $\alpha([x, 1]) = [\widehat{\alpha}(x), 1]$. Similarly, let $\breve{\alpha}$ be the diffeomorphism so that $\alpha([x, -1]) = [\breve{\alpha}(x), -1]$. For $y \in Y_n$ we have

$$\begin{aligned}
[\widehat{\alpha}(y), 1] &= \alpha([y, 1]) \\
&= \alpha([f(n)^p y, -1]) \\
&= [\breve{\alpha} f(n)^p y, -1] \\
&= [f(n)^{-q} \breve{\alpha} f(n)^p y, 1]
\end{aligned}$$

and conclude that $\widehat{\alpha}(y) = f(n)^{-q} \breve{\alpha} f(n)^p y$, so that $f(n)^q = \breve{\alpha} f(n)^p \widehat{\alpha}^{-1}$ on Y_n.

Either one of $\delta(\widehat{\alpha})$ or $\delta(\breve{\alpha})$ is equal to $+1$, or both are -1. In the former case there is an isotopy $H\colon I \times Y_n \to I \times Y_n$ so that $H_0(y) = f(n)^{q-p}(y)$ and $H_1(y) = \breve{\alpha}\widehat{\alpha}^{-1}(y)$. Now identify $X_+(n)$ with $(I \times Y_n) \cup X_+(n)$ and define

$$\xi\colon (E(2) \# 2\overline{\mathbb{C}P}^2 \backslash \mathrm{int}(X_+(n))) \cup_{f(n)^p} X_+(n) \to (E(2) \# 2\overline{\mathbb{C}P}^2 \backslash \mathrm{int}(X_+(n))) \cup_{f(n)^q} X_+(n),$$

by

$$\xi(w) = \begin{cases} w, & w \in E(2) \# 2\overline{\mathbb{C}P}^2 \setminus \operatorname{int}(X_+(n)) \\ H(t,y), & \text{for } w = (t,y) \in I \times Y_n, \\ \breve{\alpha}\widehat{\alpha}^{-1}(w), & w \in X_+(n) \end{cases} .$$

One checks that this is a well-defined diffeomorphism to conclude that $p = q$ by part (1) of Proposition 2.1 . In the second case there is an isotopy $H\colon I \times Y_n \to I \times Y_n$ so that $H_0(y) = f(n)^{q+p}$ and $H_1(y) = \breve{\alpha}\widehat{\alpha}^{-1}(y)$. The argument from the first case then implies that $p = -q$.

If ψ is orientation reversing, pre-compose with $\Phi_{p,n}\rho\Phi^{-1}_{-p,n}$ where

$$\rho\colon X_+(n) \cup_{f(n)^{-p}} -X_+(n) \to X_+(n) \cup_{f(n)^p} -X_+(n)$$

is the map given by $\rho([x,\epsilon]) = [x,-\epsilon]$. This produces an orientation preserving diffeomorphism $\psi'\colon S^4 \to S^4$ so that $\psi' \circ j_{-p,n} = j_{q,n} \circ f(n)^{-p}$. Indeed,

$$\begin{aligned} \psi' \circ j_{-p,n}(y) &= \psi\Phi_{p,n}\rho\Phi^{-1}_{-p,n}(y) \\ &= \psi\Phi_{p,n}([y,-1]) \\ &= \psi\Phi_{p,n}i_{p,n}f(n)^{-p}y \\ &= \psi j_{p,n}f(n)^{-p}y = j_{q,n} \circ f(n)^{-p}(y). \end{aligned}$$

It follows that $\alpha' = \Phi^{-1}_{q,n} \circ \psi' \circ \Phi_{-p,n}$ is an orientation-preserving diffeomorphism taking the $X_+(n)$-factor to the $X_+(n)$-factor. Defining $\widehat{\alpha}'$ and $\breve{\alpha}'$ as in the orientation-preserving case, we find $\widehat{\alpha}'(y) = f(n)^{-q}\breve{\alpha}f(n)^{-p}(y)$ and conclude $p = \pm q$. Thus, if we restrict our embeddings to the ones with $p > 0$ there can be no intertwining orientation preserving or reversing diffeomorphism. □

3. Stabilization and sums

In preparation for the proof of Theorem 1.3, we now review some techniques for constructing interesting families of diffeomorphisms and embeddings. The general principle underlying [**7**] and the present paper is that stabilization (as defined below) can convert exotic smooth structures on 4-manifolds into other kinds of exotic phenomena. This method can be iterated to build exotic families of diffeomorphisms, parameterized by spheres of increasing dimension. In turn, a fixed embedding of a submanifold can be composed with such a family of diffeomorphisms, yielding a family of embeddings that can often be shown to be exotic.

The simplest instance of this strategy (leading to exotic diffeomorphisms) is to start with X and X', distinct smooth structures on the same underlying topological 4-manifold that are detected by a gauge theoretic invariant. Assume that $X \# U \cong X' \# U$ for some simple manifold U, typically $S^2 \times S^2$ or $\mathbb{C}P^2$. (At present, this is known to hold with $U = S^2 \times S^2$ for many pairs of homeomorphic smooth closed 4-manifolds, although not for all manifolds with boundary [**36**].) After picking a specific identification of these two stabilized manifolds, an embedding or diffeomorphism that makes use of the stabilizing summand U can often be shown to be exotic provided there is a way to recover information about smooth invariants of X and X' from the additional structure.

Here is a first basic example, implicit in the early work of Donaldson [**20**], using $U = \mathbb{C}P^2$. If $X \# \mathbb{C}P^2 \cong X' \# \mathbb{C}P^2$, then after picking a specific identification of the stabilized manifolds, one may conclude that the sphere arising from the $\mathbb{C}P^1 \subset (\mathbb{C}P^2 \setminus \mathrm{pt}) \subset X \# \mathbb{C}P^2$ is topologically isotopic to, but smoothly distinct from the sphere arising from $\mathbb{C}P^1 \subset (\mathbb{C}P^2 \backslash \mathrm{pt}) \subset X' \# \mathbb{C}P^2$. (The topological isotopy is deduced from [**56, 57**].) Indeed, if there was a diffeomorphism carrying one sphere to the other, blowing down the spheres would result in a diffeomorphism between X and X'. A similar construction and argument may be made assuming $X\#(S^2\times S^2)\cong X'\#(S^2\times S^2)$, via surgery on embedded spheres of self-intersection 0.

An extension of this technique was introduced in [**58**] to study diffeomorphisms. The idea is that a self-diffeomorphism of U gives rise to a diffeomorphism of the stabilized manifold $X \# U$ supported on U. For certain choices of U and self-diffeomorphism, a gauge theoretic invariant of diffeomorphisms shows that if the diffeomorphism associated to the decomposition $X \# U$ were isotopic to the one associated to the decomposition $X' \# U$, then X and X' would have the same gauge theoretic invariants. As in the previous example, these smoothly non-isotopic diffeomorphisms are topologically isotopic.

3.1. Submanifold sums. Many of the manifolds that go into our proof of Theorem 1.3 are constructed using *submanifold sum* along a submanifold Σ, which we briefly review. Generally, we will write $N(\Sigma, X)$ for the normal bundle of Σ in X, omitting the ambient manifold when it's clear. Assume that Σ is embedded in X and V in such a way that the normal bundles are oppositely oriented, or in symbols $N(\Sigma, V) = N(\Sigma, X)^\vee$ where $N^\vee$ denotes the bundle with orientation opposite to N. The submanifold sum is given by

$$X \#_\Sigma V = (X \backslash N(\Sigma)) \cup (V \backslash N(\Sigma))/\sim, \tag{3.1}$$

where $N(\Sigma, X)\backslash\Sigma$ is identified to $N(\Sigma, V)\backslash\Sigma$ via a map that acts as $v \mapsto |v|^{-2}v$ on the fibers.

The submanifold sum is a fairly general operation and depends only on the embedding of the normal bundle up to isotopy. In dimension 4, this operation is most often done along a surface, typically denoted by Σ, but we will also consider sums along 3-manifolds (typically denoted by Y). For sums along 3-manifolds, (as pictured in Figure 6 below), the isotopy type is determined by orientations and the isotopy type of the embedding of the submanifold. In section 4, we will consider sums along families and so in the following remark discuss the potential dependence on choices. The bottom line is that in the family setting, the case of codimension one submanifold sums is a little simpler than codimension two.

In codimension one, since Y and V are both oriented, the normal bundle is $N(Y) = Y\times[-1,1]$ and this identification is unique up to a choice in the orientation-preserving automorphism group of the normal bundle, which is contractible. Therefore $\mathrm{Emb}(Y, V)$ is homotopy equivalent to $\mathrm{Emb}(N(Y), V)$. On the other hand, if Σ has codimension two, the embeddings of the normal bundle extending a given embedding of the submanifold differ up to isotopy by a bundle automorphism. The space of bundle automorphisms may be identified with maps from Σ into the circle. Its path components, each of which are contractible, are given by $H^1(\Sigma; \mathbb{Z})$.

There is a minor variation of the submanifold sum in codimension one that we will use, and we explain the notation here. If we write $X \#_Y (-V)$, then this should be taken as an oriented manifold where the orientation agrees with the given orientation on X. More formally, when Y separates both X and V, this can be written as follows. Label the components of $X\backslash Y$ as $X_\pm$ so that $Y \times [-1, 0) \subset X_-$, and do the same for V. Then $X \#_Y (-V)$ would be written as

$$((X_- \cup V_+)/\sim)\coprod((X_+ \cup V_-)/\sim)$$

where '$\sim$' indicates the same inversion on $N(Y)$ as in the usual submanifold sum. This process is depicted in Figure 6; we also do this in families.

Using these ideas, here is how to build many collections of exotic embeddings. The construction, and the related construction of higher-parameter families of embeddings, uses as building blocks some 4-manifolds with specific properties. We describe a particular collection of such manifolds below in Section 5, but it's easier to see the overall principle if one doesn't worry too much about the specific ingredients.

Let X_p be a collection of exotic 4-manifolds that are distinguished by some gauge theoretic invariant I (say $I = D$ for Donaldson or $I = \mathrm{SW}$ for Seiberg-Witten) so that

$$I(X_p) = I(X_q) \text{ implies } p = q.$$

Suppose that each X_p has an embedding $\nu_p \colon N(\Sigma) \hookrightarrow X_p$, here Σ is a surface. By convention, exotic manifolds are homeomorphic; in this case we further assume that there are homeomorphisms that respect the embedding of $N(\Sigma)$, i.e.,

$$\psi_p \colon X_p \to X_0, \text{ so that } \psi_p \circ \nu_p = \nu_0.$$

We now let U be a stabilizing manifold so that $X_p \# U$ is diffeomorphic to $X_0 \# U$, where the diffeomorphism respects the embeddings ν_p. It is not hard to find such diffeomorphisms; see for example [**5**]. The point of the embedded copy of Σ is to allow a submanifold sum of a 'buffer' manifold V containing a copy of Σ, giving the construction greater flexibility.

Suppose further that the collection $\{X_p \#_\Sigma V\}$ is exotic, and continues to be distinguished by the invariant I. This would be the case for $V = \overline{\mathbb{C}P}^2$ when the submanifolds sum is just connected sum (along a 3-sphere) and I is a Seiberg-Witten or Donaldson invariant. Let

$$\iota \colon Y \hookrightarrow V \backslash \Sigma$$

be a smooth, separating embedding of a 3-manifold so that $V \backslash \iota(Y) = V_0 \amalg V_1$ with $\iota(\Sigma) \subset V_0$. Set $\widehat{V} = V \# U$ where the connected sum takes place in V_1, so that $\widehat{V} = \overline{V_0} \cup_Y (\overline{V_1 \# U})$. The assumption that the X_p become diffeomorphic after connected sum with U, respecting the embedding of Σ, implies that there is a diffeomorphism

$$\varphi_p \colon X_p \#_\Sigma \widehat{V} \to X_0 \#_\Sigma \widehat{V}.$$

We also take $\varphi_0 = 1$ and $\psi_0 = 1$. It will be convenient to have φ_p be homotopic to $\psi_p \#_\Sigma 1_{\widehat{V}}$ for all p. Since we are dealing with simply connected 4-manifolds, two maps are homotopic if and only if they induce the same isomorphism on homology [**17**]. By a classic theorem of Wall [**67**] one can compose an initial choice of φ_p with a diffeomorphism to ensure that this is the case if X_p is indefinite and $U = S^2 \times S^2$. In this case a result of Kreck [**42**] states that $\varphi_p(\psi_p \#_\Sigma 1_{\widehat{V}})^{-1}$ is pseudoisotopic to the identity, and so work of Perron [**56**] and Quinn [**57**] implies that it is topologically isotopic to the identity. Notice that both φ_p and $(\psi_p \#_\Sigma 1_{\widehat{V}})^{-1}$ restrict to smooth maps on $\widehat{V} \backslash \Sigma$ and in particular are smooth on Y. Thus we can define *smooth* embeddings

$$j_p \colon Y \hookrightarrow X_0 \#_\Sigma \widehat{V},$$

by $j_p = \varphi_p(\psi_p \#_\Sigma 1_{\widehat{V}})^{-1} \circ \iota$, and immediately conclude that the maps j_p are topologically isotopic to one another.

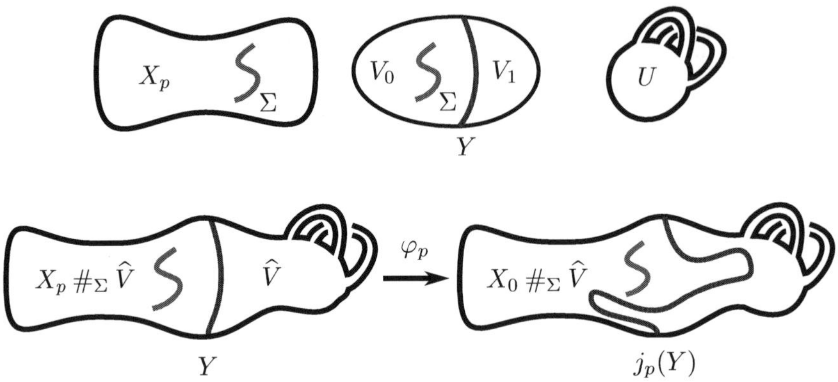

FIGURE 4. Construction of the embedding j_p

It remains to show that the embeddings j_p are smoothly distinct. Assuming that there is a diffeomorphism $f\colon X_0 \#_\Sigma \widehat{V} \to X_0 \#_\Sigma \widehat{V}$ so that $f \circ j_p = j_q$, one would conclude that

$$\overline{\left((X_0 \#_\Sigma \widehat{V})\backslash j_p(Y)\right)_0 \bigcup_Y \overline{V_1}} \cong \overline{\left((X_0 \#_\Sigma \widehat{V})\backslash j_q(Y)\right)_0 \bigcup_Y \overline{V_1}}.$$

Here the 0-subscript refers to the component disjoint from $U\backslash\mathrm{pt}$. Using extensions of suitable restrictions of φ_p and φ_q one would further conclude that $X_p \#_\Sigma V \cong X_q \#_\Sigma V$, implying that $p = q$.

REMARK 3.1. Theorem 1.1 draws a finer distinction between smoothly knotted embeddings. The complements of the embeddings in that theorem are independent of the choice of embedding. The fact that the embeddings are not smoothly equivalent follows from the existence of non-extendible diffeomorphisms on a manifold with boundary. The paper [**41**] constructs a number of other interesting examples of smoothly knotted embeddings based on a different exotic phenomenon, namely the non-smoothablility of certain homeomorphisms.

3.2. Higher parameter families. The construction of exotic diffeomorphisms via stably equivalent smooth structures combines with further stabilizations by a manifold U (such as $\mathbb{C}P^2$ or $S^2 \times S^2$, often combined with a buffer manifold as in the previous subsection) to yield exotic spherical families of diffeomorphisms. The construction is recursive, raising the dimension of the parameterizing sphere by one at each step. At the base level, let X and X' be distinct smooth manifolds with a diffeomorphism $\varphi\colon X \# U \to X' \# U$ that induces the same isomorphism on homology as $\psi \# 1_U$ for a homeomorphism $\psi\colon X \to X'$ and preserves the embeddings of Σ, and write Z_0 for $X \# U$. Let $f\colon U \to U$ be a diffeomorphism acting non-trivially on $H_2(U)$, defined using 2-spheres embedded in U. A prototype might be $U = \mathbb{C}P^2$, with f defined as complex conjugation, or equivalently as a reflection in the 2-sphere $\mathbb{C}P^1$. In examples, the construction is slightly more elaborate.

In this setting, it is reasonable to conjecture that

$$\alpha^0 = [\varphi, 1 \# f] = \varphi \circ (1_{X'} \# f)\varphi^{-1}(1_X \# f)^{-1}\colon Z_0 \to Z_0, \tag{3.2}$$

is an exotic diffeomorphism. Indeed the original construction of [**58**] was of this form, and there are now a number of additional examples that use the same technique. It is worth noting that the commutator in Equation (3.2) has embedded in it a crucial switch from X to X'; other commutator-like constructions of this sort will be used without further comment. A particular choice of α^0 and U is given in Section 5. To describe the recursion, imagine that $\alpha^k\colon S^k \to \mathrm{Diff}(Z_k)$ represents a non-trivial element of $\pi_k(\mathrm{Diff}(Z_k))$, and that $\alpha^k \# 1_U$ represents the trivial element of $\pi_k(\mathrm{Diff}(Z_k \# U))$. In what follows, we will be assuming that all of the families (and homotopy of families) of diffeomorphisms are relative to $N(\Sigma)$, but we omit it from the notation for sanity's sake.

The presence of the summand of U in $\hat{V}$ implies that $\alpha^k \#_\Sigma 1_{\hat{V}}$ represents the trivial element of $\pi_k(\mathrm{Diff}(Z_k \#_\Sigma \hat{V}))$. Let $F\colon I \times S^k \to \mathrm{Diff}(Z_k \#_\Sigma \hat{V})$ be the corresponding null-homotopy. Set $Z_{k+1} = Z_k \#_\Sigma \hat{V}$ and let $\beta_0\colon \hat{V} = V \# U \to \hat{V}$ be the diffeomorphism constructed as in the base case, in other words, $\beta_0 = 1_V \#_\Sigma f$. We then define $\alpha^{k+1}\colon I \times S^k \to \mathrm{Diff}(Z_{k+1})$ by $\alpha^{k+1} = [F, 1_{Z_k} \#_\Sigma \beta_0]$. In other words given $(t,v) \in I \times S^k$ and $z \in Z_{k+1}$ we have

$$\alpha^{k+1}(t,v)(z) = F(t,v)(1_{Z_k} \#_\Sigma \beta_0)F(t,v)^{-1}(1_{Z_k} \#_\Sigma \beta_0)^{-1}(z). \tag{3.3}$$

Since $F(0,v) = 1_{Z_k}$, and $F(1,v) = \alpha^k \#_\Sigma 1_{\hat{V}}$ commutes with $1_{Z_k} \#_\Sigma \beta_0$, we see that $\alpha^{k+1}|_{\partial I \times S^k} = 1_{Z_{k+1}}$ so it induces a map on the suspension of S^k, i.e., on S^{k+1}. In fact it also restricts to the identity on $I \times \{v_0\}$ so it also induces a map on the reduced suspension.

This process, which gives rise to families of diffeomorphisms α^k as well as associated families of embeddings, is represented schematically in Figure 5. In this figure, we have represented a diffeomorphism or family of diffeomorphisms by a red curve and a blue curve. One can imagine that the diffeomorphism takes the red to the blue (or that a spherical family moves these curves around in some fashion.) The smooth isotopy F does not exist without the U summand in $\hat{V}$, so one imagines the isotopy would give a family moving the blue curve to the red. However, this must pass through the right-most copy of U, as represented by a pair of handles on the far right of the picture, and this in some sense is what makes it an 'interesting' isotopy.

In section 4, we will discuss invariants that can be used to show that families constructed in this way from suitable initial collections of 4-manifolds are non-trivial.

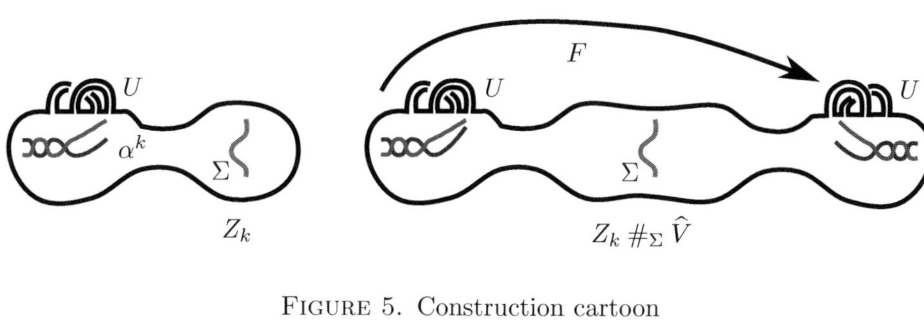

FIGURE 5. Construction cartoon

4. Families and invariants

Families of diffeomorphisms, manifolds, and embeddings are closely related. For instance, a family of manifolds diffeomorphic to Z is, by definition, a bundle over a base B with fiber diffeomorphic to Z and structure group $\mathrm{Diff}(Z)$, and hence is equivalent to a map from B to the classifying space $B\,\mathrm{Diff}(Z)$. We discuss some of the relations between these three types of families here and then turn to the definition of various invariants that are useful to distinguish these families. The connection between families of diffeomorphisms and bundles is most visible in the case of spherical families.

DEFINITION 4.1. If $\alpha\colon S^k \to \mathrm{Diff}(Z)$ is a spherical family of diffeomorphisms, the family of spaces associated to α via the *clutching construction* is

$$\mathcal{Z}[\alpha] = \{\pm 1\} \times D^{k+1} \times Z/\sim,$$

where $(1, v, z) \sim (-1, v, \alpha(v)(z))$ for $|v| = 1$. If we view $S^{k+1} = \{\pm 1\} \times D^{k+1}/\sim$, then there is a natural fiber bundle projection $\mathcal{Z}[\alpha] \to S^{k+1}$.

We have the following facts about the clutching construction.

LEMMA 4.2.

(a) *Let $\alpha\colon S^k \to \mathrm{Diff}(Z)$ be a spherical family of diffeomorphisms and $F\colon I \times S^k \to \mathrm{Diff}(Z)$ a contraction to the constant family $\mathbf{1}_Z$ given by 1_Z at each point of S^k. Then there is a family isomorphism $\Phi\colon \mathcal{Z}[\mathbf{1}_Z] \to \mathcal{Z}[\alpha]$ given by*

$$\Phi([\epsilon, v, z]) = \begin{cases} [1, v, z], & \text{if } \epsilon = 1, \\ [-1, v, F(|v|, v)(z)], & \text{if } \epsilon = -1. \end{cases}$$

(b) *Let $\beta\colon Z' \to Z$ be a diffeomorphism. Then there is a family isomorphism*

$$\xi\colon \mathcal{Z}'[\beta^{-1}\alpha\beta] \to \mathcal{Z}[\alpha],$$

given by $\xi([\epsilon, v, w]) = [\epsilon, v, \beta(w)]$.

REMARK 4.3. The lemma follows by a direct check of the definitions. The second isomorphism ξ would also work if β was a spherical family of diffeomorphisms.

We now describe how to use these ideas to construct and detect interesting families of embeddings. The construction of families begins the same way as the construction of smoothly knotted codimension one embeddings in Section 3.1. The ingredients that go into the construction are

(1) A collection of homeomorphic smooth manifolds X_p together with a family of invariants $I_{\mathbf{c}_q}$, parameterized by data on the manifold, so that $I_{\mathbf{c}_q}(X_p) = \gamma_p \delta_{pq}$ for some non-zero constants γ_p. In the Donaldson case, $\mathbf{c}_q$ describes a certain SO(3) bundle, while in Seiberg-Witten case, it specifies a spin^{c} structure.
(2) Embeddings $\nu_p\colon N(\Sigma_1 \amalg \Sigma_2) \hookrightarrow X_p$.
(3) A smooth buffer manifold $V = \overline{V_0} \cup_Y \overline{V_1}$ with $N(\Sigma_2) \hookrightarrow V_0$.
(4) A collection of manifolds Z_k together with topologically trivial families of diffeomorphisms $\alpha^k\colon S^k \to \mathrm{Diff}(Z_k, N(\Sigma_1))$ so that $\alpha^k \# 1_U$ smoothly contracts and the corresponding $I^{H_k\,\mathrm{Diff}}_{\mathbf{c}_q}(\alpha^k \#_{\Sigma_1} 1_{X_p \#_{\Sigma_2} V}) = \gamma'_p \delta_{pq}$ for some non-zero constants γ'_p and data q.

As before, we will have a manifold U (typically $S^2 \times S^2$ or $(S^2 \times S^2) \# \overline{\mathbb{CP}}^2$) and write $\widehat{V} = V \# U$ where the connected sum takes place in V_1.

Notice that we just need one spherical family of diffeomorphisms for each dimension and use a large collection of exotic embeddings of a submanifold to build a large collection of families of embeddings in each dimension.

We now provide a recipe for combining these ingredients into examples of the proper form. We will list some assumptions along the way. Let $\psi_p\colon X_p \to X_0$

be homeomorphisms so that $\psi_p \circ \nu_p = \nu_0$, and $\varphi_p\colon X_p \#_{\Sigma_2} \widehat{V} \to X_0 \#_{\Sigma_2} \widehat{V}$ be diffeomorphisms homotopic to $\psi_p \#_{\Sigma_2} 1_{\widehat{V}}$. This implies that there is a topological isotopy $G^p\colon I \to \mathrm{Homeo}^0(X_0 \#_{\Sigma_2} \widehat{V})$ with $G^p_1 = \varphi_p(\psi_p \#_{\Sigma_2} 1_{\widehat{V}})^{-1}$ and $G^p_0 = 1_{X_0\#_{\Sigma_2}\widehat{V}}$. In the examples used in this paper, the isotopy G^p comes from the 'key stable isotopy' in [**5**, §2] and so may be chosen to be the identity on $N(\Sigma_1)$. This additional property holds in many other settings; for example if the complement of Σ_1 has trivial [**64**], infinite cyclic [**18**] or in some cases finite cyclic [**37**] fundamental group. Define $j_p\colon Y \to X_0 \#_{\Sigma_2} \widehat{V}$ by $j_p = \varphi_p(\psi_p \#_{\Sigma_2} 1_{\widehat{V}})^{-1}\iota$. Notice that this is smooth because the restriction of $\varphi_p(\psi_p \#_{\Sigma_2} 1_{\widehat{V}})^{-1}$ to the image of the embedding $\iota\colon Y \to \widehat{V}\backslash\Sigma_2$ is smooth. Set

$$Z'_k = Z_k \#_{\Sigma_1} \left(X_0 \#_{\Sigma_2} \widehat{V}\right).$$

Notice that $X_0\#_{\Sigma_2}\widehat{V}$ contains a U-summand. This implies that $\alpha^k\#_{\Sigma_1}1_{X_0\#_{\Sigma_2}\widehat{V}}\colon S^k \to \mathrm{Diff}(Z_k \#_{\Sigma_1} X_0 \#_{\Sigma_2} \widehat{V})$ smoothly contracts. By Lemma 4.2 we conclude that there is a bundle isomorphism

$$\Phi^k\colon \mathcal{Z}[\alpha^k \#_{\Sigma_1} 1_{X_0\#_{\Sigma_2}\widehat{V}}] \to S^{k+1} \times Z'_k.$$

Since $\alpha^k \#_{\Sigma_1} 1_{X_0\#_{\Sigma_2}\widehat{V}}$ acts trivially on $\widehat{V}\backslash\Sigma_2$, the embedding j_p extends to a family of embeddings $\underline{j_p}\colon S^{k+1} \times Y \to \mathcal{Z}[\alpha^k \#_{\Sigma_1} 1_{X_0\#_{\Sigma_2}\widehat{V}}]$ given by $\underline{j_p}([\epsilon, v], y) = [\epsilon, v, j_p(y)]$. Now define

$$J^{k+1}_p\colon S^{k+1} \to \mathrm{Emb}(Y, Z'_k), \text{ by } J^{k+1}_p([\epsilon, v])(y) = \mathrm{pr}_{Z'_k}\Phi^k(\underline{j_p}([\epsilon, v], y))$$

LEMMA 4.4. *The classes* $[J^{k+1}_p]$ *and* $[J^{k+1}_0]\pi_{k+1}(\mathrm{Emb}^{\mathrm{Top}}(Y, Z'_k))$.

PROOF. As discussed above, we are working in a setting where the topological isotopy G^p has support in $X_0 \#_{\Sigma_2} \widehat{V}\backslash\Sigma_1$ so extends to define a homotopy

$$\underline{G}^p\colon I \to \mathrm{Emb}(Y, Z'_k), \text{ given by } \underline{G}^p_t(y) = \mathrm{pr}_{Z'_k}\Phi_k([\epsilon, v, G^p_t j_0(y)]).$$ □

There is a family version of the submanifold sum, both for sums along surfaces and along 3-manifolds. As we will see, it interacts well with our constructions of families of diffeomorphisms.

DEFINITION 4.5. Let $J\colon \Xi \to \mathrm{Emb}(N(Y), Z)$ and $K\colon \Xi \to \mathrm{Emb}(N(Y), W)$ be given and set

$$\mathcal{Z}[\![J, K]\!] = ((\Xi \times Z)\backslash J(Y)) \amalg ((\Xi \times W)\backslash K(Y)) / \sim$$

where the fibers of the normal bundles are identified via inversion in the unit sphere as in (3.1).

REMARK 4.6. Note that if Ξ is connected, then we are getting an embedding of Y in a manifold diffeomorphic to $Z \#_Y W$. But when done in the family setting, the resulting family may well be non-trivial. One could go further and define a family notion of submanifold sum for family embeddings into non-trivial families. When the K-family of embeddings is clear from the context, we will remove it from the notation.

Let Z contain a summand (along a surface) of $\widehat{V}$ as in our description of smoothly knotted embeddings in Section 3.1, and let $\iota\colon N(Y) \hookrightarrow V$ be the model embedding and $J\colon \Xi^{k+1} \to \mathrm{Emb}(N(Y), Z)$ be a family of embeddings representing a class in $H_{k+1}(\mathrm{Emb}(N(Y), Z))$. The family submanifold sum

$$(\Xi^{k+1} \times Z) \cup_J (\Xi^{k+1} \times -V)$$

will consist of two components. Let $\mathcal{Z}[\![J]\!]$ denote the component that does not include the copy of U in each fiber. We call it the preferred component. The component with U will be called the discarded component. See Figure 6, which omits the parameter space Ξ.

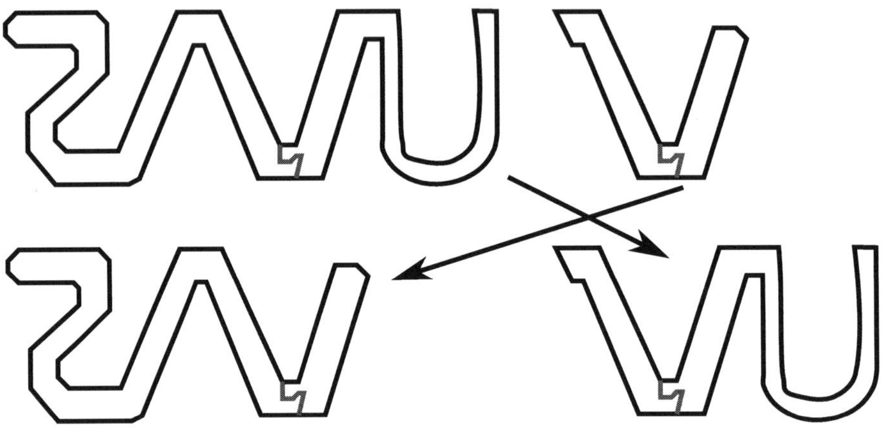

FIGURE 6. Codimension one submanifold sum

Let us assume that that the basic $I_{\mathbf{c}_q}$ extends to an invariant of families of manifolds, denoted $\mathbb{I}_{\mathbf{c}_q}$. Given this assumption, the invariant for a family of embeddings is given by

$$I_{\mathbf{c}_q}^{H_{k+1}\,\mathrm{Emb}}(J) = \mathbb{I}_{\mathbf{c}_q}(\mathcal{Z}[\![J]\!]). \tag{4.1}$$

LEMMA 4.7. *The classes* $[J_p^{k+1}]$ *are independent in* $\pi_{k+1}(\mathrm{Emb}(Y, Z_k'))$.

PROOF. Notice that $\Phi^k\colon \mathcal{Z}[\alpha^k \#_{\Sigma_1} 1_{X_0 \#_{\Sigma_2} \hat{V}}] \to S^{k+1} \times Z_k')$ satisfies

$$\Phi^k \underline{j_p}([\epsilon, v], y) = [\epsilon, v, J_p^{k+1}([\epsilon, v])(y)].$$

The diffeomorphism $1_{Z_k} \#_{\Sigma_1} \varphi_p\colon Z_k \#_{\Sigma_1} X_p \#_{\Sigma_2} \widehat{V} \to Z_k \#_{\Sigma_1} X_0 \#_{\Sigma_2} \widehat{V}$ combined with Lemma 4.2 then gives a family isomorphism

$$\xi\colon \mathcal{Z}[\alpha^k \#_{\Sigma_1} 1_{X_0 \#_{\Sigma_2} \hat{V}}] \to \mathcal{Z}[\alpha^k \#_{\Sigma_1} 1_{X_p \#_{\Sigma_2} \hat{V}}],$$

so that $\alpha^k \#_{\Sigma_1} 1_{X_0 \#_{\Sigma_2} \hat{V}} j_p = \alpha^k \#_{\Sigma_1} 1_{X_p \#_{\Sigma_2} \hat{V}} \iota$. It follows that

$$\begin{aligned}
\mathcal{Z}[\![J_p^{k+1}]\!] &= \text{preferred component}\Big((S^{k+1} \times Z_k')) \#_{J_p^{k+1}} (S^{k+1} \times -V)\Big) \\
&\cong \text{preferred component}\Big(\mathcal{Z}[\alpha^k \#_{\Sigma_1} 1_{X_0 \#_{\Sigma_2} \hat{V}}] \#_{\underline{j_p}} (S^{k+1} \times -V)\Big) \\
&\cong \text{preferred component}\Big(\mathcal{Z}[\alpha^k \#_{\Sigma_1} 1_{X_p \#_{\Sigma_2} \hat{V}}] \#_{\underline{\iota}} (S^{k+1} \times -V)\Big) \\
&\cong \mathcal{Z}[\alpha^k \#_{\Sigma_1} 1_{X_p \#_{\Sigma_2} V}]
\end{aligned}$$

Thus,

$$\begin{aligned} I^{H_{k+1}\,\mathrm{Emb}}_{\mathbf{c}_q}(J^{k+1}_p) &= \mathbb{I}_{\mathbf{c}_q}(\mathcal{Z}[\![J^{k+1}_p]\!]) \\ &= \mathbb{I}_{\mathbf{c}_q}(\mathcal{Z}[\alpha^k \#_{\Sigma_1} 1_{X_p\#_{\Sigma_2}V}]) \\ &= I^{H_k\,\mathrm{Diff}}_{\mathbf{c}_q}(\alpha^k \#_{\Sigma_1} 1_{X_p\#_{\Sigma_2}V}) = \gamma'_p\delta_{pq}. \end{aligned}$$

This is sufficient to show that the families of embeddings are independent in the homotopy group.

We can slightly recast the invariant $I^{H_{k+1}\,\mathrm{Emb}}$. By making $\mathbf{c}_q$ an argument we obtain

$$I^{H_{k+1}\,\mathrm{Emb}}_{(\cdot)} \colon H_{k+1}(\mathrm{Emb}(Y, Z'_k))) \to \mathrm{Maps}(\text{Set of } \mathbf{c}_q \text{ values}, \mathbb{Z}).$$

Thus, $I^{H_{k+1}\,\mathrm{Emb}}_{(\cdot)}\left(\ker\left[\pi_{k+1}(\mathrm{Emb}(Y, Z'_k))) \to \pi_{k+1}(\mathrm{Emb}^{\mathrm{Top}}(Y, Z'_k)))\right]\right)$ is a free group of rank equal to the number of initial exotic manifolds X_p. As it is free there is a right inverse s to $I^{H_{k+1}\,\mathrm{Emb}}_{(\cdot)}$ from this group to the kernel giving the required summand. The same argument generates the summand on the kernel at the level of homology. □

5. Specific choices

To make the outline in the prior section into a proof, one must make specific choices for the manifolds, diffeomorphisms, and invariants. We will see that there are many different choices for manifolds X_p, U, and V, and the diffeomorphism $f : U \to U$ that can be used as the basis of our constructions. In this section, we give some specific choices and specify the manifolds Z appearing in the statement of Theorem 1.3. All of them are built in some fashion out of the elliptic surface $E(2)$, which has $q_{E(2),\mathbf{c},\beta} = -1$ with the homology orientation and class $\mathbf{c}$ described in [**21**]. It also has $\mathrm{SW}(E(2), 0) = 1$.

The surface $E(2)$ has three embedded copies [**30**] of the Gompf nucleus [**27**], labeled $N^{(1)}$, $N^{(2)}$ and $N^{(3)}$, whose fibers are denoted $T^{(1)}$, $T^{(2)}$ and $T^{(3)}$. (We retain the convention from [**7**] of denoting various specific manifolds with bold letters; these are not quite the same as the manifolds appearing in this paper, which are in ordinary font.) Set $\mathbf{X} = E(2)\#\overline{\mathbb{C}P}^2$. We also define $\mathbf{Y}$ by a cork twist on $\mathbf{X}$ using the cork discussed below in Example 7.3; in particular, it decomposes as a connected sum of copies of $\mathbb{C}P^2$ and $\overline{\mathbb{C}P}^2$ and hence has trivial gauge-theoretic invariants. This is the original example of an effective cork discovered by Akbulut. It is important to notice that the cork is disjoint from the $N^{(1)}$ nucleus; see [**5**, Figure 7].

By Freedman's work there is a homeomorphism $\psi : \mathbf{X} \to \mathbf{Y}$ relative to $N^{(1)}$. Furthermore, if we set $\mathbf{M} = \mathbb{C}P^2\#2\overline{\mathbb{C}P}^2 \cong S^2 \times S^2 \# \overline{\mathbb{C}P}^2$ there is a diffeomorphism $\varphi : \mathbf{X} \# \mathbf{M} \to \mathbf{Y} \# \mathbf{M}$ relative to $N^{(1)}$ inducing the same map on homology. We use $X' = \mathbf{X}$, $X = \mathbf{Y}$, and $Z_0 = \mathbf{Y} \# \mathbf{M}$ in the construction outlined in Section 3.2.

Corresponding to the three summands in the stabilizing manifold $U = \mathbf{M}$, there are 2-spheres H, E_1, and E_2 of self-intersection 1, -1, and -1, respectively. Any sphere S of self-intersection ± 1 in a 4-manifold determines a reflection ρ^S, and we define

$$f = \rho^{E_1}\rho^{H+E_1+E_2}.$$

As described in equations (3.2) and (3.3) in Section 3.1, such a diffeomorphism gives rise to a diffeomorphism $\alpha^0 : Z_0 \to Z_0$, and by a recursive process, to spherical families α^k of diffeomorphisms on the related manifolds Z_k. This is exactly the rank one case of the construction of exotic families of diffeomorphisms presented in [7]. In addition, the commutator construction gives topological contractions and smooth stabilized contractions. The formula for each of these is

$$\begin{aligned} \alpha^{k+1} &= [F^k, 1_{\mathbf{Z}_k} \#_T \alpha] \\ G^{k+1} &= [F^k, 1_{\mathbf{Z}_k} \#_T G] \\ F^{k+1} &= [F^k, 1_{\mathbf{Z}_k} \#_T F]. \end{aligned} \tag{5.1}$$

We will only need the α^k and F^k in this paper.

The basic idea to create families of embeddings is to fiber sum a family of manifolds constructed via the clutching construction with the α^k families of diffeomorphisms with the collections of exotic embeddings described in section 3.1. Schematically we are just gluing a copy of Figure 4 onto the end of a copy of Figure 5.

The buffer manifold V is defined to be the blowup $E(2)\#\overline{\mathbb{C}P}^2$ (this is the same as $\mathbf{X}$ but we are using it differently). We take $\Sigma_2 = T^{(2)}$ the fiber in the second nucleus. Any separating 3-manifold Y that embeds in V disjoint from a fiber will satisfy the conclusions of the theorem and have large families of embeddings in the manifolds Z constructed below. Some popular choices for Y might be S^3 (as the boundary of a ball), the Brieskorn sphere $\Sigma(2,3,11)$ (boundary of a nucleus), the Brieskorn sphere $\Sigma(2,3,7)$ (boundary of a Milnor fiber [25]), or the boundary of the Akbulut-Mazur cork [2]. Note that for this last choice there are two distinct embeddings of Y, differing by reparameterization using the cork twist. In any of these cases let V_0 be the component of $V\backslash Y$ that contains $\Sigma_2 = T^{(2)}$ and let V_1 be the other component.

To construct the manifolds X_p we will use log transforms and fiber sums. Let $E(2)_{p;q;r}$ be the result of performing log transforms of multiplicity p, q, and r on the fibers in the $N^{(i)}$ nuclei for $i = 1, 2, 3$, where by convention '∞' means that no surgery has been done. $E(2)_{p;q;r}$ inherits a homology orientation from $E(2)$.

REMARK 5.1. We could define X_p to be $E(2)_{\infty;\infty;2p+1}$, and take $I_{\mathbf{c}_q}$ to be the Seiberg-Witten invariants corresponding to $2q[F_p]$, where $[F_p]$ is Poincaré dual to the multiple fiber in X_p. There are suitable family versions of this invariant [40]. In a forthcoming paper we will show how to compute the family Seiberg-Witten invariants to make this whole story work. This paper is not yet available, so we will describe a more complicated construction that is based on our computation [7] of the relevant family Donaldson invariants.

While the family version will have to wait for the computation of the family Seiberg-Witten invariants, we note that these manifolds are sufficient to establish the existence of an infinite collection of exotic codimension one embeddings as described in section 3.1 via Donaldson invariants. Indeed, by [26, Theorem 3.1] and [30, Theorem 3.1],

$$q_{X_p,\mathbf{c},\beta} = \sum_{j=-p}^{p} (-1)^{\frac{1}{2}(\mathbf{c}^2+2jc[f^{(3)}]+c[f^{(3)}])}.$$

The construction of a collection of manifolds X_p suitable for use with family Donaldson invariants generalizes $\mathbf{X} = E(2)\,\#\,\overline{\mathbb{C}P}^2$ to

$$\mathbf{X}^r = \mathbf{X}\,\#_T\, E(2)^{(\#_T)^{r+1}} \#\, \overline{\mathbb{C}P}^2.$$

The torus sum in this expression is done along the fibers labeled $T^{(1)}$ in each copy of $E(2)$. It is useful to notice that all of these are homeomorphic relative to $N^{(1)}$ to the manifold $\mathbf{Y}^r$ obtained by performing a cork twist on the cork in the zeroth copy of $\mathbf{X}$. Finally, we define the manifolds $\mathbf{X}_p^r$ by replacing the pth copy of $E(2)$ in $\mathbf{X}^r$ with $E(2)_{\infty;2;2}$. We will take $X_p = \mathbf{X}_p^r$ for our proof of Theorem 1.3. We will take Σ_1 and Σ_2 to be parallel copies of the fiber $T^{(1)}$ in the first nucleus.

We now discuss the invariants that we use to distinguish the various families that we construct. The same circle of ideas provides invariants for families of diffeomorphisms and for families of embeddings. We start with a review of the simplest sort of Donaldson invariants and then move on the family version, for the applications to families of diffeomorphisms and embeddings.

For a 4-manifold X with $b_+^2(X) > 1$, set

$$\ell_X = -\frac{3}{4}\left(\chi(X) + \sigma(X)\right),$$

and define C_X to be the set of $\eta \in H^2(X; \mathbf{Z}_2)$ such that

(i) η lifts to an integral class $c \in H^2(X)$
(ii) For any integral lift c of η we have $c \cdot c \equiv \ell_X \pmod 4$
(iii) $\eta \neq 0$.

Further, we let $\widehat{C}_X$ be the set of integral lifts $\mathbf{c}$ of elements in C_X modulo the equivalence $\mathbf{c} \sim \mathbf{c}'$ if $\mathbf{c} \equiv \mathbf{c}' \pmod 2$ and $(\frac{1}{2}(\mathbf{c} - \mathbf{c}'))^2 \equiv 0 \pmod 2$. The set C_X parameterizes a set of potential second Stiefel-Whitney classes of SO(3) bundles over X. Adding the requirement that the virtual dimension of the ASD moduli space be zero then ensures that each element of C_X corresponds to a unique bundle. The $\widehat{C}_X$ refines the choice of bundle by an equivalence class of spin^c-lift up to an equivalence relation. In addition, we take a homology orientation, i.e., $\beta \in \wedge^{top}(H^0 \oplus H_+^2(X))$. The choice of $\mathbf{c}$ and β determine a specific SO(3) bundle and the data to define an orientation on the corresponding ASD moduli space. Furthermore, for generic metrics these moduli spaces are a finite set of oriented points and the degree zero Donaldson invariant $q_{\mathbf{c},\beta}(X)$ is just the signed count of these points. These are the invariants in item (1) of the recipe from Section 3.2

For future reference, we note a special case of Donaldson's blowup formula [**20**]. If $\pi\colon X \# \overline{\mathbb{C}P}^2 \to X$ is the natural projection and $\mathbf{c} \in \widehat{C}_X$, then $q_{X\#\overline{\mathbb{C}P}^2, \pi^*\mathbf{c}} = q_{X,\mathbf{c}}$. We build many manifolds via submanifold sums, all of which are sums along tori, so we need to know how to compute invariants in this situation. Here is a specific fiber sum formula that works in this setting. Let X and Y be smooth 4-manifolds, each with a preferred embedded torus of square zero, each denoted T. Let $\mathbf{c}_X \in \widehat{C}_X$ and $\mathbf{c}_Y \in \widehat{C}_Y$ satisfy $\mathbf{c}_X[T] \equiv \mathbf{c}_Y[T] \equiv 1 \pmod 2$. Further suppose that the restriction of $\mathbf{c}_{XY} \in \widehat{C}_{X\#_T Y}$ to $X \backslash T$ (respectively $Y \backslash T$) agrees with the restriction of $\mathbf{c}_X$ (respectively $\mathbf{c}_Y$). Then by [**52**, Theorem 2.2]

$$q_{X\#_T Y, \mathbf{c}_{XY}} = q_{X,\mathbf{c}_X} q_{Y,\mathbf{c}_Y}.$$

We also need a computation of the Donaldson invariants of the $\mathbf{X}_p^r$. We identify the cohomology groups $H^*(\mathbf{X}^r)$ and $H^*(\mathbf{Y}^r)$ via a homeomorphism ψ. To specify the cohomology classes that we need for the Donaldson invariant calculation, we need a little notation. In the ith copy of $E(2)$, let the homology classes of the sections (lying in the 3 nuclei) be denoted by $\sigma^{(i,a)}$ and the homology classes of the fibers be denoted by $f^{(i,a)}$. Let $\Upsilon^{\mathbf{X}}\colon \mathbf{X} \backslash T^{(1)} \to \mathbf{X}^r$ and $\Upsilon^{(i)}\colon E(2)^{(i)} \backslash f^{(i,1)} \to \mathbf{X}^r$

be the inclusion maps. Let $\mathbf{c}_q^{\mathbf{X}}$ be cohomology classes in $H^2(\mathbf{X}^r)$ so that $(\Upsilon^{\mathbf{X}})^*\mathbf{c}_q^{\mathbf{X}}$ is the image of the Poincaré dual to the section of $\mathbf{X}$ in $H^2(\mathbf{X}\backslash F)$, $(\Upsilon^{(i)})^*\mathbf{c}_q^{\mathbf{X}}$ is the image of the Poincaré dual to $\sigma^{(i,1)} + \sigma^{(i,2)} + \sigma^{(i,3)} + f^{(i,1)} - 3f^{(i,2)} - 2f^{(i,3)}$ in $H^2(E(2)^{(i)}\backslash f^{(i,1)})$ for $i \neq q$ and is the image of the Poincaré dual to $\sigma^{(i,1)}$ for $i = q$. Set $\mathbf{c}_q = \psi^*\mathbf{c}_q^{\mathbf{X}}$. It is an elementary calculation to verify that all of the $\mathbf{c}$-classes that we have defined live in the correct $\widehat{C}$ sets. Furthermore, the Donaldson invariants satisfy the assumption that $(q_{\mathbf{X}_p^r,\mathbf{c}_q})$ is $4(-1)^r$ times the identity matrix [**7**].

In [**7**] the degree zero Donaldson invariant is extended to an invariant for families of diffeomorphisms. Assume $b_+^2(Z)$ is at least $k+3$ and that $b_+^2(Z)$ and k have the same parity. Set

$$\ell_Z^k = -\frac{3}{4}(\chi(Z) + \sigma(Z)) + \frac{1}{2}(k+1) = -\frac{1}{2}(3b_2^+(Z) - k + 2)$$

and define C_Z^k to be the set of $\eta \in H^2(Z; \mathbf{Z}_2)$ such that

(i) η lifts to an integral class $c \in H^2(Z)$
(ii) For any integral lift c of η we have $c \cdot c \equiv \ell_Z^k \pmod 4$
(iii) $\eta \neq 0$.

Further, we let $\widehat{C}_Z^k$ be the set of integral lifts $\mathbf{c}$ of elements in C_Z^k modulo the equivalence $\mathbf{c} \sim \mathbf{c}'$ if $\mathbf{c} \equiv \mathbf{c}' \pmod 2$ and $(\frac{1}{2}(\mathbf{c} - \mathbf{c}'))^2 \equiv 0 \pmod 2$.

Let $\alpha = \sum a_i\sigma_i$ be a smooth singular cycle representing a class in $H_k(\mathrm{Diff}(Z))$. For a generic metric g_* on Z one gets a smooth singular cycle in $C_k(\mathrm{Met}(Z))$ given by $\alpha^* g_* = \sum a_i\sigma_i^* g_*$. As $H_k(\mathrm{Met}(Z)) = 0$, this is the boundary of a $(k+1)$-chain $\beta = \sum b_j\tau_j$. Informally, we have an invariant

$$D_{\mathbf{c},\beta}^{H_k\,\mathrm{Diff}} \colon H^k(\mathrm{Diff}(Z)) \to \mathbb{Z},$$

given by

$$D_{\mathbf{c},\beta}^{H_k\,\mathrm{Diff}}(\alpha) = \sum b_j \# \mathcal{M}_{\mathbf{c},\beta}(\{\tau_j\}).$$

To make the definition of this invariant rigorous, a variation of smooth singular chains that incorporates suitable transversality may be used. This is the approach described in [**7**]. Composition with the Hurewicz homomorphism $\pi_k(\mathrm{Diff}(Z)) \to H_k(\mathrm{Diff}(Z))$ produces an invariant for the homotopy groups of the diffeomorphism group.

In [**40**] Konno defined a version of the degree zero Donaldson invariants for families of manifolds. His definition yields a cohomology class on $B\,\mathrm{Diff}(Z)$ (or precisely on the classifying space for a subgroup of $\mathrm{Diff}(Z)$ that preserves bundle data and homology orientations). He also describes it as a characteristic class for bundles of 4-manifolds. Konno's invariant is based on a CW representation of homology, and uses a family version of fundamental cycles. Given $\mathbf{c} \in \widehat{C}_Z^k$ and a homology orientation β on Z, the version of his invariant that we use for a $(k+1)$-dimensional family $Z \hookrightarrow \mathcal{Z} \to B$ is denoted by $\mathbb{D}_{\mathbf{c},\beta}(\mathcal{Z})$. Essentially, it assigns to each $(k+1)$-cell a count of ASD connections in a parameterized moduli space. Konno's invariants play the role of $\mathbb{I}_{\mathbf{c}_q}(\mathcal{Z}[\![J]\!])$ from (4.1). The invariants playing the role of $I_{\mathbf{c}_q}^{H_{k+1}\,\mathrm{Emb}}$ in item (1) of the recipe from Section 4 are given by

$$D_{\mathbf{c},\beta}^{H_{k+1}\,\mathrm{Emb}}(J) = \mathbb{D}_{\mathbf{c},\beta}(\mathcal{Z}[\![J]\!]).$$

The following proposition establishes one relation between the family and diffeomorphism invariants.

PROPOSITION 5.2. *For* $\alpha \in \pi_k(\mathrm{Diff}(Z))$ *one has*

$$D^{H_k\,\mathrm{Diff}}_{\mathbf{c},\beta}(\alpha) = \mathbb{D}_{\mathbf{c},\beta}(\mathcal{Z}[\alpha]).$$

REMARK 5.3. Intuitively, this proposition is clear. In simple situations one may represent a singular chain by a map from a CW complex into the space. Given transversality, the virtual neighborhood is trivial and the two definitions will agree. Drouin includes a proof of this proposition in [**22**].

To compute the invariants $D^{H_{k+1}\,\mathrm{Emb}}_{\mathbf{c},\beta}(J)$, we combine Proposition 5.2 with the homomorphism properties of $D^{H_k\,\mathrm{Diff}}$ and following result from [**7**].

THEOREM 5.4 (Anti-holomorphic blow-up). *Let* $\beta\colon S^k \to \mathrm{Diff}(Z, D)$ *be a family so that* $\beta \# 1_{\mathbf{M}}$ *smoothly contracts via a contraction* F. *We then have*

$$|D^{\pi_{k+1}}_{\mathbf{c}+\boldsymbol{\nu}}(F(1_Z \# \boldsymbol{f})(F)^{-1})| = 2|D^{\pi_k}_{\mathbf{c}}(\beta)|.$$

Putting everything together this implies that there are constants γ'_p so that

$$D^{H_{k+1}\,\mathrm{Emb}}_{\mathbf{c}_q,\beta}(J^{k+1}_p) = \gamma'_p\delta_{pq}.$$

6. Proof of Theorem 1.3

Sections 4 and 5 nearly complete the proof of Theorem 1.3. The only missing piece is that one may take one target manifold Z for which there are large rank summands in the kernels of the homology and homotopy groups for a range of indices.

As we have defined them, the manifolds Z'_k grow as k grows. Picking an upper bound n we can pick a new family of manifolds V_k so that

$$Z'_k \#_{\Sigma_1} V_k \cong Z''_n$$

independent of k for $k \leqslant n$. Provided this only changes the invariants by a non-zero factor, the same argument will work. The reason this construction only addresses homotopy or homology groups in degree of a fixed parity is that the invariant $D^{H_{k+1}\,\mathrm{Emb}}_{-,\beta}$ can only be non-zero when $b^2_+(Z) \equiv k+1 \pmod 2$. This argument is written out in more detail in [**7**]. □

REMARK 6.1. The reason we just obtain finite-rank summands is that $\widehat{C}^k_Z$ is finite. The Seiberg-Witten version of the family invariants may be parameterized by the (infinite) collection of spin^c structures. Thus, in a forthcoming paper we describe how to compute the relevant Seiberg-Witten invariants that establish the existence of infinite-rank summands of homotopy classes of embeddings into a fixed 4-manifold. It is not unreasonable to imagine that a family version of the Bauer-Furuta invariants could establish that the kernels of all of the homotopy and homology groups in a range contain infinite-rank summands.

To construct exotic embeddings of a 3-manifold Y we used an embedding of Y into a 4-manifold V with non-trivial Donaldson invariants. This leads to the following question (see also [**23**]).

QUESTION 6.2. Does every 3-manifold embed into some 4-manifold with non-trivial Donaldson invariants? Non-trivial Seiberg-Witten invariants? Non-trivial Bauer-Furuta invariants? As the referee points out, we should require that the 3-manifold separates to avoid a trivial answer such as $Y \hookrightarrow (S^1 \times Y) \# X$ where X has a non-trivial gauge-theoretic invariant.

It is known [**11**] that there are 3-manifolds that do not embed smoothly in any symplectic 4-manifold, blocking an easy route to a positive answer to this question.

7. Families of submanifolds

The previous section gave explicit families of *embeddings* of submanifolds; by definition this includes the parameterization. Our goal is to show that in some cases, these families of embeddings give rise to families of *submanifolds*. This means that we need to understand any possible dependence of the invariants from the choice of parameterization, or in other words, of automorphisms of the submanifold. In the following discussion, we formalize the family version of automorphisms of the submanifold. Let $Y \hookrightarrow \mathcal{Y} \to \Xi$ be a family of manifolds and $Z \hookrightarrow \mathcal{Z} \to \Xi$ be a family of manifolds with $\dim(Z) \geqslant \dim(Y)$. Let $\mathrm{Aut}(\mathcal{Y})$ be the group of bundle automorphisms of $\mathcal{Y}$.

DEFINITION 7.1. A family of embeddings is a bundle map $J\colon \mathcal{Y} \to \mathcal{Z}$. Define an equivalence relation on families of embeddings by $J \sim J'$ if there is $\xi \in \mathrm{Aut}(\mathcal{Y})$ so that $J\xi = J'$. A family of submanifolds is an equivalence class of a family of embeddings.

REMARK 7.2. We denote the space of family embeddings by $\mathcal{E}\mathrm{mb}(\mathcal{Y},\mathcal{Z})$. In this case the family of submanifolds with given bundles of manifolds is identified with

$$\mathcal{E}\mathrm{mb}(\mathcal{Y},\mathcal{Z})/\mathrm{Aut}(\mathcal{Y}).$$

We call a family of embeddings representing a given family of submanifolds a *marking*.

The first invariant of a family of submanifolds is the underlying bundle $\mathcal{Y}$. Such bundles are classified by homotopy classes of maps to $\mathrm{BDiff}(Y)$. The identification $\Omega\mathrm{BDiff}(Y) \simeq \mathrm{Diff}(Y)$ provides a good understanding of the homotopy type of $\mathrm{BDiff}(Y)$. For example if Y is a hyperbolic 3-manifold, $\mathrm{Diff}(Y)$ is homotopy equivalent to $\mathrm{Isom}(Y)$, so $\mathrm{BDiff}(Y)$ is a $K(\mathrm{Isom}(Y),1)$. It would be appropriate to restrict to orientation-preserving diffeomorphisms and isometries to study oriented families of submanifolds.

Now turn to the question of distinguishing families of submanifolds with the same underlying bundle of spaces. For simplicity assume this bundle is trivial, so $\Xi \times Y$. The basic invariant for families of embeddings was to consider a family of spaces constructed via family submanifold sum, $\mathcal{Z} \#_J \mathcal{V}$. The trouble with this approach for families of submanifolds is that the isomorphism class of the family of spaces may depend upon the representative of the family of submanifolds. One solution to this issue is to take the families associated to all markings of the submanifold. Since markings that differ by elements of $\mathrm{Aut}(\mathcal{Y})_0$ give rise to the same families of maps, a better option is to define a homotopy marking as an equivalence class of marking up to $\mathrm{Aut}(\mathcal{Y})_0$. The natural invariant of a family of submanifolds is the multiset (set with multiplicities) of all families of spaces obtained by taking the submanifold sum over all homotopy markings. If that multiset is finite, then one can simplify further by taking the sum of invariants over the multiset.

A similar idea works when considering higher-dimensional families of submanifolds The higher homotopy groups are groups, so it is natural to look for an invariant in the form of a group homomorphism. If the diffeomorphism group of Y has finitely many components (eg if Y is hyperbolic) then sum of the invariants over all markings will give an integer-valued homomorphism from the homology groups of the space of submanifolds.

7.1. Examples. We now consider some examples to illustrate the difference between detecting families of embeddings and families of submanifolds. In this section we will return to the choice of the manifolds X_p as the $(2p+1)$-log transforms of $E(2)$. This will simplify the discussion, and provide equally powerful results once combined with the forthcoming work on the Seiberg-Witten invariants. The first example illustrates that the invariants for $\pi_0(\mathrm{Emb})$ defined in section 3.1 can in fact depend on the marking. We address this in two ways: one is by considering the multiset of invariants, and the other is by taking the sum over all markings (when the set of markings is finite). The second example is one where the diffeomorphism group of Y is contractible, so there is only one marking. It appears as the second example because it is constructed by modifying the first example. In fact, we describe a way to replace a given embedded submanifold by one with a contractible diffeomorphism group.

EXAMPLE 7.3. The first class of examples is constructed with the submanifold $Y = \partial W$ where W is the Akbulut-Mazur cork. To start with, we will consider the dependence of our $\pi_0(\mathrm{Emb})$ invariants on the parameterization.

To understand the diffeomorphism group of Y, we use the surgery picture in Figure 7 in which all of the symmetries are visible. Combining results about surgery

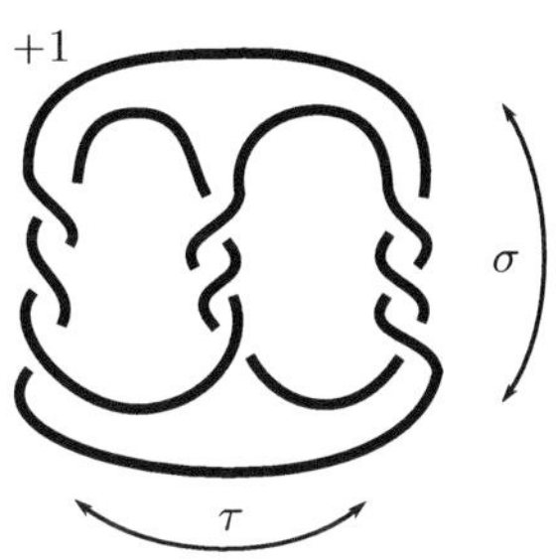

FIGURE 7. The boundary of the Akbulut-Mazur manifold

on Montesinos knots shows that Y is hyperbolic [**50, 51, 70**]. A calculation with SnapPy shows further that

$$\mathrm{Diff}(Y) \simeq \mathrm{Isom}(Y) = \langle \sigma, \tau \,|\, \sigma^2 = \tau^2 = [\sigma, \tau] = 1 \rangle$$

where σ and τ are exhibited in Figure 7. This calculation, because it takes place on a closed manifold, is not completely rigorous. Forthcoming work of the second author with N. Hoffman gives a method to transform such calculations into a calculation of the symmetry group of an associated link, which can be done rigorously.

The diffeomorphism σ extends over W, while the diffeomorphism τ is the cork twist. By [**2**], the manifold W embeds into $V = E(2) \# \overline{\mathbb{C}P}^2$ so that

$$(E(2) \# \overline{\mathbb{C}P}^2 \backslash \mathrm{int}(W)) \cup_\tau W \cong 3\mathbb{C}P^2 \# 20\overline{\mathbb{C}P}^2$$

from which it follows that τ does not extend over W. Moreover, it is known [**31**, Figure 9.4] that one can take this embedding to be disjoint from a fiber.

Set $V_1 = W$ and $U = S^2 \times S^2$. Label the three nuclei in the original $E(2)$ as $N^{(0,a)}$ for $a = 1, 2, 3$, and label the nuclei in the transformed copy of $E(2)$ by $N^{(1,a)}$

for $a = 1, 2, 3$. Denote the section classes by $\sigma^{(i,a)}$ and the fiber classes by $f^{(i,a)}$. After the log transforms to obtain $E(2)_{\infty;\infty;2p+1}$ we will have $N^{(1,3)}_{2p+1}$. Figure 8 displays the result of a $(2p+1)$-log transform on a nucleus. Let $f^{(3)}$ denote the 0-framed handle and $s^{(3)}$ denote the $(-8p^2-6p-2)$-framed handle in the displayed figure of $N^{(1,3)}_{2p+1}$.

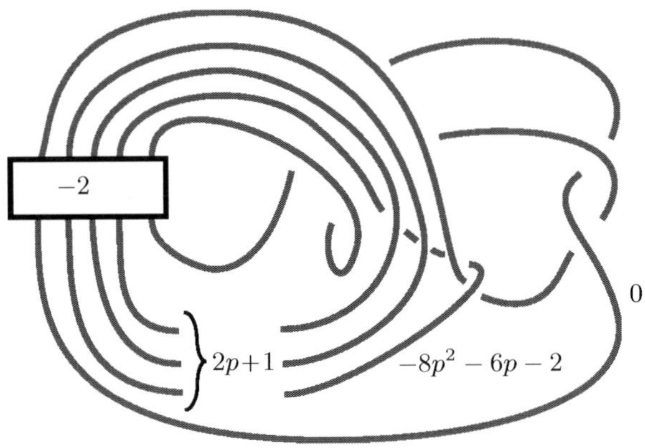

FIGURE 8. The Gompf nucleus after log transform $N^{(1,3)}_{2p+1}$

By the work of Freedman, there is a homeomorphism rel boundary $\zeta\colon N^{(1,3)}_{2p+1} \to N^{(1,3)}$ that matches the multiple fiber class $f^{(3)}$ in the transformed nucleus with the fiber class in the nucleus on the other side. Furthermore, the class $s^{(3)}$ is sent to $\sigma^{(1,3)} - (4p^2+3p)f^{(1,3)}$. Notice that after a log transform, the generic class is no longer primitive. Indeed the multiple fiber is primitive, and the generic fiber is a multiple of the multiple fiber. In this case the generic fiber is in the homology class of $(2p+1)f^{(3)}$ and $f^{(3)}$ is primitive. See [**31**, Remark 3.3.5]

In this case we have $\psi_p\colon E(2)_{\infty;\infty;2p+1} \to E(2)$ extending ζ by the identity. We also have a diffeomorphism $\varphi_p\colon E(2)_{\infty;\infty;2p+1} \#_T \widehat{V} \to E(2) \#_T \widehat{V}$ as in the introduction. We then have the embeddings $j_p\colon Y \to E(2)\#_T\widehat{V}$ with $j_p = \varphi_p(\psi_p\#_T 1_{\widehat{V}})^{-1}j$, and the submanifolds $j_p(Y)$. We also have a model embedding $k\colon Y \to V$ that we can use to form submanifold sums to distinguish the submanifolds in X.

If we now consider the submanifolds $j_p(Y)$ we can still construct the submanifold sums, but the answer depends on the choice of marking, so we compare the manifolds obtained as submanifold sum using the markings $\{j_p, j_p\sigma, j_p\tau, j_p\tau\sigma\}$ for various p. The preferred components of the submanifold sums for the markings associated to $j_p(Y)$ are:

$$\{E(2)_{\infty;\infty;2p+1} \#_T V, E(2)_{\infty;\infty;2p+1} \#_T V, 7\mathbb{C}P^2 \# 40\overline{\mathbb{C}P}^2, 7\mathbb{C}P^2 \# 40\overline{\mathbb{C}P}^2\}.$$

In all cases the discarded components are $4\mathbb{C}P^2 \# 21\overline{\mathbb{C}P}^2$.

Let $\mathbf{c}$ be the element of $\widehat{C}_{E(2)\#_T V}$ that restricts to the Poincaré duals of $\sigma^{(0,1)}$ on the $E(2)$ summands (recall that $V = E(2) \# \overline{\mathbb{C}P}^2$). Let β be the homology orientation induced by the submanifold sum. Applying the degree zero Donaldson invariant $q_{-,\mathbf{c},\beta}$ to this set of manifolds gives $\{2p+1, 2p+1, 0, 0\}$. The dependence of the multiset of Donaldson invariants on p demonstrates that the submanifolds for different values of p are not smoothly isotopic. As remarked above, since the number of homotopy markings is finite in this case, we could alternatively sum the invariants over all homotopy markings.

The general construction described in Section 6 now gives spherical families J_p^{k+1} of embeddings of Y. These in turn generate the families of submanifolds given by $J_p^{k+1}(Y)$. As in the case of π_0 the multiset of submanifold sums taken over all markings will have four elements. Applying the degree zero family Donaldson invariants $\mathbb{D}_{\mathbf{c}_q,\beta}$ will result in the multiset of values $\{\gamma_p\delta_{pq}, \gamma_p\delta_{pq}, 0, 0\}$ for exactly the same reason. The first two elements arise from $J_p^{k+1}(Y)$ and $J_p^{k+1}\sigma(Y)$. The last two zero values arise from $J_p^{k+1}\tau(Y)$ and $J_p^{k+1}\tau\sigma(Y)$. The sum over all homotopy markings then gives non-trivial homomorphisms to $\mathbb{Z}$; using different choices for $\mathbf{c}_q$ gives a surjection from the homology group of the space of submanifolds to free abelian groups of arbitrarily high (finite) rank.

The easiest case to analyze is when the diffeomorphism group of Y is contractible. This occurs when Y is hyperbolic with no symmetries (i.e. is *asymmetric.*) We start with a construction of such a Y as a modification of the Akbulut cork. The construction works in much greater generality, but we will content ourselves with a particular case. Recall that cobordism P from a manifold M to a manifold N is invertible if there is a cobordism Q from N to M with $P \cup_N Q \cong M \times I$. There is a similar notion of invertible concordance for knots, links, and tangles, where in the latter setting one requires a fixed product structure along the boundary. The relevance of invertible cobordisms to embedding problems is the following obvious lemma.

LEMMA 7.4. *A co-oriented embedding $j : N^{m-1} \hookrightarrow X^m$ and an invertible cobordism from M to N gives rise to a co-oriented embedding $M \hookrightarrow X$.*

EXAMPLE 7.5. This example exhibits a method for converting a 3-manifold N into one with a trivial symmetry group such that there is an invertible cobordism from N to the new manifold. The idea is to insert a tangle with trivial symmetry group into a surgery diagram for N. The method seems to work in some generality and can easily be applied to particular examples.

LEMMA 7.6. *Let Y be the Akbulut-Mazur cork, given as $(+1)$ surgery on the pretzel knot $P(-3,3,-3)$ as drawn in Figure 7. Then there is an invertible homology cobordism from Y to an asymmetric hyperbolic homology sphere. More precisely, there are infinitely many distinct asymmetric hyperbolic homology spheres Y_n, and invertible cobordisms from Y to Y_n.*

PROOF. The manifold Y_n is given by the following surgery diagram on the two component link $L = (K, A)$. A calculation with SnapPy [**63**] shows that L is hyperbolic with trivial symmetry group and moreover the manifold obtained by Dehn surgery on K with coefficient -1 has the same properties. The calculations are done via interval arithmetic and are therefore rigorous. A standard argument [**32**] using Thurston's Dehn surgery theorem shows that for sufficiently large n, the manifolds Y_n are asymmetric, hyperbolic, and distinct.

It remains to construct the invertible cobordism; the construction is similar to the use of the 'KT grabber' in [**12**]. Note first that doing $-1/n$ surgery on A turns K into a knot K_n that has n full twists in the region where it passes through a disk bounded by K. We will show that for all n, the pretzel knot $P(-3,3,-3)$ is invertibly concordant to K_n. Note that the knot K_n splits along the Conway

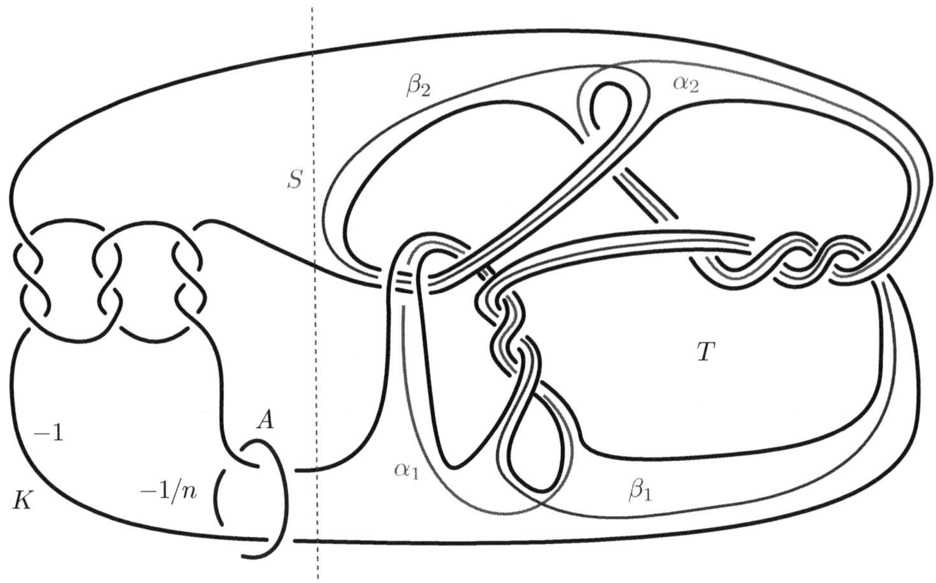

FIGURE 9. The $P(-3,3,-3)$ knot with DS tangle T and augmentation A

sphere indicated by the vertical dotted line into a sum of two tangles. On the left is a tangle whose closure is the pretzel knot $P(-3,3,-3)$; let us call the portion on the right the tangle T. If one replaces T by a trivial tangle, the resulting knot is the pretzel knot, because the extra twists come undone.

There is an invertible cobordism from the trivial tangle to T, constructed as in the proof of [**61**, Theorem 2.6]. The main points in verifying the invertibility can be seen in Figure 9, which is a variation of [**61**, Figure 1]. (Additional twists have been added to the bands to break the symmetries of the older picture.) The tangle T has two components, each of which bounds a genus 1 relative Seifert surface (with part of the boundary along T and the other part along the sphere S) and that those surfaces are disjoint. Observe that the Seifert form on homology of the lower surface has a hyperbolic basis $\{\alpha_1, \beta_1\}$, with a similar basis $\{\alpha_2, \beta_2\}$ for the upper surface.

View T as lying in $B^3 \times \{1/2\} \subset B^3 \times I$. If one does surgery on either $\{\alpha_1, \alpha_2\}$ or $\{\beta_1, \beta_2\}$, both of those surfaces become disks. Since $\{\alpha_1, \alpha_2\}$ is a trivial link one can add 2-handles in $B^3 \times [0, 1/2]$ to do the surgery, and similarly add 2-handles in $B^3 \times [1/2, 1]$ to do surgery along $\{\beta_1, \beta_2\}$. The union of the traces of the two surgeries yields two copies of $D^2 \times I$ meeting $S \times I$ in a pair of arcs $\times I$.

Gluing this cobordism to the product cobordism over the left side of the picture gives an invertible cobordism between the pretzel knot and K_n. □

EXAMPLE 7.7. Recall that the Akbulut-Mazur cork in $E(2) \# \overline{\mathbb{C}P}^2$ is disjoint from the first Gompf nucleus. As $-\partial N(2) = \Sigma(2,3,11)$, we see that the Brieskorn sphere $Y = \Sigma(2,3,11)$ embeds into $E(2) \# \overline{\mathbb{C}P}^2$ separating the torus used for fiber sums from the Akbulut-Mazur cork. By the same argument as in the application of Theorem 1.3 in Example 7.3, this embedding of Y gives rise to r distinct isotopy classes of embeddings j_p distinguished Donaldson invariants $\mathbb{D}_{\mathbf{c}_q,\beta}$. Likewise, there are rank r large summands in the homotopy and homology groups of $\mathrm{Emb}(Y, Z_r^k)$.

Now consider the extension of these results to spaces of submanifolds; the key issue again is to understand the diffeomorphism group of Y. It is known [**13**] that the mapping class group of Y is $\mathbb{Z}_2$. Moreover, the generator, say τ is represented by complex conjugation in a standard description of Y as a link of an isolated singularity. By [**49**] each component of the group $\mathrm{Diff}(\Sigma(2,3,11))$ has the homotopy type of a circle.

We first claim that the submanifolds $j_p(Y)$ are not isotopic. To show this, we note that Gompf showed [**27**, Lemma 3.7] that τ extends to a self-diffeomorphism of $N(2)$. It follows that j_p and $j_p\tau$ have the same Donaldson invariants, so the multisets of invariants distinguish the submanifolds $j_p(Y)$. With regard to the higher-dimensional families, the long exact sequence

$$\cdots \to \pi_n(\mathrm{Diff}(Y)) \to \pi_n(\mathrm{Emb}(Y, Z_k)) \to \pi_n(\mathrm{Sub}(Y, Z_k)) \to \pi_{n-1}(\mathrm{Diff}(Y)) \to \cdots$$

shows that the families $J_p^{j+1} \in \pi_{j+1}(\mathrm{Emb}(Y, Z_r^k))$ are independent.

Acknowledgments

These results were inspired in part by conversations at MSRI with Hokuto Konno, Anubhav Mukherjee, and Masaki Taniguchi. We thank Chuck Livingston and Jeff Meier for some helpful correspondence about doubly slice knots, and the referee for a careful reading of our original submission.

References

[1] Selman Akbulut and John D. McCarthy, *Casson's invariant for oriented homology 3-spheres: An exposition*, Mathematical Notes, vol. 36, Princeton University Press, Princeton, NJ, 1990, DOI 10.1515/9781400860623. MR1030042

[2] Selman Akbulut, *A fake compact contractible 4-manifold*, J. Differential Geom. **33** (1991), no. 2, 335–356. MR1094459

[3] Selman Akbulut, *On infinite order corks*, Proceedings of the Gökova Geometry-Topology Conference 2016, Gökova Geometry/Topology Conference (GGT), Gökova, 2017, pp. 151–157. MR3676087

[4] D. Auckly, *Smoothly knotted surfaces that remain distinct after many internal stabilizations*, Preprint, arXiv:2307.16266, 2023.

[5] Dave Auckly, Hee Jung Kim, Paul Melvin, and Daniel Ruberman, *Stable isotopy in four dimensions*, J. Lond. Math. Soc. (2) **91** (2015), no. 2, 439–463, DOI 10.1112/jlms/jdu075. MR3355110

[6] Dave Auckly, Hee Jung Kim, Paul Melvin, and Daniel Ruberman, *Equivariant hyperbolization of 3-manifolds via homology cobordisms*, Topology Appl. **333** (2023), Paper No. 108485, 24, DOI 10.1016/j.topol.2023.108485. MR4579944

[7] D. Auckly and D. Ruberman, *Families of diffeomorphisms and embeddings via Donaldson theory*. manuscript in preparation, 2023.

[8] D. Auckly and D. Ruberman, *Families of diffeomorphisms, embeddings, and positive scalar curvature metrics via Seiberg-Witten theory*. manuscript in preparation, 2023.

[9] David Baraglia, *Non-trivial smooth families of K3 surfaces*, Math. Ann. **387** (2023), no. 3-4, 1719–1744, DOI 10.1007/s00208-022-02508-3. MR4657435

[10] David Baraglia and Hokuto Konno, *A note on the Nielsen realization problem for K3 surfaces*, Proc. Amer. Math. Soc. **151** (2023), no. 9, 4079–4087, DOI 10.1090/proc/15544. MR4607650

[11] A. Daemi, T. Lidman, and M. M. Eismeier, *3-manifolds without any embedding in symplectic 4-manifolds*, Preprint, arXiv:2207.03631, 2022.

[12] Steven A. Bleiler, *Realizing concordant polynomials with prime knots*, Pacific J. Math. **100** (1982), no. 2, 249–257. MR669325

[13] Michel Boileau and Jean-Pierre Otal, *Scindements de Heegaard et groupe des homéotopies des petites variétés de Seifert* (French), Invent. Math. **106** (1991), no. 1, 85–107, DOI 10.1007/BF01243906. MR1123375

[14] Boris Botvinnik and Tadayuki Watanabe, *Families of diffeomorphisms and concordances detected by trivalent graphs*, J. Topol. **16** (2023), no. 1, 207–233. MR4575866

[15] Ryan Budney, *JSJ-decompositions of knot and link complements in S^3*, Enseign. Math. (2) **52** (2006), no. 3-4, 319–359. MR2300613

[16] R. Budney and D. Gabai, *Knotted 3-balls in S^4*, Preprint, `arXiv:1912.09029`, 2019.

[17] Tim D. Cochran and Nathan Habegger, *On the homotopy theory of simply connected four manifolds*, Topology **29** (1990), no. 4, 419–440, DOI 10.1016/0040-9383(90)90014-B. MR1071367

[18] Anthony Conway and Mark Powell, *Embedded surfaces with infinite cyclic knot group*, Geom. Topol. **27** (2023), no. 2, 739–821, DOI 10.2140/gt.2023.27.739. MR4589564

[19] C. L. Curtis, M. H. Freedman, W. C. Hsiang, and R. Stong, *A decomposition theorem for h-cobordant smooth simply-connected compact 4-manifolds*, Invent. Math. **123** (1996), no. 2, 343–348, DOI 10.1007/s002220050031. MR1374205

[20] S. K. Donaldson, *Irrationality and the h-cobordism conjecture*, J. Differential Geom. **26** (1987), no. 1, 141–168. MR0892034

[21] S. K. Donaldson and P. B. Kronheimer, *The geometry of four-manifolds*, Oxford Mathematical Monographs, The Clarendon Press, Oxford University Press, New York, 1990. Oxford Science Publications. MR1079726

[22] J. Drouin, *Exotic families of embedded spheres*, 2023. to appear.

[23] John B. Etnyre, Hyunki Min, and Anubhav Mukherjee, *On 3-manifolds that are boundaries of exotic 4-manifolds*, Trans. Amer. Math. Soc. **375** (2022), no. 6, 4307–4332, DOI 10.1090/tran/8586. MR4419060

[24] Michael Hartley Freedman, *The topology of four-dimensional manifolds*, J. Differential Geometry **17** (1982), no. 3, 357–453. MR679066

[25] Ronald Fintushel and Ronald J. Stern, *Homotopy $K3$ surfaces containing $\Sigma(2,3,7)$*, J. Differential Geom. **34** (1991), no. 1, 255–265. MR1114462

[26] Ronald Fintushel and Ronald J. Stern, *Donaldson invariants of 4-manifolds with simple type*, J. Differential Geom. **42** (1995), no. 3, 577–633. MR1367403

[27] Robert E. Gompf, *Nuclei of elliptic surfaces*, Topology **30** (1991), no. 3, 479–511, DOI 10.1016/0040-9383(91)90027-2. MR1113691

[28] Robert E. Gompf, *Infinite order corks*, Geom. Topol. **21** (2017), no. 4, 2475–2484, DOI 10.2140/gt.2017.21.2475. MR3654114

[29] Robert E. Gompf, *Infinite order corks via handle diagrams*, Algebr. Geom. Topol. **17** (2017), no. 5, 2863–2891, DOI 10.2140/agt.2017.17.2863. MR3704246

[30] Robert E. Gompf and Tomasz S. Mrowka, *Irreducible 4-manifolds need not be complex*, Ann. of Math. (2) **138** (1993), no. 1, 61–111, DOI 10.2307/2946635. MR1230927

[31] Robert E. Gompf and András I. Stipsicz, *4-manifolds and Kirby calculus*, Graduate Studies in Mathematics, vol. 20, American Mathematical Society, Providence, RI, 1999, DOI 10.1090/gsm/020. MR1707327

[32] Craig D. Hodgson and Jeffrey R. Weeks, *Symmetries, isometries and length spectra of closed hyperbolic three-manifolds*, Experiment. Math. **3** (1994), no. 4, 261–274. MR1341719

[33] N. Iida, H. Konno, A. Mukherjee, and M. Taniguchi, *Diffeomorphisms of 4-manifolds with boundary and exotic embeddings*, Preprint, `arXiv:2203.14878`, 2022.

[34] William H. Jaco and Peter B. Shalen, *Seifert fibered spaces in 3-manifolds*, Mem. Amer. Math. Soc. **21** (1979), no. 220, viii+192, DOI 10.1090/memo/0220. MR0539411

[35] Klaus Johannson, *Homotopy equivalences of 3-manifolds with boundaries*, Lecture Notes in Mathematics, vol. 761, Springer, Berlin, 1979. MR551744

[36] S. Kang, *One stabilization is not enough for contractible 4-manifolds*, Preprint, `arXiv:2210.07510`, 2022.

[37] Hee Jung Kim and Daniel Ruberman, *Topological triviality of smoothly knotted surfaces in 4-manifolds*, Trans. Amer. Math. Soc. **360** (2008), no. 11, 5869–5881, DOI 10.1090/S0002-9947-08-04482-6. MR2425695

[38] user101010 (`https://mathoverflow.net/users/99414/user101010`), *Non-isotopic homology spheres in S^4 with equal complements?* MathOverflow. `https://mathoverflow.net/q/412704` (version: 2021-12-29).

[39] C. Livingston and A. H. Moore, *Knotinfo: Table of knot invariants*, 2023. URL: `https://knotinfo.math.indiana.edu`.

[40] Hokuto Konno, *Characteristic classes via 4-dimensional gauge theory*, Geom. Topol. **25** (2021), no. 2, 711–773, DOI 10.2140/gt.2021.25.711. MR4251435

[41] H. Konno, A. Mukherjee, and M. Taniguchi, *Exotic codimension-1 submanifolds in 4-manifolds and stabilizations*, Preprint, `arXiv:2210.05029`, 2022.

[42] M. Kreck, *Isotopy classes of diffeomorphisms of $(k-1)$-connected almost-parallelizable $2k$-manifolds*, Algebraic topology, Aarhus 1978 (Proc. Sympos., Univ. Aarhus, Aarhus, 1978), Lecture Notes in Math., vol. 763, Springer, Berlin, 1979, pp. 643–663. MR561244

[43] P. B. Kronheimer and T. S. Mrowka, *The Dehn twist on a sum of two $K3$ surfaces*, Math. Res. Lett. **27** (2020), no. 6, 1767–1783, DOI 10.4310/MRL.2020.v27.n6.a8. MR4216604

[44] Jianfeng Lin, *Isotopy of the Dehn twist on $K3 \# K3$ after a single stabilization*, Geom. Topol. **27** (2023), no. 5, 1987–2012, DOI 10.2140/gt.2023.27.1987. MR4621924

[45] Jianfeng Lin, *The family Seiberg-Witten invariant and nonsymplectic loops of diffeomorphisms*, Preprint, `arXiv:2208.12082`, 2022.

[46] J. Lin and A. Mukherjee, *Family Bauer–Furuta invariant, Exotic Surfaces and Smale conjecture*, Preprint, `arXiv:2110.09686`, 2021.

[47] Charles Livingston and Jeffrey Meier, *Doubly slice knots with low crossing number*, New York J. Math. **21** (2015), 1007–1026. MR3425633

[48] R. Matveyev, *A decomposition of smooth simply-connected h-cobordant 4-manifolds*, J. Differential Geom. **44** (1996), no. 3, 571–582. MR1431006

[49] Darryl McCullough and Teruhiko Soma, *The Smale conjecture for Seifert fibered spaces with hyperbolic base orbifold*, J. Differential Geom. **93** (2013), no. 2, 327–353. MR3024309

[50] Jeffrey Meier, *Small Seifert fibered surgery on hyperbolic pretzel knots*, Algebr. Geom. Topol. **14** (2014), no. 1, 439–487, DOI 10.2140/agt.2014.14.439. MR3158766

[51] William W. Menasco and Morwen B. Thistlethwaite, *Surfaces with boundary in alternating knot exteriors*, J. Reine Angew. Math. **426** (1992), 47–65. MR1155746

[52] John W. Morgan and Zoltán Szabó, *Embedded tori in four-manifolds*, Topology **38** (1999), no. 3, 479–496, DOI 10.1016/S0040-9383(98)00013-5. MR1670388

[53] Robert Myers, *On mapping class groups of contractible open 3-manifolds*, Trans. Amer. Math. Soc. **335** (1993), no. 1, 1–46, DOI 10.2307/2154256. MR1069740

[54] Nobuhiro Nakamura, *The Seiberg-Witten equations for families and diffeomorphisms of 4-manifolds*, Asian J. Math. **7** (2003), no. 1, 133–138, DOI 10.4310/AJM.2003.v7.n1.a8. MR2015245

[55] Nobuhiro Nakamura, *Smoothability of $\mathbb{Z} \times \mathbb{Z}$-actions on 4-manifolds*, Proc. Amer. Math. Soc. **138** (2010), no. 8, 2973–2978, DOI 10.1090/S0002-9939-10-10413-4. MR2644908

[56] B. Perron, *Pseudo-isotopies et isotopies en dimension quatre dans la catégorie topologique* (French), Topology **25** (1986), no. 4, 381–397, DOI 10.1016/0040-9383(86)90018-2. MR862426

[57] Frank Quinn, *Isotopy of 4-manifolds*, J. Differential Geom. **24** (1986), no. 3, 343–372. MR868975

[58] Daniel Ruberman, *An obstruction to smooth isotopy in dimension 4*, Math. Res. Lett. **5** (1998), no. 6, 743–758, DOI 10.4310/MRL.1998.v5.n6.a5. MR1671187

[59] Daniel Ruberman, *A polynomial invariant of diffeomorphisms of 4-manifolds*, Proceedings of the Kirbyfest (Berkeley, CA, 1998), Geom. Topol. Monogr., vol. 2, Geom. Topol. Publ., Coventry, 1999, pp. 473–488, DOI 10.2140/gtm.1999.2.473. MR1734421

[60] Daniel Ruberman, *Positive scalar curvature, diffeomorphisms and the Seiberg-Witten invariants*, Geom. Topol. **5** (2001), 895–924, DOI 10.2140/gt.2001.5.895. MR1874146

[61] Daniel Ruberman, *Seifert surfaces of knots in S^4*, Pacific J. Math. **145** (1990), no. 1, 97–116. MR1066400

[62] Gleb Smirnov, *From flops to diffeomorphism groups*, Geom. Topol. **26** (2022), no. 2, 875–898, DOI 10.2140/gt.2022.26.875. MR4444270

[63] M. Culler, N. M. Dunfield, M. Goerner, and J. R. Weeks, *SnapPy, a computer program for studying the geometry and topology of 3-manifolds.* Available at `http://snappy.computop.org` (02/02/2023).

[64] Nathan S. Sunukjian, *Surfaces in 4-manifolds: concordance, isotopy, and surgery*, Int. Math. Res. Not. IMRN **17** (2015), 7950–7978, DOI 10.1093/imrn/rnu187. MR3404006

[65] Markus Szymik, *Characteristic cohomotopy classes for families of 4-manifolds*, Forum Math. **22** (2010), no. 3, 509–523, DOI 10.1515/FORUM.2010.027. MR2652709

[66] Motoo Tange, *Notes on Gompf's infinite order corks*, Michigan Math. J. **70** (2021), no. 1, 3–21, DOI 10.1307/mmj/1586419413. MR4255088

[67] C. T. C. Wall, *Diffeomorphisms of 4-manifolds*, J. London Math. Soc. **39** (1964), 131–140, DOI 10.1112/jlms/s1-39.1.131. MR163323

[68] T. Watanabe, *Some exotic nontrivial elements of the rational homotopy groups of* $\mathrm{Diff}(S^4)$, 2018. Preprint, arXiv:1812.02448.

[69] T. Watanabe, *Theta-graph and diffeomorphisms of some 4-manifolds*, 2020. Preprint, arXiv:2005.09545

[70] Ying-Qing Wu, *The classification of toroidal Dehn surgeries on Montesinos knots*, Comm. Anal. Geom. **19** (2011), no. 2, 305–345, DOI 10.4310/CAG.2011.v19.n2.a3. MR2835882

[71] Ming Xu, *The Bauer-Furuta invariant and a cohomotopy refined Ruberman invariant*, ProQuest LLC, Ann Arbor, MI, 2004. Thesis (Ph.D.)–State University of New York at Stony Brook. MR2706507

Department of Mathematics, Kansas State University, Manhattan, Kansas 66506
Email address: dav@math.ksu.edu

Department of Mathematics, MS 050, Brandeis University, Waltham, Massachusetts 02454
Email address: ruberman@brandeis.edu

Proceedings of Symposia in Pure Mathematics
Volume **109**, 2024
https://doi.org/10.1090/pspum/109/01991

An instanton take on some knot detection results

John A. Baldwin and Steven Sivek

ABSTRACT. We give new proofs that Khovanov homology detects the figure eight knot and the cinquefoils, and that HOMFLY homology detects 5_2 and each of the $P(-3, 3, 2n + 1)$ pretzel knots. For all but the figure eight these mostly follow the same lines as in previous work. The key difference is that in honor of Tom Mrowka's 60th birthday, the arguments here use instanton Floer homology rather than knot Floer homology.

1. Introduction

Khovanov homology has been proved to detect a handful of the simplest knots, including the unknot [**KM11**] and the trefoils $T(\pm 2, 3)$ [**BS22b**]. Recently, Dowlin [**Dow18**] constructed a spectral sequence from Khovanov homology to knot Floer homology, which made it possible to prove that Khovanov homology also detects the figure eight [**BDL**$^+$**21**], the cinquefoils $T(\pm 2, 5)$ [**BHS21**], and 5_2 [**BS22a**]. In [**BS22a**] we also used it to prove that reduced HOMFLY homology detects each of the $P(-3, 3, 2n+1)$ pretzel knots.

In this note, we give alternative arguments for most of these detection results, replacing knot Floer homology and Dowlin's spectral sequence with instanton knot homology and Kronheimer and Mrowka's spectral sequence from [**KM11**]. The figure eight detection result follows quickly from known facts about instanton L-space knots, including a criterion for their detection due to Li and Liang [**LL23**]. For $T(\pm 2, 5)$, we apply this criterion together with the classification of genus-2 instanton L-space knots given in [**FRW22**, Corollary 1.8]. Our main results are the following, which do not make any use of Heegaard Floer homology.

THEOREM 1.1. *Let K be a knot whose reduced Khovanov homology $\overline{Kh}(K)$ over some field is five-dimensional and supported in δ-grading zero. Then K is the figure eight knot.*

THEOREM 1.2. *Let K be a knot whose reduced Khovanov homology $\overline{Kh}(K)$ over some field is five-dimensional and supported in the single δ-grading ± 2. Then $K = T(\pm 2, 5)$.*

Here the δ-grading on reduced Khovanov homology is defined by $\delta = q/2 - h$, where q and h are the quantum and homological gradings, respectively.

2020 *Mathematics Subject Classification.* Primary 57K18.

The first author was supported by NSF FRG Grant DMS-1952707.

THEOREM 1.3. *Let K be a knot whose reduced HOMFLY homology $\bar{H}(K;\mathbb{Q})$ is isomorphic to $\bar{H}(5_2;\mathbb{Q})$ as triply-graded vector spaces. Then $K = 5_2$.*

THEOREM 1.4. *Let K be a knot, and suppose for some $n \in \mathbb{Z}$ that $\bar{H}(K;\mathbb{Q})$ is isomorphic to $\bar{H}(P(-3,3,2n+1);\mathbb{Q})$ as triply-graded vector spaces. Then $K = P(-3,3,2n+1)$.*

REMARK 1.5. We should emphasize that the proofs of Theorem 1.2 and 1.4 are not really new; they merely replace the parts of the arguments in [**BHS21**, **BS22a**] which involve knot Floer homology. By contrast, the proof of Theorem 1.1 is genuinely different from the one in [**BDL$^+$21**]. We also point out that we do not know how to prove that Khovanov homology detects 5_2, as in [**BS22a**], without appealing to Heegaard Floer homology.

REMARK 1.6. Beliakova, Putyra, Robert, and Wagner [**BPRW22**] recently established a new spectral sequence relating HOMFLY homology with knot Floer homology. One can use their construction in place of Dowlin's spectral sequence, together with our results in [**BS22a**], to give additional alternative proofs of Theorems 1.3 and 1.4.

Building on Theorems 1.1 and 1.2, we note that a knot K with $\dim \overline{Kh}(K) = 5$ satisfies $\det(K) = 5$ if and only if $\overline{Kh}(K)$ is supported entirely in δ-gradings of a single parity (see Lemma 2.1). We do not claim here to classify all knots K for which $\overline{Kh}(K)$ is 5-dimensional and $\det(K) = 5$, but we do come close by allowing some Heegaard Floer input.

THEOREM 1.7. *Let K be a knot other than the figure eight or $T(\pm 2,5)$, and suppose that $\dim \overline{Kh}(K;\mathbb{Z}/2\mathbb{Z}) = 5$ and $\det(K) = 5$. Then K is hyperbolic, with Seifert genus 4 and Alexander polynomial*

$$\Delta_K(t) = t^4 - t^3 + 1 - t^{-3} + t^{-4}.$$

Moreover, either K or its mirror is an instanton L-space knot, hence fibered and strongly quasipositive, with signature ± 8.

REMARK 1.8. By [**BDL$^+$21**, Theorem 1], a genus-4 knot satisfying the hypotheses of Theorem 1.7 cannot be supported in a single δ-grading.

REMARK 1.9. The Alexander polynomial appearing in Theorem 1.7 is in fact the Alexander polynomial of an instanton L-space knot, namely the $(5,2)$-cable of the right-handed trefoil. But this knot $K = 13n_{4639}$ cannot satisfy the hypotheses of Theorem 1.7, since it is a satellite. Indeed, according to [**Ras05**, §7.2], we have $\dim \overline{Kh}(K;\mathbb{Q}) = 11$, and $\dim \overline{Kh}(K;\mathbb{Z}/2\mathbb{Z})$ is even larger, since $\overline{Kh}(K;\mathbb{Q})$ has nontrivial 2-torsion.

The organization is as follows. In Section 2 we prove Theorems 1.1 and 1.2. In Section 3 we prove Theorem 1.7 by first reducing it to a question about the factorization of the Alexander polynomial $\Delta_K(t)$, and then determining enough about the factorization to rule out all cases except $g(K) = 4$. Then in Section 4 we prove Theorems 1.3 and 1.4, about the HOMFLY homology of 5_2 and the pretzels $P(-3,3,2n+1)$.

2. The figure eight and the cinquefoils

We begin with the following lemmas, which are certainly well known to experts. In what follows we let $s(K) \in 2\mathbb{Z}$ denote the Rasmussen s-invariant [**Ras10**].

LEMMA 2.1. *The determinant of a knot K is determined by its δ-graded reduced Khovanov homology $\overline{Kh}(K;\mathbb{Q})$. In particular, we have*

$$\dim \overline{Kh}(K;\mathbb{Q}) \geq \det(K),$$

and equality holds if and only if $\overline{Kh}(K;\mathbb{Q})$ is supported entirely in even δ-gradings or entirely in odd δ-gradings. If in fact it is supported in a single δ-grading σ, then $s(K) = 2\sigma$.

PROOF. The claim about $s(K)$ follows from the fact that $\overline{Kh}(K;\mathbb{Q})$ is supported in one δ-grading if and only if the unreduced Khovanov homology $Kh(K;\mathbb{Q})$ is "H-thin" [**Kho03**, Proposition 3.6], meaning supported in two δ-gradings, in which case these δ-gradings are necessarily $\sigma \pm \frac{1}{2}$. Then the part of $Kh(K;\mathbb{Q})$ in homological grading $h = 0$ must be supported in quantum gradings $q = 2(h+\delta) = 2\sigma \pm 1$, and so $s(K) = 2\sigma$ by definition.

For the claim about $\det(K)$, it follows from the fact that $\overline{Kh}(K)$ categorifies the Jones polynomial [**Kho03**]:

$$V_K(t) = \sum_{q,h\in\mathbb{Z}} (-1)^h t^{q/2} \dim \overline{Kh}^{h,q}(K).$$

Then $\det(K)$ is equal to the absolute value of

$$\begin{aligned} V_K(-1) &= \sum_{q,h\in\mathbb{Z}} (-1)^h (-1)^{q/2} \dim \overline{Kh}^{h,q}(K) \\ &= \sum_{q/2-h \text{ even}} \dim \overline{Kh}^{h,q}(K) - \sum_{q/2-h \text{ odd}} \dim \overline{Kh}^{h,q}(K) \\ &= \dim \overline{Kh}^{\delta\equiv 0 \,(\mathrm{mod}\, 2)}(K) - \dim \overline{Kh}^{\delta\equiv 1 \,(\mathrm{mod}\, 2)}(K). \end{aligned}$$

Thus $\dim \overline{Kh}(K;\mathbb{Q}) \geq \det(K)$ follows immediately from the triangle inequality. Moreover, the last expression above is equal to $\pm \dim \overline{Kh}(K)$ if and only if one of the two terms is zero, or equivalently if and only if $\overline{Kh}(K)$ is supported entirely in even δ-gradings or entirely in odd δ-gradings, as claimed. □

LEMMA 2.2. *If $\overline{Kh}(K;\mathbb{F})$ is 5-dimensional for some field $\mathbb{F}$, then so is $\overline{Kh}(K;\mathbb{Q})$, and moreover the two are supported in the same δ-gradings.*

PROOF. This follows from two applications of the universal coefficient theorem. First, we use it to show that either $\overline{Kh}(K;\mathbb{Q})$ or $\overline{Kh}(K;\mathbb{Z}/p\mathbb{Z})$ is also 5-dimensional, depending on whether $\mathbb{F}$ has characteristic 0 or $p > 0$ respectively. Then in the case $\mathbb{F} = \mathbb{Z}/p\mathbb{Z}$, the lemma follows from the universal coefficient theorem, as applied to $\overline{Kh}(K;\mathbb{Z})$, and the fact that $\dim_{\mathbb{Q}} \overline{Kh}(K;\mathbb{Q})$ is odd and neither 1 nor 3, exactly as at the start of [**BHS21**, §5]. □

We next attempt to use $\overline{Kh}(K;\mathbb{Q})$ to determine the instanton knot homology $KHI(K;\mathbb{Q})$, as defined by Kronheimer and Mrowka [**KM10b**]. This invariant comes equipped with an Alexander grading

$$KHI(K) \cong \bigoplus_{i=-g}^{g} KHI(K,i),$$

where $g = g(K)$ is the genus of K, each of whose summands are $\mathbb{Z}/2\mathbb{Z}$-graded. This decomposition recovers the Alexander polynomial of K by the relation

$$\pm\Delta_K(t) = \sum_{i=-g}^{g} \chi(KHI(K,i)) \cdot t^i, \tag{2.1}$$

as proved in [**KM10a, Lim10**]. (The sign on the left comes from differing conventions for the $\mathbb{Z}/2\mathbb{Z}$ grading.). It also satisfies the following properties:

- Symmetry: $KHI(K,i) \cong KHI(K,-i)$ as $\mathbb{Z}/2\mathbb{Z}$-graded vector spaces for all i.
- Genus detection: $\dim KHI(K,g) \geq 1$.
- Fiberedness detection: $\dim KHI(K,g) = 1$ if and only if K is fibered.

The symmetry follows from the remark after [**KM10b**, Proposition 7.1], and the genus detection and fiberedness results are [**KM10b**, Proposition 7.16] and [**KM10a**, Proposition 4.1] respectively.

PROPOSITION 2.3. *Let K be a knot for which* $\dim \overline{Kh}(K;\mathbb{Q}) = 5$. *Then* $KHI(K;\mathbb{Q})$ *has total dimension 5.*

PROOF. Kronheimer and Mrowka [**KM11**, Proposition 1.2] constructed a spectral sequence

$$\overline{Kh}(K;\mathbb{Q}) \implies I^\natural(\overline{K};\mathbb{Q})$$

which converges to the singular instanton knot homology of the mirror of K. Since the rank of the latter is invariant under mirroring, this yields a rank inequality

$$\dim I^\natural(K;\mathbb{Q}) \leq \dim \overline{Kh}(K;\mathbb{Q}) = 5.$$

Moreover, we know by [**KM11**, Proposition 1.4] that $I^\natural(K;\mathbb{Q}) \cong KHI(K;\mathbb{Q})$. Thus

$$\dim KHI(K;\mathbb{Q}) \leq 5. \tag{2.2}$$

Equation (2.1) and the fact that $\Delta_K(1) = 1$ tell us that the total rank of $KHI(K)$ must be odd. It cannot be 1, because then K would be the unknot – otherwise the summands $KHI(K,g)$ and $KHI(K,-g)$ are distinct and contribute at least 1 each to $\dim KHI(K)$ – and it cannot be 3 or else K would be a trefoil [**BS22b**, Theorem 1.6]. Thus we use (2.2) to conclude that $\dim KHI(K) = 5$. □

PROPOSITION 2.4. *Let K be a knot for which* $\dim KHI(K;\mathbb{Q}) = 5$. *Then exactly one of the following is true:*

- *K has Alexander polynomial 1, and in particular* $\det(K) = 1$.
- *K is the figure eight knot.*
- *K has genus $g \geq 2$ and instanton knot homology*

$$KHI(K;\mathbb{Q}) \cong \mathbb{Q}_g \oplus \mathbb{Q}_{g-1} \oplus \mathbb{Q}_0 \oplus \mathbb{Q}_{1-g} \oplus \mathbb{Q}_{-g},$$

  where the subscripts denote the Alexander grading of each summand. In this case either K or its mirror is an instanton L-space knot.

In particular, if $\det(K) \neq 1$ *then K is fibered.*

PROOF. Supposing first that K is not fibered, then

$$\dim KHI(K,g) + \dim KHI(K,-g) \leq 5,$$

and both terms on the left are equal and greater than 1 by the symmetry and fiberedness detection properties, so

$$\dim KHI(K,g) = \dim KHI(K,-g) = 2.$$

By symmetry the remaining $\mathbb{Q}$ summand of $KHI(K)$ can only be in Alexander grading zero, so

$$KHI(K) \cong \mathbb{Q}^2_g \oplus \mathbb{Q}_0 \oplus \mathbb{Q}^2_{-g}.$$

We apply (2.1) to determine $\Delta_K(t)$. If $KHI(K,g)$ is supported in a single $\mathbb{Z}/2\mathbb{Z}$ grading, then we have

$$\Delta_K(t) = \pm 2(t^g + t^{-g}) \pm 1$$

for some signs, and this is impossible because there is no choice of signs for which $\Delta_K(1) = 1$. So $KHI(K,g)$ must have a copy of $\mathbb{Q}$ in each $\mathbb{Z}/2\mathbb{Z}$ grading, which by (2.1) tells us that $\Delta_K(t) = 1$. Then $\det(K) = |\Delta_K(-1)| = 1$.

Next, supposing that K is fibered of genus 1, then K must be a trefoil or the figure eight. But if K is a trefoil then $\dim KHI(K) = 3$, so in fact K can only be the figure eight.

Finally, if K is fibered of genus $g \geq 2$, then we proved in [**BS22b**, Theorem 1.7] that $\dim KHI(K, g-1) \geq 1$, so $KHI(K)$ is 5-dimensional and is nonzero at least in the four distinct Alexander gradings $\pm g$ and $\pm(g-1)$. Again by symmetry each of these summands must be 1-dimensional and the remaining $\mathbb{Q}$ summand must be in degree 0, so

$$KHI(K) \cong \mathbb{Q}_g \oplus \mathbb{Q}_{g-1} \oplus \mathbb{Q}_0 \oplus \mathbb{Q}_{1-g} \oplus \mathbb{Q}_{-g}$$

as claimed. Now Li and Liang [**LL23**, Theorem 1.4] proved that any knot for which KHI has this form must be an instanton L-space knot, up to mirroring, so this completes the proof. □

PROPOSITION 2.5. *Let K be a knot satisfying* $\dim \overline{Kh}(K;\mathbb{Q}) = 5$ *and* $\det(K) \neq 1$. *Then*

$$s(K) = \begin{cases} 0 & \text{if } K \text{ is the figure eight knot} \\ 2g(K) & \text{if } K \text{ is an instanton L-space knot} \\ -2g(K) & \text{if } \overline{K} \text{ is an instanton L-space knot,} \end{cases}$$

and exactly one of these cases occurs. We also have $g(K) \geq 2$ unless K is the figure eight.

PROOF. The case where K is the figure eight is immediate, so we will suppose that K is some other knot. Proposition 2.3 says that $\dim KHI(K) = 5$, so we can apply Proposition 2.4 to see that either K or its mirror is an instanton L-space knot, and that $g(K) \geq 2$.

Let us suppose that K is an instanton L-space knot, rather than $\overline{K}$. Then K is strongly quasipositive [**BS19**, Theorem 1.15], so we know from [**Pla06**, Proposition 4] or [**Shu07**, Proposition 1.7] that $s(K) = 2g(K)$. Otherwise $\overline{K}$ is an instanton L-space knot, so the same argument says that

$$s(K) = -s(\overline{K}) = -2g(\overline{K}) = -2g(K). \qquad \square$$

We are now ready to prove Theorems 1.1 and 1.2.

Proof of Theorem 1.1. Suppose that $\overline{Kh}(K;\mathbb{F})$ is 5-dimensional and supported in the single δ-grading $\sigma = 0$. Then the same is true of $\overline{Kh}(K;\mathbb{Q})$ by Lemma 2.2. Lemma 2.1 now tells us that $\det(K) = 5$ and that $s(K) = 2\sigma = 0$, so K must be the figure eight by Proposition 2.5. □

Proof of Theorem 1.2. Suppose that $\overline{Kh}(K;\mathbb{F})$ is 5-dimensional and supported in the single δ-grading $\sigma = 2$. Then the same is true of $\overline{Kh}(K;\mathbb{Q})$ by Lemma 2.2, so Lemma 2.1 says that $\det(K) = 5$ and that $s(K) = 2\sigma = 4$. Thus Proposition 2.5 says that K must be an instanton L-space knot of genus 2. Such knots are fibered, and in [**BLSY21**, §2] we gave a partial characterization of their possible monodromies; more recently, Farber, Reinoso, and Wang [**FRW22**, Corollary 1.8] used this to show that K is necessarily $T(2,5)$. We remark that if we specifically wanted to work over $\mathbb{Z}/2\mathbb{Z}$, then we could finish the proof that $K = T(2,5)$ without recourse to [**FRW22**], using instead the arguments in [**BLSY21**].

Now if $\overline{Kh}(K;\mathbb{F})$ is 5-dimensional and supported in the δ-grading $\sigma = -2$, then we apply the above to its mirror $\overline{K}$ to conclude that $\overline{K} = T(2,5)$, and hence that $K = T(-2,5)$. This completes the proof. □

3. Other determinant-5 knots

Here we address the question of whether there are knots K other than the figure eight and cinquefoils such that $\dim \overline{Kh}(K;\mathbb{Z}/2\mathbb{Z}) = 5$ and $\overline{Kh}(K;\mathbb{Z}/2\mathbb{Z})$ is supported entirely in δ-gradings of a single parity.

In this case, Lemma 2.2 and Proposition 2.5 tell us that if K is not the figure eight, then either K or its mirror is an instanton L-space knot whose genus $g(K) = \frac{1}{2}s(K)$ is at least 2. We will see that its branched double cover is a Heegaard Floer L-space, and then use recent work of Boileau, Boyer, and Gordon [**BBG19**] to put strong conditions on the Alexander polynomial of K which rule out all cases except $g(K) = 4$.

3.1. The Alexander polynomial of a thin knot.

Proposition 3.1. *Suppose that $\overline{Kh}(K;\mathbb{Q})$ is 5-dimensional and that $\det(K) = 5$, but that K is not the figure eight or $T(\pm 2,5)$. Then either K or its mirror is an instanton L-space knot, hence fibered and strongly quasipositive, and its Alexander polynomial is*

$$\Delta_K(t) = t^g - t^{g-1} + 1 - t^{1-g} + t^{-g}$$

where $g = g(K)$ is even and at least 4.

Proof. Proposition 2.3 tells us once again that $\dim KHI(K) = 5$, so that either K or its mirror is an instanton L-space knot (hence fibered and strongly quasipositive [**BS19**, Theorem 1.15]) and

$$KHI(K;\mathbb{Q}) \cong \mathbb{Q}_g \oplus \mathbb{Q}_{g-1} \oplus \mathbb{Q}_0 \oplus \mathbb{Q}_{1-g} \oplus \mathbb{Q}_{-g}$$

by Proposition 2.4. This determines the Alexander polynomial $\Delta_K(t)$ by (2.1): applying the conditions $\Delta_K(1) = 1$ and $\Delta_K(-1) = \pm\det(K) = \pm 5$, we must have

$$\Delta_K(t) = (-1)^g \left(t^g - t^{g-1} + (-1)^g - t^{1-g} + t^{-g}\right).$$

In fact, since K is an instanton L-space knot up to mirroring, we can apply [**LY21**, Theorem 1.9] to deduce that the nonzero coefficients of $\Delta_K(t)$ alternate in sign, so

g must be even and then

$$\Delta_K(t) = t^g - t^{g-1} + 1 - t^{1-g} + t^{-g}.$$

By assumption g is not $T(\pm 2, 5)$, but there are no other instanton L-space knots of genus 2 [**FRW22**, Corollary 1.8], so then g is at least 4. □

PROPOSITION 3.2. *Suppose that $\overline{Kh}(K;\mathbb{Z}/2\mathbb{Z})$ is 5-dimensional and that $\det(K) = 5$, but that K is not the figure eight or $T(\pm 2, 5)$. Then K has signature $\sigma(K) = \pm 2g(K)$, and if we write $h = \frac{1}{2}g(K)$, then h is an integer with $h \geq 2$, and the polynomial*

$$p_h(t) = t^{4h} - t^{4h-1} + t^{2h} - t + 1$$

is a product of cyclotomic polynomials.

PROOF. Lemma 2.2 says that $\overline{Kh}(K;\mathbb{Q})$ is also 5-dimensional, so Proposition 3.1 tells us that $g(K)$ is even and at least 4, so that $h \in \mathbb{Z}$ as claimed; that K is fibered and strongly quasipositive, after possibly replacing it with its mirror; and that

$$p_h(t) = t^{g(K)}\Delta_K(t),$$

so that $p_h(t)$ is a product of cyclotomic polynomials if and only if $\Delta_K(t)$ is. Boileau, Boyer, and Gordon [**BBG19**, Corollary 1.2] proved that if the branched double cover of a fibered, strongly quasipositive knot K is a Heegaard Floer L-space, then $\Delta_K(t)$ is a product of cyclotomic polynomials; this follows from their observation [**BBG19**, Proposition 6.1] that in this case K has signature $\pm 2g(K)$. Thus it suffices to show that $\Sigma_2(K)$ is a Heegaard Floer L-space.

We now apply Ozsváth and Szabó's link surgeries spectral sequence [**OS05**], and in particular the inequality

$$\det(K) \leq \dim \widehat{HF}(\Sigma_2(K);\mathbb{Z}/2\mathbb{Z}) \leq \dim \overline{Kh}(K;\mathbb{Z}/2\mathbb{Z}),$$

to conclude that $\dim \widehat{HF}(\Sigma_2(K);\mathbb{Z}/2\mathbb{Z}) = |H_1(\Sigma_2(K))| = 5$. In other words, the branched double cover $\Sigma_2(K)$ is a Heegaard Floer L-space, and the proposition follows. □

PROOF OF THEOREM 1.7. Suppose that $\overline{Kh}(K;\mathbb{Z}/2\mathbb{Z})$ is 5-dimensional and supported in δ-gradings of a single parity. Then the same is true of $\overline{Kh}(K;\mathbb{Q})$ by Lemma 2.2, and $\det(K) = 5$ by Lemma 2.1. Moreover, since K is not the figure eight by assumption, Proposition 2.5 says that $s(K) = \pm 2g(K)$ and $g(K) \geq 2$.

We apply Proposition 3.2 and see that in fact $g(K)$ is even and at least 4, that $\sigma(K) = \pm 2g(K)$, and that the polynomial $p_h(t)$ must be a product of cyclotomic polynomials, where $h = \frac{1}{2}g(K)$. In Proposition 3.5 we will prove that this is not the case for any $h \geq 3$, so we must have $h \leq 2$. But since $2h = g(K) \geq 4$ this leaves only $h = 2$, hence $g(K) = 4$ as claimed. Proposition 3.1 establishes all of the remaining conclusions except hyperbolicity. The Alexander polynomial and genus prevent K from being a torus knot, so we need only show that it cannot be a satellite; this requires substantially different techniques, so we defer it to Proposition 3.4 below. □

We need the following lemma to prove that knots satisfying the hypotheses of Theorem 1.7 cannot be satellites.

LEMMA 3.3. *Let $P \subset S^1 \times D^2$ be a knot, and suppose that there is a branched double cover*

$$S^1 \times D^2 \to S^1 \times D^2$$

with branch locus P. Then P is isotopic to the core $S^1 \times \{0\}$.

PROOF. Fix a nontrivial torus knot $T = T(p,q)$, and consider the satellite $K = P(T)$. We form the branched double cover $\Sigma_2(K)$ by taking a double cover X of $S^3 \setminus N(T)$, which may or may not be connected a priori, and gluing it to the branched double cover Σ_P of $P \subset S^1 \times D^2$. Since the latter has connected boundary, so does X, so X is connected. We also know that X is Seifert fibered over a disk, since the same is true of $S^3 \setminus N(T)$; and that it has $r = 2$ singular fibers since it is not a solid torus. (We recall that the knot complement has base orbifold $D^2(|p|, |q|)$.)

Gluing the solid torus Σ_P to X amounts to a Dehn filling of X, so by [**Hei74**, Proposition 2] there are now two possibilities:

- the Seifert fibration extends to $\Sigma_2(K)$, with at most $r + 1 = 3$ singular fibers, or
- we have filled the fiber slope, and the resulting $\Sigma_2(K)$ is a connected sum of r nontrivial lens spaces.

In the latter case, K is a connected sum of r nontrivial knots, because the branched double cover of a prime knot is prime [**HN10**, Proposition 5.1]. But as a satellite of a torus knot, it can only be composite if the pattern P has wrapping number 1, see e.g. [**Cro04**, Theorem 4.4.1]. In this case P is a connected sum of the core $C_0 = S^1 \times \{0\}$ with some other knot $K_0 \subset S^3$, and we have

$$\Sigma_P \cong \Sigma_2(C_0) \# \Sigma_2(K_0) \cong (S^1 \times D^2) \# \Sigma_2(K_0).$$

Thus $\Sigma_2(K_0) \cong S^3$, which implies that K_0 is unknotted [**Wal69**] and hence that P is isotopic to the core C_0.

In the remaining case, we know that $\Sigma_2(K)$ is a small Seifert fibered space: it has base S^2 and at most three singular fibers. We can also arrange for $\pi_1(\Sigma_2(K))$ to be infinite by taking both p and q to be large. Then a folklore result (see e.g. [**Mot17**, Proposition 3.3] for details) says that K must be either a torus knot or a Montesinos knot with three rational tangles. These are never nontrivially satellite knots – the Montesinos case is due to Oertel [**Oer84**, Corollary 4] – so either P is contained in a ball inside $S^1 \times D^2$, in which case its branched double cover cannot actually be $S^1 \times D^2$, or P is isotopic to the core circle $S^1 \times \{0\}$, as claimed. □

The following is the last remaining claim of Theorem 1.7.

PROPOSITION 3.4. *Let K be a knot such that $\dim \overline{Kh}(K; \mathbb{Z}/2\mathbb{Z}) = 5$ and $\det(K) = 5$. Then K is not a satellite knot.*

PROOF. We first show that K is prime: if $K \cong K_1 \# K_2$, then by the Künneth formula for reduced Khovanov homology over $\mathbb{Z}/2\mathbb{Z}$ we have

$$\dim \overline{Kh}(K_1) \cdot \dim \overline{Kh}(K_2) = \dim \overline{Kh}(K) = 5,$$

so $\dim \overline{Kh}(K_i) = 1$ for some i, but then K_i must be the unknot [**KM11**]. This also implies in turn that the branched double cover $\Sigma_2(K)$ is prime, as in [**HN10**, Proposition 5.1].

Now we suppose that K is a nontrivial satellite, with pattern $P \subset S^1 \times D^2$ and companion $C \subset S^3$. By assumption P cannot be isotopic to the core $S^1 \times \{0\}$, and

C cannot be the unknot. We recall from Theorem 1.7 that K (up to mirroring) is fibered and strongly quasipositive, and from the proof of Proposition 3.2 that $\Sigma_2(K)$ is a Heegaard Floer L-space. It thus follows from [**BBG19**, Proposition 6.2 and Remark 6.3] that P must have winding number 1.

Since P has odd winding number, we can write $\Sigma_2(K)$ as a union

$$\Sigma_2(K) \cong X_2(C) \cup_{T^2} \Sigma_P,$$

where $X_2(C)$ is a connected double cover of the exterior $S^3 \setminus N(C)$, and Σ_P is a double cover of $S^1 \times D^2$ branched over P; these pieces are glued along their respective torus boundaries. We note that $\partial X_2(C)$ is incompressible since C is a nontrivial knot.

Suppose that Σ_P has incompressible boundary as well. Then this torus remains incompressible in $\Sigma_2(K)$. Hanselman, Rasmussen, and Watson [**HRW17**, Theorem 7.20] classified the prime, toroidal Heegaard Floer L-spaces Y with $|H_1(Y)| = 5$, and showed in particular that they are all built by gluing together a pair of trefoil exteriors. In each case there is a unique incompressible torus up to isotopy, so we conclude that $X_2(C)$ and Σ_P are both trefoil exteriors. In particular we have a double cover

$$S^3 \setminus N(T(\pm 2, 3)) \cong X_2(C) \to S^3 \setminus N(C).$$

Gonzalez-Acuña and Whitten [**GAW92**, Theorem 3.4] proved in this case that either

- C is not a torus knot, and then it must admit a cyclic ± 2-surgery, which contradicts the main result of [**KM04**]; or
- C is a torus knot $T(p, q)$, and we can write $2 = dpq \pm 1$ for some integer d, which is also impossible.

Thus Σ_P has compressible boundary after all, and we can write

$$\Sigma_P \cong (S^1 \times D^2) \# Z$$

for some closed 3-manifold Z, which may or may not be S^3.

Supposing that Z is different from S^3, it now follows that

$$\Sigma_2(P(C)) \cong \big(X_2(C) \cup (S^1 \times D^2)\big) \# Z$$

can only be prime if S^3 arises as a Dehn filling of $X_2(C)$, i.e., if $X_2(C)$ is the exterior of some other knot in S^3. Again this is impossible since C is nontrivial [**GAW92**, Theorem 3.4], so $\Sigma_2(K) \cong \Sigma_2(P(C))$ is not prime, which is a contradiction. So $Z \cong S^3$, and therefore Σ_P is a solid torus. Lemma 3.3 now tells us that P must be isotopic to a core of $S^1 \times D^2$. But in this case K is not a nontrivial satellite of C after all, so we are done. □

3.2. Factorization of the Alexander polynomial. In this subsection we prove the following, which completes the proof of Theorem 1.7.

PROPOSITION 3.5. *Fix an integer $h \geq 1$, and define the polynomial*

$$p_h(t) = t^{4h} - t^{4h-1} + t^{2h} - t + 1. \tag{3.1}$$

If $h \geq 3$, then $p_h(t)$ is not a product of cyclotomic polynomials.

REMARK 3.6. By contrast, we note that $p_1(t) = \Phi_{10}(t)$ and $p_2(t) = \Phi_{10}(t) \cdot \Phi_{12}(t)$.

In the proof of Proposition 3.5, we will adapt an algorithm called the "Graeffe" method [**BD89**] for recognizing cyclotomic polynomials. The idea is that if

$$p(t) = t^d + a_{d-1}t^{d-1} + \cdots + a_1 t + a_0$$

has roots $\alpha_1, \alpha_2, \ldots, \alpha_d$, then we can split p into its even and odd parts by writing

$$p(t) = p_e(t^2) + t \cdot p_o(t^2) \quad \text{where} \quad \begin{cases} p_e(t) = a_0 + a_2 t + a_4 t^2 + \ldots, \\ p_o(t) = a_1 + a_3 t + a_5 t^2 + \ldots. \end{cases}$$

Then Graeffe's root-squaring method says that the polynomial

$$q(t) = (-1)^d \left(p_e(t)^2 - t \cdot p_o(t)^2\right)$$

has roots $\alpha_1^2, \alpha_2^2, \ldots, \alpha_d^2$. For example, if $p(t) = \Phi_n(t)$ then we will have

$$q(t) = \begin{cases} \Phi_n(t), & n \text{ odd} \\ \Phi_{n/2}(t), & n \text{ even but not a multiple of 4} \\ \left(\Phi_{n/2}(t)\right)^2, & n \text{ a multiple of 4.} \end{cases} \tag{3.2}$$

We note that if $p(t) = p_h(t)$ is the polynomial given in (3.1), then we have

$$p_e(t) = t^{2h} + t^h + 1, \qquad\qquad p_o(t) = -t^{2h-1} - 1$$

and so the root-squaring method produces the polynomial

$$\begin{aligned} q_h(t) &= (t^{2h} + t^h + 1)^2 - t(-t^{2h-1} - 1)^2 \\ &= t^{4h} - t^{4h-1} + 2t^{3h} + t^{2h} + 2t^h - t + 1. \end{aligned} \tag{3.3}$$

Before we begin the proof of Proposition 3.5, we will first recall some facts about special values of cyclotomic polynomials.

LEMMA 3.7. *For all $n \geq 2$, we have*

$$\Phi_n(1) = \begin{cases} p & n = p^e \text{ is a prime power} \\ 1 & \text{otherwise} \end{cases} \quad \text{and} \quad \Phi_n(-1) = \begin{cases} 1 & n \text{ odd} \\ \Phi_{n/2}(1) & n \text{ even.} \end{cases}$$

PROOF. We evaluate both sides of

$$x^{n-1} + x^{n-2} + \cdots + x + 1 = \prod_{\substack{d|n \\ d\neq 1}} \Phi_d(x)$$

at $x = 1$ and at $x = -1$ to conclude that

$$\prod_{\substack{d|n \\ d\neq 1}} \Phi_d(1) = n, \qquad\qquad \prod_{\substack{d|n \\ d\neq 1}} \Phi_d(-1) = \begin{cases} 0 & n \text{ even} \\ 1 & n \text{ odd.} \end{cases}$$

If $n = p^e$ is a prime power, with $e \geq 1$, then the first of these implies by induction on e that $\Phi_{p^e}(1) = p$, and then it follows that $\Phi_n(1) = 1$ if n is not a prime power. The second equation similarly implies by induction that $\Phi_n(-1) = 1$ for all odd n.

In the remaining cases, we wish to evaluate $\Phi_n(-1)$ where $n = 2k$ is even. We observe that if k is odd then $\Phi_n(x) = \Phi_k(-x)$, and if k is even then $\Phi_n(x) = \Phi_k(x^2)$. In either case it follows that $\Phi_n(-1) = \Phi_k(1)$. □

We now begin to determine which cyclotomic polynmoials can divide $p_h(t)$.

LEMMA 3.8. *If n is odd, then $\Phi_n(t)$ does not divide $p_h(t)$.*

PROOF. Suppose that $\Phi_n(t)$ divides $p_h(t)$. Then the squares of the primitive nth roots of unity are also primitive nth roots of unity, so $\Phi_n(t)$ also divides the polynomial $q_h(t)$ from (3.3), and hence it divides the difference

$$q_h(x) - p_x(h) = 2t^{3h} + 2t^h = 2t^h(t^{2h}+1).$$

The roots of $\Phi_n(t)$ are all nonzero, so it divides $t^{2h}+1$ and hence $t^{4h}-1$. But this means that n is an odd divisor of $4h$, so in fact n divides h.

Letting ζ be any root of $\Phi_n(t)$, we have $\zeta^n = 1$ and therefore $\zeta^h = 1$. We compute that

$$\begin{aligned} 0 = p_h(\zeta) &= \zeta^{4h} - \zeta^{4h-1} + \zeta^{2h} - \zeta + 1 \\ &= 1 - \zeta^{-1} + 1 - \zeta + 1 \end{aligned}$$

and so $\zeta^2 - 3\zeta + 1 = 0$. But then $\zeta = \frac{1}{2}(3 \pm \sqrt{5})$ is not a root of unity, contradiction. □

LEMMA 3.9. *Let $n = 2k$ be twice an odd integer $k \geq 1$, and suppose that $\Phi_n(t)$ divides $p_h(t)$. Then $n = 10$, and h is congruent to either 1 or 2 modulo 5.*

PROOF. Since $\Phi_n(t)$ divides $p_h(t)$, the primitive kth roots of unity must be roots of $q_h(t)$, so $\Phi_k(t)$ divides $q_h(t)$. Equivalently, since $\Phi_n(t) = \Phi_k(-t)$ we see that $\Phi_n(t)$ divides $q_h(-t)$ and thus also the difference

$$\begin{aligned} q_h(-t) - p_h(t) &= 2t^{4h-1} + 2(-1)^h t^{3h} + 2(-1)^h t^h + 2t \\ &= 2t\left(t^{h-1} + (-1)^h\right)\left(t^{3h-1} + (-1)^h\right). \end{aligned}$$

Then $\Phi_n(t)$ divides either $t^{2h-2}-1$ or $t^{6h-2}-1$, and hence n divides either $2h-2$ or $6h-2$.

Suppose first that n divides $2h-2$, and let ζ be a root of $\Phi_n(t)$. Then $\zeta^{2h-2} = 1$, so

$$\begin{aligned} 0 = p_h(\zeta) &= \zeta^{4h} - \zeta^{4h-1} + \zeta^{2h} - \zeta + 1 \\ &= \zeta^4 - \zeta^3 + \zeta^2 - \zeta^1 + 1 = \Phi_{10}(\zeta) \end{aligned}$$

and since ζ is a root of the irreducible $\Phi_{10}(t)$, we must have $n = 10$. In this case $2h-2$ is a multiple of 10, so $h \equiv 1 \pmod{5}$.

Now suppose instead that n divides $6h-2$, and let ζ be a root of $\Phi_n(t)$; then $\zeta^{6h-2} = 1$. We note that n is not a multiple of 3 since $6h-2$ is not, so ζ^3 is also a primitive nth root of unity and therefore a root of $\Phi_n(t)$. Since $\Phi_n(t)$ divides $p_h(t)$, we have

$$\begin{aligned} 0 = p_h(\zeta^3) &= \zeta^{12h} - \zeta^{12h-3} + \zeta^{6h} - \zeta^3 + 1 \\ &= \zeta^4 - \zeta + \zeta^2 - \zeta^3 + 1 = \Phi_{10}(\zeta) \end{aligned}$$

and so once again we must have $n = 10$. Now $6h-2$ is a multiple of 10, and so $h \equiv 2 \pmod{5}$. □

LEMMA 3.10. *Fix $h \geq 1$. If $p_h(t)$ is a product of cyclotomic polynomials, then we have*

$$p_h(t) = \Phi_{10}(t) \cdot \prod_{j=1}^{k} \Phi_{n_j}(t) \tag{3.4}$$

where each n_j *is a multiple of 4 but not a power of 2. In particular* h *must be congruent to either* 1 *or* 2 *modulo* 5.

PROOF. By assumption we can find integers $n_0, n_1, \ldots, n_k$ $(k \geq 0)$ such that

$$p_h(t) = \Phi_{n_0}(t) \cdot \Phi_{n_1}(t) \cdot \ldots \cdot \Phi_{n_k}(t).$$

By Lemma 3.8, all of the n_j must be even, and then by Lemma 3.9 they must all be either 10 or multiples of 4. Setting $t = -1$, we have

$$5 = p_h(-1) = \prod_{j=0}^{k} \Phi_{n_j}(-1),$$

and every factor on the right is a nonnegative integer by Lemma 3.7; we order them so that $\Phi_{n_0}(-1) = 5$ and $\Phi_{n_j}(-1) = 1$ for all $j \geq 1$.

Now by Lemma 3.7, the integer $\Phi_{n_0}(-1)$ can only be 5 if $n_0 = 2 \cdot 5^e$ for some $e \geq 1$, and then n_0 is not a multiple of 4 so it must be 10. Then Lemma 3.9 guarantees that h is either 1 or 2 modulo 5. Moreover, we cannot have $n_j = 10$ for any other $j \geq 1$, because then $\Phi_{n_j}(-1)$ would not be 1, so the remaining n_j are all multiples of 4.

Finally, if instead we set $t = 1$ then $p_h(1) = 1$ implies that $\Phi_{n_j}(1) = 1$ for all $j = 1, 2, \ldots, k$, so we cannot have $n_j = 2^e$ because then Lemma 3.7 would tell us that $\Phi_{n_j}(1) = 2$ instead. □

With Lemma 3.10 in hand, we can now prove Proposition 3.5.

PROOF OF PROPOSITION 3.5. Suppose that $p_h(t)$ is a product of cyclotomic polynomials. Then Lemma 3.10 says that $h \equiv 1$ or $2 \pmod 5$, so h cannot be 3, 4, or 5. We will therefore require from now on that $h \geq 6$, so that $3h \leq 4h - 6$ and hence the polynomial $q_h(t)$ from (3.3) satisfies

$$\begin{aligned} q_h(t) &= t^{4h} - t^{4h-1} + 2t^{3h} + t^{2h} + 2t^h - t + 1 \\ &= t^{4h} - t^{4h-1} + O(t^{4h-6}). \end{aligned}$$

By assumption, $p_h(t)$ has the form (3.4). Following (3.2), the polynomial $q_h(t)$ must then equal

$$\Phi_5(t) \cdot \left(\prod_{j=1}^{k} \Phi_{n_j/2}(t) \right)^2 = t^{4h} - t^{4h-1} + O(t^{4h-6}).$$

The product being squared on the left is a monic polynomial with integer coefficients: if we write

$$\prod_{j=1}^{k} \Phi_{n_j/2}(t) = t^{2h-2} + \sum_{i=1}^{2h-2} a_i t^{2h-2-i},$$

then its square has the form

$$\left(\prod_{j=1}^{k} \Phi_{n_j/2}(t) \right)^2 = t^{4h-4} + \sum_{i=1}^{4h-4} b_i t^{4h-4-i}$$

where $b_1 = 2a_1$ and

$$b_i = 2a_i + \sum_{j=1}^{i-1} a_j a_{i-j}, \qquad i \geq 2.$$

The summands on the right occur in pairs $a_j a_{i-j} = a_{i-j} a_j$ for $1 \leq j \leq \lfloor \frac{i-1}{2} \rfloor$, so it follows that

$$b_i \equiv \begin{cases} 0, & i \text{ odd} \\ (a_{i/2})^2 \equiv a_{i/2}, & i \text{ even} \end{cases} \pmod 2.$$

Multiplying by $\Phi_5(t) = t^4+t^3+t^2+t+1$, we see that the t^{4h-1}- and t^{4h-2}-coefficients of

$$q_h(t) = (t^4 + t^3 + t^2 + t + 1)\left(t^{4h-4} + \sum_{i=1}^{4h-4} b_i t^{4h-4-i}\right)$$

are -1 and 0 respectively, so that

$$-1 = b_1 + 1 = 2a_1 + 1 \quad \Longrightarrow \quad a_1 = -1,$$

and

$$0 = b_2 + b_1 + 1 = (2a_2 + a_1^2) + (2a_1) + 1 \quad \Longrightarrow \quad a_2 = 0.$$

But then we also know that the t^{4h-5}-coefficient of $q_h(t)$ is 0 by assumption, and yet it is also

$$b_1 + b_2 + b_3 + b_4 + b_5 \equiv a_1 + a_2 \equiv -1 \pmod 2,$$

which is a contradiction. □

4. HOMFLY homology of nearly fibered knots

In this section we prove that reduced HOMFLY homology detects 5_2 and each of the $P(-3, 3, 2n + 1)$ pretzel knots, where $n \in \mathbb{Z}$.

As background, we recall that Khovanov and Rozansky [**KR08a**] defined for each integer $N \geq 1$ a bigraded $\mathfrak{sl}_N$ link homology $\bar{H}_N(K)$, where $\bar{H}_2(K)$ agrees with $\overline{Kh}(K)$ up to a change of grading. In [**KR08b**] they also defined a triply graded homology theory $\bar{H}(K)$ whose graded Euler characteristic recovers the HOMFLY polynomial. Rasmussen [**Ras15**] constructed for each $N \geq 1$ a spectral sequence

$$\bar{H}(K) \Longrightarrow \bar{H}_N(K) \tag{4.1}$$

which collapses for all large enough N.

LEMMA 4.1. *Suppose for some knot K and some*

$$J \in \{5_2\} \cup \{P(-3, 3, 2n + 1) \mid n \in \mathbb{Z}\}$$

that

$$\bar{H}(K; \mathbb{Q}) \cong \bar{H}(J; \mathbb{Q})$$

as triply-graded vector spaces. Then $\Delta_K(t) = \Delta_J(t)$, and

$$\dim \bar{H}(K; \mathbb{Q}) = \dim \overline{Kh}(K; \mathbb{Q}) = \det(K).$$

PROOF. The claim that $\Delta_K(t) = \Delta_J(t)$ follows from the fact that $\bar{H}(K)$ determines the HOMFLY polynomial of K, and hence its Alexander polynomial. We note by taking $t = -1$ that this implies that $\det(K) = \det(J)$.

We now claim that

$$\dim \bar{H}(K; \mathbb{Q}) = \dim \bar{H}(J; \mathbb{Q}) = \det(J) = \det(K). \tag{4.2}$$

It suffices to prove the middle equality. When $J = 5_2$ this follows from the fact that J is a two-bridge knot: Rasmussen proved for each $N > 4$ that J is "N-thin" [**Ras07**, Theorem 1], and hence that the $\mathfrak{sl}_N$ homology $\bar{H}_N(J;\mathbb{Q})$ has dimension $\det(J)$. We take N large enough so that the spectral sequence (4.1) collapses for J, and thus conclude that $\dim \bar{H}(J;\mathbb{Q}) = \det(J)$. The case $J = P(-3,3,2n+1)$ is [**BS22a**, Lemma 9.1], in which we use an identical argument for the two-bridge knot $P(-3,3,1) = 6_1$, and then we apply work of Wang [**Wan23**] to get the general case.

Combining (4.2) with the case $N = 2$ of (4.1) and Lemma 2.1, we now see that

$$\det(K) = \dim \bar{H}(K;\mathbb{Q}) \geq \dim \bar{H}_2(K;\mathbb{Q}) = \dim \overline{Kh}(K;\mathbb{Q}) \geq \det(K).$$

Thus equality must hold throughout, completing the proof. □

Lemma 4.1 is enough to determine the instanton knot homology of such a knot K.

LEMMA 4.2. *Suppose for some knot K and some*

$$J \in \{5_2\} \cup \{P(-3,3,2n+1) \mid n \in \mathbb{Z}\}$$

that

$$\bar{H}(K;\mathbb{Q}) \cong \bar{H}(J;\mathbb{Q})$$

as triply-graded vector spaces. Then $\dim KHI(K) = \det(K)$, *and K has genus* 1 *and*

$$\dim KHI(K,1) = 2.$$

PROOF. According to Lemma 4.1, we have

$$\Delta_K(t) = \Delta_J(t) = \begin{cases} 2t - 3 + 2t^{-1}, & J = 5_2 \\ -2t + 5 - 2t^{-1}, & J = P(-3,3,2n+1) \end{cases}$$

and

$$\dim \overline{Kh}(K;\mathbb{Q}) = \det(K) = \begin{cases} 7, & J = 5_2 \\ 9, & J = P(-3,3,2n+1). \end{cases}$$

We once again apply Kronheimer and Mrowka's spectral sequence

$$\overline{Kh}(K;\mathbb{Q}) \implies I^\natural(\overline{K};\mathbb{Q})$$

of [**KM11**, Proposition 1.2], together with the isomorphism

$$I^\natural(\overline{K};\mathbb{Q}) \cong KHI(\overline{K};\mathbb{Q})$$

of [**KM11**, Proposition 1.4] and the invariance of $\dim KHI$ under orientation reversal, to conclude that

$$\dim KHI(K;\mathbb{Q}) \leq \dim \overline{Kh}(K;\mathbb{Q}) = \det(K).$$

But the relation (2.1) implies that if we write $\Delta_K(t) = \sum_i a_i t^i$ then

$$\dim KHI(K;\mathbb{Q}) \geq \sum_{i\in\mathbb{Z}} |a_i| \geq |\Delta_K(-1)| = \det(K),$$

so if $\dim KHI(K;\mathbb{Q}) \leq \det(K)$ as well then each inequality must in fact be an equality.

In other words, we have shown that

$$\dim KHI(K,i;\mathbb{Q}) = |a_i|$$

for all $i \in \mathbb{Z}$, where the a_i are the coefficients of $\Delta_J(t)$. This is zero for all $i \geq 2$ and nonzero for $i = 1$, so the genus detection property of KHI says that $g(K) = 1$. And since $a_1 = \pm 2$ we can also conclude that $\dim KHI(K, 1) = 2$, as claimed. □

Knots satisfying the conclusion of Lemma 4.2 have been completely classified: we achieved the analogue of this with $\widehat{HFK}$ in place of KHI in [**BS22a**], and then Li and Ye [**LY22**] showed that the same conclusion holds for KHI. Specifically, for any such knot the sutured complement of a genus-1 Seifert surface must be one of two possible sutured manifolds, up to orientation: for $\widehat{HFK}$ this is [**BS22a**, Theorem 5.1], and then [**LY22**, Example 2.2] establishes the same result (with the same sutured manifolds) for KHI. Then [**BS22a**, Theorem 6.1] and [**BS22a**, Theorem 7.1] classify the knots that realize these sutured manifolds, by an argument that uses no Floer homology whatsoever.

PROPOSITION 4.3 ([**BS22a, LY22**]). *Suppose that K is a genus-1 knot, and that*

$$\dim KHI(K, 1; \mathbb{Q}) = 2.$$

Then up to mirroring, either

- $\Delta_K(t) = 2t - 3 + 2t^{-1}$, *and K is either 5_2, $15n_{43522}$, or $16n_{696530}$; or*
- $\Delta_K(t) = -2t+5-2t^{-1}$, *and K is either some $P(-3,3,2n+1)$ or $15n_{115646}$,*

where the knots $15n_{115646}$ and $16n_{696530}$ are in fact twisted Whitehead doubles $\mathrm{Wh}^{\pm}(T_{2,3}, 2)$ of the right-handed trefoil.

Using Shumakovitch's KhoHo program [**Shu18**], we can compute that

$$\dim \overline{Kh}(J; \mathbb{Q}) = \begin{cases} 17, & J = 15n_{43522} \\ 23, & J = 15n_{115646} \\ 25, & J = 16n_{696530}. \end{cases} \tag{4.3}$$

With this information at hand, we are now ready to prove the remaining results from the introduction.

PROOF OF THEOREM 1.3. Suppose that $\bar{H}(K) \cong \bar{H}(5_2)$. Then Lemmas 4.1 and 4.2 say that

$$\Delta_K(t) = 2t - 3 + 2t^{-1},$$

that $\dim \overline{Kh}(K; \mathbb{Q}) = 7$, and that K has genus one, with $\dim KHI(K, 1; \mathbb{Q}) = 2$. According to Proposition 4.3, it follows that K must be either 5_2, $15n_{43522}$, or $16n_{696530}$ up to mirroring. But the triple grading distinguishes $\bar{H}(5_2)$ from $\bar{H}(\overline{5_2})$, since they have different signatures [**Ras15**, Corollary 5.1], so K cannot be $\overline{5_2}$. It also cannot be either $15n_{43522}$ or $16n_{696530}$, or their mirrors, because by (4.3) we would then have $\dim \overline{Kh}(K; \mathbb{Q}) > 7$. Thus K must be 5_2 after all. □

PROOF OF THEOREM 1.4. Suppose that $\bar{H}(K) \cong \bar{H}(P(-3,3,2n+1))$ for some fixed n. Then Lemma 4.1 says that

$$\Delta_K(t) = -2t + 5 - 2t^{-1},$$

so $\det(K) = 9$, and that

$$\dim \bar{H}(K; \mathbb{Q}) = \dim \bar{H}_2(K; \mathbb{Q}) = \dim \overline{Kh}(K; \mathbb{Q}) = 9.$$

In particular, when $N = 2$ the spectral sequence (4.1) collapses, and so by [**Ras15**, Theorem 1] the bigrading on $\bar{H}_2(K)$ is completely determined by the triple grading on $\bar{H}(K)$. This means that

$$\overline{Kh}(K;\mathbb{Q}) \cong \overline{Kh}(P(-3,3,2n+1);\mathbb{Q})$$

as bigraded vector spaces.

Moreover, by Lemma 4.2 we know that K has genus one, with $\dim KHI(K,1;\mathbb{Q}) = 2$. Proposition 4.3 says that K must therefore be some pretzel knot $P(-3,3,2m+1)$, or $15n_{115646}$ or its mirror. (We note that $P(-3,3,2m+1)$ is the mirror of $P(-3,3,-2m-1)$.) But again by (4.3) the reduced Khovanov homologies of $15n_{115646}$ and its mirror are not 9-dimensional, so in fact K cannot be either of these knots. So now we have

$$K = P(-3,3,2m+1)$$

for some $m \in \mathbb{Z}$, and all of these pretzel knots are distinguished by the bigradings on their Khovanov homologies, by [**Sta12**, Theorem 4.1] or more generally [**HW18**, Theorem 3.2]. Thus $m = n$ after all. □

Acknowledgments

We thank the referee for helpful feedback on the original version of this paper, and for suggesting an analytic proof of Proposition 3.5.

References

[BBG19] Michel Boileau, Steven Boyer, and Cameron McA. Gordon, *Branched covers of quasi-positive links and L-spaces*, J. Topol. **12** (2019), no. 2, 536–576, DOI 10.1112/topo.12092. MR4072174

[BD89] R. J. Bradford and J. H. Davenport, *Effective tests for cyclotomic polynomials*, Symbolic and algebraic computation (Rome, 1988), Lecture Notes in Comput. Sci., vol. 358, Springer, Berlin, 1989, pp. 244–251, DOI 10.1007/3-540-51084-2_22. MR1034734

[BDL+21] John A. Baldwin, Nathan Dowlin, Adam Simon Levine, Tye Lidman, and Radmila Sazdanovic, *Khovanov homology detects the figure-eight knot*, Bull. Lond. Math. Soc. **53** (2021), no. 3, 871–876, DOI 10.1112/blms.12467. MR4275096

[BHS21] John A. Baldwin, Ying Hu, and Steven Sivek, *Khovanov homology and the cinquefoil*, arXiv:2105.12102, 2021.

[BLSY21] John A. Baldwin, Zhenkun Li, Steven Sivek, and Fan Ye, *Small Dehn surgery and $SU(2)$*, arXiv:2110.02874, 2021.

[BPRW22] Anna Beliakova, Krzysztof K. Putyra, Louis-Hadrien Robert, and Emmanuel Wagner, *A proof of Dunfield–Gukov–Rasmussen conjecture*, arXiv:2210.00878, 2022.

[BS19] John A. Baldwin and Steven Sivek, *Instantons and L-space surgeries*, J. Eur. Math. Soc. (JEMS) **25** (2023), no. 10, 4033–4122, DOI 10.4171/jems/1280. MR4634689

[BS22a] John A. Baldwin and Steven Sivek, *Floer homology and non-fibered knot detection*, arXiv:2208.03307, 2022.

[BS22b] John A. Baldwin and Steven Sivek, *Khovanov homology detects the trefoils*, Duke Math. J. **171** (2022), no. 4, 885–956, DOI 10.1215/00127094-2021-0034. MR4393789

[Cro04] Peter R. Cromwell, *Knots and links*, Cambridge University Press, Cambridge, 2004, DOI 10.1017/CBO9780511809767. MR2107964

[Dow18] Nathan Dowlin, *A spectral sequence from Khovanov homology to knot Floer homology*, arXiv:1811.07848, 2018.

[FRW22] Ethan Farber, Braeden Reinoso, and Luya Wang, *Fixed point-free pseudo-Anosovs and the cinquefoil*, arXiv:2203.01402, 2022.

[GAW92] Francisco González-Acuña and Wilbur C. Whitten, *Imbeddings of three-manifold groups*, Mem. Amer. Math. Soc. **99** (1992), no. 474, viii+55, DOI 10.1090/memo/0474. MR1117167

[Hei74] Wolfgang Heil, *Elementary surgery on Seifert fiber spaces*, Yokohama Math. J. **22** (1974), 135–139. MR375320

[HN10] Matthew Hedden and Yi Ni, *Manifolds with small Heegaard Floer ranks*, Geom. Topol. **14** (2010), no. 3, 1479–1501, DOI 10.2140/gt.2010.14.1479. MR2653731

[HRW17] Jonathan Hanselman, Jacob Rasmussen, and Liam Watson, *Bordered Floer homology for manifolds with torus boundary via immersed curves*, J. Amer. Math. Soc. **37** (2024), no. 2, 391–498, DOI 10.1090/jams/1029. MR4695506

[HW18] Matthew Hedden and Liam Watson, *On the geography and botany of knot Floer homology*, Selecta Math. (N.S.) **24** (2018), no. 2, 997–1037, DOI 10.1007/s00029-017-0351-5. MR3782416

[Kho03] Mikhail Khovanov, *Patterns in knot cohomology. I*, Experiment. Math. **12** (2003), no. 3, 365–374. MR2034399

[KM04] P. B. Kronheimer and T. S. Mrowka, *Dehn surgery, the fundamental group and $SU(2)$*, Math. Res. Lett. **11** (2004), no. 5-6, 741–754, DOI 10.4310/MRL.2004.v11.n6.a3. MR2106239

[KM10a] Peter Kronheimer and Tom Mrowka, *Instanton Floer homology and the Alexander polynomial*, Algebr. Geom. Topol. **10** (2010), no. 3, 1715–1738, DOI 10.2140/agt.2010.10.1715. MR2683750

[KM10b] Peter Kronheimer and Tomasz Mrowka, *Knots, sutures, and excision*, J. Differential Geom. **84** (2010), no. 2, 301–364. MR2652464

[KM11] P. B. Kronheimer and T. S. Mrowka, *Khovanov homology is an unknot-detector*, Publ. Math. Inst. Hautes Études Sci. **113** (2011), 97–208, DOI 10.1007/s10240-010-0030-y. MR2805599

[KR08a] Mikhail Khovanov and Lev Rozansky, *Matrix factorizations and link homology*, Fund. Math. **199** (2008), no. 1, 1–91, DOI 10.4064/fm199-1-1. MR2391017

[KR08b] Mikhail Khovanov and Lev Rozansky, *Matrix factorizations and link homology. II*, Geom. Topol. **12** (2008), no. 3, 1387–1425, DOI 10.2140/gt.2008.12.1387. MR2421131

[Lim10] Yuhan Lim, *Instanton homology and the Alexander polynomial*, Proc. Amer. Math. Soc. **138** (2010), no. 10, 3759–3768, DOI 10.1090/S0002-9939-2010-10412-4. MR2661575

[LL23] Zhenkun Li and Yi Liang, *Some computations on instanton knot homology*, J. Knot Theory Ramifications **32** (2023), no. 2, Paper No. 2350007, 14, DOI 10.1142/S0218216523500074. MR4564615

[LY21] Zhenkun Li and Fan Ye, *$SU(2)$ representations and a large surgery formula*, `arXiv:2107.11005`, 2021.

[LY22] ———, *Seifert surface complements of nearly fibered knots*, `arXiv:2208.05382`, 2022.

[Mot17] Kimihiko Motegi, *A note on L-spaces which are double branched covers of non-quasi-alternating links*, Topology Appl. **230** (2017), 172–180, DOI 10.1016/j.topol.2017.08.043. MR3702764

[Oer84] Ulrich Oertel, *Closed incompressible surfaces in complements of star links*, Pacific J. Math. **111** (1984), no. 1, 209–230. MR732067

[OS05] Peter Ozsváth and Zoltán Szabó, *On the Heegaard Floer homology of branched double-covers*, Adv. Math. **194** (2005), no. 1, 1–33, DOI 10.1016/j.aim.2004.05.008. MR2141852

[Pla06] Olga Plamenevskaya, *Transverse knots and Khovanov homology*, Math. Res. Lett. **13** (2006), no. 4, 571–586, DOI 10.4310/MRL.2006.v13.n4.a7. MR2250492

[Ras05] Jacob Rasmussen, *Knot polynomials and knot homologies*, Geometry and topology of manifolds, Fields Inst. Commun., vol. 47, Amer. Math. Soc., Providence, RI, 2005, pp. 261–280, DOI 10.1007/bf01579132. MR2189938

[Ras07] Jacob Rasmussen, *Khovanov-Rozansky homology of two-bridge knots and links*, Duke Math. J. **136** (2007), no. 3, 551–583, DOI 10.1215/S0012-7094-07-13635-4. MR2309174

[Ras10] Jacob Rasmussen, *Khovanov homology and the slice genus*, Invent. Math. **182** (2010), no. 2, 419–447, DOI 10.1007/s00222-010-0275-6. MR2729272

[Ras15] Jacob Rasmussen, *Some differentials on Khovanov-Rozansky homology*, Geom. Topol. **19** (2015), no. 6, 3031–3104, DOI 10.2140/gt.2015.19.3031. MR3447099

[Shu07] Alexander N. Shumakovitch, *Rasmussen invariant, slice-Bennequin inequality, and sliceness of knots*, J. Knot Theory Ramifications **16** (2007), no. 10, 1403–1412, DOI 10.1142/S0218216507005889. MR2384833

[Shu18] Alexander N. Shumakovitch, *KhoHo package for PARI/GP*, `https://github.com/AShumakovitch/KhoHo`, 2018.

[Sta12] Laura Starkston, *The Khovanov homology of $(p, -p, q)$ pretzel knots*, J. Knot Theory Ramifications **21** (2012), no. 5, 1250056, 14, DOI 10.1142/S0218216511010103. MR2902279

[Wal69] Friedhelm Waldhausen, *Über Involutionen der* 3*-Sphäre* (German), Topology **8** (1969), 81–91, DOI 10.1016/0040-9383(69)90033-0. MR236916

[Wan23] Joshua Wang, *Split link detection for* $\mathfrak{sl}(P)$ *link homology in characteristic* P, J. Topol. **16** (2023), no. 2, 806–821. MR4637977

Department of Mathematics, Boston College, Chestnut Hill, Massachusetts 02467
Email address: `john.baldwin@bc.edu`

Department of Mathematics, Imperial College London, London SW7 2AZ, United Kingdom
Current address: Max Planck Institute for Mathematics
Email address: `s.sivek@imperial.ac.uk`
Email address: `ssivek@mpim-bonn.mpg.de`

Proceedings of Symposia in Pure Mathematics
Volume **109**, 2024
https://doi.org/10.1090/pspum/109/02002

Examples of homology 3-spheres whose Chern-Simons function is not Morse-Bott

Hans U. Boden, Christopher M. Herald, and Paul Kirk

ABSTRACT. We construct two homology 3-spheres for which the (unperturbed) $SU(2)$ Chern-Simons function is not Morse-Bott. In one case, there is a degenerate isolated critical point. In the other, a path component of the critical set is not homeomorphic to a manifold. The examples are $+1$ surgeries on connected sums of torus knots.

1. Introduction

The purpose of this article is to address a question raised to us by D. Ruberman, namely, whether there exist examples of *homology 3-spheres* M for which the $SU(2)$ Chern-Simons function

$$c_M\colon \mathcal{B}^* \to \mathbb{R}/\mathbb{Z},$$

a circle-valued function on the space of gauge equivalence classes of irreducible $SU(2)$ connections, fails to be Morse-Bott. We construct an example of a homology 3-sphere whose Chern-Simons function has a degenerate isolated critical point; this example solves Problem 3.105(B) in [**Kir97**]. In addition, we construct a homology 3-sphere for which the critical set of the Chern-Simons function has a path component not homeomorphic to a manifold.

As is well known, holonomy identifies the critical set of c_M with the irreducible character variety (a real semi-algebraic set):

$$\chi^*(M) = \mathrm{Hom}(\pi_1(M), SU(2)) \smallsetminus \{\theta\}/_{\text{conjugation}},$$

where θ denotes the trivial homomorphism. For any homomorphism $\rho\colon \pi_1(M) \to SU(2)$ (henceforth called a representation), the cohomology group $H^1(M; su(2)_{\mathrm{ad}\,\rho})$ is called the *Zariski tangent space* of $\chi(M)$ at ρ. If M is a homology 3-sphere, the conjugacy class $[\theta]$ of θ is isolated in the character variety; it follows that $\chi^*(M)$ is compact [**AM90**]. The Hodge theorem identifies the kernel of the Hessian of c_M at ρ with the Zariski tangent space of $\chi(M)$ at ρ (e.g., see [**Tau90**]). The function c_M is Morse if all its critical points are non-degenerate; i.e., the Zariski tangent space is trivial at each critical point. It is widely known that if M is a connected

2020 *Mathematics Subject Classification.* Primary 57K18, 57K31, 57R58; Secondary 81T13.
Key words and phrases. Chern-Simons function, flat moduli space.
The first author was supported by an NSERC Discovery Grant. The second author was supported by a Simons Collaboration Grant for Mathematicians. The third author is thankful to Institut Fourier and the Max Planck Institute for support.

sum of nontrivial homology spheres, c_M is not Morse because π_1 is a nontrivial free product; there are gluing parameters (also known as bending parameters), related to conjugating a representation of one factor but not the other.

The Chern-Simons function $c_M : \mathcal{B}^* \to \mathbb{R}/\mathbb{Z}$ is Morse-Bott if every path component of the critical set is a smooth manifold and if the Hessian of c_M defines a non-degenerate bilinear form on the normal bundles of the critical submanifolds (see, for example, [**Nic11**]). This translates into the condition that, for each $[\rho] \in \chi^*(M)$, the dimension of the Zariski tangent space of $\chi^*(M)$ at $[\rho]$ equals the dimension of the path component containing $[\rho]$.

Fintushel-Stern [**FS90**] showed that if M is a Seifert-fibered homology 3-sphere, then c_M is Morse-Bott. Given two homology spheres M_1, M_2 such that c_{M_i} is Morse-Bott for $i = 1, 2$, the connected sum $M_1 \# M_2$ also has a Morse-Bott Chern-Simons function. In fact, given path components $C_1 \subset \chi^*(M_1)$ and $C_2 \subset \chi^*(M_2)$, there are three associated components in $\chi^*(M_1 \# M_2)$, diffeomorphic to $C_1 \times [\theta_2]$, $[\theta_1] \times C_2$, and $C_1 \times (SU(2)/\{\pm 1\}) \times C_2 \subset \chi^*(M_1 \# M_2)$. The latter is obtained by pairing each ρ_1 representing an equivalence class in C_1 with all $SU(2)$ conjugates of a ρ_2 representing a class in C_2.

Given relatively prime integers p, q, let $T_{p,q}$ denote the (p, q) torus knot. Consider the knot complements:

$$X = S^3 \setminus nbd(T_{3,5}), Y = S^3 \setminus nbd(T_{2,7}), \text{ and } Z = S^3 \setminus nbd\left(T_{-2,7} \# T_{-2,7}\right).$$

Equip the boundary ∂X with its natural oriented meridian-longitude pair μ_X, λ_X, and similarly μ_Y, λ_Y for Y and μ_Z, λ_Z for Z. Define $h_Y : \partial X \to \partial Y$, $h_Z : \partial X \to \partial Z$ to be (orientation-reversing) homeomorphisms inducing the maps

$$(1) \quad h_{Y*} \colon \mu_X \mapsto \mu_Y,\ \lambda_X \mapsto -\mu_Y - \lambda_Y, \qquad h_{Z*} \colon \mu_X \mapsto \mu_Z,\ \lambda_X \mapsto -\mu_Z - \lambda_Z$$

on the fundamental group. Define

$$\Sigma_1 = X \cup_{h_Y} Y \text{ and } \Sigma_2 = X \cup_{h_Z} Z.$$

It is immediate from the fact that X, Y, Z are all homology solid tori with H_1 generated by the meridians, and with the longitudes trivial in H_1, that Σ_1, Σ_2 are homology spheres.

THEOREM A.

(1) *There exists an isolated point in $\chi^*(\Sigma_1)$ with two-dimensional Zariski tangent space.*
(2) *There exists a component of $\chi^*(\Sigma_2)$ which is not homeomorphic to a manifold.*

COROLLARY B. *The critical set of c_{Σ_1} contains an isolated point at which the Hessian has a 2-dimensional kernel. The critical set of c_{Σ_2} is not homeomorphic to a manifold.*

Thus c_{Σ_1} and c_{Σ_1} are most decidedly not Morse-Bott. Taking connected sums of these with themselves and with other homology 3-spheres provides many more complicated examples.

We note that results of Kapovich and Millson [**KM17**] imply that arbitrarily bad singularities, including isolated points with nonzero Zariski tangent space and non-manifold path components, occur in $SU(2)$ character varieties of 3-manifolds. It is an open question whether their universality results hold for homology 3-spheres (see, e.g., [**KM17**, Question 8.2]). We also note that there are Seifert-fibered homology spheres for which the $SU(3)$ Chern-Simons function is not Morse-Bott [**BHK05**].

2. The character varieties of X and Y and their image in the character variety of the separating torus

For any path-connected space A let

$$\chi(A) = \mathrm{Hom}(\pi_1(A), SU(2))/_{\text{conjugation}}$$

denote its character variety. Its points are conjugacy classes, denoted $[\rho\colon \pi_1(A) \to SU(2)]$, or simply $[\rho]$. A representation $\rho\colon \pi_1(A) \to SU(2)$ is called *central, (non-central) abelian*, or *irreducible*, depending on whether the stabilizer of ρ under conjugation by $SU(2)$ is isomorphic to $SU(2)$, $U(1)$ or $\{\pm 1\}$, respectively.

When T is the 2-dimensional torus with a fixed set of generators $\mu, \lambda \in \pi_1(T)$, $\chi(T)$ is homeomorphic to a 2-sphere (usually called the pillowcase), and there is a branched covering

$$\mathbb{R}^2 \to \chi(T),\ (x,y) \mapsto [\mu \mapsto e^{x\mathbf{i}}, \lambda \mapsto e^{y\mathbf{i}}] \tag{2}$$

which can be seen as the composite of the projection $\mathbb{R}^2 \to \mathbb{R}^2/(2\pi\mathbb{Z})^2$ and the orbit map of the elliptic involution induced by $(x, y) \mapsto (-x, -y)$. Call a curve in $\chi(T)$ a *line segment* if it is the image of a line segment in $\mathbb{R}^2$. Since the slope of a line is preserved by both translations by $(2\pi\mathbb{Z})^2$ and reflections through the origin, line segments in $\chi(T)$ have well-defined slope.

For any knot K, $\chi(S^3 \smallsetminus nbd(K))$ contains an arc of (conjugacy classes of) abelian representations with central endpoints, mapping to the image of the x axis (i.e., with slope zero) in $\chi(T)$. We parameterize this arc with a path of representations $\mu \mapsto e^{a\mathbf{i}}$, $\lambda \mapsto 1$, $a \in [0, \pi]$, where μ, λ are a meridian, longitude pair.

Klassen [**Kla91**] explicitly identified the $SU(2)$ character varieties of torus knot complements. From his description of families of homomorphisms parameterizing the path components of $\chi^*(S^3 \smallsetminus nbd(T_{p,q}))$, one can readily restrict to a meridian/longitude which generate $\pi_1(T)$ to identify the image of the restriction map

$$i^*\colon \chi\left(S^3 \smallsetminus nbd(T_{p,q})\right) \to \chi(T)$$

induced by the inclusion $i\colon T = \partial\left(S^3 \smallsetminus nbd(T_{p,q})\right) \to S^3 \smallsetminus nbd(T_{p,q})$. Along with the abelian arc, $\chi(S^3 \smallsetminus nbd(T_{p,q}))$ consists of a collection of arcs of conjugacy classes of irreducible representations, mapping to $\chi(T)$ as line segments of slope $-pq$, with ends limiting to certain points on the abelian arc. The details in the case of $T_{3,5}$ are summarized in [**HHK14**]. For the purposes of this article, we require only the following part of this calculation for $T_{3,5}$, $T_{2,7}$, and $T_{-2,7}$.

PROPOSITION 1 (Klassen [**Kla91**]). *There is a path component of $\chi^*(S^3 \smallsetminus nbd(T_{3,5}))$ which is an arc mapping onto a line segment in $\chi(T)$ of slope -15, $r \in \left(\frac{\pi}{15}, \frac{11\pi}{15}\right) \mapsto (r, -15r)$, with ends limiting to the points $a = \frac{\pi}{15}$ and $a = \frac{11\pi}{15}$ on the abelian arc. Similarly, there is path component of $\chi^*(S^3 \smallsetminus nbd(T_{\pm 2,7}))$ mapping onto a line segment in $\chi(T)$ of slope ∓ 14, $r \in \left(\frac{\pi}{14}, \frac{13\pi}{14}\right) \mapsto (r, \mp 14r)$, with ends limiting to the points $a = \frac{\pi}{14}$ and $a = \frac{13\pi}{14}$ on the abelian arc.*

At each interior point on these irreducible arcs, the Zariski tangent space is 1-dimensional. For the (abelian) endpoints of either irreducible arc, the Zariski tangent space is 3-dimensional and the linearization of the restriction map to $\chi(T)$ has rank one, with horizontal image.

Figure 1 and Figure 2 illustrate neighborhoods of the left ends of the irreducible arcs described in the theorem and (lifts to $\mathbb{R}^2$ of) their images under restriction to the character variety of the boundary torus. In both cases, the neighborhoods embed into the pillowcase.

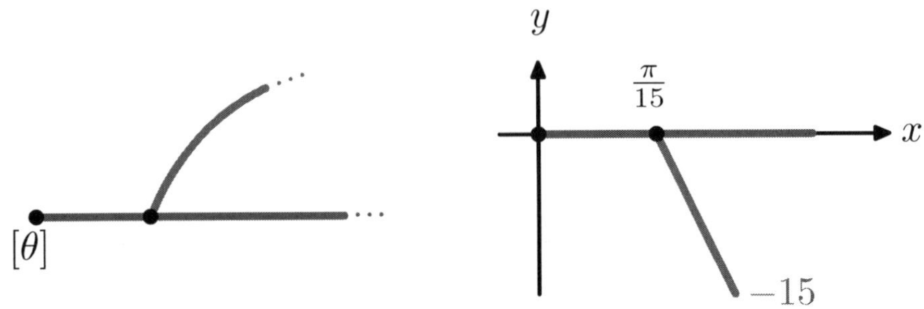

FIGURE 1. Local picture of $\chi(X) = \chi(S^3 \setminus nbd(T_{3,5}))$ near $[\theta]$ (on left) and its image under $i_X^*\colon \chi(X) \to \chi(\partial X)$ (on right)

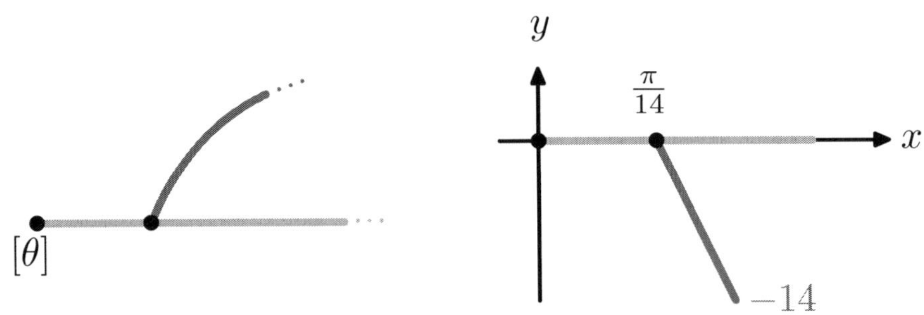

FIGURE 2. Local picture of $\chi(Y) = \chi(S^3 \setminus nbd(T_{2,7}))$ near $[\theta]$ (on left) and its image under $i_Y^*\colon \chi(Y) \to \chi(\partial Y)$ (on right)

3. Proof of Part (1)

The homeomorphism h_Y of Equation (1) induces a map $h_Y^*\colon \chi(\partial Y) \to \chi(\partial X)$ which lifts to the linear map

$$h_Y^* = \begin{pmatrix} 1 & 0 \\ -1 & -1 \end{pmatrix}$$

on $\mathbb{R}^2$, using Equation (2). Figure 3 illustrates the line segments which make up the images under the local embeddings i_X^* and $h_Y^* \circ i_Y^*$ of the portions of $\chi(X)$ and $\chi(Y)$ in Figures 1 and 2.

Consider the fiber product

$$F := \{([\rho_X], [\rho_Y]) \mid i_X^*(\rho_X) = h_Y^* \circ i_Y^*(\rho_Y)\} \subset \chi(X) \times \chi(Y).$$

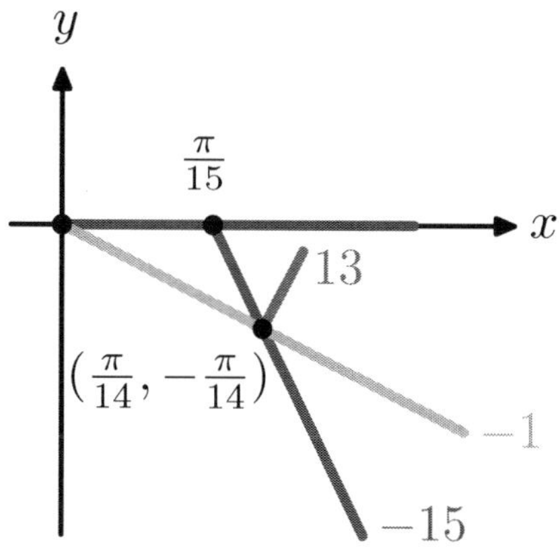

FIGURE 3. The images $i_X^*(\chi(X))$ and $h^* \circ i_Y^*(\chi(Y))$ near $[\theta]$ in $\chi(\partial X)$

The restriction map $\chi(\Sigma_1) \to \chi(X) \times \chi(Y)$ has image F and fiber over $([\rho_X], [\rho_Y])$ (known as the space of *gluing parameters*) homeomorphic to the double coset space

$$\mathrm{Stab}_{\rho_X} \backslash \mathrm{Stab}_{\rho_{\partial X}} / \mathrm{Stab}_{\rho_Y} \tag{3}$$

(see, e.g., [**HHK14**]).

From the subsets of $i_X^*(\chi(X))$, $h_Y^*(i_Y^*(\chi(Y)))$ that we have identified and sketched in Figure 3, it is clear that there are two isolated points of intersection. Specifically, $([\theta_X], [\theta_Y])$ maps to the origin in $\chi(\partial X) = \mathbb{R}^2/\sim$, and a pair $([\rho_X], [\rho_Y]) \in \chi(X) \times \chi(Y)$ which maps to $(\frac{\pi}{14}, -\frac{\pi}{14})$. Moreover, ρ_X is irreducible and ρ_Y is abelian, but non-central.

For this second pair, $\mathrm{Stab}_{\rho_{\partial X}} = U(1) = \mathrm{Stab}_{\rho_Y}$, since the representations ρ_Y and $\rho_{\partial X}$ are abelian non-central. It follows that $H^0(\partial X; su(2)_{ad\rho_{\partial X}}) = 0$. Hence the pair $([\rho_X], [\rho_Y])$ mapping to $(\frac{\pi}{14}, -\frac{\pi}{14})$ determines an *isolated point* of the character variety $\chi(\Sigma_1)$, viewed as a real semi-algebraic set.

Consider the Mayer-Vietoris sequence (with local $su(2)$ coefficients)

$$\cdots \xrightarrow{0} H^1(\Sigma_1) \longrightarrow H^1(X) \oplus H^1(Y) \xrightarrow{i_X^* - h_Y^* \circ i_Y^*} H^1(\partial X) \longrightarrow \cdots$$

We have:

- $\dim H^1(X; su(2)_{ad\rho_X}) = 1$ because $[\rho_X]$ is an interior point on the irreducible arc of $\chi(X)$ identified in Proposition 1,
- $\dim H^1(\partial X; su(2)_{ad\rho_{\partial X}}) = 2$ since the restriction $\rho_{\partial X} \colon \pi_1(\partial X) \to SU(2)$ is abelian and non-central,
- $\dim H^1(Y; su(2)_{ad\rho_Y}) = 3$ because ρ_Y is the $a = \frac{\pi}{14}$ abelian endpoint of the irreducible arc of $\chi(Y)$ identified in Proposition 1, and
- the image of the irreducible arc in $\chi^*(X)$ and the abelian arc in $h_Y^*(i_Y^*(\chi(Y)))$ have different slopes in $\chi(\partial X)$, namely -15 and -1, respectively, which shows $i_X^* - h_Y^* \circ i_Y^*$ is surjective.

Thus, the Mayer-Vietoris sequence implies that $\dim H^1(\Sigma_1; su(2)_{ad\rho}) = 2$, completing the proof of the first assertion of Theorem A.

4. Proof of Part (2)

The proof of Part (2) of Theorem A follows a similar strategy to that used to prove Part (1), but we replace Y by $Z = S^3 \setminus nbd(T_{-2,7}\#T_{-2,7})$. The exterior Z of the composite knot $T_{-2,7}\#T_{-2,7}$ may be viewed as the union of the two exteriors

$$Z_1 = Z_2 = S^3 \setminus nbd(T_{-2,7})$$

along an annulus representing a meridian. We begin by describing the relevant subset of $\chi(Z)$ and its image $i_Z^*(\chi(Z)) \subset \chi(\partial Z))$.

The fundamental group $\pi_1(Z)$ is an amalgamated free product of $\pi_1(Z_1)$ and $\pi_1(Z_2)$, where particular meridians on each of the two knot complements are identified. For any representations $\rho_i : \pi_1(Z_i) \to SU(2)$, $i = 1, 2$, which agree on the identified meridians, there is a representation of $\pi_1(Z)$ that restricts to ρ_i on $\pi_1(Z_i)$. The longitude for the composite knot is the product of the longitudes for Z_1 and Z_2 so, roughly speaking, the longitudinal coordinates in the pillowcase pictures for Z_1, Z_2 add.

The fiber product/gluing parameter results above (this time with restrictions to the annulus instead of to ∂X) demonstrate that abelian arcs and the irreducible arcs in $\chi(Z_i)$, $i = 1, 2$, described in Proposition 1 give rise to the following subsets of $\chi(Z)$:

(i) the abelian arc in $\chi(Z)$,
(ii) two *half-abelian arcs*, namely an abelian/irreducible arc and an irreducible/abelian arc, consisting of representations of $\pi_1(Z)$ that are irreducible on only one of $\pi_1(Z_i)$, and
(iii) a cylinder of irreducible/irreducible representations with S^1 gluing parameter.

All three components of $\chi(Z)$ described in (ii) and (iii) limit to the abelian points $a = \frac{\pi}{14}, \frac{13\pi}{14}$ on the abelian arc of $\chi(Z)$. Under i_Z^*, the abelian arc maps to the (image in $\chi(\partial Z)$ of) the x axis; the two half abelian arcs in (ii) map to line segments of slope 14, and the cylinder in (iii) maps onto a line segment of slope 28. This is summarized in Figure 4.

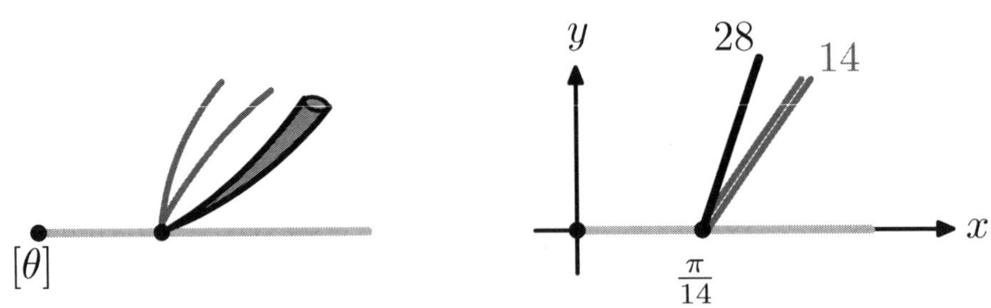

FIGURE 4. Local picture of $\chi(Z) = \chi(S^3 \setminus nbd(T_{-2,7}\#T_{-2,7}))$ near $[\theta]$ (on left) and its image under $i_Z^* \colon \chi(Z) \to \chi(\partial Z)$ (on right)

Under the map $h_Z^* \colon \chi(\partial Z) \to \chi(\partial X)$, the origin is fixed (i.e., $h_Z^*([\theta_{\partial Z}]) = [\theta_{\partial X}]$), the abelian arc in $\chi(Z)$ maps to the line segment $y = -x$, the half abelian arcs described above map onto line segments leaving the abelian arc with slope -15, and the image of the cylinder maps onto a line segment of slope -29. This is summarized in Figure 5.

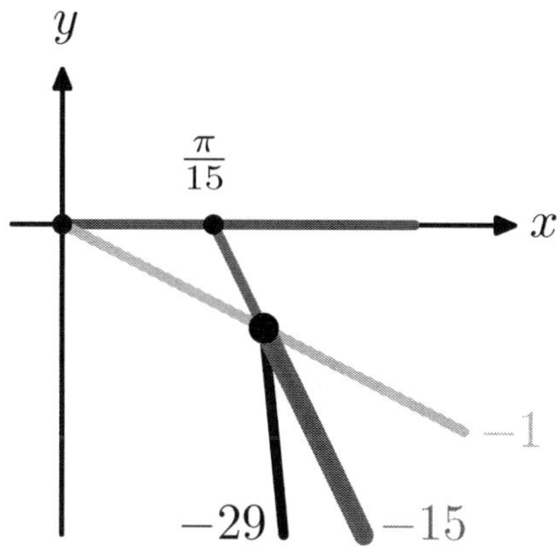

FIGURE 5. The images $i_X^*(\chi(X))$ and $h_Z^* \circ i_Z^*(\chi(Z))$ near $[\theta]$ in $\chi(\partial X)$

In Figure 5, we overlay the images of $i_X^*(\chi(X))$ and $h_Z^*(i_Z^*(\chi(Z)))$ in the same picture, so that we can apply the same sort of fiber product/gluing parameter reasoning to $\Sigma_2 = X \cup_{h_Z} Z$. We begin by noting that the abelian arc of $\chi(Z)$ meets the irreducible arc in $\chi(X)$ drawn in Figure 1 at the point $(\frac{\pi}{14}, -\frac{\pi}{14})$. This intersection corresponds to a point $[\rho_0] \in \chi(\Sigma_2)$ restricting to an irreducible representation of $\pi_1(X)$ and an abelian representation of $\pi_1(Z)$, so there is no gluing parameter. Nearby, however, the intersection includes a line segment emanating down from this point with slope -15. The preimage of that segment in $\chi(X)$ is the irreducible arc in Figure 1 and the preimage in $\chi(Z)$ is the left ends of the two half-abelian arcs on in Figure 4.

Taking gluing parameters into account, $[\rho_0]$ has a neighborhood in $\chi(\Sigma_2)$ which is a cone on two disjoint circles, so the path component containing $[\rho_0]$ is not a manifold. This proves the second assertion of Theorem A. □

5. Further discussion and other examples

We note the following fact about the spaces Σ_1, Σ_2.

PROPOSITION 2. *The homology 3-sphere Σ_1 is diffeomorphic to $+1$ surgery on $T_{3,5}\#T_{2,7}$, and Σ_2 is diffeomorphic to $+1$ surgery on $T_{3,5}\#T_{-2,7}\#T_{-2,7}$. They are both graph manifolds.* □

More lengthy calculations using similar techniques allow the analysis of the full character varieties of both of these homology spheres Σ_i, $i = 1, 2$, as well as more complicated constructions involving more torus knot complements. We highlight a few related results without proof for the interested reader.

PROPOSITION 3. *The irreducible character variety $\chi^*(\Sigma_1)$ consists of 22 isolated points with trivial Zariski tangent space, six isolated points with 2-dimensional Zariski tangent space like the one we described in detail, and a collection of Morse-Bott circle components.* □

While we have focused in this paper on the unperturbed Chern-Simons function, the effect of a small (carefully selected) holonomy perturbations on $\chi(\Sigma_1)$ is also reasonably straightforward to understand. A simple holonomy perturbation in a neighborhood of ∂X can be selected so that $i_X^*(\chi(X))$ undergoes a vertical Hamiltonian flow supported away from the central endpoints on the abelian arc, so that each of the six singular isolated points resolves into a Morse critical point and the Morse-Bott circle components remain (see, for example, [**HK18**]). Under a further perturbation using a curve that cuts through ∂X to break the symmetry giving rise to the gluing parameters, the Chern-Simons function can be made into a Morse function; the Morse-Bott circles can be seen to each contribute two isolated critical points (contributing zero points, counted algebraically, to the Casson invariant).

For clarity, the figures only show the neighborhood of the left endpoints of the irreducible arcs described in Proposition 1, but the irreducible arc parameterizations in that proposition show that the $T_{\pm 2,7}$ arcs extend further to the right than the irreducible $T_{3,5}$ arc. The following proposition is easily proved by tracking the images of the entire half-abelian arcs (see the last two paragraphs of Section 4).

PROPOSITION 4. *The path component of $\chi^*(\Sigma_2)$ containing the singular point $[\rho_0]$ is homeomorphic to a wedge of two 2-spheres.* □

PROPOSITION 5. *If one replaces Z with $S^3 \setminus nbd(3T_{-2,7}\#T_{2,7})$ in the construction of Σ_2, then the corresponding point at $(\frac{\pi}{14}, -\frac{\pi}{14})$ has a neighborhood that is a cone on the disjoint union of two circles and a 3-torus.* □

Finally, we note that the homology spheres Σ_1, Σ_2 can also be decomposed using Heegaard splittings, giving rise to different fiber product descriptions of the singularities described in Theorem A. For Heegaard splittings, the character varieties of the handlebodies are smooth submanifolds of the smooth locus of the character variety of the Heegaard surface in a neighborhood of their intersections points. Hence the local singular structure in the fiber product description associated to this decomposition is due to this pair of handlebody character varieties intersecting nontransversely.

References

[AM90] Selman Akbulut and John D. McCarthy, *Casson's invariant for oriented homology 3-spheres*, Mathematical Notes, vol. 36, Princeton University Press, Princeton, NJ, 1990. An exposition, DOI 10.1515/9781400860623. MR1030042

[BHK05] Hans U. Boden, Christopher M. Herald, and Paul A. Kirk, *The integer valued* SU(3) *Casson invariant for Brieskorn spheres*, J. Differential Geom. **71** (2005), no. 1, 23–83. MR2191768

[FS90] Ronald Fintushel and Ronald J. Stern, *Instanton homology of Seifert fibred homology three spheres*, Proc. London Math. Soc. (3) **61** (1990), no. 1, 109–137, DOI 10.1112/plms/s3-61.1.109. MR1051101

[HHK14] Matthew Hedden, Christopher M. Herald, and Paul Kirk, *The pillowcase and perturbations of traceless representations of knot groups*, Geom. Topol. **18** (2014), no. 1, 211–287, DOI 10.2140/gt.2014.18.211. MR3158776

[HK18] Christopher M. Herald and Paul Kirk, *Holonomy perturbations and regularity for traceless* SU(2) *character varieties of tangles*, Quantum Topol. **9** (2018), no. 2, 349–418, DOI 10.4171/QT/110. MR3812815

[Kir97] Rob Kirby (ed.), *Problems in low-dimensional topology*, Geometric topology (Athens, GA, 1993), AMS/IP Stud. Adv. Math., vol. 2, Amer. Math. Soc., Providence, RI, 1997, pp. 35–473, DOI 10.1090/amsip/002.2/02. MR1470751

[Kla91] Eric Paul Klassen, *Representations of knot groups in* SU(2), Trans. Amer. Math. Soc. **326** (1991), no. 2, 795–828, DOI 10.2307/2001784. MR1008696

[KM17] Michael Kapovich and John J. Millson, *On representation varieties of 3-manifold groups*, Geom. Topol. **21** (2017), no. 4, 1931–1968, DOI 10.2140/gt.2017.21.1931. MR3654101

[Nic11] Liviu Nicolaescu, *An invitation to Morse theory*, 2nd ed., Universitext, Springer, New York, 2011, DOI 10.1007/978-1-4614-1105-5. MR2883440

[Tau90] Clifford Henry Taubes, *Casson's invariant and gauge theory*, J. Differential Geom. **31** (1990), no. 2, 547–599. MR1037415

Department of Mathematics and Statistics, McMaster University, Hamilton, Ontario L8S 4K1, Canada
Email address: boden@mcmaster.ca

Department of Mathematics and Statistics, University of Nevada, Reno, Nevada 89557
Email address: herald@unr.edu

Department of Mathematics, Indiana University, Bloomington, Indiana 47405
Email address: pkirk@indiana.edu

Proceedings of Symposia in Pure Mathematics
Volume **109**, 2024
https://doi.org/10.1090/pspum/109/01992

Blowup formulas for Segre and Verlinde numbers of surfaces and higher rank Donaldson invariants

Lothar Göttsche

Dedicated to Tomasz Mrowka, on the occasion of his 60th birthday

Abstract. We formulate conjectural blowup formulas for Segre and Verlinde numbers on moduli spaces of sheaves on projective surfaces S with $p_g(S) > 0$ and $b_1(S) = 0$. As application we give a give a conjectural formula for the Donaldson invariants of S in arbitrary rank, generalizing the famous structure formula for Donaldson invariants of Kronheimer and Mrowka. Furthermore we give conjectural formulas for the K-theoretic Donaldson invariants, as well as for some Donaldson invariants with fundamental matters.

1. Introduction

In this paper we study virtual enumerative invariants of moduli spaces of sheaves on projective algebraic surfaces S, with first Betti number $b_1(S) = 0$ and geometric genus $p_g(S) > 0$. In particular we study two kinds of invariants, the Verlinde and Segre invariants of these moduli spaces. We will formulate conjectural blowup formulas relating the Verlinde and Segre invariants for the surface S and its blowup $\widehat{S}$ in a point. The analogs of many of these conjectures have already been proven by Nakajima and Yoshioka in [**NY**] in the setting of instanton counting, i.e. the equivariant setting on $\mathbb{A}^2$. We will give evidence for these conjectures and then explore some of their consequences. The main focus of this paper is on the consequences of these blowup formulas.

Let H be an ample line bundle on S. Let $M := M_S^H(\rho, c_1, c_2)$ be the moduli space of Gieseker H-semistable torsion-free sheaves E on S of rank $\mathrm{rk}(E) = \rho$ and with Chern classes $c_1(E) = c_1$, $c_2(E) = c_2$. We will assume in the whole paper that $M_S^H(\rho, c_1, c_2)$ consists only of stable sheaves, although we expect that the conjectures of this paper will also hold without this assumption, if we define the invariants as in Mochizuki [**Moc**], or via Joyce Song pairs [**JS**]. We denote by $K_0(X)$ the Grothendieck group of coherent sheaves on a scheme X, and by $K^0(X)$ the Grothendieck group of locally free sheaves. M carries a perfect obstruction theory as defined in [**Moc**], of expected dimension

$$\mathrm{vd} = \mathrm{vd}(M) = \mathrm{vd}(\rho, c_1, c_2) = 2\rho c_2 - (\rho - 1)c_1^2 - (\rho^2 - 1)\chi(\mathcal{O}_S),$$

and therefore has a virtual fundamental class $[M]^{\mathrm{vir}} \in H_{2\mathrm{vd}}(M, \mathbb{Z})$ and a virtual structure sheaf $\mathcal{O}_M^{\mathrm{vir}}$ in $K_0(M)$. To a class $V \in K^0(M)$ we can associate the virtual

2020 *Mathematics Subject Classification.* Primary 14J60; Secondary 14D21, 57K41.

holomorphic Euler characteristic $\chi^{\mathrm{vir}}(M,V) := \chi(M, V\otimes \mathcal{O}_M^{\mathrm{vir}})$. We can use this virtual structure to define and study two types of enumerative invariants of M, both associated to classes $\alpha \in K^0(S)$:

(1) the virtual Verlinde numbers $\chi^{\mathrm{vir}}(M,\lambda(\alpha))$, for $\lambda(\alpha)\in \mathrm{Pic}(M)$ the determinant line bundle associated to α,
(2) the virtual Segre invariants $\int_{[M]^{\mathrm{vir}}} c(\tau(\alpha))$, for $\tau(\alpha)\in K^0(M)$ a tautological "vector bundle" associated to α, which generalizes the tautological bundles $\alpha^{[n]}$ on Hilbert schemes of points.

We briefly review definitions and properties, both known and conjectured of these invariants, mostly based on [**GK1**]. They generalize the corresponding results on Hilbert schemes of points from [**EGL**], [**MOP1**], [**MOP2**], [**MOP3**]. More details can be found in Section 2 of this paper and in [**GK1**].

In this whole paper let S be a smooth projective surface with $b_1(S) = 0$, $p_g(S) > 0$, and let H be an ample line bundle on S. For any $\rho \in \mathbb{Z}_{>0}$, $c_1 \in H^2(S,\mathbb{Z})$ and $c_2 \in H^4(S,\mathbb{Z})$, let $M := M_S^H(\rho, c_1, c_2)$ be the moduli space of rank ρ Gieseker H-semistable torsion free sheaves on S with Chern classes c_1, c_2. We will always assume in the following that M contains no strictly semistable sheaves. For simplicity of exposition we also assume that there exists a universal sheaf $\mathbb{E}$ on $S\times M$, but in [**GK1**, Rem. 2.4, 2.5] it is explained how to remove this assumption. Let $\pi_S : S\times M \to S$ and $\pi_M : S\times M\to M$ be the projections to the factors.

Virtual Segre invariants: Consider the slant product

$$/ : H^p(S\times M,\mathbb{Q}) \times H_q(S,\mathbb{Q}) \to H^{p-q}(M,\mathbb{Q}).$$

For any $\gamma \in H^k(S,\mathbb{Q})$ let

$$\mu(\gamma) := \Big(c_2(\mathbb{E}) - \frac{\rho-1}{2\rho}c_1(\mathbb{E})^2\Big)/\mathrm{PD}(\gamma) \in H^k(M,\mathbb{Q}),$$

where $\mathrm{PD}(\gamma)$ denotes the Poincaré dual. For any class $\alpha\in K^0(S)$, we define

$$\mathrm{ch}(\tau(\alpha)) := \mathrm{ch}(-\pi_{M!}(\pi_S^*\alpha\cdot \mathbb{E}\cdot \det(\mathbb{E})^{-\frac{1}{\rho}})) \in A^*(M)_{\mathbb{Q}},$$

where $A^*(M)_{\mathbb{Q}}$ denotes the Chow ring with rational coefficients. When the root $\det(\mathbb{E})^{-1/\rho}$ does not exist, the right hand side is defined by a formal application of the Grothendieck-Riemann-Roch formula. We write $c(\tau(\alpha)) = \sum_i c_i(\tau(\alpha))$ for the Chern classes corresponding to $\mathrm{ch}(\tau(\alpha))$. Let $\mathrm{pt}\in H^4(S,\mathbb{Z})$ be the Poincaré dual of the point class, and $L\in H^2(S,\mathbb{Z})$. The virtual Segre invariant of S associated to $(\rho, c_1, c_2, \alpha, L)$ is $\int_{[M]^{\mathrm{vir}}} c(\tau(\alpha)) \exp(\mu(L)+\mu(\mathrm{pt})u)$, where u is a formal variable. In the future we will always put $s := \mathrm{rk}(\alpha)$.

Virtual Verlinde invariants: Let $c\in K(S)_{\mathrm{num}}$ be a class in the numerical Grothendieck group of S such that $\mathrm{rk}(c) = \rho$, $c_1(c) = c_1$, and $c_2(c) = c_2$. Denote

$$K_c := \{v \in K^0(S)\ :\ \chi(S, c\otimes v) = 0\}.$$

The map

$$\lambda := \lambda_{\mathbb{E}} : K_c \to \mathrm{Pic}(M), \quad \beta \mapsto \det\big(\pi_{M!}(\pi_S^*\beta\cdot[\mathbb{E}])\big)^{-1}. \tag{1.1}$$

does not depend on the choice of the universal family and can be extended to the case where the universal family does not exist. Fix $r\in\mathbb{Z}$, $L\in \mathrm{Pic}(S)\otimes\mathbb{Q}$ such that $\mathcal{L} := L\otimes \det(c)^{-\frac{r}{\rho}} \in \mathrm{Pic}(S)$ and ρ divides $\mathcal{L}c_1 + r\big(\frac{1}{2}c_1(c_1-K_S) - c_2\big)$. Take $v\in K^0(S)$ such that

- $\mathrm{rk}(v) = r$ and $c_1(v) = \mathcal{L}$,
- $c_2(v) = \frac{1}{2}\mathcal{L}(\mathcal{L} - K_S) + r\chi(\mathcal{O}_S) + \frac{1}{\rho}\mathcal{L}c_1 + \frac{r}{\rho}\big(\frac{1}{2}c_1(c_1 - K_S) - c_2\big)$.

The second condition is equivalent to $v \in K_c \subset K^0(S)$. We define $\mu(L) \otimes E^{\otimes r} := \lambda(v)$. Let $\mathcal{O}_M^{\mathrm{vir}}$ be the virtual structure sheaf of M. For $V \in K^0(M)$ let $\chi^{\mathrm{vir}}(M, V) := \chi(M, V \otimes \mathcal{O}_M^{\mathrm{vir}})$. The virtual Verlinde number of S associated to (ρ, c_1, c_2, L, r) is $\chi^{\mathrm{vir}}(M, \mu(L) \otimes E^{\otimes r})$.

If $\rho = 1$ and $c_1 = 0$, $M_S^H(1, 0, n)$ is just the Hilbert schemes $S^{[n]}$ of n points on S. Let $Z_n(S) \subset S \times S^{[n]}$ be the universal subscheme with the projections $p : Z_n(S) \to S^{[n]}$ and $q : Z_n(S) \to S$. For $\alpha \in K^0(S)$, the associated tautological bundle on $S^{[n]}$ is defined by $\alpha^{[n]} = p_* q^*(\alpha) \in K^0(S^{[n]})$. Then we have $\alpha^{[n]} = \tau(\alpha)$ in the notations above, thus the virtual Segre invariants generalize the Segre numbers $\int_{S^{[n]}} c(\alpha^{[n]})$ on Hilbert schemes of points. Similarly if $r = \mathrm{rk}(\alpha)$ and $L = \det(\alpha)$, we have that $\mu(L) \otimes E^{\otimes r} = \det(\alpha^{[n]})$; thus the virtual Verlinde invariants generalize the Verlinde numbers of line bundles on Hilbert schemes of points.

Conjectural formulas and relations. We state the conjectures of [**GK1**] expressing the Segre and Verlinde numbers in terms of Seiberg-Witten invariants. We denote the Seiberg-Witten invariant of S in class $a \in H^2(S, \mathbb{Z})$ by $SW(a)$. We follow Mochizuki's [**Moc**] convention $SW(a) = \widetilde{SW}(2a - K_S)$, where $\widetilde{SW}$ denotes the usual Seiberg-Witten invariant from differential geometry. For S a surface with $b_1(S) = 0$ and $p_g(S) > 0$, there are finitely many $a \in H^2(S, \mathbb{Z})$ with $SW(a) \neq 0$, called Seiberg-Witten basic classes.

NOTATION 1.1. We fix $\rho \in \mathbb{Z}_{>0}$, and $s, r \in \mathbb{Z}$. We denote $\varepsilon_\rho := \exp(2\pi i/\rho)$ and write $[n] := \{1, \ldots, n\}$ for any $n \in \mathbb{Z}_{\geq 0}$. For integers $n < m$ we also write $[n, m]$ for the integers l with $n \leq l \leq m$. We introduce the following changes of variables

$$z = t(1 + (1 - \tfrac{s}{\rho})t^2)^{\frac{1}{2}(1-\frac{s}{\rho})}, \quad w = v(1 + v^2)^{\frac{1}{2}(\frac{r^2}{\rho^2} - 1)},$$

and the following formal power series

$$\begin{aligned}
V_s^{(\rho)}(z) &= (1 + (1 - \tfrac{s}{\rho})t^2)^{\rho - s}(1 + (2 - \tfrac{s}{\rho})t^2)^s, \\
W_s^{(\rho)}(z) &= (1 + (1 - \tfrac{s}{\rho})t^2)^{\frac{1}{2}(s - \rho - 1)}(1 + (2 - \tfrac{s}{\rho})t^2)^{\frac{1}{2}(1-s)}, \\
X_s^{(\rho)}(z) &= (1 + (1 - \tfrac{s}{\rho})t^2)^{\frac{1}{2}(s^2 - (\rho + \frac{1}{\rho})s)}(1 + (2 - \tfrac{s}{\rho})t^2)^{-\frac{1}{2}s^2 + \frac{1}{2}} \cdot \\
&\quad \cdot (1 + (1 - \tfrac{s}{\rho})(2 - \tfrac{s}{\rho})t^2)^{-\frac{1}{2}}, \\
Q_s^{(\rho)}(z) &= \tfrac{1}{2}t^2(1 + (1 - \tfrac{s}{\rho})t^2), \quad R_s^{(\rho)}(z) = t^2, \\
T_s^{(\rho)}(z) &= \rho t^2(1 + \tfrac{1}{2}(1 - \tfrac{s}{\rho})(2 - \tfrac{s}{\rho})t^2), \\
G_r^{(\rho)}(w) &= 1 + v^2, \quad F_r^{(\rho)}(w) = \frac{(1 + v^2)^{\frac{r^2}{\rho^2}}}{(1 + \frac{r^2}{\rho^2}v^2)}.
\end{aligned}$$

We warn the reader that this notation differs from that of [**GK1**]: $z^{\frac{1}{2}}, t^{\frac{1}{2}}, w^{\frac{1}{2}}, v^{\frac{1}{2}}$ from [**GK1**] were replaced by where z, t, w, v. Also in the following conjectures $Z_s^{(\rho)}$ from [**GK1**] was replaced by $\rho Z_s^{(\rho)}$ and $B_r^{(\rho)}$ by $\rho B_r^{(\rho)}$.

CONJECTURE 1.2 ([**GK1**, Conj. 2.8, 2.9]). *For $\rho \in \mathbb{Z}_{>0}$ and $s, r \in \mathbb{Z}$, there exist $Y_s^{(\rho)}$, $Z_s^{(\rho)}$, $Y_{j,s}^{(\rho)}$, $Z_{jk,s}^{(\rho)}$, $S_s^{(\rho)}$, $S_{j,s}^{(\rho)} \in \mathbb{C}[[z]]$ and $A_r^{(\rho)}$, $B_r^{(\rho)}$, $A_{j,r}^{(\rho)}$, $B_{jk,r}^{(\rho)} \in \mathbb{C}[[w]]$*

for all $1 \le j \le k \le \rho - 1$ such that the following holds. For any $\alpha \in K^0(S)$ with $\mathrm{rk}(\alpha) = s$ *and $L \in \mathrm{Pic}(S)$, we put*

$$\phi_{S,\rho,c_1,\alpha,L} := \rho^{2-\chi(\mathcal{O}_S)} \, (V_s^{(\rho)})^{c_2(\alpha)} \, (W_s^{(\rho)})^{c_1(\alpha)^2} \, (X_s^{(\rho)})^{\chi(\mathcal{O}_S)} \, (Y_s^{(\rho)})^{c_1(\alpha)K_S} \, (Z_s^{(\rho)})^{K_S^2} \cdot e^{L^2 Q_s^{(\rho)} + (c_1(\alpha)L)R_s^{(\rho)} + (LK_S)S_s^{(\rho)} + u\, T_s^{(\rho)}}$$
$$\cdot \sum_{(a_1,\ldots,a_{\rho-1})} \prod_{j=1}^{\rho-1} \varepsilon_\rho^{ja_jc_1} \, \mathrm{SW}(a_j) \, (Y_{j,s}^{(\rho)})^{c_1(\alpha)a_j} \, e^{(a_jL)S_{j,s}^{(\rho)}} \prod_{1\le j\le k\le \rho-1} (Z_{jk,s}^{(\rho)})^{a_ja_k},$$
$$\psi_{S,\rho,c_1,L,r} := \rho^{2-\chi(\mathcal{O}_S)} \, (G_r^{(\rho)})^{\chi(L)} \, (F_r^{(\rho)})^{\frac{1}{2}\chi(\mathcal{O}_S)} \, (A_r^{(\rho)})^{LK_S} \, (B_r^{(\rho)})^{K_S^2}$$
$$\cdot \sum_{(a_1,\ldots,a_{\rho-1})} \prod_{j=1}^{\rho-1} \varepsilon_\rho^{ja_jc_1} \, \mathrm{SW}(a_j) \, (A_{j,r}^{(\rho)})^{a_jL} \prod_{1\le j\le k\le \rho-1} (B_{jk,r}^{(\rho)})^{a_ja_k}.$$

Let $M = M_S^H(\rho, c_1, c_2)$ consist only of stable sheaves. Then

$$\int_{[M]^{\mathrm{vir}}} c(\tau(\alpha)) \exp(\mu(L) + \mu(\mathrm{pt})u) = \mathrm{Coeff}_{z^{\mathrm{vd}(M)}}[\phi_{S,\rho,c_1,\alpha,L}],$$
$$\chi^{\mathrm{vir}}(M, \mu(L) \otimes E^{\otimes r}) = \mathrm{Coeff}_{w^{\mathrm{vd}(M)}}[\psi_{S,\rho,c_1,L,r}].$$

where the sums are over all $(a_1, \ldots, a_{\rho-1}) \in H^2(S, \mathbb{Z})^{\rho-1}$. Furthermore, $Y_s^{(\rho)}$, $Y_{j,s}^{(\rho)}$, $Z_s^{(\rho)}$, $Z_{jk,s}^{(\rho)}$, $S_s^{(\rho)}$, $S_{j,s}^{(\rho)}$, $A_r^{(\rho)}$, $B_r^{(\rho)}$, $A_{j,r}^{(\rho)}$, $B_{jk,r}^{(\rho)}$ are all algebraic functions.

The rest of this paper depends on Conjecture 1.2, because it states the existence of universal power series; universal in the sense that they depend only on ρ and r respectively ρ and s, and in particular are independent of S. These universal power series are the subject a number of other conjectures, which describe the properties of these universal power series and in many instances determine them, always using that they are indeed universal. In Remark 6.4 the implications between these conjectures are listed.

In this paper we formulate conjectural blowup formulas for the Verlinde and Segre invariants of moduli spaces of sheaves on surfaces. These formulas give a simple relationship between the invariants on a surface S and some of the invariants on its blowup $\widehat{S}$ in a point. These blowup formulas can be viewed as a generalization in the case of algebraic surfaces of the specialization the blowup formulas of [**FS1**] to 4-manifolds of simple type.

For moduli spaces M of sheaves of rank ρ, most of them hold for the Verlinde numbers $\chi^{\mathrm{vir}}(M, \mu(L) \otimes E^{\otimes r})$ under the condition that $|r| \le \rho$, and for the Segre invariants under the condition that $0 \le s \le 2\rho$. We have found these relations experimentally as relations among the universal functions $A_{J,r}^{(\rho)}$, $B_{J,r}^{(\rho)}$ for the Verlinde numbers and $Y_{J,r}^{(\rho)}$, $Z_{J,r}^{(\rho)}$, $S_{J,r}^{(\rho)}$ for the Segre invariants, which we will introduce in Section 2. Here we will state them directly for the generating functions of the invariants of the moduli spaces, where they take a simpler and more attractive form. We start with the blowup formulas for the Verlinde numbers.

CONJECTURE 1.3. *For any smooth projective surface S with $b_1(S) = 0$ and $p_g(S) > 0$, let $\pi : \widehat{S} \to S$ be the blowup in a point with exceptional divisor D. Let $|r| \le \rho$.*

(1) *For* $k\in[0,\rho]$ *and in addition, if* $|r|<\rho$, *for* $k\in[r,r+\rho]$ *we have*

$$\psi_{\widehat{S},\rho,\pi^*c_1,L+kD,r}=\psi_{S,\rho,c_1,L,r},\quad k\in[0,\rho]. \tag{1.2}$$

(2) *For* $\ell=1,\ldots,\rho-1$ *we have*

$$\psi_{\widehat{S},\rho,\pi^*c_1-\ell D,\pi^*L+(k+(\rho-\ell)\frac{r}{\rho})D,r}=\begin{cases}0, & k\in[1,\rho-1],\\ (-w)^{\ell(\rho-\ell)}\psi_{S,\rho,c_1,L,r}, & k{=}0,\ r\in[-\rho{+}1,\rho],\\ (-w)^{\ell(\rho-\ell)}\frac{\psi_{S,\rho,c_1,L,r}}{(1-(-v^2)^\rho)^{\rho-\ell}}, & k{=}0,\ r{=}{-}\rho,\\ w^{\ell(\rho-\ell)}\psi_{S,\rho,c_1,L,r}, & k{=}\rho,\ r{\in}[-\rho,\rho{-}1],\\ w^{\ell(\rho-\ell)}\frac{\psi_{S,\rho,c_1,L,r}}{(1-(-v^2)^\rho)^{\rho-\ell}}, & k=\rho,\ r=\rho.\end{cases} \tag{1.3}$$

For the Segre invariants, we get the following blowup formulas.

CONJECTURE 1.4. *Let* S *be a surface with* $p_g(S)>0$, $b_1(S)=0$. *Let* $L\in\mathrm{Pic}(S)$, *let* $\widehat{S}$ *be the blowup of* S *in a point with exceptional divisor* D. *Let* $\alpha\in K^0(S)$ *be a class of rank* $s\in[0,2\rho]$. *Let* $L\in H^2(S,\mathbb{Q})$.

(1) *For* $k=[0,\rho]$, *and furthermore, if* $s\in[1,2\rho-1]$ *for* $k\in[s-\rho,s]$, *we have*

$$\phi_{\widehat{S},\rho,\pi^*c_1,\pi^*\alpha+k\mathcal{O}_D(D),\pi^*L}=\phi_{S,\rho,c_1,\alpha,L}.$$

(2) *For* $k\geq 0$,

$$\phi_{\widehat{S},\rho,\pi^*c_1,\pi^*\alpha+k\mathcal{O}_D(D),\pi^*L-xD}=\phi_{S,\rho,c_1,\alpha,L}+O(x^{\rho-k+1}),$$

and furthermore, if $s\in[1,2\rho-2]$, *we have for* $k\leq s$

$$\phi_{\widehat{S},\rho,\pi^*c_1,\pi^*\alpha+k\mathcal{O}_D(D),\pi^*L-xD}=\phi_{S,\rho,c_1,\alpha,L}+O(x^{\rho+k-s+1}).$$

(3) *Let* $\widehat{c}_1\in H^2(\widehat{S},\mathbb{Z})$ *with* $\widehat{c}_1D=\ell\in[1,\rho-1]$, *let* $\widehat{\alpha}\in K^0(\widehat{S})\otimes\mathbb{Q}$ *with* $c_1(\widehat{\alpha})D=\ell\frac{s}{\rho}-s-k$ *with* $k\in\mathbb{Z}$. *Then*

$$\phi_{\widehat{S},\rho,\widehat{c}_1,\widehat{\alpha},\pi^*L-xD}=O(x^{\min(\ell(\rho-\ell+k),(\ell-k)(\rho-\ell))}).$$

In particular $\phi_{\widehat{S},\rho,\widehat{c}_1,\widehat{\alpha},\pi^*L}=0$ *for* $k\in[-\rho+\ell+1,\ell-1]$.

REMARK 1.5. Analogs of these conjectures have been proven before by Nakajima and Yoshioka in [**NY**] for equivariant intersection numbers on instanton moduli spaces on the plane and its blowup in the origin. In fact the analogue of part (1) and (2) of Conjecture 1.4 is [**NY**, Thm. 2.1], and that of part (3) of Conjecture 1.4 is [**NY**, Thm. 2.4]. Similarly [**NY**, Thm. 2.11] is a partial analogue of Conjecture 1.3.

Furthermore Tanaka and Kuhn have recently in [**KT**] proven general, but less explicit blowup formulas for the virtual invariants of the moduli spaces. As pointed out by Nakajima, it is likely that the conjectures of this paper can be proven by combining the methods and results of [**NY**] and [**KT**], and this project is currently being pursued by Kuhn and Tanaka.

Together with the known and conjectured properties of the Verlinde and Segre invariants mentioned above, these conjectural blowup formulas impose rather strong constraints. In many cases this allows to determine conjectural formulas for the universal functions $A^{(\rho)}_{J,r}$, $B^{(\rho)}_{J,r}$, $Y^{(\rho)}_{J,s}$, $Z^{(\rho)}_{J,s}$, $S^{(\rho)}_{J,s}$. This is the first step in a program to apply the blowup formulas together with other conjectured and known relations for the Segre and Verlinde invariants to completely determine their generating functions. In a work in progress, when $\rho \le 6$ and $|r| \le \rho$ or $0 \le s \le 2\rho$, we experimentally find that these relations determine the power series $A^{(\rho)}_{J,r}$, $B^{(\rho)}_{J,r}$, $Y^{(\rho)}_{J,s}$, $Z^{(\rho)}_{J,s}$, $S^{(\rho)}_{J,s}$ completely, and we determine them modulo w^{100} respectively z^{100}. This allows to find the conjectural algebraic relations that these power series fulfill. In the last section of this paper one can see the resulting formulas in case $\rho \le 4$.

In further work [**G2**] in progress we will state a conjectural virtual strange duality, and will use this to give conjectural formulas for the $A^{(\rho)}_{J,r}$, $B^{(\rho)}_{J,r}$ and $Y^{(\rho)}_{J,s}$, $Z^{(\rho)}_{J,s}$ for arbitrary values of ρ and r (respectively s). Using preliminary results from [**G2**] in [**GM1**], a complete formula for the generating functions for the Verlinde and Segre invariants of Hilbert schemes of points on a surface is given, and it is proven for surfaces with $K_S^2 = 0$.

Donaldson invariants in arbitrary rank. In this paper we will mostly concentrate on the values $r = -\rho, 0, \rho$ for the Verlinde numbers and $s = 0, \rho, 2\rho$ for the Segre invariants. In these cases we will be able to determine conjectural generating functions for the Segre and Verlinde invariants of the moduli spaces $M = M_S^H(\rho, c_1, c_2)$ of sheaves of arbitrary rank ρ. Of particular interest are the Donaldson invariants of arbitrary rank ρ, which are a special instance of the case $s = 0$ of the Segre invariants. The Segre invariants with a given s are sometimes in the physics language called Donaldson invariants with s fundamental matters. Thus we conjecturally determine also the Donaldson invariants in rank ρ with ρ or 2ρ fundamental matters. The Verlinde numbers with $r = 0$ are sometimes called K-theoretic Donaldson invariants.

Donaldson invariants are invariants of differentiable 4-manifolds X, defined using moduli spaces of anti-selfdual connections on principal $SU(2)$ and $SO(3)$ bundles on X. If X is a projective algebraic surface S, they can be computed as intersection numbers of μ-classes on moduli spaces of sheaves $M_S^H(2, c_1, c_2)$ on S (see [**Li**],[**Mor**], [**FM**]). The definition of Donaldson invariants has been generalized in [**Kro**] to higher rank bundles. A similar proof as in the rank 2 should show that they can be computed in the same way using moduli spaces of higher rank sheaves $M_S^H(\rho, c_1, c_2)$, although I am not aware of a proof in the literature.

DEFINITION 1.6. Let (S, H) be a polarized surface and fix $\rho \in \mathbb{Z}_{>0}$, $c_1 \in H^2(S, \mathbb{Z})$. Let $L \in H^2(S, \mathbb{Q})$. The rank ρ Donaldson invariants of S with respect to H, c_1 are

$$D^{S,H}_{\rho,c_1,c_2}(L + u\mathrm{pt}) = \int_{[M_S^H(\rho,c_1,c_2)]^{\mathrm{vir}}} \exp(\mu(L) + \mu(\mathrm{pt})u).$$

In the rank 2 case in [**KM1**],[**KM2**] Kronheimer and Mrowka introduced the notion of 4-manifolds of simple type, and gave a general structure theorem for the generating functions of the Donaldson invariants of 4-manifolds of simple type (see also [**FS2**]). This structure theorem expresses the generating function in terms of so-called basic classes.

With the advent of Seiberg-Witten theory these basic classes were at least conjecturally identified with the so called Seiberg-Witten basic classes. For $a \in H^2(S, \mathbb{Z})$ we put $\widetilde{a} := 2a - K_S \in H^2(S, \mathbb{Z})$. As mentioned above, we denote by $\widetilde{SW}$ the Seiberg-Witten invariants in gauge theory, which are related to our convention by $\widetilde{SW}(\widetilde{a}) = SW(a)$. The Witten conjecture [**Wit**] expresses the rank 2 Donaldson invariants in terms of the Seiberg-Witten invariants, and thus gives a new interpretation of the Kronheimer-Mrowka structure theorem. It was proven in [**GNY2**] for projective surfaces, and (modulo a technical assumption) for all differentiable 4-manifolds M with $b_1(M) = 1$, odd $b_+(M)$ and of Seiberg-Witten simple type in [**FL1, FL2**].

The Marino-Moore conjecture [**MM**] is a conjectural generalization of the Witten conjecture, and thus also of the structure theorem of Kronheimer and Mrowka to Donaldson invariants of arbitrary rank, although the formulas of [**MM**] do not appear to be completely explicit. The following conjecture can be viewed as a completely explicit version of the Marino-Moore conjecture.

NOTATION 1.7. In the whole paper let $\xi = e^{\frac{\pi i}{2\rho}}$ be a primitive 4ρ-th root of unity. For integers n, m we denote $[n, m] := \{k \in \mathbb{Z} \mid n \leq k \leq m\}$, and we put $[m] := [1, m]$. For $1 \leq i < j \leq \rho - 1$ let $\beta_{ij} := \frac{\xi^{i+j} - \xi^{-(i+j)}}{\xi^{j-i} - \xi^{i-j}}$ and $\beta_{ji} := \beta_{ij}$. For any subset $J \subset [\rho - 1]$ put

$$\beta_J := \prod_{\substack{i \in J \\ j \in [\rho-1] \setminus J}} \beta_{ij},$$

in particular $\beta_\emptyset = 1$, and let

$$B := \sum_{J \subset [\rho-1]} \beta_J.$$

CONJECTURE 1.8. *Let (S, H) be a polarized surface and fix $\rho \in \mathbb{Z}_{>0}$, $c_1 \in H^2(S, \mathbb{Z})$, $c_2 \in \mathbb{Z}$, and write* $\mathrm{vd} := \mathrm{vd}(\rho, c_1, c_2)$. *We also write $\xi = e^{\frac{\pi i}{2\rho}}$, and $\varepsilon_\rho = e^{\frac{2\pi i}{\rho}}$. Then $D^{S,H}_{\rho, c_1, c_2}(L + u\mathrm{pt})$ is the coefficient of z^{vd} of*

$$\rho^{2-\chi(\mathcal{O}_S)} B^{K_S^2} e^{(\frac{1}{2}L^2 + \rho u)z^2} \sum_{(a_1, \ldots, a_{\rho-1})} \prod_{j=1}^{\rho-1} \varepsilon_\rho^{j \cdot (a_j, c_1)} \widetilde{SW}(\widetilde{a}_j) e^{-\sin(\pi \frac{j}{\rho})(\widetilde{a}_j L) z} \cdot \prod_{1 \leq i < j \leq \rho-1} \beta_{ij}^{\frac{1}{2}\widetilde{a}_i(\widetilde{a}_j - \widetilde{a}_i)} \tag{1.4}$$

Note that by definition for $1 \leq i < j \leq \rho - 1$, we can write $\beta_{ij} = \frac{\sin(\pi \frac{i+j}{2\rho})}{\sin(\pi \frac{j-i}{2\rho})} \in \mathbb{R}^{(\rho)}_{>0}$, and we choose $\beta_{ij}^{\frac{1}{2}} \in \mathbb{R}_{>0}$.

This extends [**GK1**, Conj. 5.6], where the shape of the formula for general ρ is predicted and the explicit formula given for $\rho \leq 4$. In Section 4.2 we also indicate how this should generalize to 4-manifolds of Seiberg-Witten simple type. One should compare this formula to [**LM**] eq. (10.107), which is also a more explicit version of the Marino-Moore conjectures. There, using physics methods, the formula is determined up to an unknown multiplicative constant, which in our formula (1.4) turns out to be $\rho^{2-\chi(\mathcal{O}_S)} B^{K_S^2}$. Otherwise their formula looks very similar (1.4), but due to differences in notations and conventions I was unable check whether they agree.

Let $M = M_S^H(\rho, c_1, c_2)$, and assume that $\mathrm{vd}(\rho, c_1, c_2) = 0$. Then Conjecture 1.8 in particular gives a formula for the degree of $[M]^{\mathrm{vir}}$:

$$\int_{[M]^{\mathrm{vir}}} 1 = \rho^{2-\chi(\mathcal{O}_S)} B^{K_S^2} \sum_{(a_1,\ldots,a_{\rho-1})} \prod_{j=1}^{\rho-1} \varepsilon_\rho^{j\cdot(a_j,c_1)} \widetilde{SW}(\widetilde{a}_j) \prod_{1\le i<j\le \rho-1} \beta_{ij}^{\frac{1}{2}\widetilde{a}_i(\widetilde{a}_j-\widetilde{a}_i)}.$$

In [**GKL**] we compute generating functions of the monopole contributions to Vafa-Witten invariants using localization on Hilbert schemes of points for $\rho \le 5$, then using the S-duality conjecture we get conjectural generating functions for the virtual Euler numbers of the moduli spaces $M_S^H(\rho, c_1, c_2)$. In particular this gives us conjectural formulas for the degrees of the virtual fundamental classes of moduli spaces $M_S^H(\rho, c_1, c_2)$ of virtual dimension zero for $\rho \le 5$. These formulas also agree with the formulas of Conjecture 1.8.

2. Background material

In this section we briefly review some of the conjectures and definitions from [**GK1**], which can be viewed as generalizations of corresponding conjectures and results from [**EGL**], [**MOP1**], [**MOP2**], [**MOP3**] from the case of Hilbert schemes of points to moduli spaces of sheaves of higher rank. For more details see [**GK1**].

NOTATION 2.1. It is sometimes convenient to organize the universal functions of Conjecture 1.2 in a different form. Recall that for a positive integer n we put $[n] = \{1, \ldots, n\}$. Fix $\rho \in \mathbb{Z}_{>0}$. For any subset $J \subset [\rho - 1]$ we put

$$\begin{aligned} &Y_{J,s}^{(\rho)} := Y_s^{(\rho)} \prod_{j\in J} Y_{j,s}^{(\rho)}, \quad Z_{J,s}^{(\rho)} := Z_s^{(\rho)} \prod_{i\le j\in J} Z_{ij,s}^{(\rho)}, \quad S_{J,s}^{(\rho)} := S_s^{(\rho)} + \sum_{j\in J} S_{j,s}^{(\rho)}, \\ &A_{J,r}^{(\rho)} := A_r^{(\rho)} \prod_{j\in J} A_{j,r}^{(\rho)}, \quad B_{J,r}^{(\rho)} := B_r^{(\rho)} \prod_{i\le j\in J} B_{ij,r}^{(\rho)}. \end{aligned} \tag{2.1}$$

We note that knowing the power series $Y_{J,s}^{(\rho)}$, $Z_{J,s}^{(\rho)}$, $S_{J,s}^{(\rho)}$, $A_{J,r}^{(\rho)}$, $B_{J,r}^{(\rho)}$ is equivalent to knowing the power series $Y_s^{(\rho)}$, $Y_{i,s}^{(\rho)}$, $Z_s^{(\rho)}$, $Z_{ij,s}^{(\rho)}$, $S_s^{(\rho)}$, $S_{i,s}^{(\rho)}$, $A_r^{(\rho)}$, $A_{i,r}^{(\rho)}$, $B_r^{(\rho)}$, $B_{ij,r}^{(\rho)}$.

For a subset $|J| \subset [\rho - 1]$, we also let $|J|$ be the number of elements of J and put $\|J\| := \sum_{j\in J} j$.

Let S have an irreducible reduced canonical divisor. Then it is well known that the Seiberg-Witten classes of S are just 0 and K_S, with Seiberg-Witten invariants $SW(0) = 1$, $SW(K_S) = (-1)^{\chi(\mathcal{O}_S)}$. Thus Conjecture 1.2 implies the following.

CONJECTURE 2.2 ([**GK1**, Conj. 1.4, 1.6]).

(1) *For any $\alpha \in K^0(S)$ with $\mathrm{rk}(\alpha) = s$ and any $L \in \mathrm{Pic}(S)$, we have that $\int_{[M]^{\mathrm{vir}}} c(\tau(\alpha)) \exp(\mu(L) + \mu(\mathrm{pt})u)$ is the coefficient of $z^{\mathrm{vd}(M)}$ of*

$$\rho^{2-\chi(\mathcal{O}_S)} (V_s^{(\rho)})^{c_2(\alpha)} (W_s^{(\rho)})^{c_1(\alpha)^2} (X_s^{(\rho)})^{\chi(\mathcal{O}_S)} e^{L^2 Q_s^{(\rho)} + (c_1(\alpha)L)R_s^{(\rho)} + uT_s^{(\rho)}}$$
$$\cdot \sum_{J\subset[\rho-1]} (-1)^{|J|\chi(\mathcal{O}_S)} \varepsilon_\rho^{\|J\|K_S c_1} (Y_{J,s}^{(\rho)})^{c_1(\alpha)K_S} (Z_{J,s}^{(\rho)})^{K_S^2} e^{(K_S L)S_{J,s}^{(\rho)}}.$$

(2) *The virtual Verlinde number $\chi^{\mathrm{vir}}(M, \mu(L) \otimes E^{\otimes r})$ equals the coefficient of $w^{\mathrm{vd}(M)}$ of*

$$\rho^{2-\chi(\mathcal{O}_S)} (G_r^{(\rho)})^{\chi(L)} (F_r^{(\rho)})^{\frac{1}{2}\chi(\mathcal{O}_S)} \sum_{J\subset[\rho-1]} (-1)^{|J|\chi(\mathcal{O}_S)} \varepsilon_\rho^{\|J\|K_S c_1} (A_{J,r}^{(\rho)})^{K_S L} (B_{J,r}^{(\rho)})^{K_S^2}.$$

Furthermore, the $Y^{(\rho)}_{J,s}$, $Z^{(\rho)}_{J,s}$, $S^{(\rho)}_{J,s}$, $A^{(\rho)}_{J,r}$, $B^{(\rho)}_{J,r}$ *are all algebraic functions.*

The Segre-Verlinde correspondence, conjectured first in the case of Hilbert schemes of points in [**Joh**] and [**MOP2**] gives an explicit conjectural relation between the universal functions of the Verlinde and the Segre numbers.

CONJECTURE 2.3 ([**GK1**, Conj. 1.7]). *For any* $\rho > 0$, $r \in \mathbb{Z}$, $s := \rho + r$ *we have*

$$A^{(\rho)}_{J,r}(w) = W^{(\rho)}_s(z)\, Y^{(\rho)}_{J,s}(z), \quad B^{(\rho)}_{J,r}(w) = Z^{(\rho)}_{J,s}(z),$$

for all $J \subset [\rho - 1]$, *under the change of variables*

$$w = v(1+v^2)^{\frac{1}{2}(\frac{r^2}{\rho^2}-1)}, \quad z = t(1+(1-\tfrac{s}{\rho})t^2)^{\frac{1}{2}(1-\frac{s}{\rho})}, \quad v = t(1-\tfrac{r}{\rho}t^2)^{-\frac{1}{2}}. \tag{2.2}$$

Finally the virtual Serre duality $\chi^{\mathrm{vir}}(M, L) = (-1)^{\mathrm{vd}(M)}\chi^{\mathrm{vir}}(M, L^\vee \otimes K^{\mathrm{vir}}_M)$ (see [**FG**]), leads to the following conjecture.

CONJECTURE 2.4 ([**GK1**, Conj. 5.4]). *For any* $\rho > 0$, *we have*

$$A^{(\rho)}_{J,-r}(w) = (1+v^2)^{1-\rho} A^{(\rho)}_{J,r}(-w)^{-1},$$
$$B^{(\rho)}_{J,-r}(w) = (1+v^2)^{\binom{\rho}{2}} A^{(\rho)}_{J,r}(-w)^{\rho} B^{(\rho)}_{J,r}(-w),$$

for all $J \subset [\rho - 1]$ *and* $r \in \mathbb{Z}$.

In [**GK1**, Conj. 3.5] it is also conjectured that the coefficients of the universal functions $A^{(\rho)}_{J,r}, B^{(\rho)}_{J,r}$; $Y^{(\rho)}_{J,s}$, $Z^{(\rho)}_{J,s}$, $S^{(\rho)}_{J,s}$ are polynomials in r and s respectively. We can slightly strengthen the statement there.

CONJECTURE 2.5. *For all* $J \in [\rho - 1]$ *and all* $n \geq 0$ *we have the following.*

(1) *The coefficients of* w^n *of* $A^{(\rho)}_{J,r}$, *(respectively* $B^{(\rho)}_{J,r}$*) are polynomials in* r *of degree at most* $n-1$ *(respectively at most* n*).*
(2) *The coefficients of* z^n *of* $Y^{(\rho)}_{J,s}$, *(respectively* $Z^{(\rho)}_{J,s}$, $S^{(\rho)}_{J,s}$*) are polynomials in* s *of degree at most* $n-1$ *(respectively at most* n*).*

In the future the variables w and v, respectively z and t are always related by the change of variables (2.2). We recall the blowup relation for the virtual Verlinde numbers [**GK1**, Prop. 5.10] and state an analogous result for the Segre invariants, whose proof is an easy adaptation of the proof of the formula for Verlinde numbers. For this let S be a smooth projective surface satisfying $b_1(S) = 0$ and $p_g(S) > 0$. Let $\pi : \widehat{S} \to S$ be the blowup of S in a point with exceptional divisor D.

PROPOSITION 2.6.

$$\psi_{\widehat{S},\rho,\pi^*c_1-\ell D,L-mD,r} = \frac{\psi_{S,\rho,c_1,L,r}}{(1+v^2)^{\binom{m+1}{2}}} \sum_{J\subset[\rho-1]} \varepsilon_\rho^{\|J\|\ell} \frac{(A^{(\rho)}_{J,r})^m}{B^{(\rho)}_{J,r}}, \tag{2.3}$$

$$\begin{aligned}\phi_{\widehat{S},\rho,\pi^*c_1-\ell D,\alpha+k\mathcal{O}_D(D),L-xD} &= \phi_{S,\rho,c_1,\alpha,L} e^{-x^2 Q^{(\rho)}_s + kxR^{(\rho)}_s} \\ &\cdot \frac{(1+(1-\frac{s}{\rho})t^2)^{k(k-s+\rho)/2}}{(1+(2-\frac{s}{\rho})t^2)^{k(k-s)/2}} \sum_{J\subset[\rho-1]} \varepsilon_\rho^{\|J\|\ell} \frac{(Y^{(\rho)}_{J,s})^k}{Z^{(\rho)}_{J,s}} e^{xS^{(\rho)}_{J,s}}.\end{aligned} \tag{2.4}$$

To get the explicit formula (2.4), note that

$$c_1(\alpha+k\mathcal{O}_D(D))^2=c_1(\alpha)-k^2,\quad c_2(\alpha+k\mathcal{O}_D(D))=c_2(\alpha)-\binom{k}{2},$$
$$(V_s^{(\rho)})^{\binom{k}{2}}(W_s^{(\rho)})^{k^2}=(1+(1-\frac{s}{\rho})t^2)^{-\frac{1}{2}k(k-s+\rho)}(1+(2-\frac{s}{\rho})t^2)^{\frac{1}{2}k(k-s)}.$$

3. Virtual blowup formulas

We now state the blowup formulas for the universal functions $A^{(\rho)}_{J,r}$, $B^{(\rho)}_{J,r}$ for the Verlinde numbers and $Y^{(\rho)}_{J,s}$, $Z^{(\rho)}_{J,s}$, $S^{(\rho)}_{J,s}$ for the Segre invariants. We then show that they are equivalent to Conjectures 1.4 and 1.3 of the introduction.

3.1. Blowup formula for the Verlinde numbers.

CONJECTURE 3.1. *Let $|r|\le\rho$. With the change of variables $w=v(1+v^2)^{\frac{1}{2}(\frac{r^2}{\rho^2}-1)}$, we have the following.*

(1) *For $a\in[-\rho,0]$ and in addition, if $|r|<\rho$ for $a\in[-\rho-r,-r]$*

$$\sum_{J\subset[\rho-1]}\frac{(A^{(\rho)}_{J,r})^a}{B^{(\rho)}_{J,r}}=(1+v^2)^{\binom{a+1}{2}}. \tag{3.1}$$

(2) *For $\ell=1,\ldots,\rho-1$ and $a=i+(\ell-\rho)\frac{r}{\rho}$, with $i\in[-\rho,0]$, we have the following*

$$\sum_{J\subset[\rho-1]}\epsilon_\rho^{\ell\|J\|}\frac{(A^{(\rho)}_{J,r})^a}{B^{(\rho)}_{J,r}}=\begin{cases}0, & i\in[-\rho+1,-1],\\ (-w)^{\ell(\rho-\ell)}(1+v^2)^{\binom{a+1}{2}}, & i=0,\ r\in[-\rho+1,\rho],\\ (-w)^{\ell(\rho-\ell)}\frac{(1+v^2)^{\binom{a+1}{2}}}{(1-(-v^2)^\rho)^{\rho-\ell}}, & i=0,\ r=-\rho,\\ w^{\ell(\rho-\ell)}(1+v^2)^{\binom{a+1}{2}}, & i=-\rho,\ r\in[-\rho,\rho-1],\\ w^{\ell(\rho-\ell)}\frac{(1+v^2)^{\binom{a+1}{2}}}{(1-(-v^2)^\rho)^{\rho-\ell}}, & i=-\rho,\ r=\rho.\end{cases} \tag{3.2}$$

We now want to see that, assuming Conjecture 1.2, or equivalently Conjecture 2.2, the two Conjectures 3.1 and 1.3 are equivalent. By Conjecture 1.2, the universal power series $A^{(\rho)}_{J,r}$, $B^{(\rho)}_{J,r}$ are independent of the surface S. We can therefore assume that S is a K3-surface, and in particular that $\psi_{S,\rho,c_1,L,r}\neq 0$. We put $a=-k$, then we can rewrite (2.3) as

$$\frac{\psi_{\widehat{S},\rho,\pi^*c_1-\ell D,L+kD,r}}{\psi_{S,\rho,c_1,L,r}}=\frac{1}{(1+v^2)^{\binom{a+1}{2}}}\sum_{J\subset[\rho-1]}\varepsilon_\rho^{\|J\|\ell}\frac{(A^{(\rho)}_{J,r})^a}{B^{(\rho)}_{J,r}}. \tag{3.3}$$

Putting $\ell=0$ in (3.3) shows immediately that (3.1) is equivalent to (1.2), and similarly in case $\ell\neq 0$ (3.2) is equivalent to (1.3).

We can also directly formulate these conjectures as conjectures about the virtual holomorphic Euler characteristics of moduli spaces. If $c\in K^0(S)$ is the class of a sheaf E in $M^H_S(\rho,c_1,c_2)$, and H is an ample line bundle such that $M^H_S(\rho,c_1,c_2)$ only consists of stable sheaves, we also write $M_S(c):=M^H_S(\rho,c_1,c_2)$.

CONJECTURE 3.2. *Let S be a surface with $p_g(S)>0$, $b_1(S)=0$. Let $L\in\mathrm{Pic}(S)$, let $\pi:\widehat{S}\to S$ be the blowup of S in a point with exceptional divisor D. Let $c\in K^0(S)$ be a class of rank ρ, and let $\beta\in K^0(S)$ with $r:=rk(\beta)$ with $|r|\le\rho$.*

(1) *For $k \in [0,\rho]$ we have the following:*

$$\chi^{\mathrm{vir}}\big(M_{\widehat{S}}(\pi^*c), \lambda(\pi^*(\beta) + k\mathcal{O}_D(D))\big) = \chi^{\mathrm{vir}}\big(M_S(c), \lambda(\beta)\big),$$

and if $|r| < \rho$, then also

$$\chi^{\mathrm{vir}}\big(M_{\widehat{S}}(\pi^*c), \lambda(\pi^*(\beta) + (r+k)\mathcal{O}_D(D))\big) = \chi^{\mathrm{vir}}\big(M_S(c), \lambda(\beta)\big),$$

(2) *For $\ell = 1, \ldots, \rho - 1$, and $k = 1, \ldots, \rho - 1$ we have*

$$\chi^{\mathrm{vir}}\big(M_{\widehat{S}}(\pi^*c - \ell\mathcal{O}_D), \mu\big(L + (k + \frac{(\rho-\ell)r}{\rho})D\big) \otimes E^{\otimes r}\big) = 0.$$

We see that Conjecture 1.3 implies Conjecture 3.2: By definition we have

$$\chi^{\mathrm{vir}}(M_S(c), \mu(L) \otimes E^{\otimes r}) = \mathrm{Coeff}_{w^{\mathrm{vd}(M_S(c))}}\big[\psi_{S,\rho,c_1,L,r}\big].$$

To see part (1) of Conjecture 3.2, note that $\mathrm{vd}(M_S(c)) = \mathrm{vd}(M_{\widehat{S}}(\pi^*c))$, and if $\mu(L) \otimes E^{\otimes r} = \lambda(\beta)$ for some $\beta \in K_c(S)$ or rank r, then $\pi^*\beta + k\mathcal{O}_D(D) \in K_{\pi^*c}(\widehat{S})$, because $\chi(S, \pi^*c \otimes \mathcal{O}_D(D)) = 0$. By definition

$$\mu(\pi^*L + kD) \otimes E^{\otimes r} = \lambda(\pi^*\beta + k\mathcal{O}_D(D)).$$

Part (2) of Conjecture 3.2 follows directly from the case $k \in [1, \rho - 1]$ of (1.3).

3.2. Blowup formulas for the Segre invariants. Now we state the blowup relations for the universal functions of the Segre invariants.

CONJECTURE 3.3. *For $s \in [0, 2\rho]$ we have the following formulas.*

(1) *For $a \in [-\rho, 0]$, and furthermore, if $s \in [1, 2\rho-1]$, also for $a \in [-s, -s+\rho]$ we have*

$$\sum_{J \subset [\rho-1]} \frac{(Y_{J,s}^{(\rho)})^a}{Z_{J,s}^{(\rho)}} = \frac{(1 + (2 - \frac{s}{\rho}t^2))^{\frac{1}{2}a(a+s)}}{(1 + (1 - \frac{s}{\rho}t^2))^{\frac{1}{2}a(a+s-\rho)}}. \tag{3.4}$$

(2) *For $a \leq 0$ we have*

$$\sum_{J \subset [\rho-1]} \frac{(Y_{J,s}^{(\rho)})^a}{Z_{J,s}^{(\rho)}} e^{xS_{J,s}^{(\rho)}} = \frac{(1 + (2 - \frac{s}{\rho}t^2))^{\frac{1}{2}a(a+s)}}{(1 + (1 - \frac{s}{\rho}t^2))^{\frac{1}{2}a(a+s-\rho)}} e^{\frac{x^2}{2}\cdot t^2(1+(1-\frac{s}{\rho})t^2) + x\cdot at^2} + O(x^{\rho+a+1}), \tag{3.5}$$

furthermore, if $s \in [1, 2\rho - 2]$ we have for $a \geq -s$

$$\sum_{J \subset [\rho-1]} \frac{(Y_{J,s}^{(\rho)})^a}{Z_{J,s}^{(\rho)}} e^{xS_{J,s}^{(\rho)}} = \frac{(1 + (2 - \frac{s}{\rho}t^2))^{\frac{1}{2}a(a+s)}}{(1 + (1 - \frac{s}{\rho}t^2))^{\frac{1}{2}a(a+s-\rho)}} e^{\frac{x^2}{2}\cdot t^2(1+(1-\frac{s}{\rho})t^2) + x\cdot at^2} + O(x^{\rho-a-s+1}). \tag{3.6}$$

(3) *Let $\ell \in [1, \rho - 1]$, let $a \in \mathbb{Z}$. Then*

$$\sum_{J \subset [\rho-1]} \epsilon_\rho^{\ell\|J\|} S_{J,s}^n \frac{(Y_{J,s}^{(\rho)})^{a+(\ell-\rho)\frac{s}{\rho}}}{Z_{J,s}^{(\rho)}} = 0, \quad 0 \leq n < \min\big(\ell(\rho-\ell-a), (\ell+a)(\rho-\ell)\big). \tag{3.7}$$

In particular we have

$$\sum_{J \subset [\rho-1]} \epsilon_\rho^{\ell\|J\|} \frac{(Y_{J,s}^{(\rho)})^{a+(\ell-\rho)\frac{s}{\rho}}}{Z_{J,s}^{(\rho)}} = 0, \quad a \in [-\ell+1, -\ell+\rho-1]. \tag{3.8}$$

Like for the Verlinde formulas we want to see that Conjectures 3.3 and 1.4 are equivalent, if we assume Conjecture 1.2. We can assume by Conjecture 1.2 that S is a K3 surface, and thus that $\phi_{S,\rho,c_1,\alpha,L} \neq 0$.

We put $a = -k$, then, using the definitions of $R_s^{(\rho)}$ and $Q_s^{(\rho)}$, we get by Proposition 2.6 the formula

$$\begin{aligned}\frac{\phi_{\widehat{S},\rho,\pi^*c_1-\ell D,\alpha-a\mathcal{O}_D(D),L-xD}}{\phi_{S,\rho,c_1,\alpha,L}} &= \frac{(1+(1-\frac{s}{\rho}t^2))^{\frac{1}{2}a(a+s-\rho)}}{(1+(2-\frac{s}{\rho}t^2))^{\frac{1}{2}a(a+s)}}\\ &\quad\cdot e^{-\frac{x^2}{2}\cdot t^2(1+(1-\frac{s}{\rho})t^2)-ax\cdot t^2}\sum_{J\subset[\rho-1]}\varepsilon_\rho^{\|J\|\ell}\frac{(Y_{J,s}^{(\rho)})^a}{Z_{J,s}^{(\rho)}}e^{xS_{J,s}^{(\rho)}}.\end{aligned} \tag{3.9}$$

To see that the parts (1) of Conjectures 3.3 and 1.4 are equivalent, we put $\ell = 0$, $x = 0$ in (3.9).

By the case $\ell = 0$ of (3.9), we have for any $n > 0$ that

$$\phi_{\widehat{S},\rho,\pi^*c_1,\alpha-a\mathcal{O}_D(D),L-xD} = \phi_{S,\rho,c_1,\alpha,L} + O(x^n),$$

if and only if

$$\sum_{J\subset[\rho-1]}\frac{(Y_{J,s}^{(\rho)})^a}{Z_{J,s}^{(\rho)}}e^{xS_{J,s}^{(\rho)}} = \frac{(1+(2-\frac{s}{\rho}t^2))^{\frac{1}{2}a(a+s)}}{(1+(1-\frac{s}{\rho}t^2))^{\frac{1}{2}a(a+s-\rho)}}e^{\frac{x^2}{2}\cdot t^2(1+(1-\frac{s}{\rho})t^2)+x\cdot at^2} + O(x^n).$$

This shows that parts (2) of Conjectures 3.3 and 1.4 are equivalent.

For part (3), by (3.9) we just need to see that for a given n we have

$$\mathrm{Coeff}_{x^m}\left[e^{-\frac{x^2}{2}\cdot t^2(1+(1-\frac{s}{\rho})t^2)-ax\cdot t^2}\sum_{J\subset[\rho-1]}\varepsilon_\rho^{\|J\|\ell}\frac{(Y_{J,s}^{(\rho)})^a}{Z_{J,s}^{(\rho)}}e^{xS_{J,s}^{(\rho)}}\right] = 0$$

for all $m \leq n$, if and only if $\sum_{J\subset[\rho-1]}(S_{J,s}^{(\rho)})^m\varepsilon_\rho^{\|J\|\ell}\frac{(Y_{J,s}^{(\rho)})^a}{Z_{J,s}^{(\rho)}} = 0$ for all $m \leq n$. This is obvious.

REMARK 3.4. We can see that Conjectures 3.1 and 3.3 are compatible with the Segre-Verlinde correspondence Conjecture 2.3, and the virtual Serre duality Conjecture 2.4.

(1) Direct computation gives that replacing r by $-r$ and applying Conjecture 2.4 transforms (3.1) to itself and (3.2) to (3.2) with a replaced by $-\rho - a$.
(2) Direct computation gives that the Segre-Verlinde correspondence Conjecture 2.3 transforms (3.4) for s into (3.1) for $r = s - \rho$ and vice versa. Similarly for $a \in [-\rho + \ell - 1, \ell - 1]$ it transforms (3.8) for (a, s) into (3.2) for $(a, r = s - \rho)$ and vice versa.

REMARK 3.5. For every subset $J \subset [\rho - 1]$ and all $r, s \in \mathbb{Z}$, using (2.1), we put

$$\beta_{J,r} := \frac{B_r^{(\rho)}}{B_{J,r}^{(\rho)}} = \frac{1}{\prod_{i\leq j\in J}B_{ij,r}^{(\rho)}},$$

$$\zeta_{J,s} := \frac{Z_s^{(\rho)}}{Z_{J,s}^{(\rho)}} = \frac{1}{\prod_{i\leq j\in J}Z_{ij,s}^{(\rho)}}.$$

Then by definition $B^{(\rho)}_{J,r}=\frac{B^{(\rho)}_r}{\beta_{J,r}}$ and $Z^{(\rho)}_{J,s}=\frac{Z^{(\rho)}_s}{\zeta_{J,s}}$, and the cases $a=0$ of (3.1) and (3.4) give

$$B^{(\rho)}_r=\sum_{J\subset[\rho-1]}\beta_{J,r},\qquad r\in[-\rho,\rho],$$

$$Z^{(\rho)}_s=\sum_{J\subset[\rho-1]}\zeta_{J,s},\qquad s\in[0,2\rho].$$

This determines for $r\in[-\rho,\rho]$ the universal function $B^{(\rho)}_r$ in terms of the $B^{(\rho)}_{ij,r}$ and for $s\in[0,2\rho]$ the universal function $Z^{(\rho)}_s$ in terms of the $Z^{(\rho)}_{ij,s}$

3.3. Evidence for the conjectures. As mentioned above, the results of [**NY**] and [**KT**] provide evidence for the conjectures.

In [**GK1**] we used Mochizuki's formula [**Moc**, Thm. 7.5.2] and localization to compute the Segre and Verlinde invariants of moduli spaces $M^H_S(\rho,c_1,c_2)$ for surfaces S with $p_g(S)>0$ and $b_1(S)=0$, for $\rho=1,2,3,4$, in a number of cases of S,c_1,c_2. Assuming Conjecture 1.2, this determines the universal power series $A^{(\rho)}_{J,r}$, $B^{(\rho)}_{J,r}$, $Y^{(\rho)}_{J,s}$, $Z^{(\rho)}_{J,s}$, $S^{(\rho)}_{J,s}$ modulo certain powers of w and z. Specifically we determined them up to the following orders.

(1) $A^{(2)}_{J,r}$, $B^{(2)}_{J,r}$, with r as variable modulo w^{24} and $Y^{(2)}_{J,s}$, $Z^{(2)}_{J,s}$, $S^{(2)}_{J,s}$ with s as variable modulo z^{18}.

(2) $A^{(3)}_{J,r}$, $B^{(3)}_{J,r}$ for $-11\le r\le 3$ modulo w^{12} and $Y^{(3)}_{J,s}$, $Z^{(3)}_{J,s}$, $S^{(3)}_{J,s}$ for $-3\le s\le 12$ modulo z^{13}.

(3) Finally we determined $Y^{(4)}_{J,s}$, $Z^{(4)}_{J,s}$, $S^{(4)}_{J,s}$ for $0\le s\le 8$ modulo z^8. Using the Verlinde-Segre-correspondence Conjecture 2.3, we also determine $A^{(4)}_{J,r}$, $B^{(4)}_{J,r}$ for $|r|\le 4$ modulo w^8.

Until these orders Conjectures 3.1 and 3.3 are confirmed. In [**GK1**] we also conjecturally determined the universal functions $A^{(\rho)}_{J,r}$, $B^{(\rho)}_{J,r}$, and $Y^{(\rho)}_{J,s}$, $Z^{(\rho)}_{J,s}$, $S^{(\rho)}_{J,s}$, for $\rho=2$, $|r|\le 3$, (respectively $s=-1,\dots,5$), for $\rho=3$, $|r|\le 3$ (respectively $s=0,\dots,6$), and $\rho=4$, $r=-4$ (respectively $s=0,4$) as algebraic functions. We have confirmed modulo w^{60} and z^{60} that theses algebraic functions fulfill Conjectures 3.1 and 3.3.

In the current paper we apply some part of these blowup formulas to determine for arbitrary ranks ρ conjectural formulas for the universal functions $A^{(\rho)}_{J,r}$, $B^{(\rho)}_{J,r}$, and $Y^{(\rho)}_{J,s}$, $Z^{(\rho)}_{J,s}$, $S^{(\rho)}_{J,s}$ for $r=-\rho,0,\rho$ and $s=0,\ \rho,\ 2\rho$. We check for $\rho\le 12$ and modulo w^{60} respectively z^{60} that Conjectures 3.1 and 3.3 hold for these universal functions. In the computations the coefficients are highly overdetermined by the relations. That there is a solution for these relations, which also agrees with the previous results for $\rho=1,2,3,4$ appears quite nontrivial.

Further confirmation of Conjectures 3.1 and 3.3 comes from the results of the forthcoming paper [**G2**], which we describe now. By Conjecture 2.5 and interpolation, the conjectural formulas for $r=-\rho,0,\rho$ and $s=0,\rho,2\rho$ determine the power series $A^{(\rho)}_{J,r}$, $B^{(\rho)}_{J,r}$, and $Y^{(\rho)}_{J,s}$, $Z^{(\rho)}_{J,s}$, $S^{(\rho)}_{J,s}$ modulo w^3 (respectively z^3). Given this information, we use the blowup formulas (3.1), (3.2), (3.4), (3.8), and the coefficient of x^1 of (3.5) and (3.6) to compute for $\rho=2,3,4,5,6$ and $|r|\le\rho$ (or $0\le s\le 2\rho$) the coefficients of the power series $A^{(\rho)}_{J,r}$, $B^{(\rho)}_{J,r}$, and $Y^{(\rho)}_{J,s}$, $Z^{(\rho)}_{J,s}$, $S^{(\rho)}_{J,s}$ modulo w^{100} and z^{100} by a Pari/GP program.

Indeed applying the blowup formulas order by order in v and t gives linear equations for the coefficients of the $A^{(\rho)}_{J,r}$, $B^{(\rho)}_{J,r}$, and $Y^{(\rho)}_{J,s}$, $Z^{(\rho)}_{J,s}$, $S^{(\rho)}_{J,s}$, which at each step have a unique solution (indeed they are overdetermined). The resulting power series fulfill all the relations of Conjectures 3.1 and 3.3. Furthermore for $\rho = 2, 3, 4$, they agree, up to the order computed, with the power series determined in [**GK1**] and the algebraic functions conjectured there. We report some of the results of these computations in the last section of this paper.

Finally, also in the forthcoming paper [**G2**], we use a virtual version of strange duality that relates the power series $A^{(\rho)}_{J,r}$, $B^{(\rho)}_{J,r}$, to the power series $A^{(r)}_{J,\rho}$, $B^{(r)}_{J,\rho}$. This allows us to determine from the power series $A^{(\rho)}_{J,r}$, $B^{(\rho)}_{J,r}$ with $|r| \leq \rho \leq 6$ the $A^{(\rho)}_{J,r}$, $B^{(\rho)}_{J,r}$ with $\rho \leq |r| \leq 6$ exchanging the role of r and ρ. Again we find that the results agree with the results from [**GK1**], both for the power series and for the expressions as algebraic functions conjectured there.

Note that in the case $\rho = 1$, by identifying a zero-dimensional scheme Z with its ideal sheaf I_Z, we can identify the moduli space $M^H_S(1, 0, n)$ with the Hilbert scheme $Hilb^n(S)$ of n points on S. In the paper [**GM1**], based on preliminary results of [**G2**], complete conjectural formulas for generating functions for all the Segre and Verlinde formulas of all the Hilbert schemes of points $Hilb^n(S)$ are found. They are proven when $K_S^2 = 0$, and based on a program written with the help of Don Zagier, they are also proven for the Hilbert schemes $Hilb^n(S)$ of arbitrary surfaces for all $n < 50$.

In particular the generating functions for the invariants of Hilbert schemes of points obtained by applying the conjectural blowup formulas, the conjectures of section 2, and the conjectural virtual strange duality mentioned above are correct for surfaces with $K_S^2 = 0$ and correct for all surfaces modulo t^{100} or v^{100}. This is very strong evidence for the conjectural blowup formulas, as well as for the whole conjectural framework, including the conjectured virtual strange duality in [**G2**].

4. Verlinde formulas for $r = \pm\rho$ and Donaldson invariants in arbitrary rank

We now will use the blowup formulas to conjecturally determine the Verlinde formula for $\chi^{\mathrm{vir}}(M^H_S(\rho, c_1, c_2), \mu(L) \otimes E^{\otimes r})$ in the case $r = -\rho$ and the Segre invariants in case $s = 0$ for arbitrary rank ρ. We determine the Donaldson invariants in arbitrary rank ρ as the special case $\alpha = 0$ of the Segre invariants for $s = 0$.

4.1. Verlinde formula for $r = -\rho$.

We fix $\rho \in \mathbb{Z}_{>0}$. In order to state the conjectural Verlinde formula for $r = -\rho$, we introduce the following notations, which we will also use in the rest of this paper in the formulation of our main conjectures on Segre and Verlinde invariants.

NOTATION 4.1. Let $\xi = e^{\frac{\pi i}{2\rho}}$ be a primitive 4ρ-th root of unity. For $1 \leq i < j \leq \rho - 1$ let

$$\beta_{ij} := \frac{\xi^{i+j} - \xi^{-(i+j)}}{\xi^{i-j} - \xi^{j-i}}, \quad \beta_{ji} := \beta_{ij}.$$

We put

$$
(4.1)\qquad
\begin{aligned}
B_{ij} &:= \beta_{ij}^2 = \frac{(\xi^{i+j}-\xi^{-i-j})^2}{(\xi^{i-j}-\xi^{-i+j})^2}, \qquad 1\le i<j\le \rho-1,\\
B_{ii} &:= \prod_{j\neq i\in[\rho-1]}\frac{1}{\beta_{ij}} = (-1)^{i-1}\prod_{j\neq i}\frac{\xi^{i-j}-\xi^{-i+j}}{\xi^{i+j}-\xi^{-i-j}}, \qquad 1\le i\le \rho-1.
\end{aligned}
$$

Then we set

$$
(4.2)\qquad
\begin{aligned}
\beta_I &:= \frac{1}{\prod_{i\le j\in I} B_{ij}}, \qquad I\subset[\rho-1],\\
B &:= \sum_{J\subset[\rho-1]} \beta_J,\\
B_I &:= \frac{B}{\beta_I} = \sum_{J\subset[\rho-1]}\frac{\beta_J}{\beta_I}, \qquad I\subset[\rho-1].
\end{aligned}
$$

Note that for $I\subset[\rho-1]$ we have

$$
\beta_I = \frac{\prod_{i\in I}\prod_{j\neq i}\beta_{ij}}{\prod_{i<j\in I}\beta_{ij}^2} = \frac{\prod_{i\in I}\prod_{j\neq i}\beta_{ij}}{\prod_{i\neq j\in I}\beta_{ij}} = \prod_{\substack{i\in I\\ j\in[\rho-1]\setminus I}}\beta_{ij}.
$$

Now the Verlinde numbers with $r=-\rho$ are conjecturally determined by the following formulas.

CONJECTURE 4.2. (1) *$B^{(\rho)}_{J,-\rho} := B_J$ for all $J\subset[\rho-1]$, or equivalently $B^{(\rho)}_{-\rho} := B$ and $B^{(\rho)}_{ij,-\rho} := B_{ij}$ for all $1\le i\le j\le \rho-1$.*

(2)

$$
(4.3)\qquad A^{(\rho)}_{-\rho} := \prod_{j=1}^{\rho-1}\frac{1}{1+\xi^{\rho+2j}w}, \quad A^{(\rho)}_{j,-\rho} := \frac{1+\xi^{\rho+2j}w}{1+\xi^{\rho-2j}w}, \text{ for } j\in[\rho-1],
$$

equivalently, for each $J\subset[\rho-1]$

$$
A^{(\rho)}_{J,-\rho} = \prod_{j\in J}\frac{1}{1+\xi^{\rho-2j}w}\prod_{j\in[\rho-1]\setminus J}\frac{1}{1+\xi^{\rho+2j}w}.
$$

We outline the assumptions, arguments and calculations that lead to this conjecture. We use two working hypotheses.

(1) We expect that for $J\subset[\rho-1]$ the universal functions $B^{(\rho)}_{J,-\rho}$ are constant, i.e. independent of the variable w.
(2) We expect the formula (4.3) for the universal power series $A^{(\rho)}_{J,-\rho}$ to be true.

The motivation for these working hypotheses is the following.

(1) Note that, by the Segre-Verlinde correspondence Conjecture 2.3, if $Z^{(\rho)}_{J,0}$ is constant, then $B^{(\rho)}_{J,-\rho} = Z^{(\rho)}_{J,0}$. That $Z^{(\rho)}_{J,0}$ is constant is conjectured in [**GK1**] as part of a conjecture about the structure of the Donaldson invariants in arbitrary rank, and, if one assumes Conjecture 1.2, it also follows from the Marino-Moore conjecture [**MM**]. The conjecture of [**GK1**] is true for $\rho=1$. For $\rho=2$ and $\alpha=0$ it is the Witten conjecture for algebraic surfaces with $p_g>0$ and $b_1=0$, which was proven in [**GNY2**].

For $\rho=3$ and $\rho=4$ it was confirmed up to high expected dimension in [**GK1**].

Heuristically we can motivate that the $Z^{(\rho)}_{J,0}$ should be constant as follows. We see from the definition that $X_0^{(\rho)}=1$. Thus, if S is a surface with $p_g(S)>0$ and $b_1(S)=1$ and an irreducible reduced canonical divisor, we get for $c_1\in H^2(S,\mathbb{Z})$, that

$$\phi_{S,\rho,c_1,0,0}=\rho^{2-\chi(\mathcal{O}_S)}\sum_{J\subset[\rho-1]}(-1)^{|J|}\varepsilon_\rho^{\|J\|K_Sc_1}(Z^{(\rho)}_{J,0})^{K_S^2}.$$

So for $c_2\in\mathbb{Z}$ we get by Conjecture 2.2 that

$$\begin{aligned}\mathrm{Coeff}_{z^{\mathrm{vd}(\rho,c_1,c_2)}}\Big[\rho^{2-\chi(\mathcal{O}_S)}&\sum_{J\subset[\rho-1]}(-1)^{|J|\chi(\mathcal{O}_S)}\varepsilon_\rho^{\|J\|K_Sc_1}(Z^{(\rho)}_{J,0})^{K_S^2}\Big]\\ &=\int_{[M_S^H(\rho,c_1,c_2)]^{\mathrm{vir}}}1,\end{aligned}$$

and the right hand side is zero unless $\mathrm{vd}(\rho,c_1,c_2)=0$. This is an indication that all the $Z^{(\rho)}_{J,0}$ should be constant.

(2) For ranks $\rho=2,3,4$ the conjectural formulas for the $A^{(\rho)}_{J,-\rho}$ were found in [**GK1**] based on computations for not too large virtual dimension, using Mochizuki's formula and localization (for $\rho=4$ also using virtual Serre duality). One can see that these formulas are specializations of (4.3). As the formula is very simple and natural, we expect that it holds for arbitrary rank ρ.

For $\rho=2,\ldots 8$, we now define $A^{(\rho)}_{-\rho}$ and the $A^{(\rho)}_{j,-\rho}$ by the formulas (4.3). We choose variables for the $B^{(\rho)}_{J,-\rho}$ for $J\subset[\rho-1]$. Imposing the blowup relations (3.1), (3.2) order by order in $v=w$ gives an infinite number of linear equations for the $B^{(\rho)}_{J,-\rho}$, which we can solve successively using Pari/GP. We find that the $B^{(\rho)}_{J,-\rho}$ are uniquely determined by these equations, (and that they fulfill them to high order in w). We analyze the explicit formulas that we obtain for $B^{(\rho)}_{-\rho}$ and the $B^{(\rho)}_{ij,-\rho}$ for $\rho=2,\ldots 8$. Using also Remark 3.5, we find that, for $\rho=2,\ldots,8$, we can write

$$\frac{B^{(\rho)}_{\emptyset,-\rho}}{B^{(\rho)}_{J,-\rho}}=\prod_{\substack{i\in I\\ j\in[\rho-1]\setminus I}}\beta_{ij},$$

for suitable numbers $\beta_{ij}=\beta_{ji}$, which we finally see as given by formula (4.1). This gives the formulas of Conjecture 4.2 for $\rho=2,\ldots,8$. Using Pari/GP, we check that these universal functions satisfy the blowup formulas (3.1) and (3.2) as identities of rational functions in w for $\rho\le 7$ and $\rho\le 6$ respectively, and that they satisfy (3.1), (3.2) modulo $O(w^{100})$ for $\rho\le 12$.

If we chose ρ,c_1,c_2 with $\mathrm{vd}(\rho,c_1,c_2)=0$, then the degree of the virtual fundamental class $\int_{[M_S^H(\rho,c_1,c_2)]^{\mathrm{vir}}}1$, is both the coefficient of w^0 of $\psi_{S,\rho,c_1,L,r}$ for all $r\in\mathbb{Z}$ (using the virtual Riemann-Roch formula [**FG**]) and the coefficient of z^0 of $\phi_{S,\rho,c_1,\alpha,L}$ for any $\alpha\in K^0(S)$. Thus we conjecture that the constant coefficient of each $Z^{(\rho)}_{J,s}$ and $B^{(\rho)}_{J,r}$ for all $r,s\in\mathbb{Z}$ is B_J. In fact we can also see directly that Conjecture 2.2, Conjecture 2.3, Conjecture 2.5 and Conjecture 4.2 imply the following statement.

CONJECTURE 4.3.

(1) *For all* $J \subset [\rho-1]$ *and* $1 \le i \le j \le \rho-1$ *and all* $r \in \mathbb{Z}$, *we can write* $B^{(\rho)}_{J,r} = B^0_{J,r}B_J$, $B^{(\rho)}_r = B^0_r B$, $B^{(\rho)}_{ij,r} = B^0_{ij,r}B_{ij}$ *with* $B^0_{J,r}$, B^0_r, $B^0_{ij,r} \in 1 + w\mathbb{C}[[w]]$.

(2) *In the same way for all* $J \subset [\rho-1]$ *and* $1 \le i \le j \le \rho-1$ *and all* $s \in \mathbb{Z}$, *we can write* $Z^{(\rho)}_{J,s} = Z^0_{J,s}B_J$, $Z^{(\rho)}_s = Z^0_s B$, $Z^{(\rho)}_{ij,s} = Z^0_{ij,s}B_{ij}$ *with* $Z^0_{J,s}$, Z^0_s, $Z^0_{ij,s} \in 1 + z\mathbb{C}[[z]]$.

4.2. Segre numbers for $s = 0$ and Donaldson invariants in arbitrary rank. Applying the Segre-Verlinde correspondence Conjecture 2.3 to Conjecture 4.2 determines the universal functions $Y^{(\rho)}_{J,0}$, $Z^{(\rho)}_{J,0}$. Thus in order to get the Segre numbers for $s = 0$ in arbitrary rank we only need to determine the $S^{(\rho)}_{J,0}$. We obtain the following conjecture, which generalizes the results of [**GK1**] for $\rho \le 4$.

CONJECTURE 4.4. *Under the change of variables* $z = t(1+t^2)^{\frac{1}{2}}$, *we have*

$$Z^{(\rho)}_{J,0} := B^{(\rho)}_{J,0} \quad \textit{for all } J \subset [\rho-1],$$

$$Y^{(\rho)}_{J,0} := \frac{(1+t^2)^\rho}{(1+2t^2)^{\frac{1}{2}}} \prod_{j\in J} \frac{1}{(1+t^2)^{\frac{1}{2}} + \xi^{\rho-2j}t} \prod_{j\in[\rho-1]\setminus J} \frac{1}{(1+t^2)^{\frac{1}{2}} + \xi^{\rho+2j}t},$$

$$S^{(\rho)}_{J,0} := \Big(-\sum_{j\in J}\xi^{\rho-2j} + \sum_{j\in[\rho-1]\setminus J}\xi^{2j-\rho}\Big)z.$$

In particular, we have

$$S^{(\rho)}_0 = \Big(\sum_{j\in[\rho-1]}\xi^{2j-\rho}\Big)z, \quad S^{(\rho)}_{j,0} = (-\xi^{2j-\rho} - \xi^{\rho-2j})z = -2\sin(\pi\frac{j}{\rho})z. \tag{4.4}$$

To determine the $S^{(\rho)}_{J,0}$, we make the working assumption that $S^{(\rho)}_0$ and the $S^{(\rho)}_{i,0}$ for $i = 1, \dots, \rho-1$ are constant multiples $s_0 z$, $s_{i,0}z$ of z with $s_0 \in \mathbb{C}$, $s_{i,0} \in \mathbb{C}$. For $\rho = 2, \dots, 8$ we first let s_0 and the $s_{i,0}$ be variables. Then the coefficients of x^1 of the blowup formulas (3.5) impose, order by order in t, linear equations for s_0 and the $s_{i,0}$, which uniquely determine them by the formula of (4.4). Then for $\rho \le 12$, we define $S^{(\rho)}_0$ and the $S^{(\rho)}_{i,0}$ by (4.4) and check using Pari/GP, that for $\rho \le 12$ the blowup formulas (3.4), (3.5), (3.6), (3.7) hold modulo $O(t^{100})$.

Conjecture 1.8 follows from Conjecture 4.4 by specializing to $\alpha = 0$: By Definition $D^{S,H}_{\rho,c_1,c_2}(L + u\mathrm{pt}) = \mathrm{Coeff}_{z^{\mathrm{vd}(\rho,c_1,c_2)}}\big[\phi_{S,\rho,c_1,0,L}\big]$. As in the introduction we put $\widetilde{a} := 2a - K_S$ for $a \in H^2(S,\mathbb{Z})$, and write $\widetilde{SW}(\widetilde{a}) := SW(a)$. Note that

$$\frac{1}{2}\sum_{j=1}^{\rho-1}(\xi^{2j-\rho} + \xi^{\rho-2j}) = \frac{1}{2}\sum_{j=1}^{\rho-1}(\xi^{2j-\rho} + \xi^{2(\rho-j)-\rho}) = \sum_{j=1}^{\rho-1}\xi^{2j-\rho}.$$

Therefore we get

$$\prod_{j=1}^{\rho-1} e^{-\sin(\pi\frac{j}{\rho})(\widetilde{a}_j L)z} = e^{(K_S L)S^{(\rho)}_0 z}\prod_{j=1}^{\rho-1} e^{(a_j L)S^{(\rho)}_{j,0}z}.$$

Using $a_iK_S = a_i^2$, $a_jK_S = a_j^2$, we see that $\frac{1}{2}\widetilde{a}_i(\widetilde{a}_j - \widetilde{a}_i) = 2a_ia_j - a_i^2 - a_j^2$. By $B_{ij} = \beta_{ij}^2$ and $B_{ii} = \prod_{j\in[\rho-1]\setminus\{i\}} \beta_{ij}^{-1}$, this gives

$$\prod_{1\le i<j\le \rho-1} \beta_{ij}^{\frac{1}{2}\widetilde{a}_i(\widetilde{a}_j-\widetilde{a}_i)} = \prod_{1\le i<j\le\rho-1} B_{ij}^{a_ia_j}\cdot\prod_{i\neq j}\beta_{ij}^{-a_i^2} = \prod_{1\le i\le j\le\rho-1} B_{ij}^{a_ia_j}.$$

Therefore Conjecture 1.8 follows from Conjecture 4.4 and Conjecture 1.2.

REMARK 4.5. The Donaldson invariants are more generally defined for differentiable 4-manifolds M with $b_+(M) > 1$ and $b_1(M) = 0$. If M is a complex projective surface, they agree up to different sign conventions with the definition of the Donaldson invariants above. Let $\sigma(M) = b_+(M) - b_-(M)$ be signature of the intersection form on $H_2(M,\mathbb{Q})$, and $e(M)$ the topological Euler characteristic of M. Replacing in (1.4) K_S^2 by $3\sigma(M) + 2e(M)$ and $\chi(\mathcal{O}_S)$ by $\frac{1}{4}(\sigma(M) + e(M))$, we obtain a formula that (up to sign) is expressed in terms of invariants of the 4-manifold M. Note that $\frac{c_1^2-c_1K_S}{2} \in \mathbb{Z}$. An easy calculation shows that changing the sign from $\prod_{j=1}^{\rho-1}\varepsilon_\rho^{j\cdot(a_jc_1)}$ to $\prod_{j=1}^{\rho-1}\varepsilon_\rho^{j\cdot\frac{1}{2}(\widetilde{a}_j+c_1)c_1}$, just multiplies (1.4) by the global sign $(-1)^{(\rho-1)\frac{c_1^2-c_1K_S}{2}}$.

We conjecture that, after making these replacements, formula (1.4) gives (up to possibly different sign conventions) the Donaldson invariants of rank ρ of any simply connected 4 manifold M of Seiberg-Witten simple type.

4.3. Verlinde formula for $r = \rho$ and Segre formula for $s = 2\rho$. Using the virtual Serre duality Conjecture 2.4, we immediately get from Conjecture 4.2 a Verlinde formula for the case $r = \rho$. In addition we checked using Pari/GP that the universal functions $A_{J,\rho}^{(\rho)}$, $B_{J,\rho}^{(\rho)}$ below satisfy the blowup formulas (3.1), (3.2) as identities of rational functions in w for $\rho \le 6$ and $\rho \le 5$ respectively, and that they satisfy (3.1), (3.2) modulo $O(w^{100})$ for $\rho \le 12$.

CONJECTURE 4.6. *For all $J \in [\rho-1]$ we have*

$$A_{J,\rho}^{(\rho)} = (1+w^2)^{1-\rho}\prod_{j\in J}(1-\xi^{\rho-2j}w)\prod_{j\in[\rho-1]\setminus J}(1-\xi^{\rho+2j}w),$$

$$B_{J,\rho}^{(\rho)} = (1+w^2)^{\binom{\rho}{2}}B_J\prod_{j\in J}\frac{1}{(1-\xi^{\rho-2j}w)^\rho}\prod_{j\in[\rho-1]\setminus J}\frac{1}{(1-\xi^{\rho+2j}w)^\rho}.$$

We can apply the Segre-Verlinde correspondence to Conjecture 4.6, and then use the blowup formulas again to get the Segre invariants for $s = 2\rho$. Again the results agree with those of [**GK1**] for $\rho \le 3$.

CONJECTURE 4.7. *We have for $J \subset [\rho-1]$*

$$Y_{J,2\rho}^{(\rho)} = (1+z^2)^{\frac{1-\rho}{2}}\prod_{j\in J}(1-\xi^{\rho-2j}z)\prod_{j\in[\rho-1]\setminus J}(1-\xi^{\rho+2j}z),$$

$$Z_{J,2\rho}^{(\rho)} = (1+z^2)^{\binom{\rho}{2}}B_J\prod_{j\in J}\frac{1}{(1-\xi^{\rho-2j}z)^\rho}\prod_{j\in[\rho-1]\setminus J}\frac{1}{(1-\xi^{\rho+2j}z)^\rho},$$

$$S_{J,2\rho}^{(\rho)} = \frac{\rho+1}{1+z^2} - \frac{2\rho}{1-(-z^2)^\rho} + \sum_{j\in J}\frac{1}{1-\xi^{\rho+2j}z} + \sum_{j\in[\rho-1]\setminus J}\frac{1}{1-\xi^{\rho-2j}z}.$$

The formulas for the $Y^{(\rho)}_{J,\rho}$ and $Z^{(\rho)}_{J,\rho}$ follow from Conjecture 4.6 and the Segre-Verlinde correspondence Conjecture 2.3. In fact, note that we have for $r=\rho$ the variable change $w=v$, and in the Segre-Verlinde correspondence we have the change of variables $v=t(1-t^2)^{-\frac{1}{2}}$, and finally we have $z=t(1-t^2)^{-\frac{1}{2}}$, i.e. $t=z/(1+z^2)^{\frac{1}{2}}$. Combining these, we get the trivial change of variables $w=z$, and we have $W^{(\rho)}_{2\rho}=(1-t^2)^{\frac{1-\rho}{2}}=(1+z^2)^{\frac{1-\rho}{2}}$.

For $\rho=2,\dots,6$ let $S^{(\rho)}_{2\rho}$ and the $S^{(\rho)}_{i,2\rho}$ for $i=1,\dots,5$ be power series in z with vanishing constant terms and independent variables as coefficients. We impose order by order in z the coefficient of x^1 of the blowup formulas (3.5), (3.7). These impose linear relations on the coefficients of $S^{(\rho)}_{2\rho}$ and the $S^{(\rho)}_{i,2\rho}$, which we check have a unique solution modulo $O(z^{100})$, given by the above formulas. We then check with a Pari/GP program, that with these formulas Conjecture 3.3. holds for $\rho\le 12$ modulo $O(z^{100})$.

5. K-theoretic Donaldson invariants and Segre invariants for $s=\rho$

In this section we give a conjectural formula for the holomorphic Euler characteristics $\chi^{\mathrm{vir}}(M^H_S(\rho,c_1,c_2),\mu(L))$ for arbitrary rank $\rho>0$. These are also called the K-theoretic Donaldson invariants, and have been studied in case $\rho=2$ e.g. in [**GNY1**],[**GKW**],[**GY**],[**G1**]. Finally we use the Verlinde-Segre correspondence and the blowup formulas to also determine the Segre invariants with $s=\rho$.

5.1. K-theoretic Donaldson invariants. The universal functions for the K-theoretic Donaldson invariants are determined by the following formulas, which generalize the formulas of [**GK1**] for $\rho\le 3$. These formulas have a number of similarities with the formulas in case for $r=\pm\rho$.

CONJECTURE 5.1. *Let again $\xi=e^{\frac{\pi i}{2\rho}}$. Then we have the following.*

(5.1)
$$A^{(\rho)}_0=\prod_{i=1}^{\rho-1}\Big(\big(1+(\tfrac{\xi^{\rho-2i}-\xi^{2i-\rho}}{2})^2w^2\big)^{\frac{1}{2}}-(\tfrac{\xi^{\rho-2i}+\xi^{2i-\rho}}{2})w\Big),$$
$$A^{(\rho)}_{i,0}=\frac{\big(1+(\frac{\xi^{\rho-2i}-\xi^{2i-\rho}}{2})^2w^2\big)^{\frac{1}{2}}+(\frac{\xi^{\rho-2i}+\xi^{2i-\rho}}{2})w}{\big(1+(\frac{\xi^{\rho-2i}-\xi^{2i-\rho}}{2})^2w^2\big)^{\frac{1}{2}}-(\frac{\xi^{\rho-2i}+\xi^{2i-\rho}}{2})w},\qquad i\in[\rho-1],$$

(5.2)
$$\gamma_{i,\rho-i}=\Big(\frac{1+(\frac{\xi^{\rho-2i}-\xi^{2i-\rho}}{2})^2w^2}{1-w^2}\Big)^{\frac{1}{2}},\qquad i\in[\rho-1],$$
$$\gamma_{i,j}=\frac{(\xi^{\rho-2j}+\xi^{2j-\rho})\gamma_{i,\rho-i}+(\xi^{\rho-2i}+\xi^{2i-\rho})\gamma_{j,\rho-j}}{\xi^{\rho-2i}+\xi^{2i-\rho}+\xi^{\rho-2j}+\xi^{2j-\rho}},\qquad 1\le i\ne j\le\rho-1,$$

(5.3)
$$\gamma_I=\Big(\prod_{i\in I,\ j\notin I}\gamma_{ij}\Big)\Big(\prod_{i\in I}A^{(\rho)}_{i,0}\Big)^{\frac{\rho}{2}},\qquad I\subset[\rho-1],$$

(5.4)
$$B^{(\rho)}_0=\sum_{I\subset[\rho-1]}\beta_I\gamma_I,\qquad B^{(\rho)}_{J,0}=\frac{B^{(\rho)}_0}{\beta_J\gamma_J},\qquad J\subset[\rho-1].$$

We explain the strategy for finding these formulas: we use the virtual Serre duality relation Conjecture 2.4. In general it relates r to $-r$, but in our case $r=0$ it gives nontrivial relations among the generating functions $A^{(\rho)}_{J,0}$, $B^{(\rho)}_{J,0}$. It implies for $J \in [\rho-1]$ the relations

$$A^{(\rho)}_{J,0}(w) = \frac{(1-w^2)^{\rho-1}}{A^{(\rho)}_{J,0}(-w)}, \tag{5.5}$$

$$B^{(\rho)}_{J,0}(w) = \frac{A^{(\rho)}_{J,0}(-w)^{\rho} B^{(\rho)}_{J,0}(-w)}{(1-w^2)^{\binom{\rho}{2}}}. \tag{5.6}$$

In particular we have the weaker relation

$$\frac{B^{(\rho)}_{J,0}(w)}{B^{(\rho)}_{0}(w)} = \frac{B^{(\rho)}_{J,0}(-w)}{B^{(\rho)}_{0}(-w)} \prod_{i\in J} A^{(\rho)}_{i,0}(-w)^{\rho}. \tag{5.7}$$

In [**GK1**] the universal series $A^{(\rho)}_{J,0}$, $B^{(\rho)}_{J,0}$ have been determined for $\rho \le 3$. By Conjectures 4.2, 4.6, we have conjectural formulas for the $A^{(\rho)}_{J,r}$, $B^{(\rho)}_{J,r}$ when $r = \pm\rho$. Using Conjecture 2.5, this allows us to use interpolation to determine the $A^{(\rho)}_{J,r}$, $B^{(\rho)}_{J,r}$ up to degree 1 in w. Now for $\rho \le 4$ we take the coefficients of the power series $A^{(\rho)}_{0}$, $A^{(\rho)}_{i,0}$, $B^{(\rho)}_{0}$, $B^{(\rho)}_{ij,0}$ as indeterminants. Then the blowup relations (3.1), (3.2) give successively degree by degree in w linear relations for these coefficients. We use them to determine the power series modulo $O(w^{60})$ with a Pari/GP program. From these truncated power series and (5.5) we guess the formulas (5.1) for $\rho \le 4$. We then make the Ansatz that the formula (5.1) extends to arbitrary rank ρ. Note that this makes (5.5) automatic.

In the case $r = -\rho$ we found for $J \subset [\rho-1]$ that

$$B^{(\rho)}_{-\rho} = \sum_{I\subset[\rho-1]} \beta_I, \qquad B^{(\rho)}_{J,-\rho} = \frac{B^{(\rho)}_{-\rho}}{\beta_J}, \qquad \beta_J = \prod_{i\in J\ j\notin J} \beta_{ij},$$

where $\beta_{ij} = \beta_{ji}$ are constants. We expect the formula for the $B^{(\rho)}_{J,0}$ to have a similar structure. We make the Ansatz that (5.4) and (5.3) hold with unknown power series $\gamma_{ij} = \gamma_{ji} \in 1 + w^2\mathbb{C}[[w^2]]$. This is also motivated both by Remark 3.5, and the fact that with this Ansatz at least the relation (5.7) is automatically fullfilled. In fact this appears to be the simplest modification of formula in case $r = -\rho$ compatible with Remark 3.5 and (5.7). The relation (5.6) will not automatically follow from the Ansatz but requires the right choice of the γ_{ij}.

Now for $\rho \le 8$, we take the coefficients of positive degree of the γ_{ij} for $1 \le i < j \le \rho - 1$ as indeterminants. Then the blowup relations (3.1), (3.2) give, successively degree by degree in w, linear relations for these coefficients, which we use with a Pari/GP program to determine them modulo $O(w^{40})$. We analyze these power series, and find the formula for the $\gamma_{i,\rho-i}$, and then that the general γ_{ij} are given by the convex combination (5.2). We then use a Pari/GP program to show that with these formulas the formula (5.6) and the blowup formulas Conjecture 3.1 hold modulo $O(w^{80})$ for $\rho \le 10$.

5.2. Segre invariants for $s = \rho$. We can now apply the Segre Verlinde correspondence to determine the universal power series $Y^{(\rho)}_{J,\rho}$, $Z^{(\rho)}_{J,\rho}$ of the Segre invariants for $s = \rho$. Then the power series $S^{(\rho)}_{J,\rho}$ can be determined using the blowup formulas of Conjecture 3.3. We obtain the following formulas.

CONJECTURE 5.2. *Let again* $\xi = e^{\frac{\pi i}{2\rho}}$. *We have*

$$
\begin{aligned}
Y_\rho^{(\rho)} &= \prod_{i=1}^{\rho-1}\Big(\big(1+(\frac{\xi^{\rho-2i}+\xi^{2i-\rho}}{2})^2z^2\big)^{\frac12}+(\frac{\xi^{\rho-2i}+\xi^{2i-\rho}}{2})z\Big),\\
Y_{i,\rho}^{(\rho)} &= \Big(\big(1+(\frac{\xi^{\rho-2i}+\xi^{2i-\rho}}{2})^2z^2\big)^{\frac12}-(\frac{\xi^{\rho-2i}+\xi^{2i-\rho}}{2})z\Big)^2,\\
\zeta_{i,\rho-i} &= \big(1+(\frac{\xi^{\rho-2i}+\xi^{2i-\rho}}{2})^2z^2\big)^{\frac12}, \qquad i\in[\rho-1],\\
\zeta_{i,j} &= \frac{(\xi^{\rho-2j}+\xi^{2j-\rho})\zeta_{i,\rho-i}+(\xi^{\rho-2i}+\xi^{2i-\rho})\zeta_{j,\rho-j}}{\xi^{\rho-2i}+\xi^{2i-\rho}+\xi^{\rho-2j}+\xi^{2j-\rho}}, \qquad 1\le i\ne j\le \rho-1,\\
\zeta_I &= \Big(\prod_{i\in I,\ j\notin I}\zeta_{i,j}\Big)\Big(\prod_{i\in I}(Y_{i,\rho}^{(\rho)})^{\frac{\rho}{2}}\Big), \qquad I\subset[\rho-1],\\
Z_\rho^{(\rho)} &= \sum_{I\subset[\rho-1]}\beta_I\zeta_I, \qquad Z_{J,\rho}^{(\rho)}=\frac{Z_\rho^{(\rho)}}{\beta_J\zeta_J}, \qquad J\subset[\rho-1],\\
S_\rho^{(\rho)} &= -\sum_{i=1}^{\rho-1}\Big((\frac{\xi^{\rho-2i}+\xi^{2i-\rho}}{2})^2z^2+(\frac{\xi^{\rho-2i}+\xi^{2i-\rho}}{2})z\big(1+(\frac{\xi^{\rho-2i}+\xi^{2i-\rho}}{2})^2z^2\big)^{\frac12}\Big),\\
S_{i,\rho}^{(\rho)} &= -\frac{\xi^{2(\rho-2i)}-\xi^{2(2i-\rho)}}{2}z^2-(\xi^{\rho-2i}+\xi^{2i-\rho})z\big(1+(\frac{\xi^{\rho-2i}+\xi^{2i-\rho}}{2})^2z^2\big)^{\frac12}, \quad i\in[\rho-1].
\end{aligned}
$$

The formulas for the $Y_{J,\rho}^{(\rho)}$ and the $Z_{J,\rho}^{(\rho)}$ follow directly from Conjecture 5.1 and Conjecture 2.3. Let $S_\rho^{(\rho)}$, and $S_{i,\rho}^{(\rho)}$ for $i = 1,\dots,\rho-1$ be power series in z with constant coefficient 0 and the other coefficients as variables. Then the coefficients of x^1 of the blowup formulas (3.5) give linear relations for the coefficients of the $S_{i,\rho}^{(\rho)}$ which we use to determine them up to degree 50 in z for $\rho\le 7$. From these we can read off the formulas for $S_\rho^{(\rho)}$ and the $S_{i,\rho}^{(\rho)}$. We then check that with the formulas of Conjecture 5.2, the blowup formulas of Conjecture 1.4 hold up to degree 60 in z.

6. Generating functions for Verlinde and Segre formulas in low rank.

As mentioned in the introduction, and sketched in Section 3.3, we can use the blowup formulas to compute the Segre and Verlinde numbers for ρ not too large. We obtained the coefficients of the power series $A_{J,r}^{(\rho)}$, $B_{J,r}^{(\rho)}$ modulo v^{100} (and for some values of r and ρ for a much higher order) and of $Y_{J,s}^{(\rho)}$, $Z_{J,s}^{(\rho)}$ and $S_{J,s}^{(\rho)}$ for $\rho = 2,3,4,5,6$ and $0 < s < |2\rho|$ modulo t^{100}. Here we always use the changes of variables $w = v(1+v^2)^{\frac12(1-\frac{r^2}{\rho^2})}$ and $z = t(1+(1-\frac{s}{\rho})t^2)^{\frac12(1-\frac{s}{\rho})}$. Up to the order to which they are computed, these power series satisfy the Verlinde-Segre correspondence and the virtual Serre duality.

Furthermore we find that up to the order in v and t to which we computed, these universal power series always satisfy an algebraic equation of degree $2^{\rho-1}$ with coefficients in the rational functions $\mathbb{Q}(v^2)$ or $\mathbb{Q}(t^2)$ in v^2 respectively t^2. In fact by the above computations, assuming Conjecture 1.2, and blowup formulas Conjectures 1.4 and 1.3, the following conjecture holds for $\rho = 2,3,4,5,6$ and $|r| < \rho$ or $0 < s < 2\rho$.

Conjecture 6.1.

(1) *For all* $\rho \in \mathbb{Z} > 0$ *and all* $|r| < \rho$ *we have*

$$\prod_{I \subset [\rho-1]} (x - A_{I,r}^{(\rho)}) \in \mathbb{Q}(v^2)[x], \qquad \prod_{I \subset [\rho-1]} (x - B_{I,r}^{(\rho)}) \in \mathbb{Q}(v^2)[x].$$

(2) *For all* $\rho \in \mathbb{Z} > 0$ *and all* $0 < s < \rho$ *we have*

$$\prod_{I \subset [\rho-1]} (x - Y_{I,s}^{(\rho)}) \in \mathbb{Q}(t^2)[x], \qquad \prod_{I \subset [\rho-1]} (x - Z_{I,s}^{(\rho)}) \in \mathbb{Q}(t^2)[x],$$

$$\prod_{I \subset [\rho-1]} (x - S_{I,s}^{(\rho)}) \in \mathbb{Q}(t^2)[x].$$

For illustration and so that one gets some feeling for their shape, we list some of these polynomials in case $\rho \leq 4$. We restrict ourselves to polynomials for the $A_{I,r}^{(\rho)}$ and $B_{I,r}^{(\rho)}$. By the Verlinde-Segre correspondence Conjecture 2.3, this also determines the polynomials for the $Y_{I,s}^{(\rho)}$ and $Z_{I,s}^{(\rho)}$. Furthermore we restrict to the case $0 < r < \rho$, because the case $-\rho < r < 0$ is then determined by the virtual Serre duality relation Conjecture 2.4.

For ρ at most 3, these relations were already found by different methods in a different form in [**GK1**].

Rank 2: When $\rho = 2$, $r = 1$ we get with the change of variables $w = v(1+v^2)^{\frac{3}{8}}$

$$(x - A_{\emptyset,1}^{(2)}(w))(x - A_{\{1\},1}^{(2)}(w)) = x^2 - \frac{2+v^2}{1+v^2} + \frac{1}{(1+v^2)^2},$$

$$(x - B_{\emptyset,1}^{(2)}(w))(x - B_{\{1\},1}^{(2)}(w)) = x^2 - (1+v^2)^2(4+v^2)(x-1).$$

Rank 3: When $\rho = 3$, $r = 1$ we get with the change of variables $w = v(1+v^2)^{\frac{4}{9}}$

$$\prod_{I \subset \{1,2\}} (x - A_{I,1}^{(3)}(w)) = x^4 - \frac{4+v^2}{1+v^2}x^3 + \frac{6}{(1+v^2)^2}x^2 - \frac{4+v^2}{(1+v^2)^4}x + \frac{1}{(1+v^2)^6},$$

$$\prod_{I \subset \{1,2\}} (x - B_{I,1}^{(3)}(w)) = x^4 - (1+v^2)^3(9+v^2)(2+6v^2+v^4)x^3$$
$$+ (1+v^2)^4(9+v^2)(13+49v^2+36v^4-4v^6)x^2$$
$$- (1+v^2)^8(9+v^2)^2(x-1).$$

When $\rho = 3$, $r = 2$ we get with the change of variables $w = v(1+v^2)^{\frac{5}{18}}$

$$\prod_{I \subset \{1,2\}} (x - A_{I,2}^{(3)}(w)) = x^4 - \frac{4+3v^2}{(1+v^2)^2}x^3 + \frac{6+6v^2-v^6}{(1+v^2)^4}x^2$$
$$- \frac{4+3v^2}{(1+v^2)^6}x + \frac{1}{(1+v^2)^8},$$

$$\prod_{I \subset \{1,2\}} (x - B_{I,2}^{(3)}(w)) = x^4 - (1+v^2)^4(9+4v^2)(2+10v^2+5v^4)x^3$$
$$+ (1+v^2)^6(9+4v^2)(13+74v^2+89v^4-47v^6-95v^8-11v^{10}+17v^{12}$$
$$+ 2v^{14} - v^{16})x^2 - (1+v^2)^{12}(9+4v^2)^2(-4+v^2)(x-1).$$

Rank 4: When $\rho = 4$, we only write down the cases $r = 1$ and $r = 3$ in a form that makes them become simpler. We get with the change of variables $w = v(1+v^2)^{\frac{15}{32}}$

$$\prod_{I\subset\{1,2,3\}} (x-(1+v^2)^2 A_{I,1}^{(4)}(w)) = x^8 - (8+8v^2+v^4)x^7 + (28+48v^2+20v^4)x^6$$
$$- (7+8v^2)(8+8v^2+v^4)x^5 + (70+160v^2+104v^4+16v^6+v^8)x^4$$
$$- (7+8v^2)(8+8v^2+v^4)x^3 + (28+48v^2+20v^4)x^2 - (8+8v^2+v^4)x + 1,$$
$$\prod_{I\subset\{1,2,3\}} (x - A_{I,1}^{(4)}(w)B_{I,1}^{(4)}(w)) = x^8 - (1+v^2)^2(16+v^2)(6+21v^2+12v^4+v^6)x^7$$
$$+ (1+v^2)^2(16+v^2)(232+1287v^2+2442v^4+1874v^6+441v^8-51v^{10})x^6$$
$$- (1+v^2)^5(16+v^2)^2(296+1033v^2+1022v^4+270v^6)x^5 + (1+v^2)^4(16+v^2)^2$$
$$\cdot (3536+18030v^2+36507v^4+36600v^6+18924v^8+4980v^{10}+680v^{12}+27v^{14})x^4$$
$$- (1+v^2)^6(16+v^2)^3(32+27v^2)(50+129v^2+105v^4+22v^6)x^3$$
$$+ (1+v^2)^7(16+v^2)^3(32+27v^2)(72+130v^2+61v^4)x^2$$
$$- (1+v^2)^8(16+v^2)^4(32+27v^2)^2(x-1).$$

We get with the change of variables $w = v(1+v^2)^{\frac{7}{32}}$

$$\prod_{I\subset\{1,2,3\}} (x-(1+v^2)^3 A_{I,3}^{(4)}(w)) = x^8 - (8+8v^2+v^4)x^7 + (28+48v^2+20v^4)x^6$$
$$- (7+8v^2+v^8)(8+8v^2+v^4)x^5$$
$$+ (70+160v^2+104v^4+16v^6+17v^8+16v^{10}+v^{16})x^4$$
$$- (7+8v^2+v^8)(8+8v^2+v^4)x^3 + (28+48v^2+20v^4)x^2 - (8+8v^2+v^4)x + 1,$$
$$\prod_{I\subset\{1,2,3\}} (x - A_{I,3}^{(4)}(w)^2 B_{I,3}^{(4)}(w)) = x^8 - (16+9v^2)(6+23v^2+16v^4+v^6)x^7$$
$$+ (16+9v^2)(232+1103v^2+1604v^4+602v^6-435v^8-407v^{10}-88v^{12}+v^{14})x^6$$
$$- (16+9v^2)^2(296+1217v^2+1678v^4+725v^6+327v^8+583v^{10}+259v^{12}$$
$$- 2v^{14}+17v^{16}+12v^{18})x^5 + (16+9v^2)^2(3536+12958v^2+17251v^4+9220v^6$$
$$+ 5096v^8+6958v^{10}+1881v^{12}-648v^{14}+1296v^{16}+456v^{18}-216v^{20}+96v^{22}$$
$$+ 32v^{24}-13v^{26}+4v^{28})x^4 - (16+9v^2)^3(32-13v^2+4v^4)(50+143v^2+142v^4$$
$$+ 58v^6+17v^8+27v^{10}+19v^{12}+v^{14}+v^{18})x^3$$
$$+ (16+9v^2)^3(32-13v^2+4v^4)(216+446v^2+257v^4+28v^6+24v^8$$
$$+ 15v^{10}+v^{14})x^2 - (16+9v^2)^4(32-13v^2+4v^4)^2(x-1).$$

These formulas have a number of interesting features. The fact that in $\prod_{I\subset[\rho-1]}(x - B_{I,r}^{(\rho)})$, the coefficient of x is the negative of the constant coefficient is a direct consequence of the simplest of the blowup formulas of Conjecture 1.3: $\sum_{I\subset[\rho-1]}(B_{I,r}^{(\rho)})^{-1} = 1$ if $|r| \le \rho$.

The most obvious observation we can make from the above formulas is that the polynomials $\prod_{I\subset[\rho-1]}(x - A_{I,r}^{(\rho)})$ are almost palindromic. In fact the computations for $|r| < \rho \le 6$ confirm up to order at least 100 in v the following conjecture.

CONJECTURE 6.2. *For $|r| < \rho$ we have with $w = v(1+v^2)^{\frac{1}{2}(1-\frac{r^2}{\rho^2})}$ the following.*

(1)
$$p_{\rho,r}(x,v) := \prod_{I\subset[\rho-1]} \left(x-(1+v^2)^{\frac{r+\rho-1}{2}}A^{(\rho)}_{I,r}(w)\right)$$

is palindromic in x, *i.e.* $p_{\rho,r}(x,v)=x^{2^{\rho-1}}p_{\rho,r}(x^{-1},v)$.

(2) *More precisely we have*

$$A^{(\rho)}_{[\rho-1]\setminus I,r}(w)=\frac{1}{(1+v^2)^{r+\rho-1}A^{(\rho)}_{I,r}(w)}.$$

for all $I\subset[\rho-1]$.

Remark 6.3. Note that by the definition $A^{(\rho)}_{I,r}(w)=A^{(\rho)}_r\prod_{i\in I}A^{(\rho)}_{i,r}(w)$ part (2) of Conjecture 6.2 is equivalent to

$$A^{(\rho)}_r(w)^2=\frac{1}{(1+v^2)^{r+\rho-1}\prod_{i\in[\rho-1]}A^{(\rho)}_{i,r}(w)},$$

or equivalently

$$A^{(\rho)}_{I,r}(w)^2=\frac{\prod_{i\in I}A^{(\rho)}_{i,r}(w)}{(1+v^2)^{r+\rho-1}\prod_{i\in[\rho-1]\setminus I}A^{(\rho)}_{i,r}(w)}\qquad \text{for all } I\subset[\rho-1].$$

Thus Conjecture 6.2 impllies that $A^{(\rho)}_r$ and all the $A^{(\rho)}_{I,r}$ are determined by the $A^{(\rho)}_{i,r}$. This observation will play a role in the forthcoming work [**G2**].

Remark 6.4. This paper contains a number of conjectures, with web of implications between them. We briefly sum up the dependencies among them.

(1) Conjecture 2.2 follows from Conjecture 1.2. As mentioned in the introduction, all other conjectures depend on Conjecture 1.2, because it states the existence of universal power series; universal in the sense that they depend only on ρ and r respectively ρ and s, and in particular are independent of S. These universal power series are the subject of all the other conjectures. The blowup formulas are only useful for determining these power series because of their universality. So for the rest of this remark we assume Conjecture 1.2.
(2) Conjecture 1.3 is equivalent to Conjecture 3.3 and Conjecture 1.4 is equivalent to Conjecture 3.1.
(3) Assuming the Verlinde-Segre correspondence Conjecture 2.3, parts (1) of Conjectures 3.3 and 3.1 are equivalent, and also part (2) of Conjecture 3.1 implies (3.8) in Conjecture 3.3.
(4) Assuming Conjecture 2.3, Conjecture 4.2 is equivalent to Conjecture 4.4 (except the formulas for $S^{(\rho)}_{J,0}$, $S^{(\rho)}_0$, $S^{(\rho)}_{j,0}$). Under the same assumption, Conjecture 4.6 is equivalent to Conjecture 4.7, and Conjecture 5.1 is equivalent to Conjecture 5.2 (except the formulas for the S-series).
(5) Assuming Conjecture 2.4, Conjectures 4.2 and 4.6 are equivalent.
(6) Conjecture 2.5 and Conjecture 4.2 imply Conjecture 4.3.

Acknowledgments

This work grew out of my collaboration with Martijn Kool over the last few years, in which I benefitted from numerous discussions and insights, without which the current paper could not have been written. The results and ideas of [**GK1**] play an important role in this paper. I thank Hiraku Nakajima for pointing out and explaining the references [**NY**] and [**KT**] and their significance. I also thank the referee for his comments which helped to improve the exposition of this paper.

References

[BF] K. Behrend and B. Fantechi, *The intrinsic normal cone*, Invent. Math. **128** (1997), no. 1, 45–88, DOI 10.1007/s002220050136. MR1437495

[CFK] Ionuţ Ciocan-Fontanine and Mikhail Kapranov, *Virtual fundamental classes via dg-manifolds*, Geom. Topol. **13** (2009), no. 3, 1779–1804, DOI 10.2140/gt.2009.13.1779. MR2496057

[Don] S. K. Donaldson, *Polynomial invariants for smooth four-manifolds*, Topology **29** (1990), no. 3, 257–315, DOI 10.1016/0040-9383(90)90001-Z. MR1066174

[EGL] Geir Ellingsrud, Lothar Göttsche, and Manfred Lehn, *On the cobordism class of the Hilbert scheme of a surface*, J. Algebraic Geom. **10** (2001), no. 1, 81–100. MR1795551

[FG] Barbara Fantechi and Lothar Göttsche, *Riemann-Roch theorems and elliptic genus for virtually smooth schemes*, Geom. Topol. **14** (2010), no. 1, 83–115, DOI 10.2140/gt.2010.14.83. MR2578301

[FL1] Paul M. N. Feehan and Thomas G. Leness, *The* SO(3) *monopole cobordism and superconformal simple type*, Adv. Math. **356** (2019), 106817, 38, DOI 10.1016/j.aim.2019.106817. MR4011024

[FL2] Paul M. N. Feehan and Thomas G. Leness, *Superconformal simple type and Witten's conjecture*, Adv. Math. **356** (2019), 106821, 35, DOI 10.1016/j.aim.2019.106821. MR4013801

[FS1] Ronald Fintushel and Ronald J. Stern, *The blowup formula for Donaldson invariants*, Ann. of Math. (2) **143** (1996), no. 3, 529–546, DOI 10.2307/2118535. MR1394968

[FS2] Ronald Fintushel and Ronald J. Stern, *Donaldson invariants of 4-manifolds with simple type*, J. Differential Geom. **42** (1995), no. 3, 577–633. MR1367403

[FM] Robert Friedman and John W. Morgan, *Smooth four-manifolds and complex surfaces*, Ergebnisse der Mathematik und ihrer Grenzgebiete (3) [Results in Mathematics and Related Areas (3)], vol. 27, Springer-Verlag, Berlin, 1994, DOI 10.1007/978-3-662-03028-8. MR1288304

[G1] Lothar Göttsche, *Verlinde-type formulas for rational surfaces*, J. Eur. Math. Soc. (JEMS) **22** (2020), no. 1, 151–212, DOI 10.4171/jems/919. MR4046012

[G2] L. Göttsche, in preparation.

[GK1] Lothar Göttsche and Martijn Kool, *Virtual Segre and Verlinde numbers of projective surfaces*, J. Lond. Math. Soc. (2) **106** (2022), no. 3, 2562–2608. MR4498560

[GK2] Lothar Göttsche and Martijn Kool, *Sheaves on surfaces and virtual invariants*, Surveys in differential geometry 2019. Differential geometry, Calabi-Yau theory, and general relativity. Part 2, Surv. Differ. Geom., vol. 24, Int. Press, Boston, MA, [2022] ©2022, pp. 67–116. MR4479719

[GKL] L. Göttsche, M. Kool and T. Laarakker, *Virtual degeneracy loci, Ramanujan's continued fractions, and cosmic strings*, `arXiv:2108.13413`.

[GKW] L. Göttsche, M. Kool, and R. A. Williams, *Verlinde formulae on complex surfaces: K-theoretic invariants*, Forum Math. Sigma **9** (2021), Paper No. e5, 31, DOI 10.1017/fms.2020.50. MR4202490

[GM1] L. Göttsche, A. Mellit, *Refined Verlinde and the Segre formula for Hilbert schemes*, Preprint, `arXiv:2210.01059`, 2022.

[GNY1] Lothar Göttsche, Hiraku Nakajima, and Kōta Yoshioka, *K-theoretic Donaldson invariants via instanton counting*, Pure Appl. Math. Q. **5** (2009), no. 3, Special Issue: In honor of Friedrich Hirzebruch., 1029–1111, DOI 10.4310/PAMQ.2009.v5.n3.a5. MR2532713

[GNY2] Lothar Göttsche, Hiraku Nakajima, and Kōta Yoshioka, *Donaldson = Seiberg-Witten from Mochizuki's formula and instanton counting*, Publ. Res. Inst. Math. Sci. **47** (2011), no. 1, 307–359, DOI 10.2977/PRIMS/37. MR2827729

[GY] Lothar Göttsche and Yao Yuan, *Generating functions for K-theoretic Donaldson invariants and Le Potier's strange duality*, J. Algebraic Geom. **28** (2019), no. 1, 43–98, DOI 10.1090/jag/703. MR3875361

[Joh] Drew Johnson, *Universal series for Hilbert schemes and strange duality*, Int. Math. Res. Not. IMRN **10** (2020), 3130–3152, DOI 10.1093/imrn/rny101. MR4098636

[JS] Dominic Joyce and Yinan Song, *A theory of generalized Donaldson-Thomas invariants*, Mem. Amer. Math. Soc. **217** (2012), no. 1020, iv+199, DOI 10.1090/S0065-9266-2011-00630-1. MR2951762

[KM1] P. B. Kronheimer and T. S. Mrowka, *Recurrence relations and asymptotics for four-manifold invariants*, Bull. Amer. Math. Soc. (N.S.) **30** (1994), no. 2, 215–221, DOI 10.1090/S0273-0979-1994-00492-6. MR1246469

[KM2] P. B. Kronheimer and T. S. Mrowka, *Embedded surfaces and the structure of Donaldson's polynomial invariants*, J. Differential Geom. **41** (1995), no. 3, 573–734. MR1338483

[Kro] P. B. Kronheimer, *Four-manifold invariants from higher-rank bundles*, J. Differential Geom. **70** (2005), no. 1, 59–112. MR2192061

[KT] N. Kuhn, Y. Tanaka, *A blowup formula for virtual enumerative invariants on projective surfaces*, Preprint, arXiv:2107.08155, 2021.

[LM] Jose Labastida and Marcos Marino, *Topological quantum field theory and four manifolds*, Mathematical Physics Studies, vol. 25, Springer, Dordrecht, 2005. MR2161779

[Leh] Manfred Lehn, *Chern classes of tautological sheaves on Hilbert schemes of points on surfaces*, Invent. Math. **136** (1999), no. 1, 157–207, DOI 10.1007/s002220050307. MR1681097

[LP] Le Potier, J.: *Dualité étrange, sur les surfaces.* Preliminary version 18.11.05

[Li] Jun Li, *Algebraic geometric interpretation of Donaldson's polynomial invariants*, J. Differential Geom. **37** (1993), no. 2, 417–466. MR1205451

[MM] Marcos Mariño and Gregory Moore, *The Donaldson-Witten function for gauge groups of rank larger than one*, Comm. Math. Phys. **199** (1998), no. 1, 25–69, DOI 10.1007/s002200050494. MR1660219

[MOP1] Alina Marian, Dragos Oprea, and Rahul Pandharipande, *Segre classes and Hilbert schemes of points* (English, with English and French summaries), Ann. Sci. Éc. Norm. Supér. (4) **50** (2017), no. 1, 239–267, DOI 10.24033/asens.2320. MR3621431

[MOP2] Alina Marian, Dragos Oprea, and Rahul Pandharipande, *The combinatorics of Lehn's conjecture*, J. Math. Soc. Japan **71** (2019), no. 1, 299–308, DOI 10.2969/jmsj/78747874. MR3909922

[MOP3] Alina Marian, Dragos Oprea, and Rahul Pandharipande, *Higher rank Segre integrals over the Hilbert scheme of points*, J. Eur. Math. Soc. (JEMS) **24** (2022), no. 8, 2979–3015, DOI 10.4171/jems/1149. MR4416594

[Moc] Takuro Mochizuki, *Donaldson type invariants for algebraic surfaces*, Lecture Notes in Mathematics, vol. 1972, Springer-Verlag, Berlin, 2009. Transition of moduli stacks, DOI 10.1007/978-3-540-93913-9. MR2508583

[Mor] John W. Morgan, *Comparison of the Donaldson polynomial invariants with their algebro-geometric analogues*, Topology **32** (1993), no. 3, 449–488, DOI 10.1016/0040-9383(93)90001-C. MR1231956

[NY1] Hiraku Nakajima and Kōta Yoshioka, *Lectures on instanton counting*, Algebraic structures and moduli spaces, CRM Proc. Lecture Notes, vol. 38, Amer. Math. Soc., Providence, RI, 2004, pp. 31–101, DOI 10.1090/crmp/038/02. MR2095899

[NY2] Hiraku Nakajima and Kōta Yoshioka, *Instanton counting on blowup. I. 4-dimensional pure gauge theory*, Invent. Math. **162** (2005), no. 2, 313–355, DOI 10.1007/s00222-005-0444-1. MR2199008

[NY] Hiraku Nakajima and Kōta Yoshioka, *Perverse coherent sheaves on blowup, III: Blow-up formula from wall-crossing*, Kyoto J. Math. **51** (2011), no. 2, 263–335, DOI 10.1215/21562261-1214366. MR2793270

[Ob] Georg Oberdieck, *Universality of descendent integrals over moduli spaces of stable sheaves on K3 surfaces*, SIGMA Symmetry Integrability Geom. Methods Appl. **18** (2022), Paper No. 076, 15, DOI 10.3842/SIGMA.2022.076. MR4495687

[Voi] Claire Voisin, *Segre classes of tautological bundles on Hilbert schemes of surfaces*, Algebr. Geom. **6** (2019), no. 2, 186–195, DOI 10.14231/AG-2019-010. MR3914750

[Wit] Edward Witten, *Monopoles and four-manifolds*, Math. Res. Lett. **1** (1994), no. 6, 769–796, DOI 10.4310/MRL.1994.v1.n6.a13. MR1306021

The Abdus Salam International Centre for Theoretical Physics, Strada Costiera, 11, 34151 Trieste, Italy

Email address: gottsche@ictp.it

Proceedings of Symposia in Pure Mathematics
Volume **109**, 2024
https://doi.org/10.1090/pspum/109/01993

A note on PL-disks and rationally slice knots

Kristen Hendricks, Jennifer Hom, Matthew Stoffregen, and Ian Zemke

ABSTRACT. We give infinitely many examples of manifold-knot pairs (Y, J) such that Y bounds an integer homology ball, J does not bound a non-locally-flat PL-disk in any integer homology ball, but J does bound a smoothly embedded disk in a rational homology ball. The proof relies on formal properties of involutive Heegaard Floer homology.

Every knot K in S^3 bounds a non-locally-flat PL-embedded disk in B^4, obtained by taking the cone over K. (Throughout, we will not require PL-disks to be locally-flat.) The analogous statement does not hold for knots in more general manifolds. Adam Levine [**9**, Theorem 1.2] found examples of manifold-knot pairs (Y, J) such that Y bounds a contractible 4-manifold and J does not bound a PL-disk in any homology ball X with $\partial X = Y$; see also [**7**].

The main result of this note concerns rationally slice knots in homology spheres bounding integer homology balls:

THEOREM 1. *There exist infinitely many manifold-knot pairs (Y, J) where Y is an integer homology sphere and*

(1) *Y bounds an integer homology 4-ball,*
(2) *J does not bound a PL-disk in any integer homology 4-ball,*
(3) *J does bound a smoothly embedded disk in a rational homology 4-ball.*

Throughout, let Y be an integer homology sphere. Recall that a knot $J \subset Y$ is *rationally slice* if J bounds a smoothly embedded disk in a rational homology 4-ball W with $\partial W = Y$. Two manifold-knot pairs (Y_0, J_0) and (Y_1, J_1) are *integrally* (respectively *rationally*) *homology concordant* if J_0 and J_1 are concordant in an integral (respectively rational) homology cobordism between Y_0 and Y_1. A knot $J \subset Y$ is integrally (respectively rationally) homology concordant to a knot K in S^3 if and only if $J \subset Y$ bounds a PL-disk in an integer (respectively rational) homology ball.

Theorem 1 is an immediate consequence of the following theorem, where $\underline{V}_0$ and $\overline{V}_0$ are the involutive knot Floer homology invariants of [**5**] and V_0 the knot Floer homology invariant defined in [**10**, Section 2.2] (see also [**13**], [**11**]):

2020 *Mathematics Subject Classification.* Primary 57K10, 57K18.

The first author was partially supported by NSF grant DMS-2019396 and a Sloan Research Fellowship. The second author was partially supported by NSF grants DMS-1552285 and DMS-2104144. The third author was partially supported by NSF grant DMS-1702532. The fourth author was partially supported by NSF grant DMS-1703685.

THEOREM 2. *Let K be a negative amphichiral rationally slice knot in S^3 with $\underline{V}_0 \geq 1$ and $V_0 = \overline{V}_0 = 0$ and let μ be the core of surgery in $M = S^3_{-1/\ell}(K)$, where ℓ is an odd positive integer. Consider $J = \mu \# U \subset M \# -M$, where U denotes the unknot in $-M$. Then $(M \# -M, J)$ is rationally slice, hence rationally homology concordant to a knot in S^3, but $(M \# -M, J)$ is not integrally homology concordant to any knot in S^3.*

REMARK 3. The figure-eight satisfies the hypotheses of Theorem 2 by [**3**] (see also [**1**, Section 3] and [**5**, Theorem 1.7]). More generally, the genus one knots K_n with n positive full twists in one band and n negative full twists in the other band, n odd, also satisfy the hypotheses of Theorem 2; see Figure 1. By [**2**, Theorem 4.16], K_n is rationally slice. (Alternatively, K_n is strongly negative amphichiral, hence rationally slice [**8**, Section 2].) Furthermore, $\sigma(K_n) = 0$ since K_n is amphichiral. The knot K_n has Seifert form

$$\begin{pmatrix} n & 1 \\ 0 & -n \end{pmatrix}$$

which implies that $\operatorname{Arf}(K_n) = 1$ if and only if n is odd. Since K_n is alternating, it now follows from [**5**, Theorem 1.7] that for n odd, $\underline{V}_0 = 1$ and $V_0 = \overline{V}_0 = 0$.

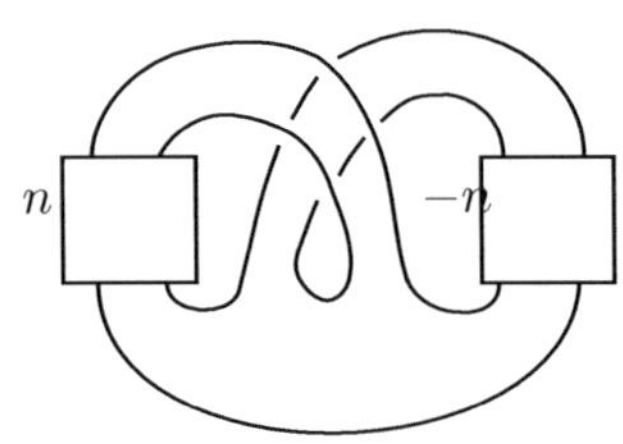

FIGURE 1. The knot K_n, where n and $-n$ denote the number of positive full twists.

REMARK 4. Note that M does not bound an integer homology ball (since, for instance, $\underline{d}(M) = 2\underline{V}_0 \neq 0$), but $M \# -M$ does.

The proof of Theorem 2 is inspired by the proof of [**7**, Theorem 1.1(1)]. Our proof relies on the following result from [**4**] relating the involutive correction term $\underline{d}$ [**5**, Section 5] with the ordinary Heegaard Floer correction term d [**12**, Section 4], for even denominator surgery on knots in S^3:

PROPOSITION 5 ([**4**, Proposition 1.7]). *Let K be a knot in S^3 and let $p, q > 0$ be relatively prime integers, with p odd and q even. Then*

$$\underline{d}(S^3_{p/q}(K), [p/2q]) = d(S^3_{p/q}(K), [p/2q])$$

where $[p/2q]$ denotes the unique self-conjugate Spin^c *structure on $S^3_{p/q}(K)$.*

The key feature from the above proposition is that for even denominator surgery on a knot in S^3, we have that $\underline{d}$ is equal to d for the unique self-conjugate Spin^c structure on the surgery. More generally, we have the following corollary of Proposition 5:

COROLLARY 6. *Let J be a knot in an integer homology sphere Y and let $p, q > 0$ be relatively prime integers, with p odd and q even. If (Y, J) is integrally homology concordant to a knot in S^3, then*

$$\underline{d}(Y_{p/q}(J), [p/2q]) = d(Y_{p/q}(J), [p/2q])$$

where $[p/2q]$ denotes the unique self-conjugate Spin^c *structure on $Y^3_{p/q}(J)$.*

PROOF. If (Y, J) is integrally homology concordant to a knot (S^3, K), then $Y^3_{p/q}(J)$ and $S^3_{p/q}(K)$ are integrally homology cobordant; the homology cobordism is given by surgering along the concordance annulus from (Y, J) to (S^3, K). Since $\underline{d}$ and d are invariants of integer homology cobordism, the result follows from Proposition 5. □

The proof of Theorem 2 relies on finding manifold-knot pairs (Y, J) where $\underline{d}$ and d of even denominator surgery along J differ; the result then follows from Corollary 6.

PROOF OF THEOREM 2. We first show that $(M \# -M, J)$ is rationally slice. Since K is rationally slice, the core of surgery in $M = S^3_{-1/\ell}(K)$ is rationally homology concordant to the core of surgery in $S^3_{-1/\ell}(U)$, which is the unknot in S^3; that is, (M, μ) is rationally slice. Hence $(M \# -M, J)$ is also rationally slice.

We now show that $(M \# -M, J)$ is not integrally homology concordant to any knot in S^3. Since μ is the core of surgery in $S^3_{-1/\ell}(K)$, we have that

$$M_{1/n}(\mu) = S^3_{1/(n-\ell)}(K).$$

Choose an even positive integer n such that $n > \ell$. Since ℓ is odd, n is even, and $n - \ell > 0$, by [**4**, Proposition 1.7] we have that

$$\underline{d}(M_{1/n}(\mu)) = \underline{d}(S^3_{1/(n-\ell)}(K)) = -2\underline{V}_0(K)$$

and

$$d(M_{1/n}(\mu)) = d(S^3_{1/(n-\ell)}(K)) = -2V_0(K) = 0.$$

Since $J = \mu \# U \subset M \# -M$, we have that

$$(M \# -M)_{1/n}(J) = M_{1/n}(\mu) \# -M.$$

Note that $-M = S^3_{1/\ell}(-K) = S^3_{1/\ell}(K)$, where the last equality follows from the fact that K is negative amphichiral. Since $\ell > 0$, [**4**, Proposition 1.7] implies that

$$\underline{d}(-M) = -2\underline{V}_0(K) \quad \text{and} \quad d(-M) = \overline{d}(-M) = 0.$$

Recall that [**6**, Proposition 1.3] states that if Y_1 and Y_2 are integer homology spheres, then

$$\underline{d}(Y_1 \# Y_2) \leq \underline{d}(Y_1) + \overline{d}(Y_2).$$

Hence $\underline{d}(M_{1/n}(\mu) \# -M) \leq -2\underline{V}_0(K)$. Since d is additive under connected sum, we have that $d(M_{1/n}(\mu) \# -M) = 0$.

We have shown that

$$\underline{d}((M \# -M)_{1/2}(J)) \leq -2\underline{V}_0(K) \quad \text{and} \quad d((M \# -M)_{1/2}(J)) = 0.$$

Recall that $\underline{V}_0(K) \geq 1$. Now by Corollary 6, it follows that $(M \# -M, J)$ is not integrally homology concordant to any knot in S^3. □

Acknowledgments

We thank JungHwan Park for helpful conversations and Chuck Livingston for thoughtful comments on an earlier draft.

References

[1] Selman Akbulut and Kyle Larson, *Brieskorn spheres bounding rational balls*, Proc. Amer. Math. Soc. **146** (2018), no. 4, 1817–1824, DOI 10.1090/proc/13828. MR3754363

[2] Jae Choon Cha, *The structure of the rational concordance group of knots*, Mem. Amer. Math. Soc. **189** (2007), no. 885, x+95, DOI 10.1090/memo/0885. MR2343079

[3] Ronald Fintushel and Ronald J. Stern, *A μ-invariant one homology 3-sphere that bounds an orientable rational ball*, Four-manifold theory (Durham, N.H., 1982), Contemp. Math., vol. 35, Amer. Math. Soc., Providence, RI, 1984, pp. 265–268, DOI 10.1090/conm/035/780582. MR780582

[4] Kristen Hendricks, Jennifer Hom, Matthew Stoffregen, and Ian Zemke, *Surgery exact triangles in involutive Heegaard Floer homology*, 2020, Preprint, arXiv:2011.00113.

[5] Kristen Hendricks and Ciprian Manolescu, *Involutive Heegaard Floer homology*, Duke Math. J. **166** (2017), no. 7, 1211–1299, DOI 10.1215/00127094-3793141. MR3649355

[6] Kristen Hendricks, Ciprian Manolescu, and Ian Zemke, *A connected sum formula for involutive Heegaard Floer homology*, Selecta Math. (N.S.) **24** (2018), no. 2, 1183–1245, DOI 10.1007/s00029-017-0332-8. MR3782421

[7] Jennifer Hom, Adam Simon Levine, and Tye Lidman, *Knot concordance in homology cobordisms*, Duke Math. J. **171** (2022), no. 15, 3089–3131, DOI 10.1215/00127094-2021-0110. MR4497224

[8] Akio Kawauchi, *Rational-slice knots via strongly negative-amphicheiral knots*, Commun. Math. Res. **25** (2009), no. 2, 177–192. MR2554510

[9] Adam Simon Levine, *Nonsurjective satellite operators and piecewise-linear concordance*, Forum Math. Sigma **4** (2016), Paper No. e34, 47, DOI 10.1017/fms.2016.31. MR3589337

[10] Yi Ni and Zhongtao Wu, *Cosmetic surgeries on knots in S^3*, J. Reine Angew. Math. **706** (2015), 1–17, DOI 10.1515/crelle-2013-0067. MR3393360

[11] Peter Ozsváth and Zoltán Szabó, *Holomorphic disks and knot invariants*, Adv. Math. **186** (2004), no. 1, 58–116, DOI 10.1016/j.aim.2003.05.001. MR2065507

[12] Peter Ozsváth and Zoltán Szabó, *Absolutely graded Floer homologies and intersection forms for four-manifolds with boundary*, Adv. Math. **173** (2003), no. 2, 179–261, DOI 10.1016/S0001-8708(02)00030-0. MR1957829

[13] Jacob Andrew Rasmussen, *Floer homology and knot complements*, ProQuest LLC, Ann Arbor, MI, 2003. Thesis (Ph.D.)–Harvard University. MR2704683

Department of Mathematics, Rutgers University, New Brunswick, New Jersey 08854
Email address: kristen.hendricks@rutgers.edu

School of Mathematics, Georgia Institute of Technology, Atlanta, Georgia 30332
Email address: hom@math.gatech.edu

Department of Mathematics, Michigan State University, East Lansing, Michigan 48824
Email address: stoffre1@msu.edu

Department of Mathematics, Princeton University, Princeton, New Jersey 08544
Email address: izemke@math.princeton.edu

Proceedings of Symposia in Pure Mathematics
Volume **109**, 2024
https://doi.org/10.1090/pspum/109/01994

Khovanov homology of strongly invertible knots and their quotients

Robert Lipshitz and Sucharit Sarkar

Abstract. We construct a spectral sequence relating the Khovanov homology of a strongly invertible knot to the annular Khovanov homologies of the two quotient knots. Using this spectral sequence, we re-prove that Khovanov homology distinguishes certain slice disks. We also give an analogous spectral sequence for $\widehat{HF}$ of the branched double cover.

Contents

1. Introduction

A knot $K \subset \mathbb{R}^3$ is called *strongly invertible* if K meets some straight line in exactly two points, and is preserved setwise by rotation by $180°$ around that line; this straight line is called the *axis*. While strongly invertible knots have been studied for decades (see [**Sak86**] and the references therein), they have recently seen a surge in interest. For example, there is a somewhat mysterious, noncommutative concordance group of strongly invertible knots [**Sak86, DP23**], as well as natural equivariant analogues of the slice genus [**BI22**]. Related to this, many of the pairs of non-isotopic slice disks or more general slice surfaces which have appeared in the literature recently come from strongly invertible knots [**Hay, SS22, HS**], a phenomenon which has led to connections with Heegaard Floer-theoretic invariants [**DMS23**]. In addition to Heegaard Floer homology, Donaldson's diagonalization theorem [**BI22**], the G-signature theorem and related equivariant signature techniques [**AB, DP**], Kojima-Yamasaki's η-function [**KY79, Sak86, BI22, DPF**],

2020 *Mathematics Subject Classification.* Primary 57K18, 55P42; Secondary 57K45, 55N91.

The first author was supported by NSF Grants DMS-1810893 (research) and 2204214 (revisions). The second author was supported by NSF Grant DMS-1905717.

and Khovanov homology [**Cou09, Wat17, Sna18, LW21**] have also been applied recently to study strongly invertible knots.

The quotient of K by its strong inversion is naturally an embedded arc with boundary on the axis. By gluing this arc to part of the axis, we obtain a quotient knot; see Figures 1.1 and 5.2. (In S^3, instead of $\mathbb{R}^3$, there are two equally natural choices of quotient knot, corresponding to the two choices of half-axis; cf. [**Sak86, BI22**].) The main goal of this paper is to construct a spectral sequence relating the Khovanov homology of a strongly invertible knot K and a variant of the Khovanov homology of its quotient.

Similar results have been proved before, for other symmetries. Stoffregen-Zhang [**SZ**] and Borodzik-Politarczyk-Silvero [**BPS21**] showed that there is a spectral sequence relating the Khovanov homology of a periodic knot (a knot preserved by rotation around an axis disjoint from it) and the annular Khovanov homology of its quotient (see also [**Cor, Zha18**]). An analogous result relating the symplectic Khovanov homology of a 2-periodic knot and of its quotient was proved earlier by Seidel-Smith [**SS10**]. Using the same technical tool, Hendricks proved a similar relationship for knot Floer homology [**Hen15**], and analogous results have been given for other symmetries in Heegaard and monopole Floer homology [**Hen12, HLS16, LT16, LM18, Lar, HLL22**].

Like Stoffregen-Zhang's and Borodzik-Politarczyk-Silvero's spectral sequence for periodic knots, the spectral sequence we construct for strongly invertible knots relates the Khovanov homology of K to the annular Khovanov homology of its quotient $\overline{K}$. If τ denotes the 180° rotation around the axis then by a slight τ-equivariant perturbation of the knot K we may assume that the projection of K to the plane perpendicular to the axis is a knot diagram. Following the literature [**Boy21, BI22**], we call such a diagram *intravergent*. The quotient knot $\overline{K}$ may be viewed as an annular knot in two natural ways, $\overline{K}_0$ and $\overline{K}_1$, corresponding to taking the quotients of the 0-resolution or the 1-resolution of the fixed crossing of the intravergent diagram K; see Figure 1.1 and 5.2. (These quotients depend on the diagram K; see Remark 4.5 and Proposition 4.6.) The annular Khovanov chain complexes of these knots are related by an *axis-moving map* $f^+\colon \Sigma^{0,0,1}ACKh(\overline{K}_1) \to ACKh(\overline{K}_0)$, where $\Sigma^{a,b,c}$ denotes a (homological, quantum, annular) trigrading shift by (a,b,c); we construct the map f^+, which is a special case of Akhmechet-Khovanov's maps associated to anchored cobordisms [**AK23**], in Section 3. By a slight abuse of notation, we define the annular Khovanov chain complex of the pair of annular knots $(\overline{K}_1, \overline{K}_0)$ to be the mapping cone of f^+,

$$ACKh(\overline{K}_1, \overline{K}_0) = \mathrm{Cone}\left(\Sigma^{0,0,1}ACKh(\overline{K}_1) \xrightarrow{f^+} ACKh(\overline{K}_0)\right),$$

and the annular Khovanov homology of the pair to be the homology of this complex which, over any field $\mathbb{F}$, is also (unnaturally) isomorphic to the homology of the mapping cone of the induced map on homology,

$$\begin{aligned} AKh(\overline{K}_1, \overline{K}_0; \mathbb{F}) &= H_*\,\mathrm{Cone}\left(\Sigma^{0,0,1}ACKh(\overline{K}_1;\mathbb{F}) \xrightarrow{f^+} ACKh(\overline{K}_0;\mathbb{F})\right) \\ &\cong H_*\,\mathrm{Cone}\left(\Sigma^{0,0,1}AKh(\overline{K}_1;\mathbb{F}) \xrightarrow{f^+} AKh(\overline{K}_0;\mathbb{F})\right). \end{aligned}$$

THEOREM 1.1. *Given a strongly invertible knot K with annular quotients $\overline{K}_0$, $\overline{K}_1$ there is a spectral sequence with the following properties:*

(Θ-1) *The E^1-page is $Kh(K;\mathbb{F}_2)\otimes\mathbb{F}_2[\theta^{-1},\theta]$ with d^1-differential the map $\theta(\mathrm{Id}+\tau_*)$,*

$$\cdots \longrightarrow \theta^{-1}Kh(K;\mathbb{F}_2) \xrightarrow{\theta(\mathrm{Id}+\tau_*)} \theta^0 Kh(K;\mathbb{F}_2) \xrightarrow{\theta(\mathrm{Id}+\tau_*)} \theta^1 Kh(K;\mathbb{F}_2) \longrightarrow \cdots,$$

where τ_* is induced by the strong inversion.

(Θ-2) *The d^r-differential preserves the quantum grading and increases the θ-power grading by r.*

(Θ-3) *The spectral sequence converges to $AKh(\overline{K}_1,\overline{K}_0;\mathbb{F}_2)\otimes\mathbb{F}_2[\theta^{-1},\theta]$. Keeping track of quantum gradings, the summand of the spectral sequence in quantum grading j converges to*

$$\bigoplus_{\substack{\bar{\imath},\bar{\jmath},k\in\mathbb{Z}\\ 2\bar{\jmath}+k=j-1+3N_- -6\overline{N}_-}} H_*\,\mathrm{Cone}\left(AKh_{\bar{\imath},\bar{\jmath},k-1}(\overline{K}_1;\mathbb{F}_2)\xrightarrow{f^+} AKh_{\bar{\imath},\bar{\jmath},k}(\overline{K}_0;\mathbb{F}_2)\right)\otimes\mathbb{F}_2[\theta^{-1},\theta] \tag{1.1}$$

where N_- (respectively $\overline{N}_-$) is the number of negative crossings of K (respectively $\overline{K}$).

Since the quantity $3N_- - 6\overline{N}_-$ comes up frequently, let

$$\Delta = N_- - 2\overline{N}_-. \tag{1.2}$$

See Remark 4.5 and Proposition 4.6 for a little further discussion of the grading shift, and Section 2 for our grading conventions for Khovanov homology. Like the periodic knot case, the proof of Theorem 1.1 uses the Khovanov stable homotopy type and Smith theory.

COROLLARY 1.2. *For any quantum grading j, we have*

$$\sum_i \dim Kh_{i,j}(K;\mathbb{F}_2) \geq \sum_{\substack{\bar{\imath},\bar{\jmath},k\in\mathbb{Z}\\ 2\bar{\jmath}+k=j-1+3\Delta}} \dim AKh_{\bar{\imath},\bar{\jmath},k}(\overline{K}_1,\overline{K}_0;\mathbb{F}_2).$$

PROOF. This follows from Theorem 1.1 by comparing the ranks of the E^1-page and E^∞-page. □

COROLLARY 1.3. *Assume that in some quantum grading j, $Kh_{*,j}(K;\mathbb{F}_2)$ is supported in a single homological grading i. Then,*

$$\dim Kh_{i,j}(K;\mathbb{F}_2) - 2\,\mathrm{rank}((\mathrm{Id}+\tau_*)_{i,j}) = \sum_{\substack{\bar{\imath},\bar{\jmath},k\in\mathbb{Z}\\ 2\bar{\jmath}+k=j-1+3\Delta}} \dim AKh_{\bar{\imath},\bar{\jmath},k}(\overline{K}_1,\overline{K}_0;\mathbb{F}_2).$$

where $(\mathrm{Id}+\tau_*)_{i,j}$ is the induced endomorphism on $Kh_{i,j}(K;\mathbb{F}_2)$.

PROOF. This also follows from Theorem 1.1 by equating the ranks of the E^2-page and E^∞-page: since θ has homological, quantum bigrading $(-1,0)$, statement (Θ-2) in the theorem implies that the d^r-differentials vanish for $r>1$. □

COROLLARY 1.4. *Assume that in some quantum grading j, $Kh_{*,j}(K;\mathbb{F}_2)$ is supported in a single homological grading i. Suppose also that*

$$\bigoplus_{\bar{\imath},\bar{\jmath},k\mid 2\bar{\jmath}+k=j-1+3\Delta} AKh_{\bar{\imath},\bar{\jmath},k}(\overline{K}_1,\overline{K}_0;\mathbb{F}_2)=0.$$

Then, the endomorphism $(\mathrm{Id}+\tau_*)_{i,j}$ on $Kh_{i,j}(K;\mathbb{F}_2)$ has rank $\frac{1}{2}\dim Kh_{i,j}(K;\mathbb{F}_2)$. In particular, if $Kh_{i,j}(K;\mathbb{F}_2)$ is 2-dimensional then, up to a change of basis, τ_* is given by the matrix $\left(\begin{smallmatrix}0&1\\1&0\end{smallmatrix}\right)$.

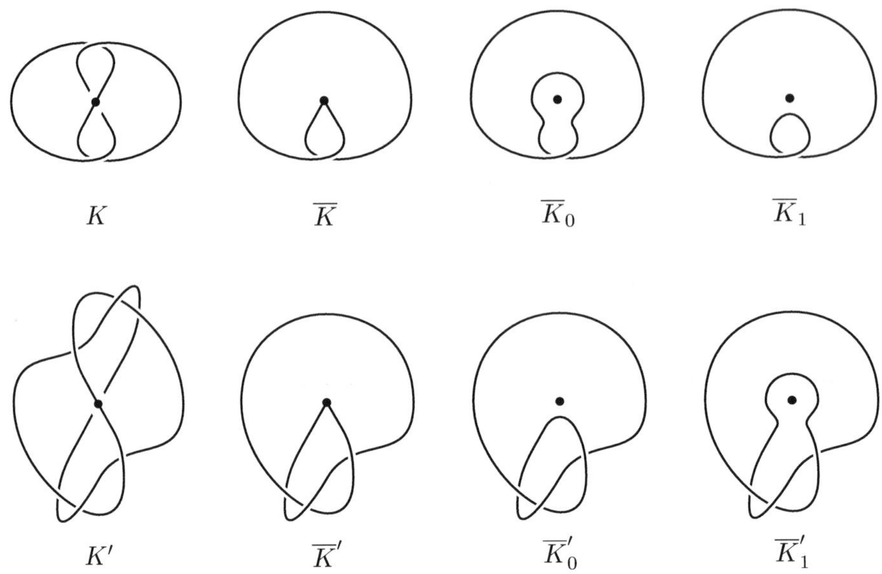

FIGURE 1.1. **Strong inversions of the trefoil.** Each row shows an intravergent diagram for a strong inversion of the trefoil, the quotient knot, and the two annular quotients. The two strong inversions are isotopic in S^3 but not $\mathbb{R}^3$; in S^3, the two different quotients correspond to the two choices of half-axis.

PROOF. The first statement is immediate from Corollary 1.3. The second follows from the first and the fact that any involution of $\mathbb{F}_2^2$ is one of $\left(\begin{smallmatrix}1&0\\0&1\end{smallmatrix}\right)$, $\left(\begin{smallmatrix}0&1\\1&0\end{smallmatrix}\right)$, $\left(\begin{smallmatrix}1&1\\0&1\end{smallmatrix}\right)$, or $\left(\begin{smallmatrix}1&0\\1&1\end{smallmatrix}\right)$, and the last two are conjugate to the second one. □

EXAMPLE 1.5. Figure 1.1 shows two intravergent diagrams K and K' for strong inversions of the positive trefoil. (See also [**BI22**, Appendix A].) The first has quotient $\overline{K}$ the unknot; the quotient $\overline{K}'$ of the second is the negative trefoil.

With our grading conventions, the E^1-page of the spectral sequences from Theorem 1.1 for both K and K' are given by the Khovanov homology of the trefoil tensored with $\mathbb{F}_2[\theta^{-1}, \theta]$, i.e.,

	0	1	2	3
9				F
7			D	E
5			C	
3	B			
1	A			

,

where each letter is a copy of $\mathbb{F}_2[\theta^{-1}, \theta]$. The homological grading is horizontal and quantum grading is vertical. The two involutions are isotopic in S^3; hence, the induced maps τ_* on Khovanov homology are the same [**MWW22**]. In fact, since τ_* is a grading-preserving involution, it must be the identity map, so the d^1-differential on the spectral sequence from Theorem 1.1 vanishes. So, the table above also gives the E^2-pages.

By Point (Θ-2), the only possibly nontrivial differential is the d^2-differential, for which we could have $d^2(D) = \theta^2 E$. It is easy to check directly that for K this differential is trivial and the spectral sequence collapses at E^2 with total rank 6 over $\mathbb{F}_2[\theta^{-1}, \theta]$. It follows from Theorem 1.1 and the computation of the E^∞-page below that this holds for K' as well.

Direct computation shows that the cones of f^+ for $\overline{K}$ and $\overline{K}'$ are given by

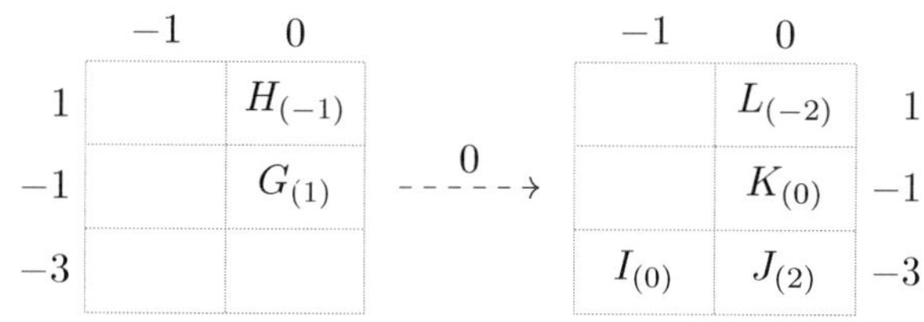

and

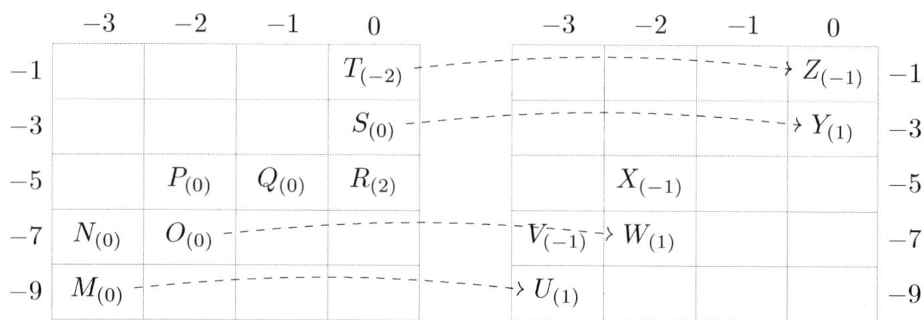

respectively. Here, the subscripts indicate the k-grading. In the first case, the map f^+ vanishes; in the second, it is given by the dashed arrows. In particular, in both cases the homology of the mapping cone has rank 6, consistent with Theorem 1.1. One can further use the grading shift to match up the gradings. For example, for K, $\Delta = -2$, while for K', $\Delta = -5$, and so the quantum gradings on K or K' corresponding to the generators of $\overline{K}$ and $\overline{K}'$ are:

G	H	I	J	K	L
7	9	1	3	5	7

N	P	Q	R	V	X
3	7	7	9	1	5

Again, this is consistent with Theorem 1.1. Alternatively, one could use Theorem 1.1 to deduce that f^+ is the map given above.

One reason strongly invertible knots have appeared recently is that they have furnished examples of non-isotopic pairs of slice disks. It turns out that Corollary 1.4 and properties of the maps on Khovanov homology can be used to prove that certain pairs of slice disks are distinguished by Khovanov homology, without explicitly computing the maps associated to the slice disks. We illustrate this phenomenon for the knot 9_{46} in Section 5.

The reduced Khovanov homology of K is closely related to the Heegaard Floer homology of the branched double cover $\Sigma(S^3, K)$ of K [**OSz05, Rob13, GW10**], and Theorem 1.1 has an analogue for Heegaard Floer homology:

THEOREM 1.6. *Given a strongly invertible knot K with quotient knot $\overline{K}$ there is an ungraded spectral sequence with E^1-page given by $\widehat{HF}(\Sigma(S^3, K)) \otimes \mathbb{F}_2[\theta^{-1}, \theta]]$ converging to $\widehat{HF}(\Sigma(S^3, \overline{K})) \otimes \mathbb{F}_2[\theta^{-1}, \theta]]$.*

As we will see, this follows from a localization result of Hendricks, Lidman, and the first author for the Heegaard Floer homology of double branched covers [**HLL22**], which in turn follows from a general localization theorem in Lagrangian intersection Floer homology of Large [**Lar**]. Note that in Theorem 1.6, there are two choices for the quotient knot $\overline{K}$, depending on which half of the axis one chooses. The statement holds for either choice. There is also an analogue for the knot Floer homology relative to a preimage of the axis, Theorem 6.3.

Another spectral sequence associated to a strongly invertible knot was constructed by Lobb-Watson [**LW21**], although they used a different kind of diagrams, transvergent rather than intravergent. (They also mention intravergent diagrams briefly, in the discussion around their Figure 7.) It might be interesting to compare their $\mathcal{F}$ spectral sequence with the one constructed here; in particular, this might give an approach to proving that their other, $\mathcal{G}$, spectral sequence collapses [**LW21**, Question 6.5].

We expect that the spectral sequence in Theorem 1.1 is an invariant of the strong inversion on K, but do not pursue this here. (In fact, we expect a somewhat stronger statement to hold: that the complex $CKh(K)$, viewed as a complex of $\mathbb{F}_2[\mathbb{Z}/2]$-modules, is invariant up to quasi-isomorphism. Invariance of the spectral sequence from Theorem 1.6 follows from [**HLL22**, Remark 4.13].) One could also ask about whether the spectral sequence is an invariant of the strongly invertible knot in S^3, i.e., does not depend on the choice of half-axis, as was the case in Example 1.5; this seems hard to prove using an intravergent diagram, but easier if one could relate it to a transvergent one. Theorem 1.6 suggests there might be an interesting reduced version of Theorem 1.1, but we do not pursue that either. One could also consider links in Theorems 1.1 and 1.6, meeting the axis in more than two points, but we also do not pursue that generalization.

This paper is organized as follows. Since there are many conventions for Khovanov homology, we review ours in Section 2. We then introduce the axis-moving maps on annular Khovanov homology and some of their basic properties in Section 3. Section 4 proves the main localization result, Theorem 1.1. We give the application to slice disks in Section 5. We end with the proof of the analogue for Heegaard Floer homology, Theorem 1.6, in Section 6.

2. Conventions for Khovanov homology

We start by describing our grading conventions for Khovanov homology (which also serves as a terse review of Khovanov homology itself). Given a knot diagram with N crossings numbered 1 through N, consider the Kauffman cube of resolutions, where the complete resolution of the diagram at the vertex $v = (v_1, \dots, v_N) \in \{0,1\}^N$ is obtained by resolving the ith crossing $\times$ by the 0-resolution $\asymp$ if $v_i = 0$ or by the 1-resolution $)($ if $v_i = 1$, for each $i \in \{1, \dots, N\}$. We will view the cube as $(1 \to 0)^N$, so each edge runs from a vector with a 1 in some coordinate to the corresponding vector with a 0 in that coordinate. Associated to each edge is an elementary saddle cobordism between the corresponding resolutions.

The Frobenius algebra $\mathbb{Z}[X]/(X^2)$ (with comultiplication given by $1 \mapsto 1 \otimes X + X \otimes 1, X \mapsto X \otimes X$) corresponds to a 2-dimensional TQFT. The Khovanov chain complex *CKh* is obtained by applying this TQFT to the cube of resolutions, and then taking the total complex. More concretely, *CKh* is freely generated by the Khovanov generators x, which consist of a choice of a vertex $v \in \{0,1\}^N$ and a labeling of the circles in the complete resolution at v by the labels $\{1, X\}$. The homological grading of x is $|v| - N_-$, and the quantum grading is $N - 3N_- + |v| + \#\{\text{circles labeled } X\} - \#\{\text{circles labeled } 1\}$. (Here N_- is the number of negative crossings in the diagram and $|v| = \sum_i v_i$ is the L^1-norm of v.) So, the quantum grading of X is two more than the quantum grading of 1.

The differential on the Khovanov chain complex is a sum of maps along the edges of the cube; it preserves the quantum grading and decreases the homological grading by 1. The component of the differential along the edge from the vertex $v = (v_1, \dots, v_{n-1}, 1, v_{n+1}, \dots, v_N)$ to the vertex $w = (v_1, \dots, v_{n-1}, 0, v_{n+1}, \dots, v_N)$ is $(-1)^{v_1 + \dots + v_{n-1}}$ times the map associated by the TQFT to the saddle cobordism. That is, if the saddle cobordism merges two circles into one, then the map is induced by the multiplication map in $\mathbb{Z}[X]/(X^2)$, and if the saddle cobordism splits a circle into two, then the map is induced by the comultiplication map in $\mathbb{Z}[X]/(X^2)$.

This convention differs from Khovanov's original [**Kho00**] in a couple of ways: the differential in Khovanov's original paper increased the homological grading; and the quantum grading of X was lower than the quantum grading of 1. However, in order to ensure that arc algebras are supported in non-negative quantum gradings, Khovanov switched the latter convention in [**Kho02**], and his subsequent papers follow the switched convention (where the quantum grading of X is higher than the quantum grading of 1). However, Khovanov's original quantum grading convention had a desirable feature that the positive knots (except the unknot) had Khovanov homologies supported in positive quantum gradings; unfortunately, with the switched convention, their Khovanov homologies were supported in negative quantum gradings. With our grading conventions—additionally making the differential decrease the homological grading—we try to tread a middle ground: arc algebras and Khovanov homologies of (non-trivial) positive knots are both supported in non-negative quantum gradings.

The Khovanov chain complex with our convention is the dual of the Khovanov chain complex from Khovanov's original convention, preserving the bigrading (this follows easily from the duality statement [**Kho00**, Proposition 32]). So, over any field, Khovanov homology with our convention is (unnaturally) bigraded isomorphic to the original Khovanov homology; over $\mathbb{Z}$, the free parts are bigraded isomorphic, and the torsion subgroup with our convention is isomorphic to the original torsion subgroup, but with its homological grading shifted down by 1 (and quantum grading unchanged).

The Khovanov complex of a link in the annulus inherits an extra *annular* or *winding number* filtration; the homology of the associated graded complex is annular Khovanov homology. We follow the usual conventions in the literature for the annular filtration. Specifically, given a labeled resolution, orient circles labeled 1 counter-clockwise (positively) and circles labeled X clockwise (negatively). Then, the annular filtration of a labeled resolution is the winding number around the axis.

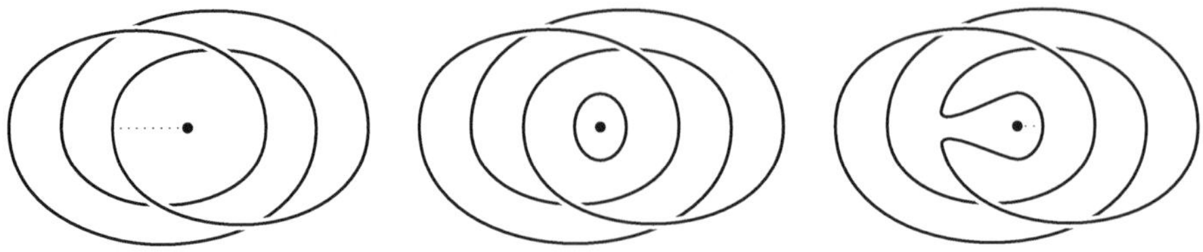

FIGURE 3.1. **The axis moving maps.** Left: the link L and the arc connecting a point p on L to the axis (dotted). Center: the link $L \amalg U$. Right: the link L' and the corresponding arc connecting p' to the axis.

Terms in the differential either preserve the annular filtration or decrease it by 2. (The latter occurs when merging a nullhomotopic circle labeled X with an essential circle labeled 1, merging two essential circles labeled 1, splitting an essential circle labeled 1 into a nullhomotopic circle labeled 1 and an essential circle labeled X, or splitting a nullhomotopic circle labeled X into two essential circles labeled X.)

3. A map on annular Khovanov homology

Let L be an annular link diagram. Fix a point p on L adjacent to the axis of the annulus. Isotoping p across the axis of the annulus gives a new link L'. (See Figure 3.1.) In this section we define and spell out basic properties of the axis moving maps

$$f^+,\ f^-: AKh(L) \to AKh(L').$$

While the rest of the results in this paper use $\mathbb{F}_2$-coefficients, in this section, we will work with $\mathbb{Z}$-coefficients.

Let $L \amalg U$ be the result of adding an essential circle U around the axis, adjacent to the axis and disjoint from L. (See Figure 3.1.) The annular Khovanov complex of $L \amalg U$ is $ACKh(L) \otimes ACKh(U) = ACKh(L) \otimes \mathbb{Z}\langle 1, X\rangle$. So, there are two inclusions $\iota_1, \iota_X : ACKh(L) \hookrightarrow ACKh(L \amalg U)$, defined by $\iota_1(y) = y \otimes 1$ and $\iota_X(y) = y \otimes X$; these have (homological, quantum, annular) trigradings $(0, -1, 1)$ and $(0, 1, -1)$, respectively. Merging L and U at the point p gives a map $m: ACKh(L \amalg U) \to L$; this map has trigrading $(0, 1, 0)$. (This is the annular merge map. So, for instance, merging two essential circles labeled 1 is the zero map.) By composing, we get trigrading-preserving maps

$$f^+ = m \circ \iota_1 : \Sigma^{0,0,1} ACKh(L) \to ACKh(L')$$
$$f^- = m \circ \iota_X : \Sigma^{0,2,-1} ACKh(L) \to ACKh(L'),$$

where $\Sigma^{a,b,c}$ denotes a trigrading shift by (a, b, c). The maps $f^{\pm}$ are compositions of chain maps, hence are chain maps. Abusing notation, we will also use $f^{\pm}$ to denote the induced map on annular homology.

When we want to indicate the dependence of $f^{\pm}$ on the point p we will write them as $f_p^{\pm}$. Also, the point p on L becomes a point p' on L' after the isotopy across the axis.

We give some basic properties of the maps $f^{\pm}$; though we will not use them in the rest of this paper, perhaps they will be useful in related applications.

PROPOSITION 3.1. *Up to homotopy and sign, the maps $f^{\pm}$ are isotopy invariants of the pair of the annular link L and the arc connecting p to the axis. Further,*

$$f^+_{p'} \circ f^+_p = f^-_{p'} \circ f^-_p = 0 \tag{3.1}$$

$$f^+_{p'} \circ f^-_p = f^-_{p'} \circ f^+_p = X\cdot_p \tag{3.2}$$

Here, $X\cdot_p$ is the basepoint action on annular Khovanov homology, the result of merging in a nullhomotopic circle labeled X at the point p, via the annular merge map. Also, the first statement uses naturality of annular Khovanov homology (see, e.g., [**GLW18**]).

PROOF. For the first statement, call an arc from an annular link to the axis *short* if its projection to the annulus is a smooth embedding and disjoint from the projection of the rest of the link to the annulus (i.e., the rest of the link diagram). Given a pair of an annular link L and a (not short) arc γ connecting that link to the axis, there is a canonical isotopy from $L \cup \gamma$ to a pair $L' \cup \gamma'$ where γ' is short: just shrink γ and pull L along with it.

Now, fix link diagrams L_i, $i = 0, 1$, and points p_i on L_i adjacent to the axis, and let γ_i be the short arc from p_i to the axis. Assume that $L_0 \cup \gamma_0$ is isotopic to $L_1 \cup \gamma_1$, via an isotopy $L_t \cup \gamma_t$. Applying the canonical isotopy $L_{t,s} \cup \gamma_{t,s}$ from the previous paragraph to each L_t gives a new isotopy $L_{t,1} \cup \gamma_{t,1}$ from $L_0 \cup \gamma_0$ to $L_1 \cup \gamma_1$ so that the arc $\gamma_{t,1}$ is short for all t. Perturb this isotopy so that the projection to the annulus is generic. This gives a sequence of Reidemeister moves (in the annulus) connecting L_0 to L_1 and disjoint from the arc. Each Reidemeister move induces a map on the annular Khovanov complex, and it is immediate from the definitions that these maps commute with the maps $f^{\pm}$.

Equations (3.1) and (3.2) are clear from the definitions. □

REMARK 3.2. These maps are a special case of maps associated to *anchored cobordisms* by Akhmechet-Khovanov [**AK23**]. Specifically, the trace of an isotopy moving p across the axis is an anchored cobordism. To define a map, we must also label the intersection point between this cobordism and the axis (the point where p crosses the axis) by 1 or 2. If the winding number of K around the axis is odd then f^+ corresponds to labeling the intersection point 1 and f^- corresponds to labeling it 2; if the winding number is even then f^+ corresponds to labeling the intersection point 2 and f^- corresponds to labeling it 1 (compare [**AK23**, Proof of Theorem 2.19]).

Equations (3.1) and (3.2) follow from Akhmechet-Khovanov's curtain relation [**AK23**, Corollary 2.7].

REMARK 3.3. Given (L, p) as above, let L'' be the result of performing a Reidemeister I move at p across the axis, changing the winding number by ± 1 (i.e., a Markov 2 move). Choose the Reidemeister move so that the 1-resolution of the new crossing is the disjoint union of L and a new (essential) circle. Then, up to some overall grading shift, the mapping cone of $f^+ \colon \Sigma^{0,0,1} ACKh(L) \to ACKh(L')$ (respectively $f^- \colon \Sigma^{0,2,-1} ACKh(L) \to ACKh(L'))$ is a subcomplex of $ACKh(L'')$: it is the subcomplex where either the new crossing is 0-resolved or the new crossing is 1-resolved and the new essential circle is labeled 1 (respectively X).

4. The localization theorem for strongly invertible knots

In this section, we prove Theorem 1.1. The spectral sequence is constructed as follows. Given an intravergent diagram for K, rotation τ by 180° induces a $\mathbb{Z}/2$ action, which we still call τ, on $CKh(K;\mathbb{F}_2)$. Consider the Tate complex for this $\mathbb{Z}/2$-action, which is given by $CKh(K;\mathbb{F}_2)\otimes\mathbb{F}_2[\theta^{-1},\theta]$ with differential $d+\theta(\mathrm{Id}+\tau)$, where d is the Khovanov differential:

$$(4.1)\qquad \cdots \xrightarrow{\theta(\mathrm{Id}+\tau)} \theta^0\, CKh(K;\mathbb{F}_2) \circlearrowleft^{d} \xrightarrow{\theta(\mathrm{Id}+\tau)} \theta^1\, CKh(K;\mathbb{F}_2) \circlearrowleft^{d} \xrightarrow{\theta(\mathrm{Id}+\tau)} \cdots.$$

If we give θ the (homological, quantum) bigrading $(-1,0)$ then this complex decomposes according to quantum gradings and the differential decreases the homological grading by 1. The complex has a filtration by the θ-power. This filtration induces the spectral sequence in Theorem 1.1.

The main work is to compute the E^∞-page of the spectral sequence (Item (Θ-3)); the other properties are immediate. The strategy is similar to Stoffregen-Zhang's [**SZ**] and Borodzik-Politarczyk-Silvero's [**BPS21**]: we prove that the fixed points of the $\mathbb{Z}/2$-action induced by the strong inversion on a CW complex representing the Khovanov stable homotopy type of K is related to the annular Khovanov stable homotopy type of $(\overline{K}_1,\overline{K}_0)$, and then apply classical Smith theory.

Before embarking on the proof, we briefly summarize the relevant aspects of the Khovanov stable homotopy type [**LS14**], following the more recent box map construction [**LLS20, LLS17**] (see also [**HKK16**]). As sketched in Section 2, in a fixed quantum grading j, the Khovanov chain complex $CKh_{*,j}(K)$ is obtained as follows:

(Kh-1) First construct a cube-shaped diagram of abelian groups, $F_j^{Kh}\colon (1\to 0)^N \to \mathsf{Ab}$, which associates to each vertex v the free abelian group generated by all Khovanov generators x at v with quantum grading j, and associates to each edge $v\to w$ the saddle cobordism map from the TQFT corresponding to the Frobenius algebra $\mathbb{Z}[X]/(X^2)$.

(Kh-2) Viewing F_j^{Kh} as a cube-shaped diagram of chain complexes, take the mapping cone N times, in the N directions of the cube, to obtain a single chain complex. (The signs, like $(-1)^{v_1+\dots+v_{n-1}}$ from Section 2, appear during this iterated mapping cone construction; the precise signs depend on the order in which the mapping cones are done.)

(Kh-3) Finally, shift the homological grading of the chain complex down by N_- to get the Khovanov chain complex $CKh_{*,j}(K)$. (The total Khovanov complex over all quantum gradings is given by $\bigoplus_j CKh_{*,j}(K)$.)

In order to remove the choice about the order in which to take the mapping cones (which amounts to choosing an ordering of the N crossings of the knot diagram), one can replace Step (Kh-2) by the following:

(Kh-2′) Extend F_j^{Kh} trivially to a diagram $(F_j^{Kh})_+$ from a slightly larger category $(1\to 0)_+^N$ which has an additional object and a unique morphism from every $v\neq 0$ to it, and then take the homotopy colimit to obtain a single chain complex. (The functor $(F_j^{Kh})_+$ sends the new object to the trivial group.)

The lift to a Khovanov stable homotopy type follows the same outline, replacing the cube of abelian groups by a cube of topological spaces. More concretely:

($\mathscr{X}$-0) Fix an integer $D \geq N$.

($\mathscr{X}$-1) Construct a cube-shaped diagram $F_j^{\mathscr{X}} \colon (1 \to 0)^N \to \mathsf{Top}$ of based CW complexes which associates to each vertex a wedge sum of D-dimensional spheres, so that its composition with the reduced homology functor, $\widetilde{H}_D \circ F_j^{\mathscr{X}}$, equals the cube-shaped diagram F_j^{Kh} from Step (Kh-1). (It suffices to construct a homotopy coherent diagram instead of a strictly commuting one.)

($\mathscr{X}$-2) Extend (trivially) to a diagram $(F_j^{\mathscr{X}})_+$ from the larger category $(1 \to 0)_+^N$ mapping the new object to a 1-point space, and then take homotopy colimit to obtain a single topological space.

($\mathscr{X}$-3) Finally, formally desuspend $(D + N_-)$ times to get the Khovanov spectrum $\mathscr{X}_j(K)$ in quantum grading j. (The total Khovanov spectrum over all quantum gradings is given by $\bigvee_j \mathscr{X}_j(K)$.)

Of these, the hardest step is Step ($\mathscr{X}$-1), which we undertake by first constructing a lax diagram in a third category—the Burnside 2-category $\mathscr{B}$ of finite sets, finite correspondences, and bijections between correspondences. Specifically, letting $\mathbb{Z}\langle\cdot\rangle \colon \mathscr{B} \to \mathsf{Ab}$ denote the functor that replaces a finite set by the free abelian group generated by it, we do the following:

($\mathscr{B}$-1) Construct a cube-shaped 2-functor $F_j^{\mathscr{B}} \colon (1 \to 0)^N \to \mathscr{B}$ which associates to each vertex v the set of Khovanov generators x at v with quantum grading j, to each edge $v \to w$ a correspondence $F_j^{\mathscr{B}}(v) \xleftarrow{s} F_j^{\mathscr{B}}(v \to w) \xrightarrow{t} F_j^{\mathscr{B}}(w)$, so that $\mathbb{Z}\langle\cdot\rangle \circ F_j^{\mathscr{B}} = F_j^{Kh}$, and to each 2-dimensional face $u \rightrightarrows {}^{v}_{v'} \rightrightarrows w$ a 2-morphism (which is an isomorphism of correspondences) $F_j^{\mathscr{B}}(v \to w) \circ F_j^{\mathscr{B}}(u \to v) \xrightarrow{\cong} F_j^{\mathscr{B}}(v' \to w) \circ F_j^{\mathscr{B}}(u \to v')$ satisfying a coherence relation for every 3-dimensional face. Note that this data specifies, for any $v > w$ in the poset $(1 > 0)^N$, a correspondence $F_j^{\mathscr{B}}(v \to w) \colon F_j^{\mathscr{B}}(v) \to F_j^{\mathscr{B}}(w)$, by composing the correspondences along a sequence of oriented edges connecting v to w; any two such sequences specify canonically isomorphic correspondences. (We can also define the total Burnside functor over all quantum gradings as $\coprod_j F_j^{\mathscr{B}}$ where $\amalg$ is defined by taking disjoints unions of sets and correspondences at vertices and edges, respectively.)

Most of the construction of $F_j^{\mathscr{B}}$ is forced. The only choices are the isomorphisms of correspondences for certain 2-dimensional faces (which we call ladybugs), and we explicitly choose the isomorphisms for those faces (which we call ladybug matchings) [**LS14**, Section 5.4], [**LLS20**, Section 8.1]. Once we have the 2-functor $F_j^{\mathscr{B}} \colon (1 \to 0)^N \to \mathscr{B}$, we can carry out Step ($\mathscr{X}$-1) by the *box map construction*, as follows:

($\mathscr{X}$-1′) For each vertex v, define $F_j^{\mathscr{X}}(v) = \left(\coprod_{x \in F_j^{\mathscr{B}}(v)} B_x\right)/\partial$, where B_x is a D-dimensional rectangular prism (*box*) associated to the Khovanov generator x. For each edge $v \to w$, choose disjoint D-dimensional sub-boxes $\{B_b\}_{b \in F_j^{\mathscr{B}}(v \to w)}$ inside $\coprod_{x \in F_j^{\mathscr{B}}(v)} B_x$ so that each B_b lies in $B_{s(b)}$; define the map $F_j^{\mathscr{X}}(v \to w) \colon F_j^{\mathscr{X}}(v) \to F_j^{\mathscr{X}}(w)$ by sending each sub-box $B_b \subset B_{s(b)}$ to $B_{t(b)}$ by scaling and translation, and the complement of all these sub-boxes to the basepoint. Call maps as in the previous sentence *box maps*. To extend this to a homotopy coherent diagram $F_j^{\mathscr{X}}$ on the entire cube, we specify, for every chain $v_\ell > \cdots > v_0$ in the poset $(1 > 0)^N$, a $[0,1]^{\ell-1}$-parameter family of box maps $F_j^{\mathscr{X}}(v_\ell) \to F_j^{\mathscr{X}}(v_0)$ satisfying certain coherence conditions on its boundary, and refining the correspondence $F_j^{\mathscr{B}}(v_\ell \to v_0)\colon F_j^{\mathscr{B}}(v_\ell) \to F_j^{\mathscr{B}}(v_0)$. By induction on ℓ, and using the coherence conditions and the 2-morphisms in the Burnside category, such a family of maps is already defined on the boundary $\partial[0,1]^{\ell-1}$. Extend it to the entire cube $[0,1]^{\ell-1}$ using $(D-2)$-connectedness of the space of labeled sub-boxes and the assumption that $D \geq N$.

The constructions for annular Khovanov complexes and annular Khovanov homotopy types mirror these definitions. There is an extra annular grading, and we only consider maps that preserve that grading. Therefore, in each (quantum, annular) bigrading (j,k), we get diagrams $AF_{j,k}^{Kh}\colon (1 \to 0)^N \to \mathsf{Ab}$, $AF_{j,k}^{\mathscr{B}}\colon (1 \to 0)^N \to \mathscr{B}$, and $AF_{j,k}^{\mathscr{X}}\colon (1 \to 0)^N \to \mathsf{Top}$; their extensions $(AF_{j,k}^{Kh})_+\colon (1 \to 0)_+^N \to \mathsf{Ab}$ and $(AF_{j,k}^{\mathscr{X}})_+\colon (1 \to 0)_+^N \to \mathsf{Top}$; and the chain complex $ACKh_{*,j,k}$ and the spectrum $A\mathscr{X}_{j,k}$. For the pair $(\overline{K}_1, \overline{K}_0)$ of annular knots whose annular Khovanov chain complex $ACKh(\overline{K}_1, \overline{K}_0)$ is defined as the mapping cone of $\Sigma^{0,0,1} ACKh(\overline{K}_1) \xrightarrow{f^+} ACKh(\overline{K}_0)$, it still may be viewed as a subcomplex of another annular Khovanov chain complex—see Remark 3.3—and therefore, all these constructions work for the pair $(\overline{K}_1, \overline{K}_0)$ as well.

Given an intravergent diagram of K, the strong inversion induces $\mathbb{Z}/2$-actions on these various objects as follows.

LEMMA 4.1. *The 180° rotation on the intravergent diagram of K induces a $\mathbb{Z}/2$-action on the cube $(1 \to 0)^N$ and an* external $\mathbb{Z}/2$-action *on the 2-functor $F_j^{\mathscr{B}}(K)\colon (1 \to 0)^N \to \mathscr{B}$ in each quantum grading j, in the sense of Stoffregen-Zhang* [**SZ**, Definition 3.4].

PROOF. The proof is similar to Stoffregen-Zhang's corresponding result for 2-periodic knots [**SZ**, Proposition 6.4]. The 180° rotation τ around the axis induces a $\mathbb{Z}/2$-action on the N crossings of K, which in turn induces a $\mathbb{Z}/2$-action (also denoted τ) on the cube category $(1 \to 0)^N$ after identifying it with $(1 \to 0)^{\{\text{crossings of } K\}}$ by ordering the crossings. It also induces a $\mathbb{Z}/2$-action (still denoted τ) on the set of all Khovanov generators in quantum grading j, sending $F_j^{\mathscr{B}}(v)$ to $F_j^{\mathscr{B}}(\tau v)$. Moreover, for each edge $v \to w$, it induces an isomorphism of correspondences

$$\begin{array}{ccccc}
F_j^{\mathscr{B}}(v) & \xleftarrow{s} & F_j^{\mathscr{B}}(v \to w) & \xrightarrow{t} & F_j^{\mathscr{B}}(w) \\
\downarrow{\scriptstyle\tau} & & \downarrow{\scriptstyle\tau} & & \downarrow{\scriptstyle\tau} \\
F_j^{\mathscr{B}}(\tau v) & \xleftarrow[s]{} & F_j^{\mathscr{B}}(\tau v \to \tau w) & \xrightarrow[t]{} & F_j^{\mathscr{B}}(\tau w)
\end{array}$$

since for any Khovanov generators $x \in F_j^{\mathscr{B}}(v), y \in F_j^{\mathscr{B}}(w)$, the set $s^{-1}(x) \cap t^{-1}(y)$ has 0 or 1 elements. So the only thing to check is that τ respects the ladybug matchings across 2-dimensional faces, which holds since the ladybug matching is invariant under planar isotopy, and in particular, the 180° rotation τ. □

This $\mathbb{Z}/2$-action on $F_j^{\mathscr{B}}(K)\colon (1\to 0)^N \to \mathscr{B}$ has a *fixed point functor* $(F_j^{\mathscr{B}}(K))^\tau$ [**SZ**, Definition 3.11]. The functor $(F_j^{\mathscr{B}}(K))^\tau$ is defined on the fixed subcategory of the cube category $(1\to 0)^N$, which is itself isomorphic to the cube category $(1\to 0)^{\overline{N}+1}$ where $\overline{N}=(N-1)/2$ is the number of crossings of the quotient diagram $\overline{K}$; $(F_j^{\mathscr{B}}(K))^\tau$ assigns to vertices and edges the fixed subset of the τ-action on the sets and correspondences, respectively. It turns out that these fixed point functors are related to the Burnside 2-functors associated to the pair of annular knots $(\overline{K}_1,\overline{K}_0)$:

LEMMA 4.2. *For any quantum grading j, the fixed point functor $(F_j^{\mathscr{B}}(K))^\tau$ is isomorphic to $\coprod_{\overline{\jmath},k|2\overline{\jmath}+k=j-1+3\Delta} AF^{\mathscr{B}}_{\overline{\jmath},k}(\overline{K}_1,\overline{K}_0)$, where Δ is as in Equation (1.2).*

PROOF. The proof is similar to the 2-periodic case [**SZ**, Theorem 6.7]. First, order the $\overline{N}$ crossings of $\overline{K}$ arbitrarily. Then, at any vertex $v=(v_1,\dots,v_{\overline{N}+1})\in(1\to 0)^{\overline{N}+1}$, the set $AF^{\mathscr{B}}_{\overline{\jmath},k}(\overline{K}_1,\overline{K}_0)(v)$ is defined to be the set of Khovanov generators of $\overline{K}_{v_{\overline{N}+1}}$ over the vertex $(v_1,\dots,v_{\overline{N}})$ in (quantum, annular) bigrading $(\overline{\jmath},k-v_{\overline{N}+1})$.

Order the N crossings of K such that the crossing on the axis is ordered last and the quotient map $\{$other crossings of $K\}\to\{$crossings of $\overline{K}\}$ is order-preserving. Then, there is an inclusion of cube categories $\iota\colon(1\to 0)^{\overline{N}+1}\to(1\to 0)^N$ which sends the vertex $v=(v_1,\dots,v_{\overline{N}},v_{\overline{N}+1})$ to the vertex $\iota(v)=(v_1,v_1,\dots,v_{\overline{N}},v_{\overline{N}},v_{\overline{N}+1})\in(1\to 0)^N$; the image is precisely the fixed subcategory of $(1\to 0)^N$.

We construct a natural isomorphism

$$\eta\colon\Big(\coprod_{\substack{\overline{\jmath},k\\ 2\overline{\jmath}+k=j-1+3\Delta}} AF^{\mathscr{B}}_{\overline{\jmath},k}(\overline{K}_1,\overline{K}_0)\Big)\longrightarrow\Big((F_j^{\mathscr{B}}(K))^\tau\circ\iota\Big)$$

between the two Burnside functors. For any vertex $v\in(1\to 0)^{\overline{N}+1}$ and any Khovanov generator $x\in AF^{\mathscr{B}}_{\overline{\jmath},k}(\overline{K}_1,\overline{K}_0)(v)$, let $\eta(x)$ be the Khovanov generator of K over the vertex $\iota(v)$ which labels each circle in the $\iota(v)$-resolution of K by the same label that x labels its quotient circle in the $(v_1,\dots,v_{\overline{N}})$-resolution of $\overline{K}_{v_{\overline{N}+1}}$. It is a straightforward calculation that $\eta(x)$ has quantum grading j if and only if $(\overline{\jmath},k)$ satisfies $2\overline{\jmath}+k=j-1+3\Delta$. For the reader's convenience, we summarize the calculation below.

In the $(v_1,\dots,v_{\overline{N}})$-resolution of the annular knot $\overline{K}_{v_{\overline{N}+1}}$, let a_1 and a_X be the numbers of essential circles that x labels by 1 and X, respectively, and let b_1 and b_X be the numbers of non-essential circles that x labels by 1 and X, respectively. Then,

$$\begin{aligned}\overline{\jmath}&=\overline{N}-3\overline{N}_-+(|v|-v_{\overline{N}+1})+(a_X+b_X)-(a_1+b_1)\\ k-v_{\overline{N}+1}&=a_1-a_X\\ 2\overline{\jmath}+k+6\overline{N}_-+1&=(2\overline{N}+1)+(2|v|-v_{\overline{N}+1})+(a_X+2b_X)-(a_1+2b_1).\end{aligned}$$

In the $\iota(v)=(v_1,v_1,\dots,v_{\overline{N}},v_{\overline{N}},v_{\overline{N}+1})$-resolution of K, the number of circles labeled $1,X$ by the generator $\eta(x)$ is (a_1+2b_1) and (a_X+2b_X), respectively. Then,

$$\begin{aligned}j+3N_-&=N+|\iota(v)|+(a_X+2b_X)-(a_1+2b_1)\\ &=(2\overline{N}+1)+(2|v|-v_{\overline{N}+1})+(a_X+2b_X)-(a_1+2b_1).\end{aligned}$$

Therefore, $2\overline{\jmath}+k = j-1+3(N_- - 2\overline{N}_-) = j-1+3\Delta$.

To make the notation less cumbersome, let $AF^{\mathscr{B}}_{[\overline{\jmath},k]} = \coprod_{\overline{\jmath},k|2\overline{\jmath}+k=j-1+3\Delta} AF^{\mathscr{B}}_{\overline{\jmath},k}$.

Next, we must specify the natural isomorphism η on the 1-morphisms, that is, for any edge $v \to w$ in $\{1 \to 0\}^{\overline{N}+1}$, we must specify an isomorphism between correspondences:

$$AF^{\mathscr{B}}_{[\overline{\jmath},k]}(\overline{K}_1, \overline{K}_0)(v \to w) \xrightarrow{\eta} F^{\mathscr{B}}_j(K)(\iota(v) \to \iota(w)).$$

As in the proof of Lemma 4.1, for any generators $x \in AF^{\mathscr{B}}_{[\overline{\jmath},k]}(\overline{K}_1, \overline{K}_0)(v)$, $y \in AF^{\mathscr{B}}_{[\overline{\jmath},k]}(\overline{K}_1, \overline{K}_0)(w)$, the set $s^{-1}(x) \cap t^{-1}(y) \subset AF^{\mathscr{B}}_{[\overline{\jmath},k]}(\overline{K}_1, \overline{K}_0)(v \to w)$ has either 0 or 1 elements, and the set $s^{-1}(\eta(x)) \cap t^{-1}(\eta(y)) \subset F^{\mathscr{B}}_j(K)(\iota(v) \to \iota(w))$ also has either 0 or 1 elements, correspondingly; therefore, the isomorphism η between the correspondences is forced. This is checked by a direct case analysis: When $v_{\overline{N}+1} = w_{\overline{N}+1}$, then this is Stoffregen-Zhang's case analysis for the 2-periodic link $K_{v_{\overline{N}+1}}$ [**SZ**, Theorem 6.8]. When $v_{\overline{N}+1} > w_{\overline{N}+1}$, the proof is the following similar (but shorter) case analysis. Consider the axis-moving isotopy from the $(v_1, \dots, v_{\overline{N}})$-resolution of the annular knot $\overline{K}_1$ to the corresponding resolution of the annular knot $\overline{K}_0$. There are two cases, depending on whether an essential circle becomes inessential or an inessential circle becomes essential. In the first (respectively second) case, $s^{-1}(x) \cap t^{-1}(w) \subset AF^{\mathscr{B}}_{[\overline{\jmath},k]}(\overline{K}_1, \overline{K}_0)(v \to w)$ is non-empty (and has only one element) if and only if x and y label the moving circle by X (respectively 1) and all other circles by the same labels. In the picture for K, we get a corresponding saddle cobordism from the $\iota(v)$-resolution of K to the $\iota(w)$-resolution of K. In the first (respectively second) case, the saddle is a split (respectively merge) and $s^{-1}(\eta(x)) \cap t^{-1}(\eta(y)) \subset F^{\mathscr{B}}_j(K)(\iota(v) \to \iota(w))$ is non-empty (and has only one element) if and only if $\eta(x)$ and $\eta(y)$ label all the circles involved in the saddle by X (respectively 1), and all other circles by the same labels.

Finally, we have to check that compatibility of these isomorphisms across 2-dimensional faces. That is, given a 2-dimensional face $u \rightrightarrows^{v}_{v'} \rightrightarrows w$ in $\{1 \to 0\}^{\overline{N}+1}$, we have to check that the following diagram commutes:

$$\begin{array}{ccc}
\begin{array}{c} AF^{\mathscr{B}}_{[\overline{\jmath},k]}(\overline{K}_1, \overline{K}_0)(v \to w) \\ \circ AF^{\mathscr{B}}_{[\overline{\jmath},k]}(\overline{K}_1, \overline{K}_0)(u \to v) \end{array} & \longrightarrow & \begin{array}{c} F^{\mathscr{B}}_j(K)(\iota(v) \to \iota(w)) \\ \circ F^{\mathscr{B}}_j(K)(\iota(u) \to \iota(v)) \end{array} \\
\downarrow & & \downarrow \\
\begin{array}{c} AF^{\mathscr{B}}_{[\overline{\jmath},k]}(\overline{K}_1, \overline{K}_0)(v' \to w) \\ \circ AF^{\mathscr{B}}_{[\overline{\jmath},k]}(\overline{K}_1, \overline{K}_0)(u \to v') \end{array} & \longrightarrow & \begin{array}{c} F^{\mathscr{B}}_j(K)(\iota(v') \to \iota(w)) \\ \circ F^{\mathscr{B}}_j(K)(\iota(u) \to \iota(v')) \end{array}
\end{array}$$

where the horizontal arrows are induced by the isomorphisms that we just constructed, and the vertical arrows are induced by the isomorphisms that are part of the data for the respective Burnside 2-functors. Unless the 2-dimensional face is a ladybug, for any pair of generators $x \in AF^{\mathscr{B}}_{[\overline{\jmath},k]}(\overline{K}_1, \overline{K}_0)(u), y \in AF^{\mathscr{B}}_{[\overline{\jmath},k]}(\overline{K}_1, \overline{K}_0)(w)$, each of $s^{-1}(x) \cap t^{-1}(w) \subset AF^{\mathscr{B}}_{[\overline{\jmath},k]}(\overline{K}_1, \overline{K}_0)(u \to w)$ and $s^{-1}(\eta(x)) \cap t^{-1}(\eta(y)) \subset F^{\mathscr{B}}_j(K)(\iota(u) \to \iota(w))$ has 0 or 1 elements, and so the check is automatic. Thus, the only case remaining is when the 2-dimensional face is a ladybug. However, recall from Remark 3.3 that the functor $AF^{\mathscr{B}}_{[\overline{\jmath},k]}(\overline{K}_1, \overline{K}_0)$ may be viewed as a subfunctor of the Burnside functor associated to a different annular knot obtained from $\overline{K}_1$ by performing a Reidemeister I move. However, such a crossing (coming from a Reidemeister I move) cannot be involved in a ladybug configuration. Therefore, in order to be a ladybug, we must have $u_{\overline{N}+1} = v_{\overline{N}+1} = v'_{\overline{N}+1} = w_{\overline{N}+1}$; and in that case, commutativity of the above diagram follows from the analogue for the 2-periodic link $K_{u_{\overline{N}+1}}$ [**SZ**, Lemma 6.15]. □

Now, given the $\mathbb{Z}/2$-action on the Burnside functor $F_j^{\mathscr{B}}(K)$, and the above identification of its fixed point functor with those of the quotient annular knots $(\overline{K}_1, \overline{K}_0)$, all that remains is to refine these actions and the fixed points to the category of topology spaces. This is precisely Stoffregen-Zhang's central thesis:

PROPOSITION 4.3 ([**SZ**, Proposition 5.10]). *Let $F^{\mathscr{B}}\colon (1 \to 0)^N \to \mathscr{B}$ be a Burnside 2-functor with an external $\mathbb{Z}/2$-action τ and $(F^{\mathscr{B}})^\tau$ denote the fixed point functor. Then, the homotopy coherent diagram $F^{\mathscr{X}} : (1 \to 0)^N \to \mathsf{Top}$ refining $F^{\mathscr{B}}$ using the box map construction, as in Step ($\mathscr{X}$-1′), may be chosen $\mathbb{Z}/2$-equivariantly so that the fixed point homotopy coherent diagram $(F^{\mathscr{X}})^\tau$ refines $(F^{\mathscr{B}})^\tau$ using box maps.*

PROOF. For the reader's convenience, we sketch the proof (summarizing the proofs of [**SZ**, Lemma 4.7 and Proposition 5.10]).

Fix $D_1, D_2 \geq N$. For every $x \in \coprod_v F^{\mathscr{B}}(v)$, associate a $(D_1 + D_2)$-dimensional box $B_x \cong [0,1]^{D_1+D_2}$; endow it with the $\mathbb{Z}/2$-action τ which reflects the first D_1-coordinates and is the identity along the last D_2-coordinates, that is,

$$\tau(x_1, \dots, x_{D_1+D_2}) = (1 - x_1, \dots, 1 - x_{D_1}, x_{D_1+1}, \dots, x_{D_1+D_2}).$$

As in Step ($\mathscr{X}$-1′), for chains $v_\ell > \cdots > v_0$ in the poset $(1 > 0)^N$, we will construct a $[0,1]^{\ell-1}$-parameter family of box maps $F_j^{\mathscr{X}}(v_\ell) \to F_j^{\mathscr{X}}(v_0)$ refining the correspondence $F^{\mathscr{B}}(v_\ell \to v_0)\colon F^{\mathscr{B}}(v_\ell) \to F^{\mathscr{B}}(v_0)$ by induction on ℓ. These maps will already be specified on the boundary $\partial[0,1]^{\ell-1}$ by the compatibility condition. There are two cases:

- If the entire chain $v_\ell > \cdots > v_0$ is not fixed by τ, choose one of the two chains $\mathfrak{c} = (v_\ell > \cdots > v_0)$ or $\tau\mathfrak{c} = (\tau v_\ell > \cdots > \tau v_0)$ arbitrarily; without loss of generality, say we pick $\mathfrak{c}$. Construct the $[0,1]^{\ell-1}$-parameter family of box maps for $\mathfrak{c}$, refining the correspondence $F^{\mathscr{B}}(v_\ell \to v_0)$, arbitrarily using the $(D_1 + D_2 - 2)$-connectedness of the space of labeled sub-boxes. Then define the $[0,1]^{\ell-1}$-parameter family of box maps for the other chain $\tau\mathfrak{c}$, refining the correspondence $F^{\mathscr{B}}(\tau v_\ell \to \tau v_0)$, by pre-composing and post-composing by τ, as well as relabeling the sub-boxes by the map $\tau\colon F^{\mathscr{B}}(v_\ell \to v_0) \to F^{\mathscr{B}}(\tau v_\ell \to \tau v_0)$.
- If the entire chain $v_\ell > \cdots > v_0$ is fixed by τ, construct the $[0,1]^{\ell-1}$-parameter family of box maps refining the correspondence $F^{\mathscr{B}}(v_\ell \to v_0)$ as follows.
  - Let $A \subset F^{\mathscr{B}}(v_\ell \to v_0)$ be the subset not fixed by τ. From every pair $\{a, \tau a\} \subset A$, choose one element arbitrarily. Let $B \subset A$ be the subset of chosen elements. Pick the $[0,1]^{\ell-1}$-parameter family of sub-boxes labeled by B in the complement of the τ-fixed subspace of the boxes using the $(D_1 - 2)$-connectedness of that space. Construct the $[0,1]^{\ell-1}$-parameter of sub-boxes labeled by $A \setminus B$ by applying τ.
  - Let $C \subset F^{\mathscr{B}}(v_\ell \to v_0)$ be the subset fixed by τ. Pick the $[0,1]^{\ell-1}$-parameter family of sub-boxes labeled by C symmetrically with respect to τ. (First choose a $[0,1]^{\ell-1}$-family of D_2-dimensional boxes inside the fixed subset $\{\frac{1}{2}\}^{D_1} \times$

$[0,1]^{D_2}$ using the (D_2-2)-connectedness of that space, and then thicken the boxes τ-equivariantly to get (D_1+D_2)-dimensional boxes, while staying disjoint from the sub-boxes labeled by A.)

This produces a homotopy coherent diagram $F^{\mathscr{X}}$ refining $F^{\mathscr{B}}$ using (D_1+D_2)-dimensional box maps, and the fixed point functor $(F^{\mathscr{X}})^\tau$ is also a homotopy coherent diagram refining $(F^{\mathscr{B}})^\tau$ using D_2-dimensional box maps. □

Combining these ingredients, we get:

PROPOSITION 4.4. *The strong inversion of K induces a $\mathbb{Z}/2$-action on the Khovanov spectrum $\mathscr{X}(K)$ whose geometric fixed point set is $A\mathscr{X}(\overline{K}_1,\overline{K}_0)$ up to some formal (de)suspension. Keeping track of quantum gradings, the geometric fixed point set of the $\mathbb{Z}/2$-action on $\mathscr{X}_j(K)$ is the spectrum $\bigvee_{\overline{\jmath},k|2\overline{\jmath}+k=j-1+3\Delta} A\mathscr{X}_{\overline{\jmath},k}(\overline{K}_1,\overline{K}_0)$, up to some formal (de)suspension.*

PROOF. Choose the homotopy coherent diagram $F_j^{\mathscr{X}}(K)$ $\mathbb{Z}/2$-equivariantly, as in Proposition 4.3. Up to some (de)suspension, the Khovanov spectrum $\mathscr{X}_j(K)$ is the homotopy colimit of the extended diagram $(F_j^{\mathscr{X}}(K))_+$. The geometric fixed point set of this homotopy colimit is the homotopy colimit of the extended fixed point functor $(F_j^{\mathscr{X}}(K))_+^\tau$. But by Lemma 4.2 and Proposition 4.3, the fixed point functor $(F_j^{\mathscr{X}}(K))^\tau$ refines Burnside functor $\coprod_{\overline{\jmath},k|2\overline{\jmath}+k=j-1+3\Delta} AF^{\mathscr{B}}_{\overline{\jmath},k}(\overline{K}_1,\overline{K}_0)$, and therefore, up to some (de)suspension, the homotopy colimit of $(F_j^{\mathscr{X}}(K))_+^\tau$ is simply

$$\bigvee_{\overline{\jmath},k|2\overline{\jmath}+k=j-1+3\Delta} A\mathscr{X}_{\overline{\jmath},k}(\overline{K}_1,\overline{K}_0),$$

as claimed. □

PROOF OF THEOREM 1.1. Since the spectral sequence is induced by the θ-filtration on the Tate complex, Formula (4.1), we only need to prove Item (Θ-3): the other parts are immediate from the definition. However, this is simply the classical Smith inequality applied to the Proposition 4.4, stated in the language of spectral sequences. To wit, the Tate complex from Equation (4.1) computes the localized equivariant homology of $\mathscr{X}(K)$, which by the classical localization theorem, equals the localized equivariant homology of the geometric fixed point set $A\mathscr{X}(\overline{K}_1,\overline{K}_0)$, which simply equals its homology $AKh(\overline{K}_1,\overline{K}_0)$, tensored with $\mathbb{F}_2[\theta^{-1},\theta]$. (It is also easy to keep track of the quantum gradings using Proposition 4.4.) □

REMARK 4.5. The expression $\Delta = N_- - 2\overline{N}_-$ appears as grading shifts in Theorem 1.1, but it is not an invariant of the knot K and its strong inversion. Geometrically, the 2-periodic annular links K_0, K_1 obtained by resolving the crossing of K on the axis, and their quotient annular knots $\overline{K}_0, \overline{K}_1$, are only well-defined up to how many times they wind around the axis, and Δ captures information about this winding number. In more detail, if B (respectively T) denotes the underpass (respectively overpass) of K near its crossing on the axis, then orient the quotient knot $\overline{K}$ by orienting the quotient arc $\overline{B}$ (respectively $\overline{T}$) towards (respectively away from) the axis. This induces orientations of the two annular knots $\overline{K}_0$ and $\overline{K}_1$, as well as their pre-images K_0 and K_1 (but not of the original knot K). Let W be the winding number of K_0 (equivalently, $\overline{K}_0$) around the axis; this is one higher than the winding number of K_1 (equivalently, $\overline{K}_1$) around the axis. Then $W - \Delta$ is an

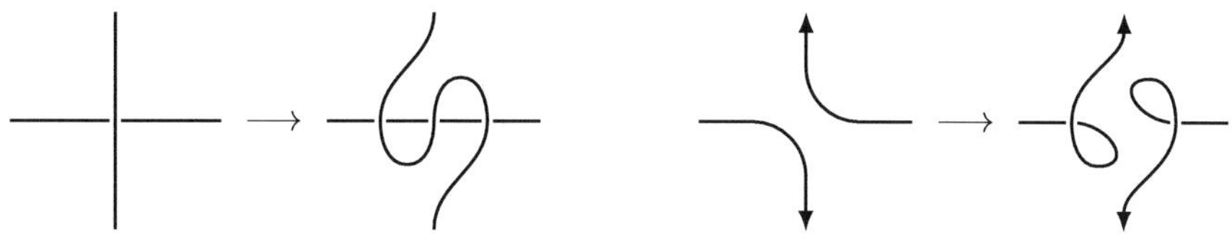

FIGURE 4.1. **An additional move.** The left half shows a new move for intravergent diagrams, corresponding to rotating the overpass T counter-clockwise over the underpass B. The right half shows the corresponding change for the 2-periodic link K_0 (with the orientation from Remark 4.5).

invariant of the knot K and its strong inversion; we prove this as Proposition 4.6 below.

PROPOSITION 4.6. *The quantity $W - \Delta$ is independent of the choice of the intravergent diagram, and in fact equals twice the axis linking number invariant* [**BI22**, Definition 4.6] *of the knot K and its strong inversion.*

PROOF. We first prove that $W - \Delta$ is an invariant. Given two intravergent diagrams for K and its strong inversion, connect them by a generic $\mathbb{Z}/2$-equivariant isotopy in $\mathbb{R}^3$; this produces a generic isotopy (in $\mathbb{R}^3$) connecting the quotient diagrams for $\overline{K}$. As in the proof of Reidemeister's theorem, this implies that the two diagrams for $\overline{K}$ are related by a finite sequence of Reidemeister moves (away from the axis), as well as a new move, corresponding to the situation when during the isotopy of K, the projection of the underpass B becomes tangent to the projection of the overpass T at the axis. The Reidemeister moves of $\overline{K}$ lift to $\mathbb{Z}/2$-equivariant pairs of Reidemeister moves for K, while this new move lifts to the move shown in the left half of Figure 4.1. (Actually, there are two moves, depending on whether the overpass T rotates clockwise or counter-clockwise over B. Figure 4.1 shows the move for the counter-clockwise rotation; the other move can be obtained by performing a $\mathbb{Z}/2$-equivariant pair of Reidemeister II moves of T over B near the axis, and then the above move in reverse.)

The Reidemeister moves for $\overline{K}$—lifting to a $\mathbb{Z}/2$-equivariant pair of Reidemeister moves for K—do not change W, the winding number of K_0. For the Reidemeister I move, depending on the shape of the clasp, either N_- increases by 2 and $\overline{N}_-$ increases by 1, or both N_- and $\overline{N}_-$ are unchanged. For the Reidemeister II move, N_- increases by 2 and $\overline{N}_-$ increases by 1, and for the Reidemeister III move, both N_- and $\overline{N}_-$ are unchanged. So, in each case $\Delta = N_- - 2\overline{N}_-$ does not change. Finally, for the new move from Figure 4.1, N_- increases by 1 and $\overline{N}_-$ also increases by 1, so Δ decreases by 1, but the winding number W decreases by 1, so the quantity $W - \Delta$ is preserved.

Next we will prove that this invariant $W - \Delta$ equals twice Boyle-Issa's axis linking number invariant [**BI22**, Definition 4.6]. Fix an orientation of the knot K. By performing the move from Figure 4.1 once if necessary, we may assume the crossing of K on the axis is a positive crossing. Then K_0 is a 2-component link, and it inherits an orientation from K. To avoid confusion, let o_{can} denote the canonical orientation of K_0 from Remark 4.5, and let o_{ind} denote the induced orientation from K. These two orientations agree on one of the components of K_0, and disagree on the other.

The number of negative crossings of K_0 with orientation o_{ind} is N_- and the number of the negative crossings of K_0 with orientation o_{can} is $2\overline{N}_-$. Therefore, $\Delta = N_- - 2\overline{N}_-$ is twice the linking number between the two components of K_0 (with orientation o_{can}); in particular, it is an even number.

Now perform the move from Figure 4.1 Δ times. (If $\Delta < 0$, then perform the reverse move $-\Delta$ times.) In the new diagram, the crossing on the axis is still positive, so the above discussion applies. Now $\Delta = 0$, and so the invariant is simply the new winding number W. Also, the linking number between the two components of K_0 in the new diagram is zero, so K_0 is the 2-component butterfly link [**BI22**, Definition 4.1], and by definition its winding number W is twice the axis linking number invariant. □

REMARK 4.7. Given a theorem about Khovanov homology, it is natural to wonder if it lifts a result about the Jones polynomial. Let $V_K(q)$ be the unreduced Jones polynomial, that is, the graded Euler characteristic of $Kh(K)$. Let $J_K(q) = V_K(q)/(q+q^{-1})$ denote the reduced Jones polynomial. For an annular knot K, let $AV_K(q,a)$ be the graded Euler characteristic of annular Khovanov homology, which was studied briefly by Roberts [**Rob13**, Section 2]. By Theorem 1.1,

$$V_K(q) \equiv q^{1-3\Delta}(qAV_{\overline{K}_1}(q^2,q) + AV_{\overline{K}_0}(q^2,q)) \pmod 2.$$

It is easy to see from Kauffman's state sum formula that if we quotient by $(q^2 + q^{-2}) - (q+q^{-1})$ then $AV(q^2,q) \equiv V(q^2)$ and $V(q) \equiv V(q^2)$. Therefore, we have

$$V_K(q) \equiv q^{-3\Delta}(q^2+q)V_{\overline{K}}(q^2) \equiv q^{-3\Delta}(q^2+q)V_{\overline{K}}(q) \pmod{2, q^2-q-q^{-1}+q^{-2}},$$

where $\overline{K}$ denotes either $\overline{K}_0$ or $\overline{K}_1$, viewed as an ordinary, not annular, knot. Since $q^2 - q - q^{-1} + q^{-2} = (q+q^{-1})(q-1+q^{-1})$ over $\mathbb{F}_2[q^{-1},q]$, we may divide by $(q+q^{-1})$ to get the equation for reduced Jones polynomial

$$J_K(q) \equiv q^{-3\Delta}(q^2+q)J_{\overline{K}}(q^2) \equiv q^{-3\Delta}(q^2+q)J_{\overline{K}}(q) \equiv J_{\overline{K}}(q) \tag{4.2}$$

modulo $(2, q-1+q^{-1})$. An analogous result can also be obtained for 2-periodic knots using [**SZ**, Theorem 1.3], giving the 2-periodic case of a formula of Murasugi's [**Mur88**, Theorem 1] and, using the fact that $J_K(i) \equiv 1 \pmod 2$, the 2-periodic case of Yokota's refinement [**Yok91**, Theorem 2]. However, Formula (4.2) is actually vacuous, since if we quotient by $(q+q^{-1})-1$, in Kauffman's state sum formula each circle contributes 1, and so for any knot or link diagram K with N crossings, N_- of which are negative, we get

$$J_K(q) \equiv \sum_{v\in\{0,1\}^N} q^{N+|v|-3N_-} = (1+q)^N q^N q^{-3N_-} \equiv 1 \pmod{2, q-1+q^{-1}}.$$

(Murasugi's and Yokota's formulas are also vacuous for 2-periodic knots, though interesting for higher periods. For Murasugi, this is [**Mur88**, Proposition 7]; Yokota only states his results for odd primes, presumably for this reason.)

5. An application to slice disks

Consider the knot $K = 9_{46}$. It bounds two slice disks as illustrated in Figure 5.1; denote them D_1 and D_2, and view them as cobordisms in $[0,1]\times\mathbb{R}^3$ from K to the empty link $\varnothing$. Let $\widehat{D}_i$ denote the image of D_i under the map $(t,x,y,z)\mapsto(1-t,x,y,z)$, so $\widehat{D}_i$ is a cobordism from $\varnothing$ to K. Sundberg-Swann showed that the disks D_1 and D_2 are distinguished by their induced maps on Khovanov homology.

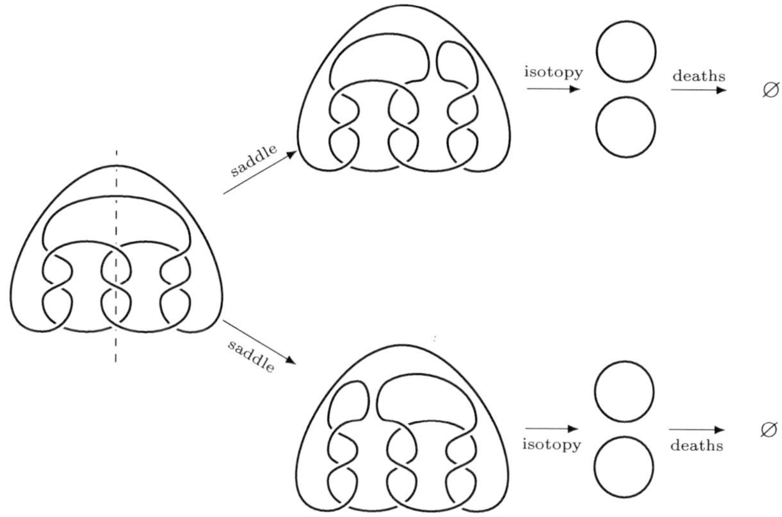

FIGURE 5.1. **The knot $K = 9_{46}$ and a pair of slice disks for it.** The knot is on the left, and the two movies on the two rows represent its two slice disks. Note that the two movies are related by a 180° rotation around the dashed line.

We will recover this result using Theorem 1.1. In fact, we get a little more; see Porism 5.2 below. The argument is reminiscent of the recent work of Dai-Mallick-Stoffregen using Heegaard Floer homology [**DMS23**].

THEOREM 5.1 ([**SS22**]). *The slice disks D_1 and D_2 induce different maps on Khovanov homology $Kh(9_{46}; \mathbb{F}_2) \to Kh(\varnothing; \mathbb{F}_2) = \mathbb{F}_2$.*

PROOF. For any cobordism F, let F_* denote the induced map on Khovanov homology. We will find an element $\gamma \in Kh_{0,1}(K; \mathbb{F}_2)$ satisfying $(D_1)_*(\gamma) \neq (D_2)_*(\gamma)$.

The Khovanov homology of K in quantum grading ± 1, retrieved from the Knot Atlas [**BM**] and converted to the conventions of Section 2, is

	-1	0
1		$\mathbb{Z}^2$
-1	$\mathbb{F}_2$	$\mathbb{Z}$

and therefore the Khovanov homology over $\mathbb{F}_2$ of K in these quantum gradings is

	-1	0
1		$\mathbb{F}_2^2$
-1	$\mathbb{F}_2$ ←	$\mathbb{F}_2^2$

where the arrow indicates a rank one Bockstein homomorphism associated to the coefficient sequence $0 \to \mathbb{Z}/2 \to \mathbb{Z}/4 \to \mathbb{Z}/2 \to 0$.

Let $\alpha \in Kh_{0,-1}(K; \mathbb{F}_2)$ be the generator of the kernel of the Bockstein homomorphism, and let $\beta \in Kh_{0,1}(K; \mathbb{F}_2)$ be the image of α under the X-action (the

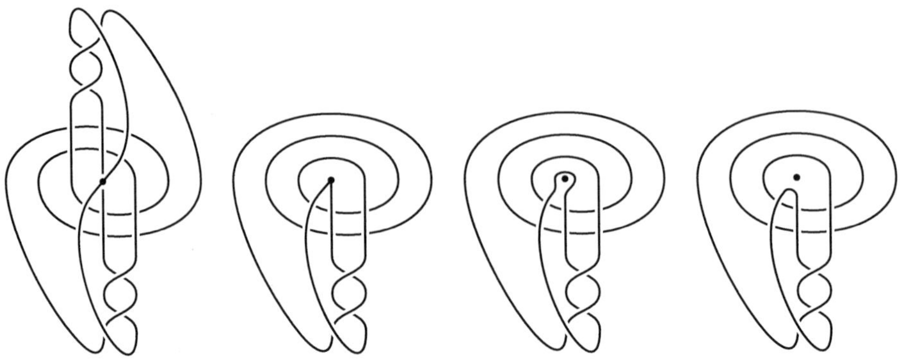

K $\overline{K}$ $\overline{K}_0$ $\overline{K}_1$

FIGURE 5.2. **The strongly invertible knot $K = 9_{46}$, its quotient knot $\overline{K}$, and the two induced annular trefoil knots $\overline{K}_0$ and $\overline{K}_1$.** Here K has been redrawn as an intravergent diagram with the axis coming straight out of the page through the marked point; it is also the axis for the two annular trefoils.

basepoint map). If $\mathrm{Id}^\bullet$ denotes the identity cobordism from K to itself decorated with a single dot, then this X-action is the map $\mathrm{Id}^\bullet_*$ induced on Khovanov homology by $\mathrm{Id}^\bullet$.

For $i \in \{1,2\}$, consider the cobordism $F_i = D_i \circ \mathrm{Id}^\bullet \circ \widehat{D}_i$ from $\varnothing$ to $\varnothing$. Since F_i is a (knotted) dotted sphere, by a result of Rasmussen and Tanaka [**Ras, Tan06**] (or, more precisely, a trivial extension of it [**LS22**, Lemma 6.16]), $(F_i)_* = (D_i)_* \circ \mathrm{Id}^\bullet_* \circ (\widehat{D}_i)_*$ is the identity map on Khovanov homology $\mathbb{F}_2 = Kh(\varnothing;\mathbb{F}_2) \to Kh(\varnothing;\mathbb{F}_2) = \mathbb{F}_2$. Therefore, both $(\widehat{D}_1)_*$ and $(\widehat{D}_2)_*$ map the generator of $Kh(\varnothing;\mathbb{F}_2)$ (in bigrading $(0,0)$) to α, which is the unique non-zero element of $Kh_{0,-1}(K;\mathbb{F}_2)$ in the kernel of the Bockstein. Therefore, both $\mathrm{Id}^\bullet_* \circ (\widehat{D}_1)_*$ and $\mathrm{Id}^\bullet_* \circ (\widehat{D}_2)_*$ map the generator of $Kh(\varnothing;\mathbb{F}_2)$ to $\beta \in Kh_{0,1}(K;\mathbb{F}_2)$, and both $(D_1)_*$ and $(D_2)_*$ map β to the generator of $Kh(\varnothing;\mathbb{F}_2)$. Choose some $\gamma \in Kh_{0,1}(K;\mathbb{F}_2)$ so that $\{\beta,\gamma\}$ is a basis of $Kh_{0,1}(K;\mathbb{F}_2)$.

The knot K is strongly invertible with respect to the 180° rotation around the dashed vertical line in Figure 5.1; call the involution τ and the induced map on Khovanov homology $\tau_*\colon Kh(K;\mathbb{F}_2) \to Kh(K;\mathbb{F}_2)$. The two slice disks D_1 and D_2 are related by the involution (Id,τ) of $[0,1] \times \mathbb{R}^3$. Therefore, $\tau_* \circ \mathrm{Id}^\bullet_* \circ (\widehat{D}_1)_* = \mathrm{Id}^\bullet_* \circ (\widehat{D}_2)_*$, which implies $\tau_*(\beta) = \beta$.

The annular quotient knots $\overline{K}_0$ and $\overline{K}_1$ are shown in Figure 5.2. We consider their annular Khovanov homology in gradings corresponding to the quantum grading $j = 1$ on K. (The grading correction term $\Delta = 4$.) Computer computation, using code by Davis [**Dav**], gives that the annular Khovanov homology of $\overline{K}_1$ in gradings with $2\jmath + k + 1 = 12$ is $\mathbb{F}_2^2$, supported in gradings $(2,7,-3)$ and $(3,5,1)$, while the annular Khovanov homology of $\overline{K}_0$ in gradings with $2\jmath + k = 12$ is also

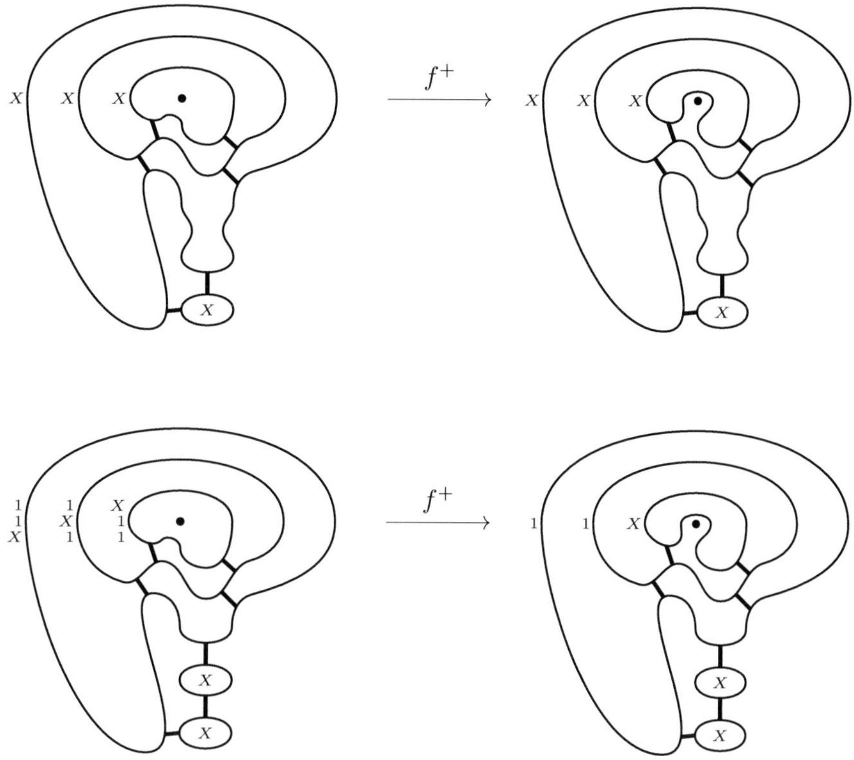

FIGURE 5.3. **Cycles in and their images under f^+.** Top left: a generator for $AKh_{2,7,-3}(\overline{K}_1;\mathbb{F}_2)$. Top right: its image in $AKh_{2,7,-2}(\overline{K}_0;\mathbb{F}_2)$ under f^+. Bottom left: an element of $AKh_{3,5,1}(\overline{K}_1;\mathbb{F}_2)$ which is the sum of three terms, where the two inessential circles and one essential circle are labeled X and the other two essential circles are labeled 1. (We have labeled the inessential circles and listed all possible labels of the three essential circles.) Bottom right: its image in $AKh_{3,5,2}(\overline{K}_0;\mathbb{F}_2)$. Crossings that were 1-resolved, i.e., where there is a Khovanov differential out of this state, are indicated with thick line segments.

$\mathbb{F}_2^2$, supported in gradings $(2,7,-2)$ and $(3,5,2)$. It is not hard to find representatives of these cycles in $AKh(\overline{K}_1;\mathbb{F}_2)$ by hand; see Figure 5.3. Their images under f^+ are (distinct) nontrivial elements of $AKh(\overline{K}_0;\mathbb{F}_2)$, so f^+ is an isomorphism. (Verifying that the image is nontrivial by hand is straightforward for the cycle in grading $(3,5,1)$, but is quite tedious for the cycle in grading $(2,7,-3)$, and might be better done by computer.) Thus, $\bigoplus_{\bar{\imath},\bar{\jmath},k|2\bar{\jmath}+k=1-1+3\Delta} AKh_{\bar{\imath},\bar{\jmath},k}(\overline{K}_1,\overline{K}_0;\mathbb{F}_2)=0$ so, by Corollary 1.4 the map $\tau_*\colon Kh_{0,1}(K;\mathbb{F}_2)\to Kh_{0,1}(K;\mathbb{F}_2)$ is given by $\left(\begin{smallmatrix}0&1\\1&0\end{smallmatrix}\right)$ with respect to an appropriate basis.

Since $\tau_*(\beta) = \beta$, it follows that the basis must be $\{\gamma, \gamma+\beta\}$ and $\tau_*(\gamma) = \gamma+\beta$. So,

$$(D_1)_*(\gamma) + (D_2)_*(\gamma) = (D_1)_*(\gamma) + (D_1)_*(\tau_*(\gamma)) = (D_1)_*(\beta) = 1 \in \mathbb{F}_2 = Kh(\varnothing; \mathbb{F}_2),$$

and hence $(D_1)_*(\gamma) \neq (D_2)_*(\gamma)$, as claimed. □

PORISM 5.2. *The knot 9_{46} does not admit an equivariant slice disk, with respect to the $\mathbb{Z}/2$-action on $[0,1] \times \mathbb{R}^3$ by (Id, τ). In fact, for any slice disk D for 9_{46}, D and $(\mathrm{Id}, \tau)(D)$ are distinguished by the induced maps on Khovanov homology.*

PROOF. The proof is the same as the proof of Theorem 5.1, with D and $(\mathrm{Id}, \tau)(D)$ in place of D_1 and D_2. □

Note that the first half of the porism also follows from Sakuma's work [**Sak86**]. The second half implies that D and $(\mathrm{Id}, \tau)(D)$ are not even related by taking connected sum with a knotted 2-sphere, since that operation does not change the map on Khovanov homology.

REMARK 5.3. For the last portion of the proof of Theorem 5.1, one could instead compute the map τ_* directly; arguably, that involves less work than the argument given. On the other hand, the argument above only requires computing the dimensions of certain annular Khovanov homology groups (including $AKh(\overline{K}_1, \overline{K}_0)$), for which there is a well-known, fast divide-and-conquer algorithm [**BN07**]; computing the action of τ_* efficiently would require further conceptual work.

6. An analogue in Heegaard Floer homology

The main ingredient in the proof of Theorem 1.6 is a recent localization theorem for the Heegaard Floer homology of branched double covers:

THEOREM 6.1 ([**HLL22**, Theorem 4.1]). *Let Y be a closed 3-manifold and $K \subset Y$ an oriented, nullhomologous knot with Seifert surface F. Let $\Sigma(Y, K)$ be the double cover branched along K induced by the Seifert surface F. Then, there is a spectral sequence with E^1-page given by*

$$\widehat{HF}(\Sigma(Y, K)) \otimes \mathbb{F}_2[\theta^{-1}, \theta]]$$

converging to

$$\widehat{HF}(Y) \otimes \mathbb{F}_2[\theta^{-1}, \theta]].$$

Consider a strongly invertible knot $K \subset \mathbb{R}^3 \subset S^3$ with axis $A \subset S^3$ (a circle). The intersection of K with A decomposes A into two intervals; let A_1 be the interval contained in $\mathbb{R}^3$ and A_2 the other interval. Let $\tau\colon S^3 \to S^3$ be rotation by $180°$ around A and let $\overline{K} \subset S^3 = S^3/\tau$ be the image of $K \cup A_1$ under the quotient map.

PROOF OF THEOREM 1.6. Let K' be the preimage of A_2 in $\Sigma(S^3, \overline{K})$. We claim that $\Sigma(S^3, K)$ is the double cover of $\Sigma(S^3, \overline{K})$ branched along K'. Theorem 6.1 then gives the result.

The claim follows from [**AB**, Lemma 3.1]; we explain this case of their proof. Let $q\colon S^3 \to S^3/\tau$ be the quotient map. The map q is the double cover branched along $q(A)$. Fix a Seifert surface $\overline{F}$ for $\overline{K}$ meeting $q(A_2)$ transversely, and let $F = q^{-1}(\overline{F})$. Let Y be the result of cutting S^3 along F, so $\partial Y = F_+ \cup_K F_-$. Since F is taken to itself by τ, τ induces an involution of Y. The fixed set of this involution is a copy of (the preimage of) A_2; the two copies of A_1 (one in F_+ and the other in F_-) are exchanged by the involution.

We can form $\Sigma(S^3, K)$ by gluing two copies of Y together. The involution τ induces an involution $\widetilde{\tau}$ of $\Sigma(S^3, K)$ with fixed set the preimage of A_2. The deck transformation of $\Sigma(S^3, K)$ gives another involution $\widetilde{\sigma}$, exchanging the two copies of Y and commuting with $\widetilde{\tau}$. So, $\widetilde{\sigma}$ descends to an involution σ of $\Sigma(S^3, K)/\widetilde{\tau}$. The quotient $\big(\Sigma(S^3, K)/\widetilde{\tau}\big)/\sigma$ is S^3, and the fixed set of σ is the preimage of $\overline{K}$. Thus, $\Sigma(S^3, K)/\widetilde{\tau} = \Sigma(S^3, \overline{K})$ and

$$\Sigma(S^3, K) = \Sigma(\Sigma(S^3, \overline{K}), K'), \tag{6.1}$$

as claimed. □

EXAMPLE 6.2. For the knots in Example 1.5, since the double cover of S^3 branched over the trefoil is a lens space, the spectral sequence associated to K has E^1-page $\mathbb{F}_2[\theta^{-1}, \theta]]^3$. The involution acts nontrivially on the set of Spin^C-structures on $\Sigma(S^3, K)$, so two of the three generators cancel under the d^1-differential and the spectral sequence collapses at the E^2-page to $\widehat{HF}(S^3) \otimes \mathbb{F}_2[\theta^{-1}, \theta]] = \mathbb{F}_2[\theta^{-1}, \theta]]$. By contrast, for K', the E^1-page and E^∞ page are both given by $\widehat{HF}(\Sigma(S^3, K')) \otimes \mathbb{F}_2[\theta^{-1}, \theta]] = \mathbb{F}_2[\theta^{-1}, \theta]]^3$, and d^i is trivial for all $i \geq 1$. (The action of the involution on the set of Spin^C-structures is trivial in this case.) In particular, unlike the spectral sequences for Khovanov homology, the spectral sequences in Heegaard Floer homology distinguish K and K'.

We conclude with a relative version of Theorem 1.6, which follows from a theorem of Large [**Lar**] (see also [**Hen12**]). As in the proof of Theorem 1.6, the preimage of A_2 is a knot K' inside $\Sigma(S^3, \overline{K})$; let $\widetilde{K}'$ be the preimage of K' in $\Sigma(S^3, K)$, which is also the preimage of A_2 in $\Sigma(S^3, K)$.

THEOREM 6.3. *With notation as in Theorem* 1.6, *there is a spectral sequence with* E^1*-page given by* $\widehat{HFK}(\Sigma(S^3, K), \widetilde{K}') \otimes \mathbb{F}_2[\theta^{-1}, \theta]]$ *and which converges to* $\widehat{HFK}(\Sigma(S^3, \overline{K}), K') \otimes \mathbb{F}_2[\theta^{-1}, \theta]]$.

PROOF. Large proved that given a nullhomologous knot L in a 3-manifold Y and a branched double cover $\Sigma(Y, L)$ of (Y, L) there is a spectral sequence

$$\widehat{HFK}(\Sigma(Y, L), \widetilde{L}) \otimes \mathbb{F}_2[\theta^{-1}, \theta]] \Rightarrow \widehat{HFK}(Y, L) \otimes \mathbb{F}_2[\theta^{-1}, \theta]],$$

where $\widetilde{L}$ is the preimage of L [**Lar**, Theorem 1.5]. (He states the result as a rank inequality.) By Formula (6.1), Large's theorem with $Y = \Sigma(S^3, \overline{K})$ and $L = K'$ gives the result. □

REMARK 6.4. The spectral sequence in Theorem 1.6 decomposes along Spin^C-structures on $\Sigma(\overline{K})$ (see [**HLL22**, Theorem 1.1]), and the spectral sequence in Theorem 6.3 decomposes along Spin^C-structures on $\Sigma(\overline{K})$ and Alexander gradings, i.e., relative Spin^C-structures on $(\Sigma(\overline{K}), K')$.

Acknowledgments

We thank Champ Davis for helpful conversations and computer code and Keegan Boyle, Kristen Hendricks, Mikhail Khovanov, Tye Lidman, Andrew Lobb, Matthew Stoffregen, and Liam Watson for further helpful discussions. We also thank the referee for suggestions and corrections.

References

[AB] Antonio Alfieri and Keegan Boyle, *Strongly invertible knots, invariant surfaces, and the Atiyah-Singer signature theorem*, `arXiv:2109.09915`.

[AK23] Rostislav Akhmechet and Mikhail Khovanov, *Anchored foams and annular homology*, Algebr. Geom. Topol. **23** (2023), no. 7, 3129–3204, DOI 10.2140/agt.2023.23.3129. MR4647674

[BI22] Keegan Boyle and Ahmad Issa, *Equivariant 4-genera of strongly invertible and periodic knots*, J. Topol. **15** (2022), no. 3, 1635–1674. MR4461855

[BM] Dror Bar-Natan, Scott Morrison, et al., *The Knot Atlas*, `http://katlas.org/`.

[BN07] Dror Bar-Natan, *Fast Khovanov homology computations*, J. Knot Theory Ramifications **16** (2007), no. 3, 243–255, DOI 10.1142/S0218216507005294. MR2320156

[Boy21] Keegan Boyle, *Involutions of alternating links*, Proc. Amer. Math. Soc. **149** (2021), no. 7, 3113–3128, DOI 10.1090/proc/15405. MR4257819

[BPS21] Maciej Borodzik, Wojciech Politarczyk, and Marithania Silvero, *Khovanov homotopy type, periodic links and localizations*, Math. Ann. **380** (2021), no. 3-4, 1233–1309, DOI 10.1007/s00208-021-02157-y. MR4297186

[Cor] James Cornish, *Sutured annular Khovanov homology and two periodic braids*, `arXiv:1606.03034`.

[Cou09] Olivier Couture, *Khovanov homology for signed divides*, Algebr. Geom. Topol. **9** (2009), no. 4, 1987–2026, DOI 10.2140/agt.2009.9.1987. MR2550464

[Dav] Champ Davis, *Annular Khovanov homology calculator*, `http://pages.uoregon.edu/champd/Annular.py`.

[DMS23] Irving Dai, Abhishek Mallick, and Matthew Stoffregen, *Equivariant knots and knot Floer homology*, J. Topol. **16** (2023), no. 3, 1167–1236, DOI 10.1112/topo.12312. MR4638003

[DP] Alessio Di Prisa, *Equivariant algebraic concordance of strongly invertible knots*, `arXiv:2303.11895`, 2023

[DP23] Alessio Di Prisa, *The equivariant concordance group is not abelian*, Bull. Lond. Math. Soc. **55** (2023), no. 1, 502–507. MR4568356

[DPF] Alessio Di Prisa and Giovanni Framba, *A new invariant of equivariant concordance and results on 2-bridge knots*, `arXiv:2303.08794`, 2023

[GLW18] J. Elisenda Grigsby, Anthony M. Licata, and Stephan M. Wehrli, *Annular Khovanov homology and knotted Schur-Weyl representations*, Compos. Math. **154** (2018), no. 3, 459–502, DOI 10.1112/S0010437X17007540. MR3731256

[GW10] J. Elisenda Grigsby and Stephan M. Wehrli, *Khovanov homology, sutured Floer homology and annular links*, Algebr. Geom. Topol. **10** (2010), no. 4, 2009–2039, DOI 10.2140/agt.2010.10.2009. MR2728482

[Hay] Kyle Hayden, *Corks, covers, and complex curves*, `arXiv:2107.06856`, 2021

[Hen12] Kristen Hendricks, *A rank inequality for the knot Floer homology of double branched covers*, Algebr. Geom. Topol. **12** (2012), no. 4, 2127–2178, DOI 10.2140/agt.2012.12.2127. MR3020203

[Hen15] Kristen Hendricks, *Localization of the link Floer homology of doubly-periodic knots*, J. Symplectic Geom. **13** (2015), no. 3, 545–608, DOI 10.4310/JSG.2015.v13.n3.a2. MR3412087

[HKK16] Po Hu, Daniel Kriz, and Igor Kriz, *Field theories, stable homotopy theory, and Khovanov homology*, Topology Proc. **48** (2016), 327–360. MR3465966

[HLL22] Kristen Hendricks, Tye Lidman, and Robert Lipshitz, *Rank inequalities for the Heegaard Floer homology of branched covers*, Doc. Math. **27** (2022), 581–612. MR4424030

[HLS16] Kristen Hendricks, Robert Lipshitz, and Sucharit Sarkar, *A flexible construction of equivariant Floer homology and applications*, J. Topol. **9** (2016), no. 4, 1153–1236, DOI 10.1112/jtopol/jtw022. MR3620455

[HS] Kyle Hayden and Isaac Sundberg, *Khovanov homology and exotic surfaces in the 4-ball*, `arXiv:2108.04810`, 2022

[Kho00] Mikhail Khovanov, *A categorification of the Jones polynomial*, Duke Math. J. **101** (2000), no. 3, 359–426, DOI 10.1215/S0012-7094-00-10131-7. MR1740682

[Kho02] Mikhail Khovanov, *A functor-valued invariant of tangles*, Algebr. Geom. Topol. **2** (2002), 665–741, DOI 10.2140/agt.2002.2.665. MR1928174

[KY79] Sadayoshi Kojima and Masayuki Yamasaki, *Some new invariants of links*, Invent. Math. **54** (1979), no. 3, 213–228, DOI 10.1007/BF01390230. MR553219

[Lar] Tim Large, *Equivariant Floer theory and double covers of three-manifolds*, arXiv:1910.12119, 2019

[LLS17] Tyler Lawson, Robert Lipshitz, and Sucharit Sarkar, *The cube and the Burnside category*, Categorification in geometry, topology, and physics, Contemp. Math., vol. 684, Amer. Math. Soc., Providence, RI, 2017, pp. 63–85. MR3611723

[LLS20] Tyler Lawson, Robert Lipshitz, and Sucharit Sarkar, *Khovanov homotopy type, Burnside category and products*, Geom. Topol. **24** (2020), no. 2, 623–745, DOI 10.2140/gt.2020.24.623. MR4153651

[LM18] Tye Lidman and Ciprian Manolescu, *Floer homology and covering spaces*, Geom. Topol. **22** (2018), no. 5, 2817–2838, DOI 10.2140/gt.2018.22.2817. MR3811772

[LS14] Robert Lipshitz and Sucharit Sarkar, *A Khovanov stable homotopy type*, J. Amer. Math. Soc. **27** (2014), no. 4, 983–1042, DOI 10.1090/S0894-0347-2014-00785-2. MR3230817

[LS22] Robert Lipshitz and Sucharit Sarkar, *A mixed invariant of nonorientable surfaces in equivariant Khovanov homology*, Trans. Amer. Math. Soc. **375** (2022), no. 12, 8807–8849, DOI 10.1090/tran/8736. MR4504654

[LT16] Robert Lipshitz and David Treumann, *Noncommutative Hodge-to-de Rham spectral sequence and the Heegaard Floer homology of double covers*, J. Eur. Math. Soc. (JEMS) **18** (2016), no. 2, 281–325, DOI 10.4171/JEMS/590. MR3459952

[LW21] Andrew Lobb and Liam Watson, *A refinement of Khovanov homology*, Geom. Topol. **25** (2021), no. 4, 1861–1917, DOI 10.2140/gt.2021.25.1861. MR4286365

[Mur88] Kunio Murasugi, *Jones polynomials of periodic links*, Pacific J. Math. **131** (1988), no. 2, 319–329. MR922222

[MWW22] Scott Morrison, Kevin Walker, and Paul Wedrich, *Invariants of 4-manifolds from Khovanov-Rozansky link homology*, Geom. Topol. **26** (2022), no. 8, 3367–3420, DOI 10.2140/gt.2022.26.3367. MR4562565

[OSz05] Peter Ozsváth and Zoltán Szabó, *On the Heegaard Floer homology of branched double-covers*, Adv. Math. **194** (2005), no. 1, 1–33, DOI 10.1016/j.aim.2004.05.008. MR2141852

[Ras] Jacob Rasmussen, *Khovanov's invariant for closed surfaces*, arXiv:math/0502527, 2005

[Rob13] Lawrence P. Roberts, *On knot Floer homology in double branched covers*, Geom. Topol. **17** (2013), no. 1, 413–467, DOI 10.2140/gt.2013.17.413. MR3035332

[Sak86] Makoto Sakuma, *On strongly invertible knots*, Algebraic and topological theories (Kinosaki, 1984), Kinokuniya, Tokyo, 1986, pp. 176–196. MR1102258

[Sna18] Michael Snape, *Homological invariants of strongly invertible knots*, Ph.D. thesis, University of Glasgow, December 2018.

[SS10] Paul Seidel and Ivan Smith, *Localization for involutions in Floer cohomology*, Geom. Funct. Anal. **20** (2010), no. 6, 1464–1501, DOI 10.1007/s00039-010-0099-y. MR2739000

[SS22] Isaac Sundberg and Jonah Swann, *Relative Khovanov-Jacobsson classes*, Algebr. Geom. Topol. **22** (2022), no. 8, 3983–4008, DOI 10.2140/agt.2022.22.3983. MR4562563

[SZ] Matthew Stoffregen and Melissa Zhang, *Localization in Khovanov homology*, arXiv:1810.04769, 2022

[Tan06] Kokoro Tanaka, *Khovanov-Jacobsson numbers and invariants of surface-knots derived from Bar-Natan's theory*, Proc. Amer. Math. Soc. **134** (2006), no. 12, 3685–3689, DOI 10.1090/S0002-9939-06-08397-3. MR2240683

[Wat17] Liam Watson, *Khovanov homology and the symmetry group of a knot*, Adv. Math. **313** (2017), 915–946, DOI 10.1016/j.aim.2017.04.003. MR3649241

[Yok91] Yoshiyuki Yokota, *The Jones polynomial of periodic knots*, Proc. Amer. Math. Soc. **113** (1991), no. 3, 889–894, DOI 10.2307/2048629. MR1064908

[Zha18] Melissa Zhang, *A rank inequality for the annular Khovanov homology of 2-periodic links*, Algebr. Geom. Topol. **18** (2018), no. 2, 1147–1194, DOI 10.2140/agt.2018.18.1147. MR3773751

Department of Mathematics, University of Oregon, Eugene, Oregon 97403
Email address: lipshitz@uoregon.edu

Department of Mathematics, University of California, Los Angeles, California 90095
Email address: sucharit@math.ucla.edu

Proceedings of Symposia in Pure Mathematics
Volume **109**, 2024
https://doi.org/10.1090/pspum/109/01995

On almost complex embeddings of rational homology balls

Paolo Lisca and Andrea Parma

ABSTRACT. We use elementary arguments to prove that none of the Stein rational homology 4-balls shown by the authors and Brendan Owens to embed smoothly but not symplectically in the complex projective plane admit such almost complex embeddings. In particular, we show that those rational balls admit no symplectic embeddings in the complex projective plane without appealing to the work of Evans and Smith.

1. Introduction

Let $p > q \geq 1$ be coprime integers and $B_{p,q}$ the Stein rational homology ball smoothing of the quotient singularity $\frac{1}{p^2}(pq-1,1)$. In [**LP1**] the authors extended work of Brendan Owens [**Ow20**] by exhibiting a subfamily

$$\{B(k,m),\ k \geq 0, m \geq 1\} \subset \{B_{p,q}\}$$

such that each $B(k,m)$ smoothly embeds in the complex projective plane. We later realized that the smooth embeddings constructed in [**LP1**] were obtained using certain special handlebody decompositions of $\mathbb{CP}^2$ called *horizontal* and that every smooth, closed, orientable 4-manifold admits horizontal decompositions [**LP2**]. In [**LP3**] we use horizontal decompositions to prove the existence of many more smooth embeddings of the rational balls $B_{p,q}$ into $\mathbb{CP}^2$.

Work of Evans and Smith [**ES18**] – based on Weimin Chen's adjunction formula for pseudoholomorphic curves in almost complex orbifolds [**Ch04**] – implies that $B(k,m)$ admits no symplectic embedding in $\mathbb{CP}^2$. In fact, Evans and Smith show that $B_{p,q}$ embeds symplectically in $\mathbb{CP}^2$ if and only if there are integers s and t such that

$$\text{(ES)} \qquad p^2 + s^2 + t^2 = 3pst \quad \text{and} \quad \pm\, q \equiv 3s/t \bmod p.$$

Note that the above sign ambiguity is irrelevant because (ES) holds for (p,q) if and only if it holds for $(p,p-q)$. In fact, $B_{p,q}$ and $B_{p,p-q}$ are symplectomorphic [**ES18**, Remark 2.8].

The main purpose of this note is to show by elementary arguments independent of [**ES18**] that the smooth embeddings constructed in [**Ow20, LP1**] are not homotopic to almost complex embeddings. In particular, we deduce that the rational balls $B(k,m)$ admit no symplectic embedding in $\mathbb{CP}^2$ without appealing to [**ES18**].

2020 *Mathematics Subject Classification.* Primary 57R40; Secondary 57K43, 57R17.
The present work is part of the MIUR-PRIN project 2017JZ2SW5.

We achieve this by associating to each collared, orientation-preserving topological embedding $j : B_{p,q} \hookrightarrow \mathbb{CP}^2$ an integer $c(j)$ which identifies the homotopy class of j and a sign $h(j) \in \{\pm 1\}$ which, together with $c(j)$, determines whether the pull-back by j of the standard complex structure of $\mathbb{CP}^2$ is homotopic to the Stein structure on $B_{p,q}$. Recall that, given a closed, topological 4-manifold X, a topological embedding $j : B_{p,q} \hookrightarrow X$ is *collared* if the restriction $j|_{\partial B_{p,q}}$ extends to a topological embedding $[-1,1] \times \partial B_{p,q} \hookrightarrow X$. The following is our main result.

THEOREM 1.1. *There is a collared, orientation-preserving topological embedding of $B_{p,q}$ into a closed, oriented topological 4-manifold X homotopy equivalent to $\mathbb{CP}^2$ if and only if $pq-1$ is a quadratic residue modulo p^2. Moreover, to each collared topological embedding $j : B_{p,q} \hookrightarrow \mathbb{CP}^2$ one can associate an integer $0 < c(j) < p^2$ such that $c(j)^2 + 1 \equiv pq \bmod p^2$ and a sign $h(j) \in \{\pm 1\}$ having the following properties:*

- *if $j_1, j_2 : B_{p,q} \hookrightarrow \mathbb{CP}^2$ are two collared topological embeddings, then j_1 is homotopic to j_2 if and only if $c(j_1) \equiv c(j_2) \mod p$;*
- *let J_0 denote the standard complex structure on $\mathbb{CP}^2$. Then, the pulled-back almost complex structure $j^*(J_0)$ is homotopic to the Stein structure on $B_{p,q}$ if and only if $h(j)q \equiv 3c(j) \bmod p$ and*

$$\frac{c(j)^2(p^2 - pq - 1) - 1}{p^2} \equiv \frac{h(j)c(j)q + 3}{p} \bmod 2. \tag{1}$$

REMARKS 1.2.

(a) In view of Theorem 1.1 it is natural to ask whether there is a rational ball $B_{p,q}$ and two non-homotopic orientation-preserving embeddings $j_1, j_2 : B_{p,q} \hookrightarrow \mathbb{CP}^2$.

(b) Theorem 1.1 is consistent with the results of [**ES18**] in the following sense. For any symplectic embedding $j : B_{p,q} \hookrightarrow \mathbb{CP}^2$ the almost complex structure $j^*(J_0)$ is homotopic to the Stein structure on $B_{p,q}$ and therefore if (ES) holds the conditions of Theorem 1.1 must be satisfied by some integer $0 < c < p^2$ and some sign $h \in \{\pm 1\}$. Lemma 4.3 implies that this is indeed the case.

COROLLARY 1.3. *Let $j : B_{p,q} \hookrightarrow \mathbb{CP}^2$ be a collared, orientation-preserving topological embedding. If $j^*(J_0)$ is homotopic to the Stein structure on $B_{p,q}$ then $q^2 + 9 \equiv 0 \bmod p$. In particular, $B(k,m)$ does not smoothly embed in $\mathbb{CP}^2$ as an almost complex manifold. A fortiori, $B(k,m)$ does not admit symplectic embeddings in $\mathbb{CP}^2$.*

PROOF. Let $j : B_{p,q} \hookrightarrow \mathbb{CP}^2$ be an embedding as in the statement such that $j^*(J_0)$ is homotopic to the Stein structure on $B_{p,q}$ and let $c(j)$ and $h(j)$ be the integers provided by Theorem 1.1. Since $c(j)^2 + 1 \equiv pq \bmod p^2$, we have $c(j)^2 \equiv -1 \bmod p$, hence $h(j)q \equiv 3c(j) \bmod p$ implies $q^2 \equiv -9 \bmod p$. On the other hand, the rational balls $B(k,m)$ were shown [**Ow20, LP1**] to be of the form $B_{p,q}$ with $q^2 + 9$ not divisible by p for each $k \geq 0$ and $m \geq 1$. The statement follows. □

REMARK 1.4. By recent work of Gompf [**Go23**, Corollary 1.2] the existence of a topological embedding $B_{p,q} \subset \mathbb{CP}^2$ implies that the (image of the) interior of $B_{p,q}$ is topologically isotopic to a Stein open subset $U \subset \mathbb{CP}^2$. By Corollary 1.3, in the case of the smooth embeddings $B(k,m) \subset \mathbb{CP}^2$ of [**Ow20, LP1**] the Stein structure which exists on $\mathrm{int}(B(k,m))$ as a smoothing of a quotient singularity is not homotopic to the Stein structure pulled-back from U by the time-1 map of the isotopy.

Earlier work of Gompf [**Go13**] implies that $B_{p,q}$ embeds holomorphically in $\mathbb{CP}^2$ if there is a smooth embedding $j : B_{p,q} \hookrightarrow \mathbb{CP}^2$ such that $j^*(J_0)$ is homotopic to the Stein structure on $B_{p,q}$. In fact, it follows by [**Go13**, Theorem 2.1] that after a smooth ambient isotopy of $\mathbb{CP}^2$ the induced complex structure on the image of j makes it a holomorphically embedded Stein handlebody. One can combine Theorem 1.1 with [**Go13**] to obtain the following.

COROLLARY 1.5. *A smooth, orientation-preserving embedding* $j : B_{p,q} \hookrightarrow \mathbb{CP}^2$ *is homotopic to a holomorphic embedding if and only if* $h(j)q \equiv 3c(j) \bmod p$ *and Equation* (1) *holds.*

PROOF. We have observed above that by [**Go13**] if $j : B_{p,q} \hookrightarrow \mathbb{CP}^2$ is a smooth, orientation-preserving embedding, then j is homotopic to a holomorphic embedding if and only if $j^*(J_0)$ is homotopic to the Stein structure on $B_{p,q}$. On the other hand, by Theorem 1.1 the latter condition on j is equivalent to the stated congruences. □

REMARK 1.6. We do not know which $B_{p,q}$'s smoothly embed in $\mathbb{CP}^2$, although many pairs (p, q) are obstructed by Donaldson's Theorem A [**Do83**]. Indeed, assuming $B_{p,q} \subset \mathbb{CP}^2$ one can construct a positive definite 4-manifold of the form $W = P \cup (\mathbb{CP}^2 \setminus B_{p,q})$ for a suitable 4-dimensional plumbing P with positive definite intersection lattice Λ_P. By Donaldson's theorem the intersection lattice Λ_W is standard, and arguing as in [**Ow20**, Section 3] one can find a contradiction. On the other hand, it is not difficult to find $B_{p,q}$'s such that p is not a Markov number, p divides $q^2 + 9$ but Donaldon's theorem does not obstruct the existence of a smooth embedding. For instance, $B_{10,1}$ and $B_{73,8}$ are such rational balls. In fact, it is easy to check that for $(c, h) \in \{(3, -1), (97, 1)\}$ in the first case and $(c, h) \in \{(1998, 1), (3331, -1)\}$ in the second case the conditions of Theorem 1.1 are satisfied, so that the balls $B_{10,1}$ and $B_{73,8}$ could conceivably admit holomorphic embeddings into $\mathbb{CP}^2$, although by [**ES18**] they admit no such symplectic embeddings. These examples have led us to the following question.

QUESTION 1.7. Is there a $B_{p,q}$ which admits a holomorphic embedding but no symplectic embedding into $\mathbb{CP}^2$?

2. Embeddings into homotopy complex projective planes

If $pq - 1$ is a quadratic residue mod p^2 then the linking form of $-\partial B_{p,q}$ is realized by the matrix (p^2) [**Ed05**, Theorem 3.1], so it follows from work of Boyer and Stong [**Bo93, St93**] that $-\partial B_{p,q}$ is the boundary of a an oriented, simply connected topological 4-manifold V with intersection form (p^2). Gluing $B_{p,q}$ and V along their boundaries produces a closed, oriented topological 4-manifold X homotopy equivalent to $\mathbb{CP}^2$ and containing a topologically embedded collared copy of $B_{p,q}$. This establishes one direction of the first sentence of Theorem 1.1. To prove the other direction we will use the following

PROPOSITION 2.1. *Let X be a closed, oriented topological 4-manifold homotopy equivalent to* $\mathbb{CP}^2$. *Let* $j : B_{p,q} \hookrightarrow X$ *be a collared, orientation-preserving topological embedding and set* $V := X \setminus \overline{j(B_{p,q})}$. *Then,* $H_1(V;\mathbb{Z}) = 0$ *and* $H_2(V;\mathbb{Z}) \cong \mathbb{Z}$. *Moreover, if* $i : V \hookrightarrow X$ *is the inclusion map, the subgroup* $i_*(H_2(V;\mathbb{Z})) \subset H_2(X;\mathbb{Z}) \cong \mathbb{Z}$ *has index* p.

PROOF. A Mayer-Vietoris argument [**Ow20**, Lemma 3.1] applied to the decomposition

$$\mathbb{CP}^2 = j(B_{p,q}) \cup V$$

gives $H_1(V;\mathbb{Z}) = 0$ and $H_2(V;\mathbb{Z}) \cong \mathbb{Z}$. Let $g \in H_2(V;\mathbb{Z})$ be a generator and

$$\alpha \in H_2(V, \partial V;\mathbb{Z}) \cong H^2(V;\mathbb{Z}) \cong \mathrm{Hom}(H_2(V;\mathbb{Z}),\mathbb{Z})$$

a relative homology class such that $\alpha \cdot g = 1$. Recall [**ES18**, § 2.3] that $H_1(B_{p,q}) \cong \mathbb{Z}/p\mathbb{Z}$. Then, $p\partial_*\alpha \in H_1(\partial V;\mathbb{Z})$ has zero image in $H_1(B_{p,q};\mathbb{Z})$. This implies that $p\,\mathrm{PD}(\alpha) \in H^2(V;\mathbb{Z})$ is the restriction of a class in $H^2(X;\mathbb{Z})$, i.e. k times a generator $\Lambda \in H^2(X;\mathbb{Z})$ for some $k \in \mathbb{Z}$. Let $\ell := \mathrm{PD}(\Lambda) \in H_2(X;\mathbb{Z})$. Then,

$$p = p\langle \mathrm{PD}(\alpha), g\rangle = k\langle \Lambda, g\rangle = k(\ell \cdot i_* g).$$

Thus, we have $i_*g = d\ell$, where d divides p. Exactness of the sequence

$$\begin{aligned} H_2(\partial V;\mathbb{Z}) = 0 \to H_2(V;\mathbb{Z}) \cong \mathbb{Z} \xrightarrow{(g\cdot g)\cdot} H_2(V,\partial V;\mathbb{Z}) \cong \mathbb{Z} \\ \to H_1(\partial V;\mathbb{Z}) \cong \mathbb{Z}/p^2\mathbb{Z} \to H_1(V;\mathbb{Z}) = 0 \end{aligned}$$

shows that $d^2 = p^2$, and the statement follows. □

COROLLARY 2.2. *Let X be a closed, oriented topological 4-manifold homotopy equivalent to $\mathbb{CP}^2$ and $j : B_{p,q} \hookrightarrow \mathbb{CP}^2$ a collared, orientation-preserving topological embedding. Then, $pq-1$ is a quadratic residue* mod p^2.

PROOF. Proposition 2.1 implies that the intersection form of V is represented by the matrix (p^2), which therefore presents the linking form on $\partial V = -\partial B_{p,q} = L(p^2, p^2 - pq + 1)$. It easily follows (cf. [**Ed05**, Theorem 3.1]) that $pq - 1$ is a quadratic residue modulo p^2. □

3. An auxiliary 4-manifold and its intersection lattice

In this section we establish some facts which will be used in Section 4 to prove the second part of Theorem 1.1. Let $j : B_{p,q} \hookrightarrow \mathbb{CP}^2$ be a collared, orientation-preserving topological embedding and let $V \subset \mathbb{CP}^2$ be the topological 4-manifold of Proposition 2.1. Let $R_{p,q}$ be the minimal resolution of the cyclic quotient singularity of type $\frac{1}{p^2}(pq-1, 1)$. Note that there is a canonical identification $\partial R_{p,q} = \partial B_{p,q}$ because $B_{p,q}$ is a smoothing of the same singularity. We use this identification to define the oriented topological 4-manifold

$$\widehat{X} := R_{p,q} \cup V$$

by gluing $\partial R_{p,q}$ and ∂V along their boundaries. Let $n = b_2(R_{p,q})$. Using the Mayer-Vietoris sequence it is easy to check that $H_1(\widehat{X};\mathbb{Z}) = 0$ and $b_2(\widehat{X}) = n+1$. By Poincaré duality and the Universal Coefficients Theorem we have

$$\mathrm{Tor}(H_2(\widehat{X};\mathbb{Z})) \cong \mathrm{Tor}(H_1(\widehat{X};\mathbb{Z})) = 0.$$

Denote by Λ_R, Λ_V and $\Lambda_{\widehat{X}}$, respectively, the free Abelian groups $H_2(R_{p,q};\mathbb{Z})$, $H_2(V;\mathbb{Z})$ and $H_2(\widehat{X};\mathbb{Z})$ viewed as intersection lattices. Recall that, as a smooth manifold, $R_{p,q}$ is the 4-dimensional plumbing of 2-disk bundles over spheres associated to a string of integers $(-a_1, \ldots, -a_n)$, where $a_i \geq 2$ for $i = 1, \ldots, n$ and

$$[a_1, \ldots, a_n] := a_1 - \cfrac{1}{a_2 - \cfrac{1}{\cdots - \cfrac{1}{a_n}}} = \frac{p^2}{pq-1} \tag{2}$$

The core 2-spheres of the plumbing $S_1, \ldots, S_n \subset R_{p,q}$ can be chosen to be smooth complex curves, and with their complex orientation they define the *vertex basis*

$$\{v_1 = [S_1], \ldots, v_n = [S_n]\} \subset \Lambda_R.$$

Let $D_n \subset R_{p,q}$ be a smoothly and properly embedded 2-disk normal to S_n, oriented so that $S_n \cdot D_n = +1$. Note that "the last sphere" S_n is well-defined unless $n > 1$ and

$$(a_1, a_2, \ldots, a_n) = (a_n, a_{n-1}, \ldots, a_1),$$

which by [**OW77**, Lemmas A.1 and A.2] holds if and only if $pq-1$ equals its inverse mod p^2, which is p^2-pq-1. Thus, $(a_1, \ldots, a_n)$ is palindromic if and only if $p = 2q$. But, since p and q are coprime, this can happen only if $(p,q) = (2,1)$, and then $\dfrac{p^2}{pq-1} = 4 = [4]$, so $n = 1$ and S_1 is well-defined.

It is well-known and easy to check that the homology class $[\partial D_n]$ is a generator of $H_1(\partial R_{p,q}; \mathbb{Z}) \cong \mathbb{Z}/p^2\mathbb{Z}$. Recall that by Proposition 2.1 we have $H_2(V; \mathbb{Z}) \cong \mathbb{Z}$.

DEFINITION 3.1. Let $g \in H_2(V; \mathbb{Z})$ be the generator such that $i_* g \in H_2(\mathbb{CP}^2; \mathbb{Z})$ is p times the class of a complex line, where i_* is the inclusion-induced map. Moreover, let

$$\alpha \in H_2(V, \partial V; \mathbb{Z}) \cong H^2(V; \mathbb{Z}) \cong \operatorname{Hom}(H_2(V; \mathbb{Z}), \mathbb{Z}) \cong \mathbb{Z}$$

be the (unique) relative homology class such that $\alpha \cdot g = 1$.

Note that, since $g \cdot g = i_* g \cdot i_* g = p^2$ and we are assuming $p > 1$, we must have $\partial\alpha \neq 0 \in H_1(V; \mathbb{Z}) = H_1(\partial R_{p,q}; \mathbb{Z})$, otherwise α would be a multiple of g and therefore $\alpha \cdot g \neq 1$.

DEFINITION 3.2. Let $0 < c(j) < p^2$ be the unique integer such that

$$c(j)[\partial D_n] = \partial\alpha \in H_1(\partial R_{p,q}; \mathbb{Z}).$$

LEMMA 3.3. *There is an an element $v_{n+1} \in \Lambda_{\widehat{X}}$ such that $v_{n+1} \cdot i_* g = 1$ and $\{v_1, \ldots, v_n, v_{n+1}\} \subset \Lambda_{\widehat{X}}$ is a basis with associated Gram matrix $G = (v_i \cdot v_j)$ given by*

$$G = \begin{pmatrix} -a_1 & 1 & 0 & 0 & \cdots & 0 \\ 1 & -a_2 & 1 & 0 & \cdots & 0 \\ \vdots & \vdots & \vdots & \vdots & \vdots & \vdots \\ 0 & \cdots & 1 & -a_{n-1} & 1 & 0 \\ 0 & \cdots & \cdots & 1 & -a_n & c(j) \\ 0 & \cdots & \cdots & 0 & c(j) & -a_{n+1} \end{pmatrix}.$$

Moreover, $c(j)^2 \equiv pq - 1 \bmod p^2$ and $a_{n+1} = \dfrac{c(j)^2(p^2-pq-1)-1}{p^2}$.

PROOF. By construction, the pair $(c(j)\,\mathrm{PD}([D_n]), \mathrm{PD}(\alpha)) \in H^2(R_{p,q};\mathbb{Z}) \oplus H^2(V;\mathbb{Z})$ is mapped to zero by the difference of the restriction maps

$$H^2(R_{p,q};\mathbb{Z}) \oplus H^2(V;\mathbb{Z}) \to H^2(\partial R_{p,q};\mathbb{Z}) \cong H^2(\partial V;\mathbb{Z}).$$

Therefore, by the cohomology Mayer-Vietoris sequence there is a homology class $v_{n+1} \in H_2(\widehat{X};\mathbb{Z})$ such that $\mathrm{PD}(v_{n+1}) \in H^2(\widehat{X};\mathbb{Z})$ is sent to $c(j)\,\mathrm{PD}([D_n])$ by the restriction map $H^2(\widehat{X};\mathbb{Z}) \to H^2(R_{p,q};\mathbb{Z})$ and to $\mathrm{PD}(\alpha)$ by the restriction map $H^2(\widehat{X};\mathbb{Z}) \to H^2(V;\mathbb{Z})$.

We claim that the classes $v_1, \ldots, v_n, v_{n+1}$ are a basis $\Lambda_{\widehat{X}}$. Since by construction $v_{n+1} \cdot g = 1$, the lattice $\langle g \rangle \subset H_2(V;\mathbb{Z})$ generated by g is *primitive* in $\Lambda_{\widehat{X}}$, i.e. the quotient $\Lambda_{\widehat{X}}/\langle g \rangle$ is torsion-free. Since $g \cdot g = p^2$, the lattice $\langle g \rangle$ is nondegenerate. Applying [**Di92**, Lemma A34] we obtain

$$\det(\langle g \rangle^\perp) = \det(\langle g \rangle) = p^2.$$

Since $\langle v_1, \ldots, v_n \rangle$ is a finite-index sublattice of the non-degenerate lattice $\langle g \rangle^\perp$ having the same determinant, by [**Di92**, Lemma A5] we have $\langle v_1, \ldots, v_n \rangle = \langle g \rangle^\perp$. Thus, for each $\lambda \in \Lambda_{\widehat{X}}$ the class

$$\lambda - (\lambda \cdot g) v_{n+1} \in \langle g \rangle^\perp$$

is a integral linear combination of $v_1, \ldots, v_n$. This shows that $\langle v_1, \ldots, v_n, v_{n+1} \rangle = \Lambda_{\widehat{X}}$, and since the classes v_i are clearly independent the claim holds.

Since $[D_n] \cdot v_n = 1$, by the definition of v_{n+1} we have $v_{n+1} \cdot v_n = c(j)$. Setting $a_{n+1} := -v_{n+1} \cdot v_{n+1}$, we see that the Gram matrix $G = (v_i \cdot v_j)$ has the stated form. Using e.g. [**OW77**, Lemmas A1 and A2] one can check that the numerator of the fraction $[a_1, \ldots, a_{n-1}]$ is the integer between 0 and p^2 which is the inverse mod p^2 of $pq - 1$, i.e. $p^2 - pq - 1$. Moreover, since $\Lambda_{\widehat{X}}$ is unimodular of signature $(1, n)$ and the submatrix of G given by the first $n - 1$ rows and columns is negative definite, we have

$$\det(G) = (-1)^n = -(-1)^n p^2 a_{n+1} - c(j)^2 (-1)^{n-1} (p^2 - pq - 1).$$

Thus, $c(j)^2(-pq - 1) \equiv 1 \bmod p^2$, i.e. $c(j)^2 \equiv pq - 1 \bmod p^2$, and the formula for a_{n+1} follows. □

As before, let $g \in H_2(V;\mathbb{Z})$ be the generator of Definition 3.1. Then, by Lemma 3.3 we can write

$$i_* g = \sum_{i=1}^{n+1} b_i v_i \in \Lambda_{\widehat{X}}$$

for some $b_1, \ldots, b_{n+1} \in \mathbb{Z}$.

LEMMA 3.4. *There is a unique integer $h(j) \in \{\pm 1\}$ such that $b_1, \ldots, b_n$ are given by the recursive rule:*

$$b_1 = h(j) c(j), \quad b_2 = a_1 b_1, \quad b_s = a_{s-1} b_{s-1} - b_{s-2}, \ s = 3, \ldots, n.$$

Moreover, $b_n = b_1(p^2 - pq - 1)$ and $h(j) b_{n+1} = (a_n b_n - b_{n-1})/b_1 = p^2$.

PROOF. Since $g \cdot g = p^2$, the lattice $\langle g \rangle$ is a finite-index sublattice of the non-degenerate lattice $\Lambda_R^\perp = \langle v_1, \ldots, v_n \rangle^\perp$ having the same determinant, therefore by [**Di92**, Lemma A5] we have $\langle g\rangle = \Lambda_R^\perp$. By Lemma 3.3, $\mathbf{b} := (b_1,\dots,b_{n+1})^t \in \mathbb{Z}^{n+1}$ generates the kernel of the $n\times(n+1)$ matrix

$$M := \begin{pmatrix} -a_1 & 1 & 0 & 0 & \cdots & 0\\ 1 & -a_2 & 1 & 0 & \cdots & 0\\ \vdots & \vdots & \vdots & \vdots & \vdots & \vdots\\ 0 & \cdots & 1 & -a_{n-1} & 1 & 0\\ 0 & \cdots & \cdots & 1 & -a_n & c(j)\end{pmatrix}$$

viewed as a homomorphism $\mathbb{Z}^{n+1}\to\mathbb{Z}^n$. The system $M\mathbf{b}=0$ consists of the n equations

$$\begin{cases} -a_1b_1+b_2=0,\\ b_{s-2}-a_{s-1}b_{s-1}+b_s=0,\ \ s=3,\dots,n,\\ b_{n-1}-a_nb_n+c(j)b_{n+1}=0.\end{cases}\tag{3}$$

Since $\mathbf{b}$ is primitive the b_i's are coprime, so the first $n-1$ equations of (3) imply that b_1 divides $b_2,\dots,b_n$. On the other hand, the last equation implies that b_1 divides $c(j)b_{n+1}$ and therefore, since the b_i's are coprime, that $c(j)=h(j)b_1$ for some $h(j)\in\mathbb{Z}$. Since b_1 divides b_i for $i=1,\dots,n$ and $c(j)\neq 0$, we can define $d_1,\dots,d_n$ by the equations $b_i=b_1d_i$, $i=1,\dots,n$. Hence $d_1=1$ and the d_i's satisfy the n equations

$$\begin{cases} -a_1d_1+d_2=0,\\ d_{s-2}-a_{s-1}d_{s-1}+d_s=0,\ \ s=3,\dots,n,\\ d_{n-1}-a_nd_n+h(j)b_{n+1}=0,\end{cases}\tag{4}$$

which we may write in the form

$$\frac{d_{i+1}}{d_i}=[a_i,\dots,a_1],\ i=1,\dots,n-1,\qquad \frac{h(j)b_{n+1}}{d_n}=[a_n,\dots,a_1].\tag{5}$$

Note that (4) and $d_1=1$ imply that $h(j)$ and d_n are coprime. Hence, it follows from (2) and the last equation of (5) that $h(j)$ divides p^2. But $h(j)$ also divides $c(j)$, which is coprime with p because by Lemma 3.3 we have $c(j)^2\equiv -1 \bmod p$, therefore $h(j)=\pm 1$. It is easy to check by induction that, since $a_i>1$ for each i, we have

$$1=d_1<\cdots<d_n,$$

and the last equation of (4) yields $h(j)b_{n+1}=(a_nb_n-b_{n-1})/b_1$. Finally, the last equality of (5) implies that $h(j)b_{n+1}=p^2$ and by e.g. [**OW77**, Appendix] $d_n(pq-1)\equiv 1 \bmod p^2$, therefore $d_n=p^2-pq-1$. This concludes the proof. □

4. Embeddings into the complex projective plane

We continue using the notation of the previous sections.

LEMMA 4.1. *Let $j_1, j_2 : B_{p,q}\hookrightarrow \mathbb{CP}^2$ two collared topological embeddings. Then, j_1 is homotopic to j_2 if and only if $c(j_1)\equiv c(j_2) \bmod p$.*

PROOF. Recall that $\mathbb{CP}^\infty$ is a $K(\mathbb{Z},2)$ and observe that, since $B_{p,q}$ is homotopy equivalent to a 2-dimensional CW-complex, the set $[B_{p,q},\mathbb{CP}^\infty]=H^2(B_{p,q};\mathbb{Z})$ of homotopy classes of maps $B_{p,q}\to\mathbb{CP}^\infty$ is in $1-1$-correspondence with $[B_{p,q},\mathbb{CP}^2]$, the correspondence being given by composing a map $B_{p,q} \to \mathbb{CP}^2$ with the inclusion $\mathbb{CP}^2 \subset \mathbb{CP}^\infty$. Hence the homotopy class of a map $j : B_{p,q} \to \mathbb{CP}^2$ is determined by the pull-back $j^*(\mathrm{PD}(\ell))$, where $\ell \in H_2(\mathbb{CP}^2;\mathbb{Z})$ is the class of a complex line. Therefore, j_1 is homotopic to j_2 if and only if $j_1^*(\mathrm{PD}(\ell)) = j_2^*(\mathrm{PD}(\ell))$. On the other hand, the cohomology exact sequence of the pair $(B_{p,q}, \partial B_{p,q})$ shows that the inclusion-induced map $H^2(B_{p,q};\mathbb{Z}) \to H^2(\partial B_{p,q};\mathbb{Z})$ is injective. Therefore $j_1^*(\mathrm{PD}(\ell)) = j_2^*(\mathrm{PD}(\ell))$ if and only if

$$j_1^* \mathrm{PD}(\ell)|_{\partial B_{p,q}} = j_2^* \mathrm{PD}(\ell)|_{\partial B_{p,q}}.$$

Observe that, for each $k = 1, 2$, we have $\mathrm{PD}(\ell)|_{\partial V_k} = p\,\mathrm{PD}(\alpha)|_{\partial V_k}$, where $V_k := \overline{\mathbb{CP}^2 \setminus j_k(B_{p,q})}$. Since by definition of $c(j_k)$

$$j_k^* \mathrm{PD}(\ell)|_{\partial V_k} = p j_k^* \mathrm{PD}(\alpha))|_{\partial V_k} = pc(j_k)\,\mathrm{PD}([\partial D_n]), \quad k = 1, 2,$$

we conclude that j_1 is homotopic to j_2 if and only if $pc(j_1) \equiv pc(j_2) \bmod p^2$, i.e. if and only if $c(j_1) \equiv c(j_2) \bmod p$. □

LEMMA 4.2. *Let $j : B_{p,q} \hookrightarrow \mathbb{CP}^2$ be a collared, orientation-preserving topological embedding. Then, the pulled-back almost complex structure $j^*(J_0)$ is homotopic to the Stein structure on $B_{p,q}$ if and only if $h(j)q \equiv 3c(j) \bmod p$ and Equation* (1) *holds.*

PROOF. Let $s_0(\mathbb{CP}^2) \in \mathrm{Spin}^c(\mathbb{CP}^2)$ be the Spin^c-structure associated to the standard complex structure on $\mathbb{CP}^2$, and $s_0(B_{p,q}) \in \mathrm{Spin}^c(B_{p,q})$ the Spin^c-structure associated to the Stein structure on $B_{p,q}$. Recall that, given any map $j : B_{p,q} \to X$ there is an induced pull-back map $j^\sharp : \mathrm{Spin}^c(X) \to \mathrm{Spin}^c(B_{p,q})$ between the sets of Spin^c-structures on X and $B_{p,q}$. Since $B_{p,q}$ is a 2-complex, homotopy classes of almost complex structures on $B_{p,q}$ are in 1-1 correspondence with Spin^c structures, with the correspondence given by sending an almost complex structure to the associated Spin^c structure [**Go97**, Remark(a), p. 48]. Therefore, $j^*(J_0)$ is homotopic to the Stein structure on $B_{p,q}$ if and only if $j^\sharp s_0(\mathbb{CP}^2) = s_0(B_{p,q})$. Therefore, to prove the theorem it suffices to show that $j^\sharp s_0(\mathbb{CP}^2) = s_0(B_{p,q})$ if and only if the stated congruences hold.

If $j^\sharp s_0(\mathbb{CP}^2) = s_0(B_{p,q})$ then, since $s_0(R_{p,q})|_{\partial R_{p,q}} = s_0(B_{p,q})|_{\partial B_{p,q}}$, $s_0(\mathbb{CP}^2)|_V$ extends to a Spin^c-structure $\bar{s}_0 \in \mathrm{Spin}^c(\widehat{X})$. Let $\beta = c_1(\bar{s}_0) \in H^2(\widehat{X};\mathbb{Z})$. The class β restricts to $R_{p,q}$ as $c_1(R_{p,q})$, hence we have $\langle \beta, i_*g\rangle = \langle c_1(\mathbb{CP}^2), i_*g\rangle = 3p$ because $c_1(\mathbb{CP}^2) = 3\,\mathrm{PD}(\ell)$, where $g \in H_2(V;\mathbb{Z})$ is the generator of Proposition 2.1. Let $\{v_1^\#, \ldots, v_{n+1}^\#\} \subset H^2(\widehat{X};\mathbb{Z})$ be the basis dual to $\{v_1, \ldots, v_{n+1}\}$. Since $v_1, \ldots, v_n \in H_2(R_{p,q};\mathbb{Z})$ are homology classes of smooth complex curves of genus zero with their canonical orientation, by the classical adjunction formula we have

$$\langle \beta, v_i\rangle = \langle c_1(R_{p,q}), v_i\rangle = 2 - a_i \quad \text{for} \quad i = 1, \ldots, n.$$

We can write

$$\beta = \sum_{i=1}^{n} (2 - a_i) v_i^\# + x v_{n+1}^\# \in H^2(\widehat{X};\mathbb{Z})$$

for some $x \in \mathbb{Z}$ with $x = \langle \beta, v_{n+1}\rangle \equiv a_{n+1} \bmod 2$ because β is characteristic. Since $i_*g = \sum_{i=1}^{n+1} b_i v_i$, we have

$$3p = \langle \beta, i_*g\rangle = \sum_{i=1}^{n} (2 - a_i) b_i + x b_{n+1}.$$

By Lemma 3.4

$$\sum_{i=2}^{n} b_i = \sum_{i=1}^{n-1} a_i b_i - \sum_{i=1}^{n-2} b_i \quad \Longrightarrow \quad \sum_{i=1}^{n} b_i = \sum_{i=1}^{n} a_i b_i - \sum_{i=1}^{n} b_i - b_1 pq,$$

where in the last equality we used that $b_1 p^2 = a_n b_n - b_{n-1}$ and $b_n = b_1(p^2 - pq - 1)$. Therefore

$$\sum_{i=1}^{n} (2 - a_i) b_i = -b_1 pq, \tag{6}$$

so by Lemma 3.4 it follows that

$$3p = -b_1 pq + x b_{n+1} = h(j) p(-c(j) q + xp),$$

therefore $h(j)q \equiv 3c(j) \bmod p$ and $x = \frac{c(j)q + 3h(j)}{p}$. Since $x = \langle \beta, v_{n+1} \rangle$ must be congruent to $v_{n+1} \cdot v_{n+1} = -a_{n+1} \bmod 2$ because β is characteristic, Equation (1) follows from Lemma 3.3, and the first half of the proof is concluded.

Conversely, let $j : B_{p,q} \hookrightarrow \mathbb{CP}^2$ be a collared, orientation-preserving topological embedding and suppose that $h(j)q \equiv 3c(j) \bmod p$ and (1) holds. Then, we can write $h(j)c(j)q + 3 = h(j)yp$, where y is an integer such that

$$y \equiv \frac{c(j)^2(p^2 - pq - 1) - 1}{p^2} \bmod 2.$$

By Lemma 3.3 the class

$$\beta := \sum_{i=1}^{n} (2 - a_i) v_i^{\#} + y v_{n+1}^{\#} \in H^2(\widehat{X}; \mathbb{Z})$$

satisfies $\langle \beta, v_i \rangle \equiv v_i \cdot v_i \bmod 2$ for each $i = 1, \ldots, n+1$, and therefore it is a characteristic class on $\widehat{X}$. Since $H_1(\widehat{X}; \mathbb{Z}) = 0$ we have $\beta = c_1(s)$ for a unique Spin^c structure $s \in \mathrm{Spin}^c(\widehat{X})$, and since $\beta|_{R_{p,q}} = c_1(R_{p,q})$ the restriction $s|_{R_{p,q}}$ coincides with the Spin^c-structure $s_0(R_{p,q})$ induced by the complex structure. Now observe that, by Lemma 3.4 and our choice of y,

$$\langle \beta, i_* g \rangle = \sum_{i=1}^{n} (2 - a_i) b_i + y b_{n+1} = -b_1 pq + y h(j) p^2 = p(-h(j)c(j)q + h(j)yp) = 3p.$$

In view of Proposition 2.1 this implies $\beta|_V = c_1(\mathbb{CP}^2)|_V$ and therefore, since $H_1(\mathbb{CP}^2; \mathbb{Z}) = 0$, we have $s|V = s_0(\mathbb{CP}^2)|_V$. In particular,

$$s|_{\partial V} = s_0(\mathbb{CP}^2)|_{\partial V = j(\partial B_{p,q})}.$$

Hence,

$$j^{\sharp} s_0(\mathbb{CP}^2)|_{\partial B_{p,q}} = s|_{\partial R_{p,q}} = s_0(R_{p,q})|_{\partial R_{p,q}} = s_0(B_{p,q})|_{\partial B_{p,q}}.$$

As observed in the proof of Lemma 4.1, since the map $H^2(B_{p,q}; \mathbb{Z}) \to H^2(\partial B_{p,q}; \mathbb{Z})$ induced by the inclusion $\partial B_{p,q} \subset B_{p,q}$ is injective, so is the inclusion-induced map $\mathrm{Spin}^c(B_{p,q}) \to \mathrm{Spin}^c(\partial B_{p,q})$. Therefore

$$j^{\sharp} s_0(\mathbb{CP}^2)|_{\partial B_{p,q}} = s_0(B_{p,q})|_{\partial B_{p,q}} \quad \Longrightarrow \quad j^{\sharp} s_0(\mathbb{CP}^2) = s_0(B_{p,q}).$$

This concludes the proof. □

PROOF OF THEOREM 1.1. The first sentence in the statement of Theorem 1.1 was established in Section 2. The second part of the statement follows from the combination of Lemmas 4.1 and 4.2. □

We close the paper with the lemma referred to in Remark 1.2(b).

LEMMA 4.3. *If p and q satisfy Equation* (ES) *for some s and t then there exists an integer c and a sign $h \in \{\pm 1\}$ satisfying the conditions of Theorem* 1.1.

PROOF. Given a triple (p, s, t) satisfying (ES) we can choose an integer c_0 with $0 < c_0 < p^2$, such that $c_0 \equiv \frac{s}{t} \bmod p$, therefore the condition $hq \equiv 3c \bmod p$ is satisfied for some $h \in \{\pm 1\}$ by any c of the form $c_0 + kp$. It suffices to show that c and h satisfy the remaining congruences of Theorem 1.1 for some choice of k. Since $c_0^2 \equiv -1 \bmod p$ – which implies that c_0 is coprime with p – we can write $c_0^2 = ap - 1$, so that $c^2 + 1 \equiv p(a + 2c_0 k) \bmod p^2$. The right-hand side is congruent to pq if and only if $2c_0 k \equiv q - a \bmod p$, which becomes $2k \equiv (a - q)c_0 \bmod p$ after multiplying both sides by c_0. If p is odd we can find k by simply inverting 2, while if p is even there is a unique possibility for k modulo $\frac{p}{2}$ and two possibilities modulo p because $p \equiv 0 \bmod 2$ implies that $(a - q)c_0$ is even. In fact, q and c_0 are both odd because they are coprime with p. In particular, $c_0^2 + 1$ is congruent to $2 \bmod 4$. Since $ap = c_0^2 + 1$, this shows that a must be odd – and $p \equiv 2 \bmod 4$ – so that $(a - q)c_0$ is even. We are left with verifying that c satisfies (1). If p is odd, replacing in (1) all occurrencies of p with 1 we obtain the equivalent congruence $c^2 q + 1 \equiv cq + 1 \bmod 2$, which holds because $c^2 \equiv c \bmod 2$. If p is even, by the argument given in the first part of the proof $\frac{p}{2}$ is odd. Moreover, replacing c with $c + \frac{p^2}{2}$ changes the right-hand side mod 2 but not the left-hand side, so that exactly one between c and $c + \frac{p^2}{2}$ satisfies the congruence. In fact, the right-hand side changes by $\dfrac{hpq}{2}$, which is odd, while the left-hand side changes by

$$\frac{\left((c + \frac{p^2}{2})^2 - c^2\right)(p^2 - pq - 1)}{p^2} = \left(c + \frac{p^2}{4}\right)(p^2 - pq - 1),$$

which is even because both c and $\frac{p^2}{4}$ are odd. □

Acknowledgment

The authors wish to thank the referee for their accurate and helpful report.

References

[Bo93] Steven Boyer, *Realization of simply-connected 4-manifolds with a given boundary*, Comment. Math. Helv. **68** (1993), no. 1, 20–47, DOI 10.1007/BF02565808. MR1201200

[Ch04] Weimin Chen, *Orbifold adjunction formula and symplectic cobordisms between lens spaces*, Geom. Topol. **8** (2004), 701–734, DOI 10.2140/gt.2004.8.701. MR2057778

[Di92] Alexandru Dimca, *Singularities and topology of hypersurfaces*, Universitext, Springer-Verlag, New York, 1992, DOI 10.1007/978-1-4612-4404-2. MR1194180

[Do83] S. K. Donaldson, *An application of gauge theory to four-dimensional topology*, J. Differential Geom. **18** (1983), no. 2, 279–315. MR710056

[Ed05] Allan L. Edmonds, *Homology lens spaces in topological 4-manifolds*, Illinois J. Math. **49** (2005), no. 3, 827–837. MR2210261

[ES18] Jonathan David Evans and Ivan Smith, *Markov numbers and Lagrangian cell complexes in the complex projective plane*, Geom. Topol. **22** (2018), no. 2, 1143–1180, DOI 10.2140/gt.2018.22.1143. MR3748686

[Go23] Robert E. Gompf, *Creating Stein surfaces by topological isotopy*, J. Differential Geom. **125** (2023), no. 1, 121–171, DOI 10.4310/jdg/1695236593. MR4643804

[Go97] Robert E. Gompf, Spin^c*-structures and homotopy equivalences*, Geom. Topol. **1** (1997), 41–50, DOI 10.2140/gt.1997.1.41. MR1475553

[Go13] Robert E. Gompf, *Smooth embeddings with Stein surface images*, J. Topol. **6** (2013), no. 4, 915–944, DOI 10.1112/jtopol/jtt017. MR3145144

[LP1] Paolo Lisca and Andrea Parma, *On Stein rational balls smoothly but not symplectically embedded in* $\mathbb{CP}^2$, Bull. Lond. Math. Soc. **54** (2022), no. 3, 949–960. MR4453750

[LP2] Paolo Lisca and Andrea Parma, *Horizontal decompositions, I*, `arXiv:2205.00482`, Algebraic Geom. Topol., to appear, 2022.

[LP3] Paolo Lisca and Andrea Parma, *Horizontal decompositions, II*, `arXiv:2302.14606`, Algebraic Geom. Topol., to appear, 2023.

[OW77] P. Orlik and P. Wagreich, *Algebraic surfaces with* k^**-action*, Acta Math. **138** (1977), no. 1-2, 43–81, DOI 10.1007/BF02392313. MR460342

[Ow20] Brendan Owens, *Smooth, nonsymplectic embeddings of rational balls in the complex projective plane*, Q. J. Math. **71** (2020), no. 3, 997–1007, DOI 10.1093/qmathj/haaa013. MR4142720

[St93] Richard Stong, *Simply-connected* 4*-manifolds with a given boundary*, Topology Appl. **52** (1993), no. 2, 161–167, DOI 10.1016/0166-8641(93)90034-B. MR1241191

Dipartimento di Matematica, Largo Bruno Pontecorvo 5, 56127 Pisa, Italy
Email address: `paolo.lisca@unipi.it`

Dipartimento di Matematica, Largo Bruno Pontecorvo 5, 56127 Pisa, Italy
Email address: `andrea.parma94@gmail.com`

Proceedings of Symposia in Pure Mathematics
Volume **109**, 2024
https://doi.org/10.1090/pspum/109/01996

Explicitly describing fibered 3-manifolds through families of singularly fibered surfaces

Maggie Miller

ABSTRACT. We give an explicit description of a fibration of the complement of the closure of a homogeneous braid, understanding how each fiber intersects every cross-section of S^3.

1. Introduction

In this note, we give an explicit description of a fibration of the complement of the closure of a homogeneous braid (e.g. a torus knot). These links have long been known to be fibered, since Stallings [**3**] observed that a standard Seifert surface for such a link decomposes into a Murasugi sum of fiber surfaces for torus knots.

Abstractly proving the existence of a fibration on a knot or link complement is generally straightforward, since (for example) fiberedness is deteected by knot or link Floer homology [**2**]. However, the author of this paper usually finds it difficult to conceptualize the whole fibration of such a link complement in total, rather than a single fiber. While building fibrations in a 4-dimensional setting, we found that a deeper understanding of fibrations of classical knot complements seemed necessary in order to extrapolate to higher dimensions. We thus worked to explicitly describe how the fibers of certain link complements meet every cross-section of a standard Morse function on S^3.

We first give an imprecise statement of the main theorem, meant to illustrate the aim of the paper.

THEOREM 1.1. *Let L be the closure of a homogenous braid β. Then L is a fibered link and we can explicitly describe a fibration $F : S^3 \setminus \nu(L) \to S^1$. The level sets of F are simple, in the sense that there is a natural Morse function h on (S^3, L) so that for every $\theta \in S^1$, the restriction of h to the interior of $F^{-1}(\theta)$ has no local minima or maxima.*

The proof of Theorem 1.1 is an explicit construction, which we are sharing with the belief that this can motivate arguments in 4-dimensional (or higher) manifolds where the usual techniques in 3-dimensional topology fail. (In the author's mind, this paper is mostly an exposition on the structure of a fibration of a 3-manifold over S^1.) We focus on closures of homogeneous braids as a natural, well-understood family of fibered links. We obtain a very restrictive statement about the position of

---

2020 *Mathematics Subject Classification.* Primary 57K30.

The author was supported by a Clay Research Fellowship and a Stanford Science Fellowship.

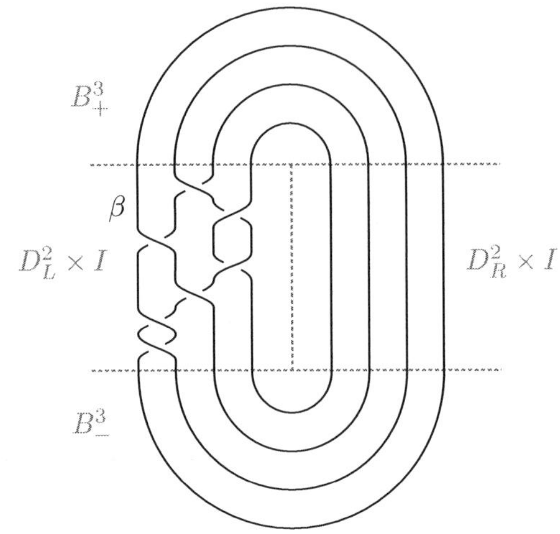

FIGURE 1. A link L in braid position as the braid closure of β. Here, β is a homogeneous braid since its given diagram is described by the braid word $\sigma_1^{-1}\sigma_1^{-1}\sigma_2^{-1}\sigma_3\sigma_1^{-1}\sigma_3\sigma_2^{-1}$, which is a word in the letters $\sigma_1^{-1}, \sigma_2^{-1}, \sigma_3$ with each letter appearing at least once.

the fibers of such a link that might be of interest to knot theorists and 3-dimensional topologists.

We give the precise statement of Theorem 1.1 now, which we will prove constructively in the following sections of this paper.

THEOREM 1.2. *Let L be the closure of a homogenous braid β on b strands, braided with respect to the standard Morse function $h : S^3 \to \mathbb{R}$, with the minima and maxima of h also minima and maxima of $h|_L$. Then there is a fibration $F : S^3 \setminus \nu(L) \to S^1$ so that simultaneously for all θ, the restriction of h to the interior of $F^{-1}(\theta)$ is Morse with no local minima nor maxima and the restriction of h to the boundary of $F^{-1}(\theta)$ is Morse with b local minima and b local maxima.*

2. Braid conventions

We choose some conventions when discussing links in braid position.

DEFINITION 2.1. Decompose $S^3 = B^3_+ \cup (S^2 \times I) \cup B^3_-$. Let $h : S^3 \to \mathbb{R}$ be Morse so that h has one local maximum in B^3_+, one local minimum in B^3_- and the restriction $h|_{S^2 \times I}$ is projection to I.

Further decompose $S^2 \times I = (D^2_L \times I) \cup (D^2_R \times I)$. Let β be a braid in $D^2_L \times I$ on b strands. Let L be a link with

$$L \cap (D^2_L \times I) = \beta \qquad \text{and} \qquad L \cap (D^2_R \times I) = \{b \text{ points}\} \times I,$$

and so that $L \cap B^3_+$ and $L \cap B^3_-$ each consist of b strands, on each of which h has one local critical point, with two of these extrema being the two critical points of h. We say L is in *braid position with respect to h* and is the *braid closure of β*. See Figure 1.

Recall that a braid diagram of a braid β on b strands can be written as a word in $\sigma_1, \sigma_1^{-1}, \dots, \sigma_{b-1}, \sigma_{b-1}^{-1}$, where σ_i is a diagram involving one positive half-twist between the i-th and $(i+1)$-th strands and no other crossings, and σ_i^{-1} is similar but with a negative half-twist. We will use the convention that the letters in a word w from left to right describe crossings in a diagram of β from bottom to top. This means that if w_1, w_2 are words for β_1, β_2, then $w_1 w_2$ is the word for the braid obtained by stacking β_2 above β_1.

DEFINITION 2.2. A braid is *homogeneous* if it can be written as a word in $\sigma_1^{s_1}, \ldots, \sigma_{b-1}^{s_{b-1}}$ for some $s_1, \ldots, s_{b-1} \in \{-1, 1\}$, with each letter appearing at least once.

In simpler terms, a braid is homogeneous if it can be written as a braid word so that for each i, the word includes one of σ_i and σ_i^{-1}, but not both. See Figure 1.

3. Background: constructing fibrations

In this section, we discuss local constructions of fibrations in dimension three. Our plan is to fix a height function $h : M^3 \to \mathbb{R}$ whose regular level sets are surfaces and whose singular level sets are surfaces with well-understood singularities. (In the case that M is closed, h is simply a Morse function.) We will then construct a family of functions f_t mapping each level set $h^{-1}(t)$ to the circle, with rules on how f_t changes with t to ensure that the function $F : M \to S^1$ given by $F(x) = f_{h(x)}(x)$ is a fibration. The advantage of this approach is that while drawing a collection of surfaces in a 3-manifold may seem daunting, drawing a collection of curves (level sets of f_t) in a surface (a level set of h) may seem more approachable.

CONVENTION 3.1. In this paper, we use the terms "singular fibration" and "height function" rather than "circular Morse function" or "Morse function" on manifolds with boundary to make it clear that we do not expect functions to be locally constant on boundary. For us, a *height function* $h : M \to \mathbb{R}$ is a smooth function that has the following properties.

- $h|_{\partial M}$ is Morse.
- h is Morse on the interior of M.
- The level sets of h meet ∂M transversely away from critical points of $h|_{\partial M}$.

For example, if L is a link in S^3 and $h : S^3 \to \mathbb{R}$ is Morse with $h|_L$ also Morse, then h restricts to a height function on $M := S^3 \setminus \nu(L)$ for $\nu(L)$ a suitable tubular neighborhood of L.

We first observe that given a fibration $F : M^3 \to S^1$ and a height function $h : M^3 \to \mathbb{R}$, we generically expect F to induce a singular fibration on each level set of h.

DEFINITION 3.2. A *singular fibration* on a compact surface Σ is a smooth map $f : \Sigma \to S^1$ with the following properties.

- The restriction $f|_{\partial\Sigma}$ is Morse.
- For all but finitely many θ, $f^{-1}(\theta)$ is a compact, properly embedded 1-manifold.
- For all θ, $f^{-1}(\theta)$ is a compact, properly embedded 1-manifold away from a finite number of singularities of the four types shown in Figure 2. We will often refer to interior wedgepoints (first image in Figure 2) as $\times$-*singularities*.

We refer to each $f^{-1}(\theta)$ as a *leaf* of f.

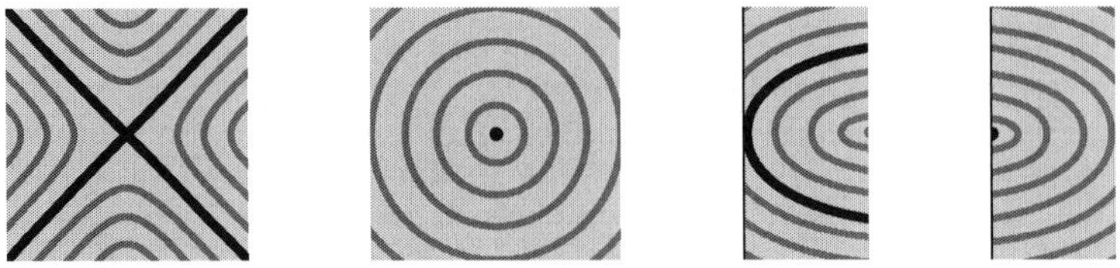

FIGURE 2. Local models near singular points of a leaf of a singular fibration $f : \Sigma \to S^1$ (away from singularities of Σ). **First:** A leaf includes an $\times$-singularity. **Second:** A leaf component is a single point in the interior of Σ. **Third:** A leaf has an isolated tangency to $\partial\Sigma$. **Fourth:** A leaf component is a single point in the boundary of Σ.

We may extend Definition 3.2 to apply to some singular surfaces. Here, by *singular surface* we mean a compact topological space embeddable in $\mathbb{R}^4$ containing a finite singularity set S whose complement is a surface.

DEFINITION 3.3. Let Σ be a compact, singular surface with finite singularity set $S = \{p_1, \dots, p_m, q_1, \dots, q_n\}$.

In words, the points $p_1, \dots, p_m$ are the singularities on the boundary of Σ. These points come in three types: some are seemingly interior points that we have artificially declared to be boundary, some are wedge points in $\partial\Sigma$, and others are isolated points that we declare to be in the boundary of Σ. The points $q_1, \dots, q_n$ are the singularities in the interior of Σ. These points come in two types: the singular points of double cones and isolated points that we declare to be in the interior of Σ.

To be precise, we require each p_i to have a neighborhood that is homeomorphic to one of: $\mathbb{R}^2$ or $\{(x, y) \mid |x| \geq |y|\}$ or $\{(0, 0)\}$, with $p_i \mapsto (0, 0)$. We require each q_i to have a neighborhood that is homeomorphic to one of: $\{(x, y, z) \mid x^2 + y^2 = |z|\}$ or $(0, 0)$, with $q_i \mapsto (0, 0)$.

A smooth (away from S) function $f : \Sigma \to S^1$ is a *singular fibration* if f restricts to a singular fibration on $\Sigma \setminus \nu(S)$ as in Definition 3.2 and in each component of $\nu(S)$, f restricts to one of the local models shown in Figure 3.

In Definition 3.3, we restrict to the given types of singularities precisely because these are the singularities that appear in level sets of a height function $h : M \to \mathbb{R}$. Interior singular points arise in $h^{-1}(t)$ when the Morse function $h|_{\mathring{M}}$ has a critical point at height t. Boundary singular points arise in $h^{-1}(t)$ when the Morse function $h|_{\partial M}$ has a critical point at height t.

We use singular fibrations of surfaces to construct a fibration of a 3-manifold by considering the level surfaces of a height function.

DEFINITION 3.4. Given a height function $h : M^3 \to \mathbb{R}$, a *movie of singular fibrations* on M^3 is a family of smooth maps $f_t : h^{-1}(t) \to S^1$ with each f_t a singular fibration on the singular surface $h^{-1}(t)$ such that the functions f_t vary smoothly with t, i.e. $F(x) = f_{h(x)}(x)$ is a smooth map from M^3 to S^1. We refer to F as the *total function* of the movie f_t.

By constructing a movie of singular fibrations on M^3 and keeping track of how singular leaves of f_t vary as t increases or decreases, we can arrange for the total function $F : M \to S^1$ to be a fibration. In [**1**] we referred to the "type" of

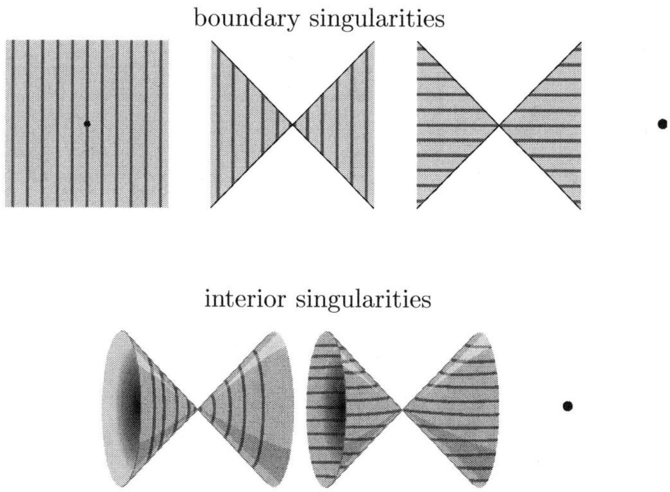

FIGURE 3. Near a singular point of a surface Σ, we require the leaves of a singular fibration f to have one of these pictured local models.

singularities in leaves of f_t, determined by the sign of $\frac{d}{dt}f_t$ at that singularity. This language is less useful in this dimension due to symmetry of $\times$-singularities. (Or in other words, this notation is less useful because an index-1 critical point of a Morse function on a surface remains an index-1 point when turned upside down.) Instead, we add arrow decorations to contour maps of f_t, as shown in Figures 4 and 5. In Figure 5 we draw only boundary singularities; models of interior singularities can be obtained by doubling.

PROPOSITION 3.5. *The total function F of a movie of singular fibrations f_t is a fibration as long as the following are true.*

- *Near singularities in leaves of f_t that are not also singularities of the surface $h^{-1}(t)$, $\frac{d}{dt}f_t$ is nonvanishing. (That is, we may always draw arrows on contour sets of f_t as in Figures 4 and 5.*
- *At singularities of $h^{-1}(t)$, $\frac{d}{dt}f_t$ vanishes and the arrows decorating $f_{t\pm\epsilon}$ are as in one of the models of Figure 5.*
- *The maps f_0 and f_1 are both fibrations.*

PROOF. Assume the listed conditions hold. This ensures that the leaves of F are nonsingular, so $F : M \to S^1$ is a fibration. □

REMARK 3.6. When the total function F of f_t is a fibration, then h restricts to a height function on each fiber of F. Figures 4 and 5 are simply models of interior and boundary critical points of $h|_{F^{-1}(\theta)}$.

NOTATION 3.7. Assume f_t has arrow decorations. A leaf of f_t containing an $\times$-singularity divides a small neighborhood of that singularity into four regions. Two of these regions (opposite to each other) locally have arrows pointing into them; we call these the *inward regions*. The other two regions locally have arrows pointing away from them; we call these the *outward regions*. See Figure 6.

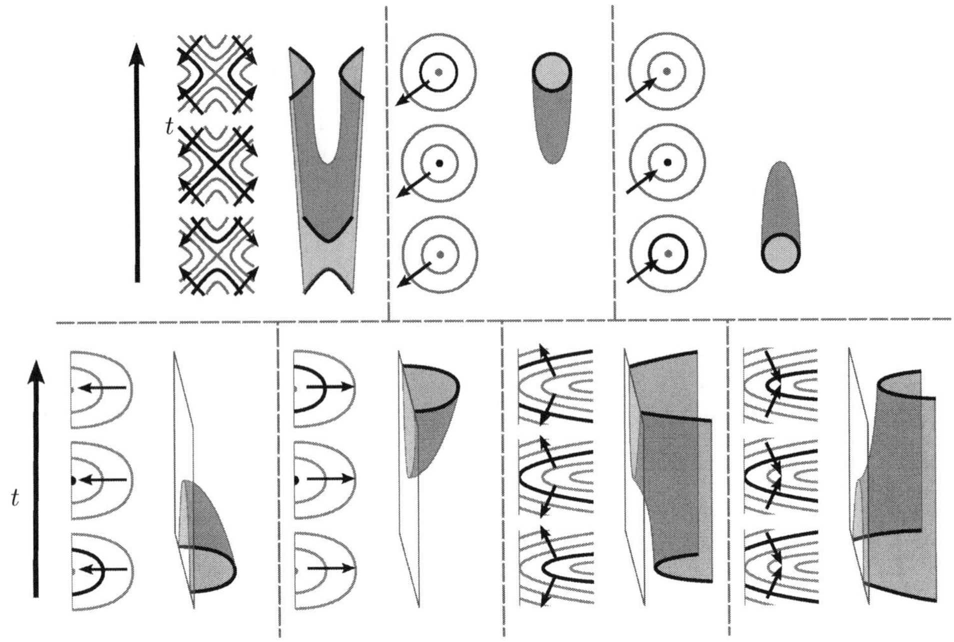

FIGURE 4. Local models near singularities of leaves of f_t that are not also singular points of $h^{-1}(t)$. In all of these models, $\frac{d}{dt}f_t$ is nonzero near the singular point. In each cell, we highlight cross-sections of $F^{-1}(\theta)$ as t increases and give a schematic of $F^{-1}(\theta)$. We include arrows near singularities of leaves of f_t that indicate the behavior of $f_{t\pm\epsilon}$ near the singularity.

4. Small movies

Here we give three main examples of movies of singular fibrations that we will use to construct more interesting movies. To prove Theorem 1.2, we will only need to consider singular fibrations on surfaces with only boundary singularities. From now on, we will only consider singular surfaces with no interior singularities.

EXAMPLE 4.1. Parametrize a 3-ball B^3 as $I \times I \times I$, and let $\pi : B^3 \to I$ be projection to the third factor. Fix smooth $f_0 : \pi^{-1}(0) \to S^1$ so that the level sets of f_0 are lines of the form $0 \times \{I\} \times \{0\}$.

Let γ be an arc properly embedded in B^3 so that $\pi(\partial\gamma) = 1$, and $\pi|_\gamma$ is Morse with exactly one local minimum that is contained in $\pi^{-1}(1/2)$, and projecting γ to $I \times I \times 0$ yields an arc contained in a level set of f_0.

Let $M^3 = B^3 \setminus \nu(\gamma)$. We choose the thickening of γ so that $h := \pi|_M$ has two singular level sets. See Figure 8. In the top row of Figure 7 we illustrate an extension of f_0 to a movie of singular fibrations $f_t : h^{-1}(t) \to S^1$. This is a concatenation of movies (iv) and (viii) of Figure 5, so the arrows are consistent with the requirements of Proposition 3.5. The leaves of the resulting total function $F : M \to S^1$ are nonsingular in their interior, but there are two $\times$-singularities in leaves of f_1.

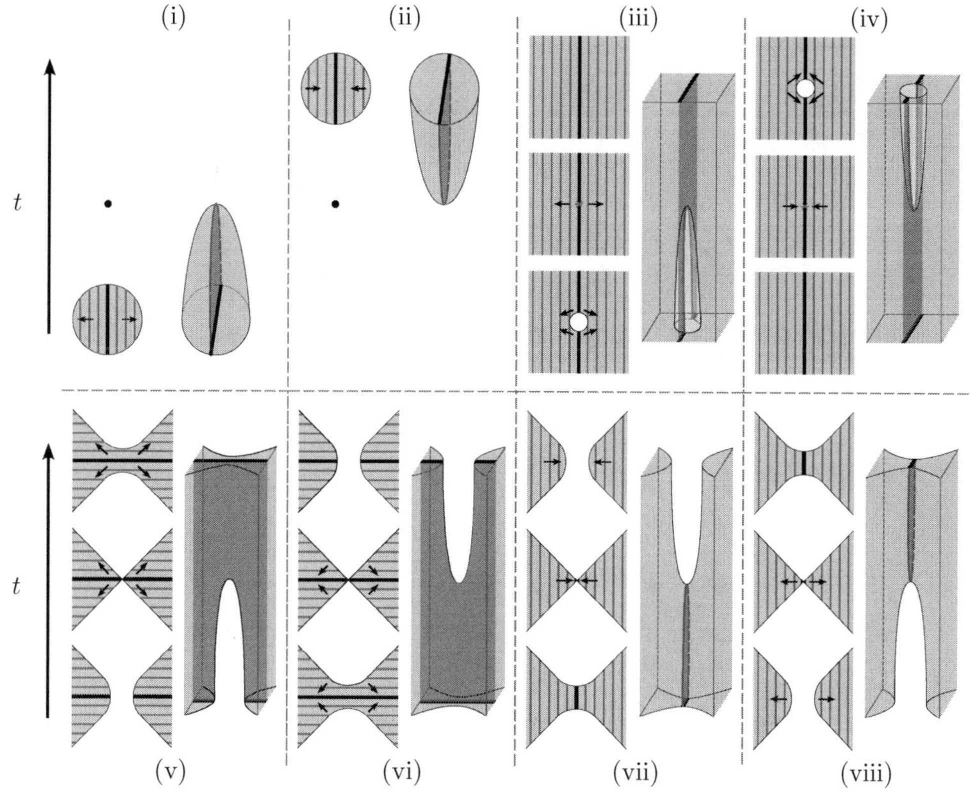

FIGURE 5. Local models near singular leaves of f_t that are also singular points of $h^{-1}(t)$. In each cell, we highlight cross-sections of $F^{-1}(\theta)$ as t increases and give a schematic of $F^{-1}(\theta)$.In each of these models, $\frac{d}{dt} f_t$ is zero at the singular point of $h^{-1}(t)$. We also include arrows in nonsingular cross-sections as in Figure 4. These arrows are drawn assuming that the non-highlighted leaves of F are also locally nonsingular. We draw each singularity as a boundary singularity; doubling these figures yields corresponding local models of singular leaves of f_t with interior singular points.

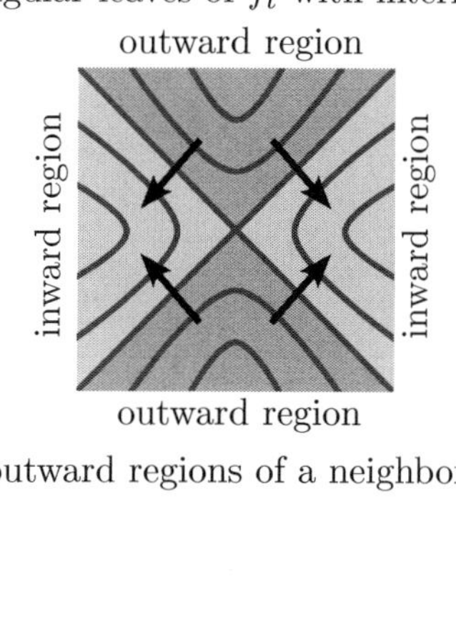

FIGURE 6. Inward and outward regions of a neighborhood of an $\times$-singularity.

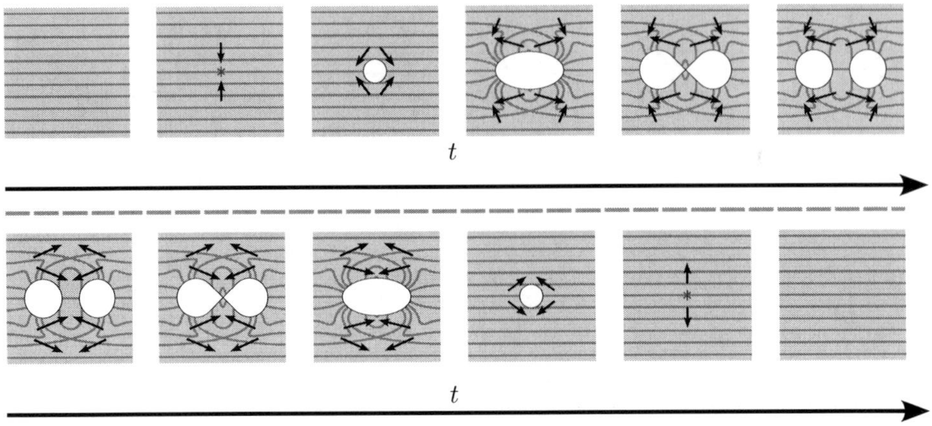

FIGURE 7. **Top row:** the movie of Example 4.1. **Bottom row:** the movie of Example 4.2.

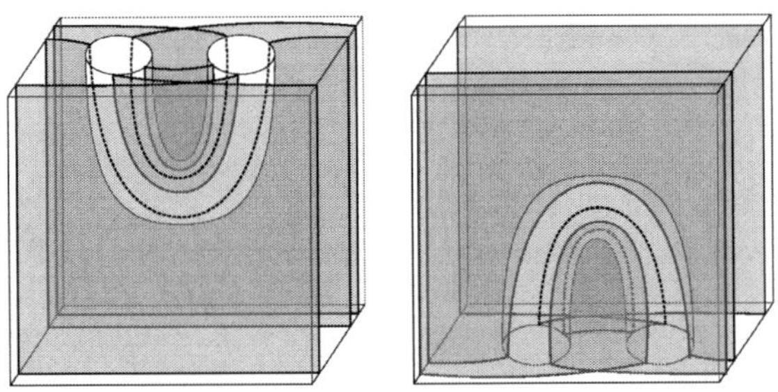

FIGURE 8. Some leaves of the total function of the movies described in Example 4.1 (left) and Example 4.2 (right).

When attempting to fiber the complement of a knot or link L, we can use the movie in Example 4.1 as a model of a fibration near a minimum of L.

EXAMPLE 4.2. Now we will essentially turn Example 4.1 upside down. Again parametrize a 3-ball B^3 as $I \times I \times I$, and let $\pi : B^3 \to I$ be projection to the third factor. Let $f_0 : h^{-1}(0) \to S^1$ agree with the function f_1 of Example 4.1. Fix an arc γ properly embedded in B^3 so that $\pi(\partial\gamma) = 0$, and $\pi|_\gamma$ is Morse with exactly one local maximum that is contained in $\pi^{-1}(1/2)$, and projecting γ to $I \times I \times 0$ yields an arc contained in a level set of f_0. In the bottom row of Figure 7 we illustrate an extension of f_0 to a movie of singular fibrations $f_t : h^{-1}(t) \to S^1$. This is a concatenation of movies (vii) and (iii) of Figure 5, so the arrows are consistent with the requirements of of Proposition 3.5. The leaves of the resulting total function $F : M \to S^1$ are nonsingular in their interior, but there are two $\times$-singularities in leaves of f_0.

When building a fibration of a knot or link in braid position, we will use the movie in the following example as a model of the fibration near a crossing. It may be helpful to think of a braid closure as obtained by performing band surgeries on an unlink, with bands corresponding to crossings in the braid. In the following example, we show how to change $f_t^{-1}(\theta)$ by band surgery to the leaves along an arc

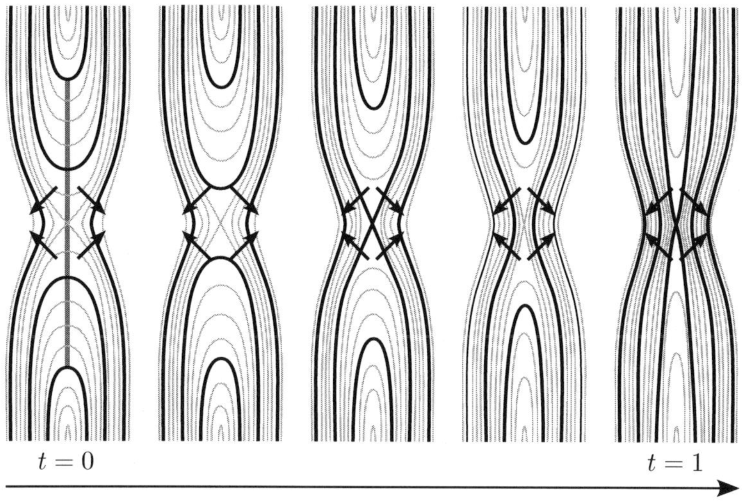

FIGURE 9. The movie of Example 4.3 (band movie). **Leftmost frame:** A neighborhood of the arc η (bold vertical arc) in the domain of f_0. The $\times$-singularity p of f_0 lies at the center of η. Also in bold, we draw the leaf $f_0^{-1}(\theta)$ that contains the endpoints of η. The leaf $f_0^{-1}(\theta)$ may intersect η in more points as shown (or even include the point p). The circular length condition on η ensures that η intersects $f_0^{-1}(\theta)$ the same number of times on either side of p. **From left to right:** f_t for increasing t.

η (the core of a band) as t increases from 0 to 1, assuming that f_0 is sufficiently nice. Since we will use this model several times, we will refer to the movie of Example 4.3 as a *band movie*. We suggest that the reader consider Figure 9 before reading the text of Example 4.3: we draw an arc η connecting one leaf of f_0 to itself, such that the arc crosses over exactly one $\times$-singularity, meeting the outward regions. We then build a movie of singular fibrations in which the arc is contracted, so that in f_1, each leaf of f_0 that met η has been changed by some number of band surgeries.

EXAMPLE 4.3. Parametrize a 3-ball M^3 as $I \times I \times I$, and let $h : M^3 \to I$ be projection to the third factor. Let $f_0 : \pi^{-1}(0) \to S^1$ be smooth so that f has one $\times$-singularity at a point p. Fix arrow decorations at p. Let η be an arc in $I \times I = h^{-1}(0)$ with $\eta(1/2) = p$, meeting the outward regions near p. (Recall arrows near $\times$-singularities point out of outward regions and into inward regions.) Take η to be transverse to all leaves of f_0 away from p. Assume $f_0(\eta(0)) = f_0(\eta(1)) = \theta$ for some θ and that in the intervals $[0, 1/2]$ and $[1/2, 1]$, $f_0 \circ \eta$ winds in equal magnitude about S^1. That is, assume there is an $l \in \mathbb{R}$ with

$$l = \int_0^{1/2} f_0(\eta(t))dt = -\int_{1/2}^1 f_0(\eta(t))dt.$$

Let $b : I \times I \to I$ be a bump function supported on a small neighborhood of η, with $b(\eta) = 1$. Then set

$$f_t(x) = f_0(x) - t \cdot l \cdot b(x).$$

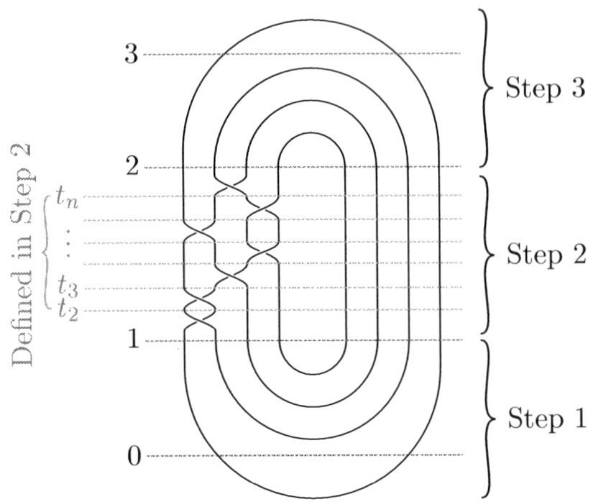

FIGURE 10. A schematic of the values of $h|_M$ relevant to each step of the construction in Theorem 1.2.

We draw f_t in Figure 9. Note that $f_1(p) = f_0(p) - l = \theta$. Our requirement that η pass through the outward regions near p ensures that the decoration arrows on f_0 actually correspond to the movie f_t.

Since this movie satisfies the first two conditions of Proposition 3.5, the total function F has leaves that are nonsingular away from $h^{-1}(\{0,1\})$.

5. Proof of Theorem 1.2

Finally, we will re-prove the result of Stallings [**3**] by explicitly constructing a fibration on the complement of any homogenous braid closure. From now on, let L be a link in braid position as the closure of a homogeneous braid β on b strands. Let $w = w_1 \cdots w_n$ be a homogenous braid word for β, so $w_1, \ldots, w_n \in \{\sigma_1^{s_1}, \ldots, \sigma_{b-1}^{s_{b-1}}\}$ for some $s_1, \ldots, s_{b-1} \in \{-1, 1\}$, with each $\sigma_i^{s_i}$ appearing at least once in w.

Since L is in braid position we have a decomposition $S^3 = B^3_+ \cup (S^2 \times I) \cup B^3_-$, where L meets $S^2 \times I$ in β and a trivial b-stranded braid, while L meets each of B^3_+ and B^3_- in a trivial b-stranded tangle. The unique minimum of the standard Morse function $h : S^3 \to \mathbb{R}$ is in B^3_- and is also a minimum of L, while the unique maximum is in B^3_+ and is a maximum of L.

Isotope as necessary so that the minima of L are in ascending order according to the projection of β described by w. That is, from bottom to top, the i-th minimum of L lies below the i-th endpoint of $\beta \cap \partial B^3_-$. Similarly isotope the maxima to be in descending order, so from top to bottom the i-th maximum of L is in the strand above the i-th endpoint of $\beta \cap \partial B^3_+$. (This is illustrated in Figure 10.)

Reparametrize h so that $h^{-1}((-\infty, 0])$ and $h^{-1}([3, \infty))$ are both balls meeting L in a single arc, and so that $h^{-1}(1) = \partial B^3_+$ while $h^{-1}(2) = \partial B^3_+$. (Again, see Figure 10.)

We will build a movie of singular fibrations on $M^3 := S^3 \setminus \nu(L)$ with respect to $h|_M$ whose total function has nonsingular leaves as in Proposition 3.5 and satisfies the conditions of Theorem 1.2, thus proving Theorem 1.2. In Figure 10, we give a schematic of the position of L and which portions of M are fibered in each step of the construction.

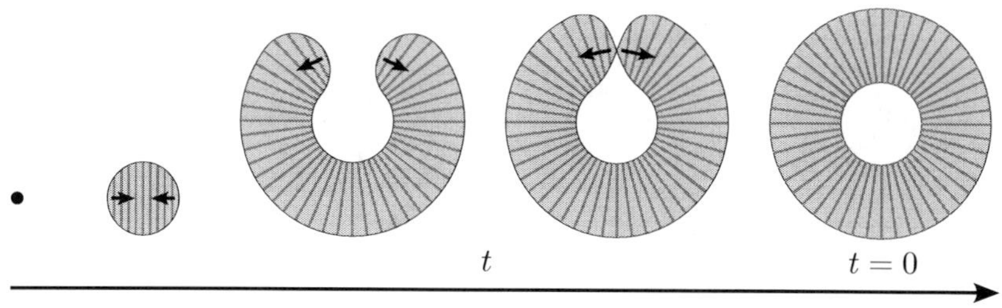

FIGURE 11. The movie f_t for $t \leq 0$. The domain of f_1 is the annulus $h|_M^{-1}(1) \setminus \nu(L)$, i.e. a 2-sphere with neighborhoods of two points in L deleted. Every leaf of f_0 is an arc between these two points (restricted to M). This movie is a concatenation of movies (ii) and (viii) of Figure 5.

5.1. Step 1: fibering $B^3_- \setminus \nu(L)$. Since $h|_M^{-1}((-\infty, 0])$ is a solid torus, it can be fibered by meridional disks. Define f_t for $t \leq 0$ so that the total function from $-\infty < t \leq 0$ is this meridional disk fibration, as in Figure 11. We call the two boundary components of $h|_M^{-1}(0)$ by the names C_1 and C_1'.

Now we extend f_t across all of $B^3_- \setminus \nu(L)$. The reader may find it helpful to begin by consulting Figure 12, in which we show how to adapt a singular fibration of a planar surface (in this case, some $h^{-1}(t)$ slightly below a local minimum of $h|_L$) to a planar surface with two additional boundary components (in this case, $h^{-1}(t')$ slightly above a local minimum of $h|_L$).

Note $L \cap B^3_-$ has b components, which we call $A_1, \ldots, A_b$ in ascending order. The arc A_1 is the only arc that meets $h^{-1}(0)$. From $t = 0$ to $t = \epsilon$, we extend f_t via isotopy, rotating an arc beneath A_2 by 180° as depicted in Figure 12. The sign of this rotation depends on s_1. If $s_1 = 1$ (i.e. σ_1 appears in w), then we take the rotation to be counterlockwise, as in the top row of Figure 12. If $s_1 = -1$ (i.e. σ_1^{-1} appears in w), then we take the rotation to be clockwise, as in the bottom row of Figure 12.

We now extend f_t above the minimum of A_2 via the movie of Example 4.1. This adds two new boundary components to the domain of f_t (for the greatest value of t for which f_t is currently defined), which we call C_2 and C_2'.

Now we can explain why we first performed a 180° rotation. Roughly, we want all boundary components of $h^{-1}(t)$ in the "left" half of the page to correspond to meridians of strands in the braid β, and the boundary components in the "right" half of the page to correspond to vertical strands, as suggested by Figure 10. The 180° rotation causes the "left" boundaries in Figure 12 to have one sign and the "right" boundaries in Figure 12 to have the opposite. The purpose of having the direction of rotation determined by s_1 will be made clear in the next step.

Repeat for A_i, $i = 3, \ldots, b$ (using a rotation of sign s_{i-1}) so that f_t is defined for all $t \leq 1$ and $h_M^{-1}(1)$ has boundary components $C_1, C_1', \ldots, C_b, C_b'$. We illustrate the leaves of f_1 and arrow decorations near the $2(b-1)$ resulting $\times$-singularities in the left half of Figure 13.

In the right half of Figure 13 we draw an alternative view of the leaves of f_1; the first view will be more useful but it might be easier for a reader to see that the second arises from applying $b-1$ instances of the movie of Example 4.1 to f_0. The

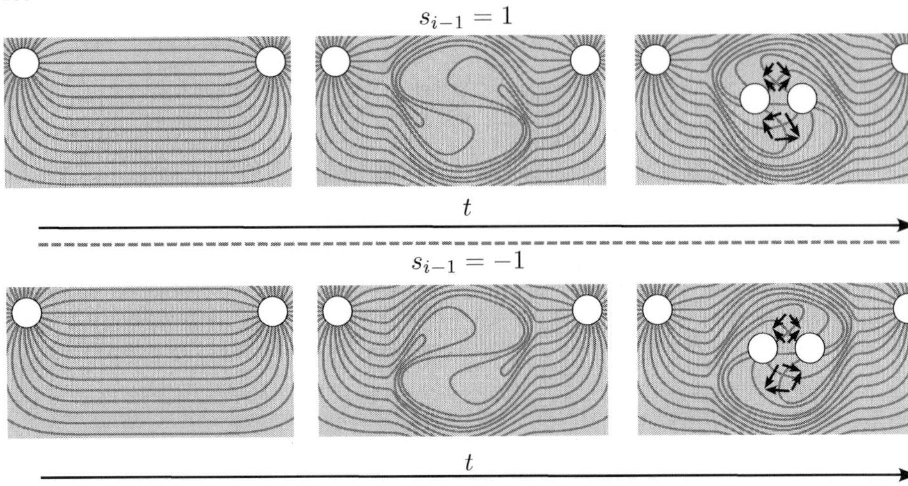

FIGURE 12. An illustration of Step 1. In both rows, we obtain f_t by isotoping f_0 and extending above the minimum of A_i via as band movie (as in Example 4.3). **Top row:** $s_{i-1} = 1$, so we rotate counterclockwise. **Bottom row:** $s_{i-1} = -1$, so we rotate clockwise.

two images of Figure 13 are related by an isotopy that rotates horizontal pairs of boundary components through 180°, exchanging the two boundaries.

5.2. Step 2: fibering $h|_M^{-1}[1,2]$. Let $1 = t_1 < t_2 < \cdots < t_n < t_{n+1} = 2$ so that β meets $h^{-1}[t_i, t_{i+1}]$ in a braid described by the i-th letter w_i of w. Our goal is to extend f_t across each $[t_i, t_{i+1}]$ by viewing the i-th crossing of β as a band move and applying the band movie of Example 4.3.

Say $w_1 = \sigma_j^{s_j}$. In the left of Figure 14, we illustrate an open set of $h_M^{-1}(1)$ that has boundary circles C_{j-1}, C_{j-1}, C_j, C'_j, C_{j+1}, and C'_{j+1}. (If $j = 1$, then ignore the top portion including C_{j-1}, C_{j-1}.) We assume $s_j = 1$; the $s_j = -1$ case is similar (mirror along a horizontal axis and take the included boundary components to be $C_j, C'_j, C_{j+1}, C'_{j+1}, C_{j+2}, C'_{j+2}$). This subspace is easily visible in the left of Figure 13. We give two cases, depending on the sign of s_{j-1}. In Figure 14, we illustrate how to extend f_t across $t \in [1, t_2]$. We first perform a band movie along the illustrated arc η. (One can check that the endpoints of η can be taken to have the same image under f_1, as required.) Then C_j, C_{j+1} exchange positions as t increases, after which we exchange their labels. We see that the contour set of f_{t_2} agrees with f_1 away from a disk containing $C_j, C'_j, C_{j+1}, C'_{j+1}$.

We similarly extend f_t along each interval $t \in [t_i, t_{i+1}]$ unless the letter w_i has already appeared in w. In this case, suppose $w_i = \sigma_j^{s_j}$ and $w_k = \sigma_j^{s_j}$ for some $k < i$. Then we must have performed the above operation in a neighborhood of $C_j, C'_j, C_{j+1}, C'_{j+1}$ when extending f_t along $t \in [t_k, t_{k+1}]$. Therefore, the leaves of f_{t_i} have the local model of Figure 14 (right), again assuming $s_j = 1$ (the case $s_j = -1$ is similar). In Figure 15, we show how to extend f_t to $t \in [t_i, t_{i+1}]$. We perform the indicated band move and obtain a singular fibration that agrees with the middle stage of Figure 14, and then proceed as in Figure 14.

Thus, we may inductively extend f_t over $t \in [t_i, t_{i+1}]$ for increasing i. If w_i is the first instance of $\sigma_j^{s_j}$ in w, then we use the movie of Figure 14 (mirrored along

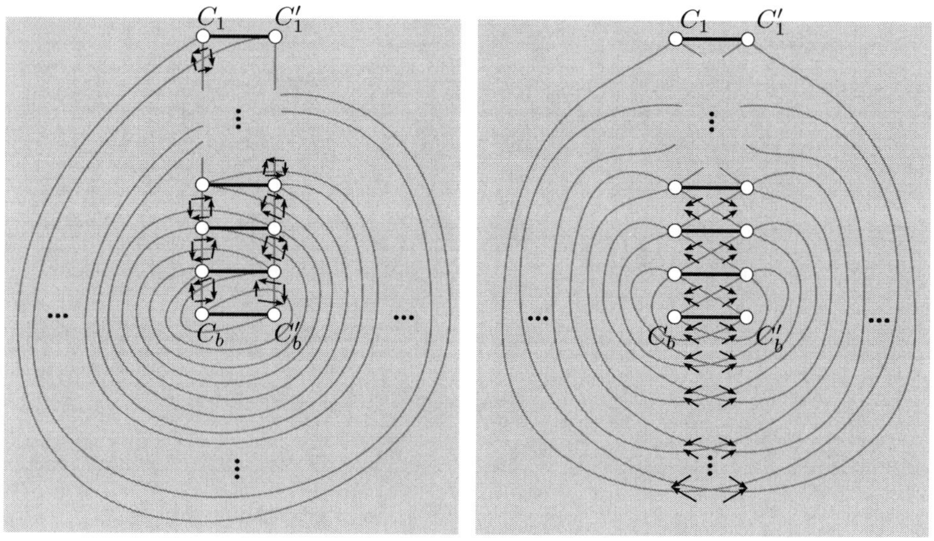

FIGURE 13. **Left:** A contour map of f_1 as obtained in Step 1. From top to bottom, the horizontal pairs of boundary components are near $A_1, \ldots, A_b$. In this figure, we have assumed $s_1 = 1$, $s_{b-4} = 1$, $s_{b-3} = -1$, $s_{b-2} = -1$, $s_{b-1} = 1$. The left column of boundary components from top to bottom are $C_1, \ldots, C_b$, while the right column of boundary components are from top to bottom $C_1', \ldots, C_b'$. In bold, we draw the projections of the arcs $L \cap B^3_-$ to $h|_M^{-1}(1)$. Each of these arcs is in a level set of f_1. **Right:** Another contour map obtained by isotoping the lefthand diagram. This isotopy rotates each pair C_i, C_i' $(i = 2, \ldots, b)$ through 180° in the direction determined by s_i, exchanging the two boundary components.

a horizontal axis if $s_j = -1$). If w_i is not the first instance of $\sigma_j^{s_j}$, then we use the movie illustrated in Figure 15 (again mirrored if $s_j = -1$).

Because every $\sigma_j^{s_j}$ appears at least once in w, the singular fibration f_2 is as in Figure 16.

5.3. Step 3: fibering $B^3_+ \setminus \nu(L)$. We have so far extended f_t to a movie of singular fibrations on M for $t \leq 2$.

While f_1 and f_2 (Figures 13 and 16) seem similar, there is a key difference between these two singular fibrations. Projections of the arcs $L \cap B^3_-$ to $h|_M^{-1}(1)$ form a level set of f_1 that lies toward *outward* regions of the $\times$-singularities, as in the movie of Example 4.1. However, projections of arcs in $L \cap B^3_+$ to $h|_M^{-1}(2)$ form a level set of f_2 that lies toward *inward* regions of the $\times$-singularities, as in the movie of Example 4.2. We may thus extend f_t across $t \in [2, 3]$ by applying $b-1$ instances of the movie of Example 4.2. (Note this is exactly the same as our procedure to extend f_t along $t \in [0, 1]$ from Step 1, turned upside down, so we abbreviate the discussion.) Then f_3 is a fibration of an annulus over S^1. We extend f_t across $t \in [3, \infty)$ via the movie of Figure 11 used in Step 1, turned upside down with

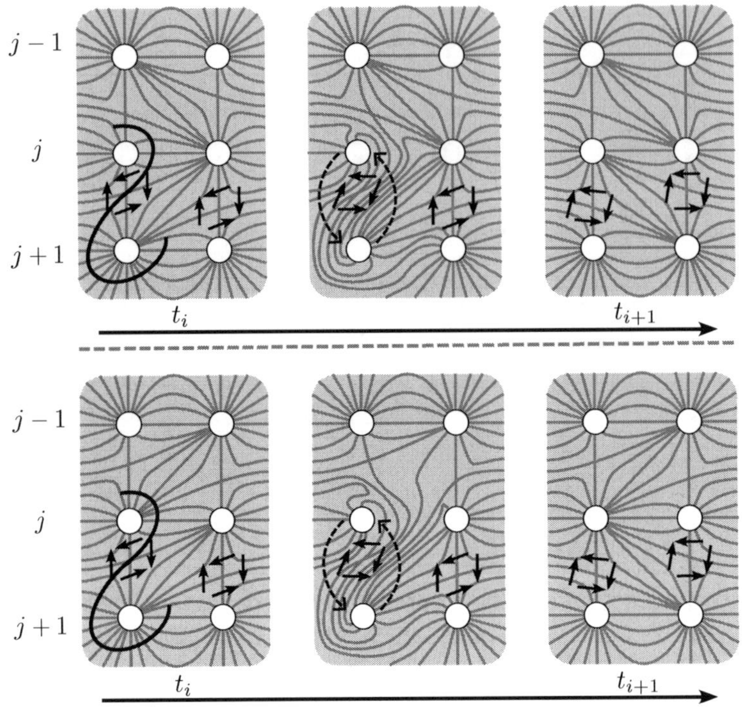

FIGURE 14. Here we assume $s_j = 1$. We give two possible local models depending on the behavior of f_{t_i} in a box containing $C_{j-1}, C'_{j-1}, C_j, C'_j$. This depends on the sign of s_{j-1} and whether we have already extended f_t across some interval corresponding to $\sigma_{j-1}^{s_{j-1}}$. **Left:** An open subset of $h_M^{-1}(t_i)$. We include an arc η satisfying the conditions of the band movie. **Middle:** We perform a band movie and indicate with dashed arrows the path of C_j, C_{j+1} as we move above the crossing of β corresponding to w_i. **Right:** An open subset of $h_M^{-1}(t_{i+1})$.

respect to t. This concludes the construction of f_t and hence the proof of Theorem 1.2.

6. Example of a fibration

In Figures 17 and 18, we perform the above algorithm to construct a fibration $F : S^3 \setminus (\text{right-handed trefoil}) \to S^1$. In Figure 17 we draw the contour set of each $f_t := F|_{h^{-1}(t)}$. In Figure 18 we highlight $F^{-1}(\theta)$ for three values of θ. Each of these leaves can be easily seen to be the standard braid surface (as expected). Note that, as stated in Theorem 1.2, h restricts to the interior of each $F^{-1}(\theta)$ as a Morse function with no local minima or maxima.

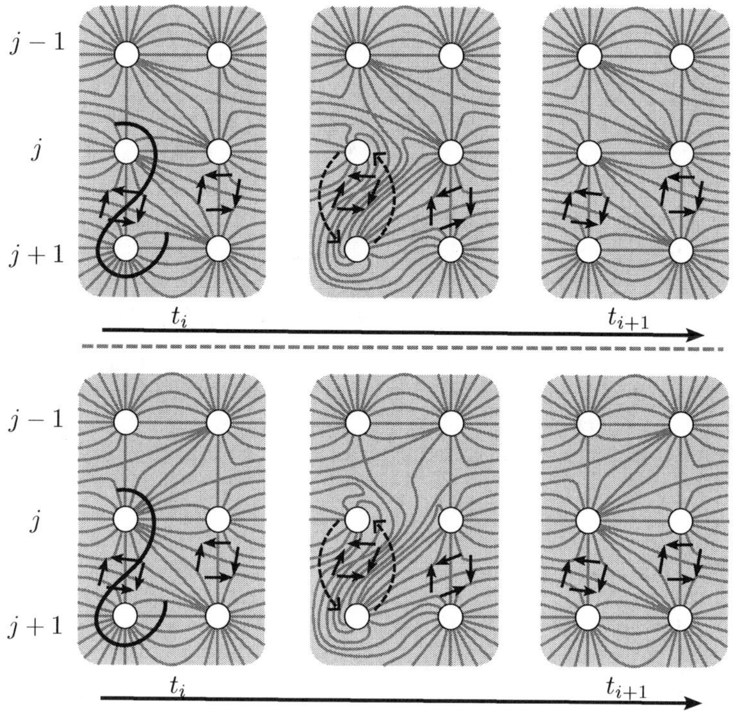

FIGURE 15. Here, $s_j = 1$ and $w_i = \sigma_j^{s_j}$ is not the first instance of $\sigma_j^{s_j}$ in w. We give two cases (top row and bottom row), depending on the behavior of f_{t_i} in a disk containing $C_{j-1}, C'_{j-1}, C_j, C'_j$. (That is, depending on the sign of s_{j-1} and whether $\sigma^{s_{j-1}}$ appears in the first $i-1$ letters of w.) **Left:** a local model of f_{t_i}. We indicate an arc along which we may perform a band movie. **Middle:** We perform the band movie. Note that this singular fibration agrees with that of the middle of 14. **Right:** An open subset of $h_M^{-1}(t_{i+1})$.

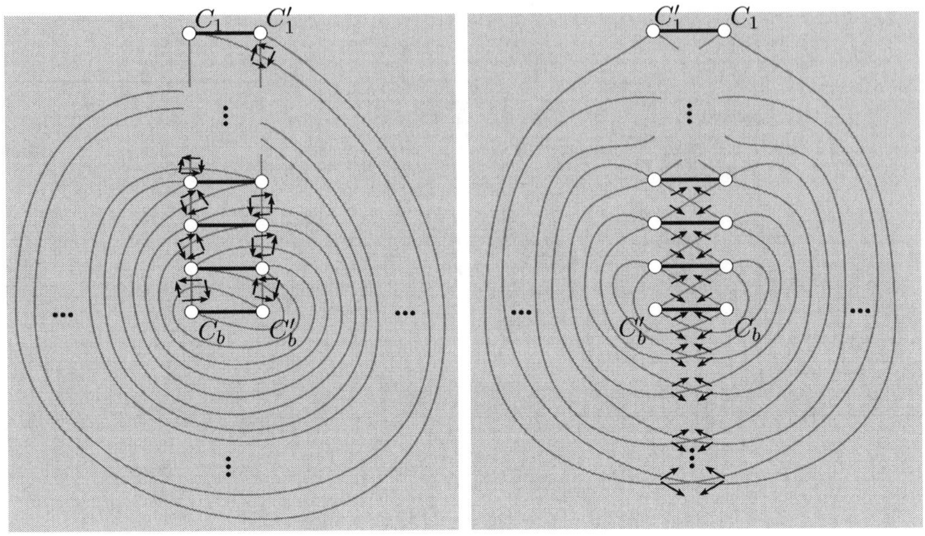

FIGURE 16. **Left:** A contour map of f_2 as obtained in Step 2. From top to bottom, the horizontal pairs of boundary components are $C_1, C_1'; \ldots, C_b, C_b'$. As in Figure 13, in this figure we have assumed $s_1 = 1$, $s_{b-4} = 1$, $s_{b-3} = -1$, $s_{b-2} = -1$, $s_{b-1} = 1$. In bold, we draw the projections of the arcs $L \cap B^3_+$ to $h|_M^{-1}(2)$. **Right:** Another contour map obtained by isotoping the lefthand diagram, included only for visualization purposes. This isotopy rotates each pair C_i, C_i' ($i = 2, \ldots, b$) through $180°$ in the direction determined by s_i, exchanging the two boundary components.

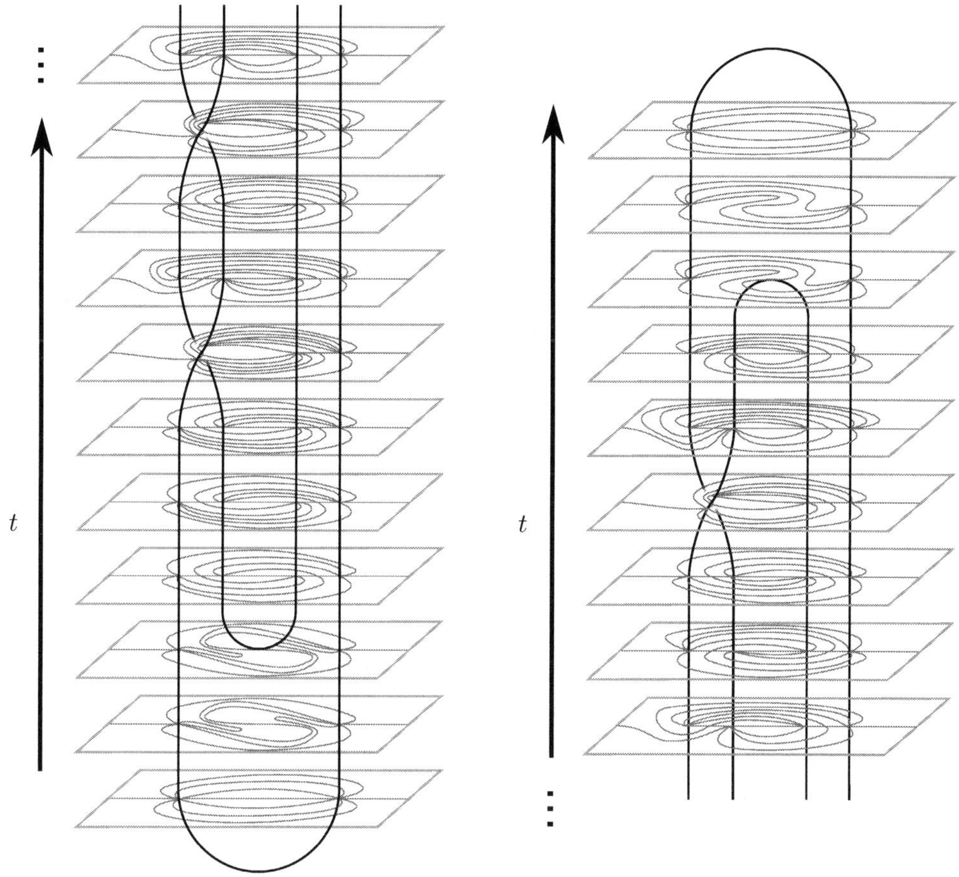

FIGURE 17. A contour set for f_t for increasing t, where f_t is a movie of singular fibrations on the trefoil complement with total function F a fibration. This movie has been constructed via the algorithm of Theorem 1.2.

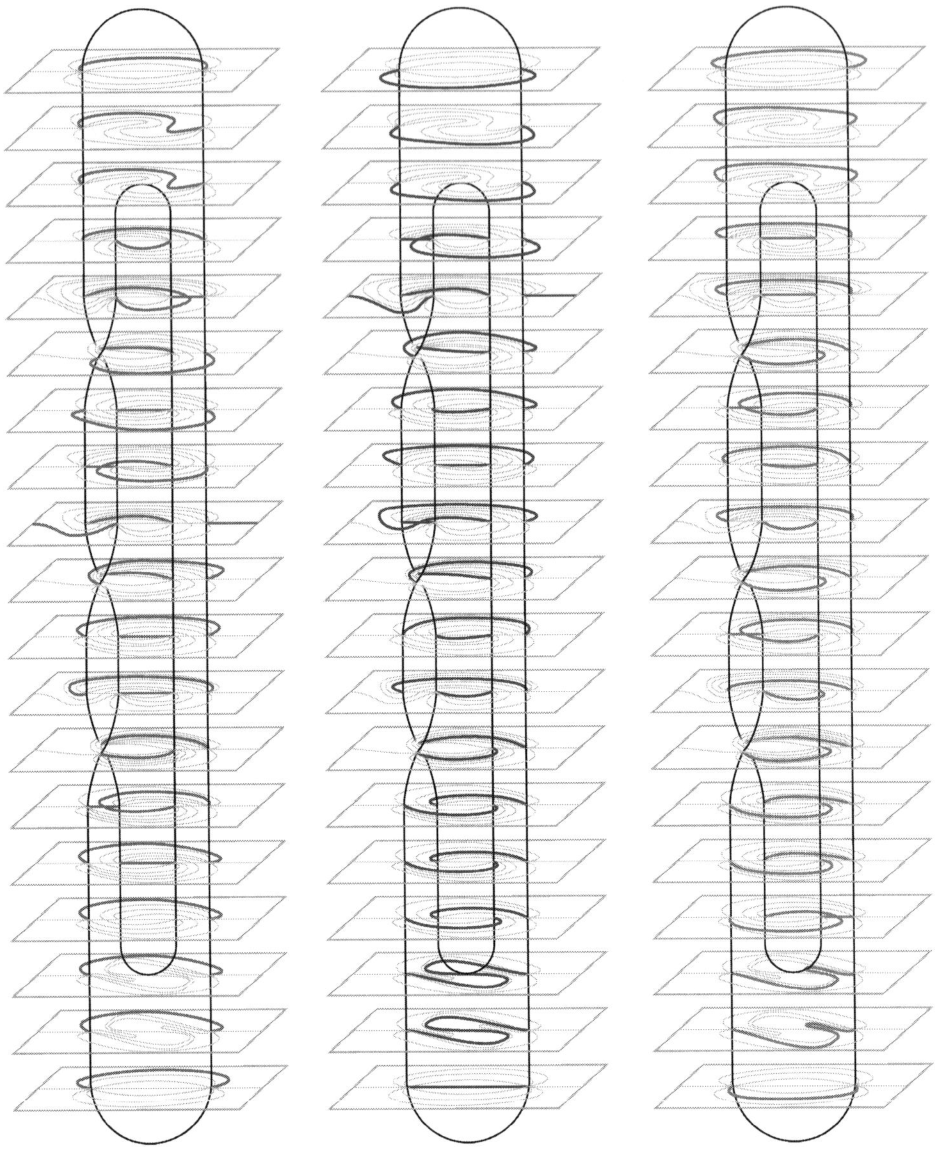

FIGURE 18. Three fibers $F^{-1}(\theta)$ for the trefoil, constructed as in Theorem 1.2.

References

[1] Maggie Miller, *Extending fibrations of knot complements to ribbon disk complements*, Geom. Topol. **25** (2021), no. 3, 1479–1550, DOI 10.2140/gt.2021.25.1479. MR4268166

[2] Yi Ni, *Knot Floer homology detects fibred knots*, Invent. Math. **170** (2007), no. 3, 577–608, DOI 10.1007/s00222-007-0075-9. MR2357503

[3] John R. Stallings, *Constructions of fibred knots and links*, Algebraic and geometric topology (Proc. Sympos. Pure Math., Stanford Univ., Stanford, Calif., 1976), Proc. Sympos. Pure Math., XXXII, Amer. Math. Soc., Providence, RI, 1978, pp. 55–60. MR520522

Department of Mathematics, Stanford University, Stanford, California 94301
Email address: maggie.miller.math@gmail.com

Proceedings of Symposia in Pure Mathematics
Volume **109**, 2024
https://doi.org/10.1090/pspum/109/01997

On the minimal genus problem in four-manifolds

András I. Stipsicz and Zoltán Szabó

ABSTRACT. We study the minimal genus problem for some smooth four-manifolds.

1. Introduction

For a closed, oriented, smooth four-manifold X the *genus function* (or minimal genus function) $g_X \colon H_2(X;\mathbb{Z}) \to \mathbb{N}$ assigns to each homology class the minimal genus of an oriented smooth surface representative of the class; that is, if $\alpha \in H_2(X;\mathbb{Z})$ then

$$g_X(\alpha) = \min\{g(\Sigma) \mid i\colon \Sigma \hookrightarrow X,\ i_*([\Sigma]) = \alpha\},$$

where Σ is an oriented, closed, connected surface of genus $g(\Sigma)$ and i is a C^∞ embedding. It is a standard fact, that every second homology class in a four-manifold can be represented by an embedded, oriented, closed, connected surface, hence the above minimum provides a finite value.

For some examples of exotic pairs of smooth four-manifolds the genus functions show very different behavior, so it is expected that this function codes important geometric information about the smooth structure of the four-manifold. In order to formalize this notion we have the following definition:

DEFINITION 1.1. *Let X and Y be smooth, closed, oriented four-manifolds with $H_1(X;\mathbb{Z}) = H_1(Y;\mathbb{Z}) = 0$. We say that X and Y have* **equivalent minimal genus functions** *if there exists an isomorphism $f\colon H_2(X;\mathbb{Z}) \longrightarrow H_2(Y;\mathbb{Z})$ with the properties that*

- *f preserves the intersection form:*
$$Q_X(a,b) = Q_Y(f(a), f(b))$$
for $a, b \in H_2(X;\mathbb{Z})$, and
- *f also preserves the minimal genus function:*
$$g_X(a) = g_Y(f(a))$$
for all $a \in H_2(X;\mathbb{Z})$.

A natural question regarding the above function is the following:

The first author was partially supported by the *Élvonal (Frontier) grant* KKP126683 of the NKFIH.The second author was partially supported by NSF grant DMS-1904628.

QUESTION 1.2. *Let X and Y be smooth, closed, oriented and simply connected four-manifolds and assume that X and Y have equivalent minimal genus functions. Is there an orientation preserving diffeomorphism between X and Y?*

REMARK 1.3. *Note that the above question only asks about the existence of an orientation preserving diffeomorphism $G\colon X \to Y$, and in particular does not require that the induced action $G_*\colon H_2(X) \to H_2(Y)$ agrees with a fixed f. Indeed, for $X = Y = K3$ the map $f\colon H_2(K3) \to H_2(K3)$ given by $f(a) = -a$ preserves both the intersection form and the minimal genus function, but (by work of Donaldson, see* [**5**, Corollary 9.1.4]*), f cannot be induced by a diffeomorphism.*

Despite its importance, very little is known in general about g_X. There are only a handful of simply connected examples where the entire genus function is known. When $X = S^4$ (or more generally, X is a homotopy 4-sphere) the genus function is trivial; in fact, the positive answer for Question 1.2 in this special case would imply the Smooth 4-dimensional Poincare Conjecture (S4PC).

For $\mathbb{CP}^2$ (and similarly, for $\overline{\mathbb{CP}}^2$) the resolution of the Thom conjecture by Kronheimer and Mrowka [**13**] provides the answer: suppose that $h \in H_2(\mathbb{CP}^2;\mathbb{Z}) \cong \mathbb{Z}$ is a generator and $d \neq 0$, then

$$g_{\mathbb{CP}^2}(dh) = \frac{1}{2}(|d|-1)(|d|-2).$$

Along similar lines, Ruberman gave a description of the minimal genus function for $\mathbb{CP}^2\#\overline{\mathbb{CP}}^2$ and for $S^2 \times S^2$ (the two S^2-bundles over S^2) in [**20**]: if $(a,b) \in H_2(\mathbb{CP}^2\#\overline{\mathbb{CP}}^2;\mathbb{Z}) \cong \mathbb{Z}^2$ (where the $\mathbb{Z}$-factors are generated by the projective lines in the factors) is a nontrivial homology class with $|a| > |b|$, then

$$g_{\mathbb{CP}^2\#\overline{\mathbb{CP}}^2}(a,b) = \frac{1}{2}(|a|-1)(|a|-2) - \frac{1}{2}|b|(|b|-1);$$

if $|a| < |b|$ then the same formula applies with the roles of a and b reversed, and $g_{\mathbb{CP}^2\#\overline{\mathbb{CP}}^2}(a,\pm a) = 0$. In a similar manner, for $(u,v) \in H_2(S^2 \times S^2;\mathbb{Z}) \cong \mathbb{Z}^2$ (where the $\mathbb{Z}$ factors are generated by the S^2-factors in $S^2 \times S^2$) we have

$$g_{S^2\times S^2}(u,v) = (|u|-1)(|v|-1)$$

if $uv \neq 0$ and $g_{S^2\times S^2}(u,v) = 0$ if $uv = 0$.

Although these four-manifolds seem to be the simplest ones (definitely with smallest Euler characteristics among simply connected closed four-manifolds), it is not known if they support distinct smooth structures. Application of a suitable sequence of Luttinger surgeries provide (not necessarily simply connected) smooth four-manifolds X_n ($n \in \mathbb{N}$) with homology isomorphic to $H_*(S^2 \times S^2;\mathbb{Z})$, first constructed in [**7**] by Fintushel, Park and Stern (see also [**2**]). In addition, the construction can be carried out so that X_1 carries a symplectic structure. We will review the construction of these manifolds in Section 2 and prove the following:

THEOREM 1.4. *Suppose that X_n is one of the homology $S^2 \times S^2$'s constructed in* [**7**] *(see also Section 2). Using the isomorphism $H_2(X_n;\mathbb{Z}) = \mathbb{Z}^2$, where the two generators x and y satisfy $Q_{X_n}(x,x) = Q_{X_n}(y,y) = 0$, $Q_{X_n}(x,y) = 1$, the minimal genus on the non-zero homology class $\alpha = ax + by$ is given by*

$$g_{X_n}(\alpha) = (|a|+1)(|b|+1).$$

In particular, for all $n, k \in \mathbb{N}$ the manifolds X_n and X_k have equivalent minimal genus functions.

Notice that (as Proposition 2.2 will show), the Seiberg-Witten values do distinguish these four-manifolds; it is still an open problem whether this family contains simply connected examples. It is also open if there is a pair $n, k \in \mathbb{N}$ so that X_n and X_k are homeomorphic — it might happen that their fundamental groups distinguish them.

Regarding Question 1.2, as partial evidence for the opposite answer, we show that the minimal genus function does distinguish (even infinite collections of) exotic smooth four-manifolds. For such example we will rely on two constructions on elliptic surfaces. First, recall that for given $n \geq 1$, $p, q > 0$, $(p, q) = 1$, the manifolds $E(n)_{p,q}$ are constructed from a simply connected genus 1 Lefschetz fibration $E(n)$ over the sphere with Euler characteristics $\chi(E(n)) = 12n$ using logarithmic transformations of multiplicity p and q. (For more on elliptic surfaces see [**11**, Chapter 3].) For $q = 1$ the manifold $E(n)_{p,q}$ will be denoted by $E(n)_p$.

A further construction of new four-manifolds out of old one is knot surgery [**8**]: suppose that X contains an embedded torus T with trivial normal bundle and with simply connected complement. For a knot $K \subset S^3$ replace the tubular neighbourhood $\nu(T) = T^2 \times D^2$ of T with $(S^3 \setminus \nu(K)) \times S^1$ (using an appropriate gluing map along the T^3 boundaries, where the meridian of the torus T will be matched with the longitude of the knot $K \subset \partial((S^3 \setminus \nu(K)) \times S^1))$. We denote the result by X_K. For an elliptic surface $E(n)$ we can choose T to be the generic fiber of the fibration; the result of the above construction then will be denoted by $E(n)_K$. We will prove the following

THEOREM 1.5. *For every $n \geq 2$ there are infinitely many smooth four-manifolds all homeomorphic to $E(n)$, distinguished by their genus function. Indeed, if K_m denotes the $(2, 2m+1)$ torus knot $T_{2,2m+1}$, then $\{E(n)_{K_m} | m \in \mathbb{N}\}$ have all distinct genus functions.*

In the closing sections of the paper we investigate the minimal genus problem for some simply connected four-manifolds with small Euler characteristics. These include smooth four-manifolds that are homeomorphic to $\mathbb{CP}^2 \# 2\overline{\mathbb{CP}}^2$ or $\mathbb{CP}^2 \# 3\overline{\mathbb{CP}}^2$, constructed by Akhmedov and Park [**1**] and by Baldridge and Kirk [**4**], respectively. For surveys on the genus function, see also [**6, 17**].

1.1. Tools used in the proofs. In determining the value of the genus function on a homology class, the Symplectic Thom Conjecture (a theorem proved in [**19**]) turns out to be rather convenient.

THEOREM 1.6. *(Ozsváth-Szabó, [**19**]) Suppose that (X, ω) is a closed symplectic four-manifold and $\Sigma \subset X$ is a connected, closed symplectic surface of genus $g(\Sigma)$. Then $g_X([\Sigma]) = g(\Sigma)$, i.e. symplectic surfaces minimize genus in their homology classes.*

The following inequality — called the *adjunction inequality* — does not determine the genus function on a specific homology class, but provides bounds and is applicable in a greater generality. Recall that the Seiberg-Witten function of a smooth, closed, oriented, simply connected four-manifold X is a map $SW_X \colon \mathrm{Spin}^c(X) \to \mathbb{Z}$ which counts solutions of a certain geometric partial differential equation. (Here $\mathrm{Spin}^c(X)$ denoted the set of spin^c structures on X.) For manifolds with $b_2^+ > 1$ (or $b_2^+ = 1$ and $b_2^- \leq 9$ and with the addition that the equations are considered with small perturbation terms) this map is a smooth invariant. SW_X has finite support, and $\mathbf{s} \in \mathrm{Spin}^c(X)$ with $SW_X(\mathbf{s}) \neq 0$ is called a **Seiberg-Witten basic class**. The four-manifold X is of Seiberg-Witten simple type if for a Seiberg-Witten basic class $\mathbf{s}$ we have that $c_1^2(\mathbf{s}) = 3\sigma(X) + 2\chi(X)$ (where $\sigma(X)$ and $\chi(X)$ are the signature and the Euler characteristic of the underlying four-manifold).

THEOREM 1.7. *(Adjunction inequality,* [**19**, Corollary 1.7]*) Suppose that X is a smooth four-manifold of Seiberg-Witten simple type with $b_1(X) = 0$ and either with $b_2^+(X) > 1$ or with $b_2^+(X) = 1$ and $b_2^-(X) \leq 9$. Then for an embedded surface $\Sigma \subset X$ with genus $g(\Sigma) > 0$ and a Seiberg-Witten basic class $\mathbf{s} \in \mathrm{Spin}^c(X)$ (i.e. with $SW_X(\mathbf{s}) \neq 0$) we have*

$$|\langle c_1(\mathbf{s}), [\Sigma]\rangle| + [\Sigma]^2 \leq 2g(\Sigma) - 2. \tag{1.1}$$

In our arguments we will frequently use the fact that for a self-diffeomorphism $f\colon X \to X$ we have $g_X(\alpha) = g_X(f_*(\alpha))$ for all $\alpha \in H_2(X;\mathbb{Z})$. Furthermore, if $S \subset X$ is an embedded sphere of self-intersection (± 1) or (± 2) in X, then it is easy to construct a self-diffeomorphism $f\colon X \to X$, for which the induced map f_* is just reflection in the class represented by S. Consequently, if $S \subset X$ is an embedded sphere with self-intersection (± 1) or (± 2), then the map

$$\alpha \mapsto \alpha - 2\frac{[S]\cdot\alpha}{[S]\cdot[S]}[S]$$

on $H_2(X;\mathbb{Z})$ is induced by a diffeomorphism.

The paper is organized as follows. In Section 2 we examine the homology $S^2 \times S^2$'s and verify Theorem 1.4. In Section 3 we study the genus function of elliptic surfaces and of the manifolds we get by logarithmic transformations or by knot surgeries on them. We also prove Theorem 1.5. Section 4 is devoted to the partial understanding of the genus function on the twice and three-times blown up complex projective plane. The exotic manifolds AP and BK (and the further infinite families of exotic smooth structures on $\mathbb{CP}^2\#2\overline{\mathbb{CP}}^2$ and on $\mathbb{CP}^2\#3\overline{\mathbb{CP}}^2$), and their genus functions are studied in Sections 5 and 6.

2. Homology $S^2 \times S^2$'s

The application of torus (and in particular, Luttinger) surgery, or some variation of it, leads to the construction of four-manifolds which have the same homology as $S^2 \times S^2$, but potentially different fundamental group. Such examples were constructed in [**7**] (see also [**2**]).

First we briefly recall the concept of surgery along a torus in a four-manifold from [**7**]. Assume that X is a smooth four-manifold and $T \subset X$ is an oriented torus with trivial normal bundle (that is, $[T]^2 = 0$). Fix a framing f on T, i.e., a trivialization of its normal bundle. For an oriented simple closed curve $\gamma \subset T$ therefore there is a unique push-off $\overline{\gamma}_f$ of γ to the boundary T^3 of a small tubular neighbourhood of T. For a rational number $\frac{p}{q}$, the manifold $(X, T, f, \gamma, \frac{p}{q})$ is defined by deleting a small tubular neighbourhood of T, and gluing back $T^2 \times D^2$ (which can be viewed as the union of a four-dimensional 2-handle, two 3-handles and a 4-handle) by attaching the 2-handle along a simple closed curve representing the homology class $p[\mu_T] \pm q[\overline{\gamma}_f]$ (where μ_T is a meridional curve to T, regarded as a subset of the boundary of the trivial neighbourhood, oriented so that it has linking number 1 with T).

In the special case, when (X, ω) is a symplectic four-manifold and $T \subset (X, \omega)$ is a Lagrangian torus (that is, $\omega|_T = 0$), the torus T comes with a canonical framing. Indeed, by the Lagrangian neighbourhood theorem an appropriate small normal neighbourhood of T is symplectomorphic to a neighbourhood of the zero-section of the cotangent bundle $T^*T^2 \to T^2$ of the 2-torus T^2 (when the total space of this bundle is equipped with the tautological symplectic form coming from its Liouville 1-form). As T^*T^2 is trivialized, this identification equips $T \subset X$ with its *Lagrangian framing.* (Pushing T along this framing we get further Lagrangian tori in (X, ω).) As is explained in [**3**], using this framing and applying $\frac{1}{k}$ surgery along T (for any choice of γ) the resulting four-manifold $(X, T, \gamma, \frac{1}{k})$ will admit a symplectic structure which is equal to ω outside of a suitably chosen neighbourhood of T. This symplectic construction is frequently referred to as *Luttinger surgery* [**3, 14**].

2.1. The construction. For the construction of the homology $S^2 \times S^2$'s in [**7**, Section 4], consider two genus 2 surfaces Σ_2^1, Σ_2^2, and equip their first integral homologies with the symplectic bases a_1, b_1, a_2, b_2 and c_1, d_1, c_2, d_2, respectively. Represent these homology classes by oriented simple closed curves (denoted by the same letters). Equip the direct product $\Sigma_2^1 \times \Sigma_2^2$ with the direct product symplectic form. Notice that in this symplectic four-manifold the product of an oriented simple closed curve γ_1 in Σ_2^1 and γ_2 in Σ_2^2 provides an oriented Lagrangian torus $T_{\gamma_1,\gamma_2} \subset X$, hence using the chosen bases we see 16 tori which are in 8 hyperbolic pairs. In addition, by considering small disjoint displacements of the curves in Σ_2^i, tori with algebraic intersection number 0 can be chosen to be disjoint. Apply now the following eight Luttinger surgeries, all with the choice $k = \pm 1$:

- $(T_{a_1,c_1}, a_1), (T_{b_1,c_1}, b_1), (T_{a_2,c_2}, a_2), (T_{b_2,c_2}, b_2)$,
- $(T_{a_2,c_1}, c_1), (T_{a_2,d_1}, d_1), (T_{a_1,c_2}, c_2), (T_{a_1,d_2}, d_2)$;

the first four with coefficient $k = -1$ and the last four with $k = 1$, resulting in the four-manifold X. As each torus comes with a dual torus, the meridional curves for each are homologically trivial. Therefore after the Luttinger surgery the chosen curve will represent the trivial element in first homology, hence $H_1(X; \mathbb{Z}) = 0$. By [**3**, Proposition 2.2] X will admit a symplectic structure so that the surfaces $\Sigma_2^1 \times \{p\}, \{q\} \times \Sigma_2^2$ are symplectic submanifolds. The intersection form (given in the basis $\Sigma_2^1 \times \{p\}, \{q\} \times \Sigma_2^2$) is a hyperbolic pair, hence the manifold is spin and its cohomology ring is isomorphic to $H^*(S^2 \times S^2; \mathbb{Z})$. By changing the last Luttinger surgery (on (T_{a_1,d_2}, d_2)) to a torus surgery with coefficient n we get a four-manifold X_n, which for $n > 1$ does not carry a natural symplectic structure.

We claim that the four-manifold $-X$ (the manifold X constructed above, with its orientation reversed) also admits a symplectic structure. Indeed, consider the map $f\colon \Sigma_2^1 \times \Sigma_2^2 \to \Sigma_2^1 \times \Sigma_2^2$ which is given by the identity on the first factor and by the orientation reversing diffeomorphism on the second factor given by the reflection which leaves c_1, c_2 fixed, and sends d_1 and d_2 to $-d_1$ and $-d_2$. The map f is an orientation reversing self-diffeomorphism of $\Sigma_2^1 \times \Sigma_2^2$, and all tori in X we used in the Luttinger surgeries above are invariant under f: the tori not involving d_1 and d_2 are fixed pointwise, while the tori T_{a_2,d_1} and T_{a_1,d_2} stay the same as subsets. Let W denote the complement of the tubular neighborhood of the eight tori in X along which we perform Luttinger surgeries, and note that the above map f restricted to W gives a map form W to $-W$. Let δ_i for $1 \leq i \leq 8$ denote the curves in the boundary of W that define the Luttinger surgeries on X. Then using surgeries along $f(\delta_i)$ in $-W$ will give $-X$. Since the map f carries the meridians of the tori to meridians (with potentially reversing their orientation) it follows that the surgeries on $-W$ are still Luttinger surgeries, and f extends to those surgeries (as an orientation reversing map). In particular $-X$ can be presented as a sequence of Luttinger surgeries along Lagrangian tori in $\Sigma_2^1 \times \Sigma_2^2$, therefore it is also symplectic.

REMARK 2.1. *Note that the above argument shows that the manifold X admits a symplectic structure with both orientations, yet it is not diffeomorphic to any surface bundle. Such examples were already known to exist.*

In the same manner, it follows that $-X_n$ is given from the product of two genus-2 surfaces by a sequence of seven Luttinger surgeries and a torus surgery with some integral coefficient, similarly to the construction of X_n.

The property of admitting symplectic structure with both orientation allows us to compute the genus function g_X of X: note that the Luttinger surgeries are in the complement of the two genus 2 surfaces $h_1 = \Sigma_2^1 \times \{p\}$ and $h_2 = \{q\} \times \Sigma_2^2$ of $\Sigma_2^1 \times \Sigma_2^2$ (for suitably chosen $p \in \Sigma_2^2$ and $q \in \Sigma_2^1$), hence provide a basis of $H_2(X;\mathbb{Z})$ with symplectic representatives.

PROOF OF THEOREM 1.4 FOR $X = X_1$. The two generators h_1, h_2 can be represented by symplectic genus 2 surfaces. If for $\alpha = ah_1 + bh_2$ both a and b are positive, then by smoothing the intersection points of push-offs, we get a symplectic surface of the claimed genus. If both a and b are negative, we can take the negative of the homology class. If one coefficient is positive and the other is negative, then by reversing the orientation on the four-manifold we are in one of the previous cases. (Notice that here we use the fact that $-X$ is constructed in a similar fashion; in particular, it is symplectic.) Finally if one of the coefficients is zero, we represent the homology class near the other surface with an appropriate cover, and choosing it within a small neighbourhood, the cover can be assumed to be symplectic. Then simple Euler characteristic computation verifies the claimed formula. □

By determining the Seiberg-Witten basic classes, the same result can be shown to hold for all X_n.

PROPOSITION 2.2. *The smooth four-manifold X_n admits two Seiberg-Witten basic classes, and so does $-X_n$ obtained by reversing the orientation on X_n.*

PROOF. The argument uses repeatedly a surgery formula for the Seiberg-Witten invariants, see [**16**, **22**].

For the version we need, we fix a smooth closed oriented compact four-manifold M with three-torus boundary $\partial M = T^3$, and a spin^c structure $\mathbf{s}$ whose first Chern class restricts trivially to ∂M. Fix two homology classes $a, b \in H_1(T^3)$ given by $a = [S^1 \times pt \times pt]$ and $b = [pt \times S^1 \times pt]$. Given a pair of integers p, q, where $(p,q) \in \mathbb{Z} \times \mathbb{Z}$ is a primitive element, let $M_{p,q}$ denote the effect of torus surgery, where we glue M and $D^2 \times T^2$ along ϕ to get $M_{p,q} = M \cup_\phi D^2 \times T^2$, so that in homology ϕ_* maps $[\partial(D^2)]$ to $pa + qb$. Let $S(p,q)$ denote the set of spin^c structures on $M_{p,q}$ whose restriction to M agrees with $\mathbf{s}$ and define $F(p,q)$ as the sum of the Seiberg-Witten invariants of classes in $S(p,q)$. Then, according to [**16**]

$$F(p,q) = p \cdot F(1,0) + q \cdot F(0,1). \tag{2.1}$$

The manifolds X_n in the theorem are all constructed from $\Sigma_2 \times \Sigma_2$ by using torus surgery eight times. Let s_0 denote the spin^c structure on $\Sigma_2 \times \Sigma_2$, with $c_1(s_0)$ evaluating as 2 on both $[\Sigma_2 \times pt]$ and $[pt \times \Sigma_2]$. Note that the Seiberg-Witten invariant of s_0 is ± 1. Now we use Formula (2.1) inductively. During these surgeries we use a to be the meridional circle. So by choosing b appropriately we can assume the effect of the torus surgery is given by $M(p,1)$, where $p = \pm 1$ in the first seven steps and $p = \pm n$ in the last step. The crucial observation is that in all these steps the manifolds $M(0,1)$ contain an embedded torus with self-intersection 0 with the property that the first Chern class of s evaluates as 2 on it. It follows from the adjunction inequality that $F(0,1) = 0$, so

$$F(p,1) = p \cdot F(1,0).$$

Since all the 8 surgeries are along tori that represent primitive second homology classes that are killed under the surgery, it follows that there is a unique spin^c structure in $S(p,1)$. This means that if we look at the nine different 4-manifolds Y_i, $i = 0,..,8$ starting with $\Sigma_2 \times \Sigma_2$ and ending in X_n, there is always a unique spin^c structure s_i that corresponds to s_0. During the inductive step we see that $Y_i = M(1,0)$ and $Y_{i+1} = M(p,1)$. It follows that the Seiberg-Witten invariant of s_8 in X_n is equal to $\pm n$.

We proved that s_8 and $-s_8$ are two non-trivial Seiberg-Witten classes on X_n. Since the surgery tori are all disjoint from $pt \times \Sigma_2$ and $\Sigma_2 \times pt$, it follows from the adjunction inequality that there are no other Seiberg-Witten basic classes. □

PROOF OF THEOREM 1.4 FOR X_n. The argument of the proof for $X = X_1$ applies almost verbatim: we consider the same surfaces constructed in that proof, and apply the adjunction inequality of Theorem 1.7 (providing lower bounds for values of the genus function) to verify that the constructions indeed provide surfaces minimizing genus. As the Seiberg-Witten calculations apply for X_n and $-X_n$ as well, minimality of the genus of the constructed representatives follows from the respective adjunction formulae. □

3. Elliptic surfaces

There are various ways to get invariants for an oriented, closed, connected four-dimensional manifold X from g_X which are easier to compare than the genus function itself. Below we consider one such example. Let Q_X denote the intersection form of X and $k = b_2^+(X)$.

DEFINITION 3.1. *A k element sequence $\mathbf{y} = (y_1, \ldots, y_k)$ with $y_i \in H_2(X, \mathbf{Z})$ is called an* **allowed sequence** *if the span of $y_1, \ldots, y_k$ is a positive definite k-dimensional subspace of $H_2(X, \mathbf{Z})$, and for all $i \neq j$ we have $Q_X(y_i, y_j) = 0$ and $Q(y_i, y_i)$ is even.*

For an allowed sequence we associate a tuple of k integers

$$Pr(X, \mathbf{y}) = (a_1, \ldots, a_k)$$

where $a_i = g_X(y_i)$.

DEFINITION 3.2. *The* **genus profile** *of X is defined as the smallest among all the corresponding tuples of k (with $k = b_2^+(X)$) integers in the lexicographical ordering:*

$$Pr(X) = min_{\mathbf{y}} Pr(X, \mathbf{y}).$$

As an example, the genus profile of the $K3$ surface is $(2, 2, 2)$. A further example is given by $X = E(2)_p$ with p odd, the $K3$ surface after a multiplicity p logarithmic transformation has been applied. The minimal genus profile of X will be $(2, 2, f(p))$, since we can find two genus 2 surfaces that are disjoint from each other and from the fiber of the fibration and have self-intersection 2. The function $f(p)$ can be estimated as follows: There is a Seiberg-Witten basic class of the form $(p-1) \cdot z$ where $z \in H^2(X)$ has square zero and $z \neq 0$. Let $\mathbf{y} = (y_1, y_2, y_3)$ be an allowed sequence for which

$$Pr(X) = Pr(X, \mathbf{y}).$$

Using the adjunction inequalities we see that $z \cdot y_1 = z \cdot y_2 = 0$, implying $z \cdot y_3 \neq 0$. Let g denote the minimal genus of y_3. Using the the adjunction inequality we have

$$2g - 2 \geq Q_X(y_3, y_3) + p - 1 \geq p + 1.$$

It follows that the minimal genus is at least $(p+3)/2$. As $f(p)$ goes to infinity as $p \to \infty$, this argument already verifies that there are infinitely many $E(2)_p$'s with distinct genus functions. A similar, more explicit computation of the genus profile provides the proof of Theorem 1.5:

PROOF OF THEOREM 1.5. Consider $E(n)_{K_m}$, the result of knot surgery along the generic fiber of the elliptic fibration on $E(n)$ (a torus with trivial normal bundle), with $K_m = T_{2,2m+1}$ the alternating torus knot with crossing number $2m+1$ ($m \geq 1$). As the knot K_m admits a Seifert surface of genus m, there is a surface in $E(n)_{K_m}$ of genus m intersecting the generic fiber once and with self-intersection $-n$: we get this surface by gluing the section of $E(n)$ with the Seifert surface of K_m. By the computation of the Seiberg-Witten invariants of $E(n)_{K_m}$ in [**8**], the adjunction inequality implies that this surface is of minimal genus in its homology class. The same holds for the surface we get by adding $k = \lceil \frac{n}{2} \rceil$ copies of the fiber, so the result will have genus $m+k$ and positive self-intersection. This then shows that $Pr(E(n)_{K_m}) = (2, \ldots, 2, k+m)$ (with $2n-2$ copies of 2), verifying the result. □

4. The genus function for $\mathbb{CP}^2\#2\overline{\mathbb{CP}}^2$ and $\mathbb{CP}^2\#3\overline{\mathbb{CP}}^2$

The genera of homology classes in $\mathbb{CP}^2\#2\overline{\mathbb{CP}}^2$ have been studied in [**10**] and [**18**], see also [**23**]. Indeed, the genera of classes in $H_2(\mathbb{CP}^2\#2\overline{\mathbb{CP}}^2; \mathbb{Z})$ with non-negative self-intersection can be determined by a simple algorithm. To explain this algorithm, we introduce the following conventions. A homology class $\alpha \in H_2(\mathbb{CP}^2\#2\overline{\mathbb{CP}}^2; \mathbb{Z})$ will be denoted by a triple (a, b_1, b_2), reflecting the decomposition

$$H_2(\mathbb{CP}^2\#2\overline{\mathbb{CP}}^2; \mathbb{Z}) = H_2(\mathbb{CP}^2; \mathbb{Z}) \oplus H_2(\overline{\mathbb{CP}}^2; \mathbb{Z}) \oplus H_2(\overline{\mathbb{CP}}^2; \mathbb{Z}) \cong \mathbb{Z}^3.$$

In $H_2(\mathbb{CP}^2\#2\overline{\mathbb{CP}}^2; \mathbb{Z})$ the classes with one coordinate ± 1 and the others 0 can be obviously represented by spheres. Therefore we can assume that in computing $g_{\mathbb{CP}^2\#2\overline{\mathbb{CP}}^2}$, we have $a, b_1, b_2 \geq 0$: otherwise we apply a reflection to the coordinate in question (which is generated by a self-diffeomorphism) to reverse its sign. In the same way, by applying a diffeomorphism interchanging the two $\overline{\mathbb{CP}}^2$ components if necessary, we can assume $b_1 \geq b_2$.

Genera of classes with $\alpha^2 \geq 0$. Suppose that $x = (a, b_1, b_2)$ has $x^2 = a^2 - b_1^2 - b_2^2 \geq 0$, and assume that $a, b_1, b_2 \geq 0$. If $a \geq b_1 + b_2$ (and $a > 0$), then the homology class can be represented by a symplectic surface. Indeed, take b_1 lines in $\mathbb{CP}^2$ passing through a given point P_1, take b_2 lines passing through another point P_2 ($P_1 \neq P_2$), and take $a - b_1 - b_2$ further generic lines, all disjoint from P_1 and P_2. Blowing up P_1 and P_2 we get a complex curve, which (after smoothing the double points in a symplectic way) provides a smooth symplectic surface representing the homology class in question. The surface will be connected, unless $a = b_1$ (and so $b_2 = 0$). By the Symplectic Thom Conjecture (Theorem 1.6) this minimizes genus in its homology class. The genus of this curve is easy to determine: for a triple (a, b_1, b_2) with $a \geq b_1 + b_2$, $a > 0$ and $a > b_1$ the genus function $g(a, b_1, b_2)$ can be given by

$$g(a, b_1, b_2) = \frac{1}{2}((a-1)(a-2) - b_1(b_1 - 1) - b_2(b_2 - 1)); \tag{4.1}$$

and $g(a, a, 0) = 0$. (In this last case, we connect sum the spherical components of the symplectic representative of the class $(a, a, 0)$.)

The following idea, determining the genus function for all classes with nonnegative self-intersection, already appeared in [**10, 23**]: Consider a class $\alpha = (a, b_1, b_2)$ and let c denote $a - b_1 - b_2$. For $c \geq 0$ the genus of α has been described above. Suppose now that $c < 0$. Consider the sphere representing the homology class $(1, 1, 1)$ (which has self-intersection (-1)) we get by connecting the projective lines in the projective plane factors, and apply a diffeomorphism f inducing reflection to this sphere. The induced map f_* takes (a, b_1, b_2) to $(a + 2c, b_1 + 2c, b_2 + 2c)$. By applying further diffeomorphisms we can assume that these integers are nonnegative, and (as $c < 0$) it is easy to see that $|a + 2c| < a$. Therefore after finitely many steps of the above algorithm we get a vector $\alpha' = (a', b_1', b_2')$ with $c' = a' - b_1' - b_2' \geq 0$ and with $g(\alpha) = g(\alpha')$. For α' the formula of Equation (4.1) above applies and determines the value of the genus function.

If $\alpha^2 = 0$ (i.e., when $a^2 - b_1^2 - b_2^2 = 0$, that is, (a, b_1, b_2) is a Pythagorean triple), the assumption $a^2 - b_1^2 - b_2^2 = 0$ together with $a - b_1 - b_2 \geq 0$ implies that one of b_1 or b_2 is equal to zero, hence the above algorithm stops at a vector of the form $(a, a, 0)$ (or $(a, 0, a)$). These classes can be represented by complex curves which consist of a copies of $\mathbb{CP}^1$. By smoothly connect summing them, we get

COROLLARY 4.1. *If $\alpha \in H_2(\mathbb{CP}^2 \# 2\overline{\mathbb{CP}}^2; \mathbb{Z})$ has $\alpha^2 = 0$ then α can be represented by a sphere (i.e. $g(\alpha) = 0$).*

Classes with $\alpha^2 < 0$. For classes of negative self-intersection we do not have a complete picture. There are a number of cases when one can find a sphere representing the class, while in some cases we have a lower bound for the genus. This latter case emerges mainly for characteristic elements.

A similar algorithm as before applies for classes of negative square and (after applying diffeomorphisms) we need to examine only those triples (a, b_1, b_2) which satisfy $2a \geq b_1 + b_2$. Indeed, by taking the sphere of self-intersection 2 representing the class $(2, 1, 1)$, we can apply reflection to a class with $d = 2a - b_1 - b_2$ negative (and further diffeomorphisms to regain $a, b_1, b_2 \geq 0$), and this process will stop at a representative of the diffeomorphism orbit satisfying $2a \geq b_1 + b_2$. This information is, however, not sufficient to determine the genus function on the homology class. Indeed, in the above form (as $b_1 \geq b_2$) we can assume that $a \geq b_2$, so the second blow-up can be performed in a point of $\mathbb{CP}^2$ where b_2 lines pass through, but for the rest of the surface representing the class (a, b_1, b_2) we need to apply ad hoc methods, which sometimes give better results for those representatives of the diffeomorphism orbit of the class which fail to satisfy $2a \geq b_1 + b_2$.

There are simple examples of homology classes which can be represented by an embedded sphere; for example:

LEMMA 4.2. *Consider the homology classes $(a, a+1, c)$ with $c = 1, 2$. These classes can be represented by embedded spheres.*

PROOF. The class $(a, a+1) \in H_2(\mathbb{CP}^2 \# \overline{\mathbb{CP}}^2; \mathbb{Z})$ can be represented by a sphere. Connect summing with the sphere representing $c = 1, 2$ we find the sphere representative of the class. □

Lower bounds for the genus function for homology classes with negative self-intersection are scarce; there are some bounds for characteristic homology classes, though, which we present below. Note that the class

$$\alpha = (a, b_1, b_2) \in H_2(\mathbb{CP}^2 \# 2\overline{\mathbb{CP}}^2; \mathbb{Z})$$

is characteristic if and only if the coordinates a, b_1, b_2 are all odd. Suppose now that the characteristic homology class α can be represented by an embedded sphere. Consider the $(|\alpha^2|-1)$-fold antiblow-up (i.e. connected sum with $\mathbb{CP}^2$) of $\mathbb{CP}^2 \# 2\overline{\mathbb{CP}}^2$, where the (still characteristic) homology class $(a, b_1, b_2, 1, \ldots, 1)$ can be now represented by an embedded sphere of self-intersection (-1). We can blow that sphere down, and as the class was characteristic, its complement (and hence the result of the blow-down) is a spin four-manifold. Applying Furuta's $\frac{10}{8}$-theorem [**9**] giving $b_2(X) \geq \frac{5}{4}|\sigma(X)| + 2$ and the divisibility of $\sigma(X)$ by 16 to this spin four-manifold, we get

LEMMA 4.3. *Suppose that $\alpha = (a, b_1, b_2) \in H_2(\mathbb{CP}^2 \# 2\overline{\mathbb{CP}}^2; \mathbb{Z})$ is a characteristic homology element with negative self-intersection $\alpha^2 = -n$ $(n \in \mathbb{N})$. If α can be represented by an embedded sphere then n is 1.* □

A more general lower bound of the genus function on characteristic elements can be derived as follows. Suppose that α is a characteristic element in $H_2(\mathbb{CP}^2 \# 2\overline{\mathbb{CP}}^2; \mathbb{Z})$ with $\alpha^2 = -8n - 1$. (Recall that the square of a characteristic element is congruent to the signature of the ambient four-manifold mod 8.) We claim that

PROPOSITION 4.4. *The smooth representative of a characteristic element $\alpha \in H_2(\mathbb{CP}^2 \# 2\overline{\mathbb{CP}}^2; \mathbb{Z})$ with $\alpha^2 = -8n - 1$ has genus at least $\frac{n}{2}$.*

PROOF. As the characteristic element represents the Poincare dual of the second Stiefel-Whitney class w_2 of the ambient manifold $\mathbb{CP}^2 \# 2\overline{\mathbb{CP}}^2$, the complement of the normal disk bundle admits a spin structure.

Suppose now that the smooth representative of α is of genus $g < \frac{n}{2}$. Consider the connected sum of $8n + 1$ copies of $\mathbb{CP}^2$ when n is even, and $8n - 7$ copies if n is odd. Below we concentrate on the case of even n: choose in every summand the projective line, connect sum them, and add g fake handles, resulting in a surface S in $\#_{8n+1}\mathbb{CP}^2$ of genus g and self-intersection $8n + 1$, which represents a characteristic class (hence its complement is spin).

In constructing the normal connected sum of the two four-manifolds along these surfaces we can choose the gluing map so that the result is a spin manifold Z. Indeed, the spin structures on a genus g surface come in two types depending on their Arf invariants (see the discussion in [**12**, page 103]), and two spin structures are diffeomorphic iff they have equal Arf invariants. The spin three-manifold Y we get as the boundary of the complement of the open neighborhood of the characteristic surface is a circle bundle over a surface (of odd Euler number). Deleting a small tubular neighbourhood of a fiber, we get a trivial bundle over a surface with boundary. On this surface the exact same classification of spin structures holds as for the closed case, implying that on the circle bundle there are two orbits of spin structures under the diffeomorphism group. The two structures can be given either as above, or considering the surface in the $(8n-7)$-fold connected sum of $\mathbb{CP}^2$, but choosing the cubic curve in one of the components as opposed to the projective line. As gluing with this latter spin structure we would get a four-manifold violating Rokhlin's theorem about signatures of closed spin manifolds, it follows that the two spin structures on the boundaries in our construction are diffeomorphic, hence (by choosing the right gluing map) provide a spin structure on Z.

As S has simply connected complement, the resulting manifold will have $b_1(Z) = 0$, hence its signature $\sigma(Z) = 8n$ and its second Betti number $b_2(Z) = \chi(Z) - 2 = 8n + 4g + 2$ satisfies the $\frac{10}{8}$-theorem of Furuta [**9**], implying

$$b_2(Z) \geq \frac{5}{4}\sigma(Z) + 2,$$

eventually providing $g \geq \frac{n}{2}$. (An almost identical argument gives the bound when n is odd.) □

An upper bound for the genus can be given by the following simple algorithm. First transform the given homology class to (a, b_1, b_2) with all nonnegative and $b_1 \geq b_2$, and in addition $2a \geq b_1 + b_2$. In particular, we can assume that $a \geq b_2$.

As $a - b_2 < b_1$ (which follows from $a^2 - b_1^2 - b_2^2 < 0$) and $a \geq b_2$ for $\alpha = (a, b_1, b_2)$ with $\alpha^2 < 0$ and $2a \geq b_1 + b_2$, we can consider b_2 lines passing through a given point P_1 and the remaining $a - b_2$ through another point P_2. Blowing up the two points, we get a symplectic representative of the class $(a, a - b_2, b_2)$. Addig $b_1 + b_2 - a$ copies of the first exceptional divisor to it, we get the right homology class, while by resolving the negative intersection points the result will not necessarily be symplectic anymore. The genus of the resulting surface gives an upper bound for the minimal genus $g(\alpha)$.

4.1. The case of $\mathbb{CP}^2\#3\overline{\mathbb{CP}}^2$. Most of the above arguments provide (after some minor modifications) results for the genus function of $\mathbb{CP}^2\#3\overline{\mathbb{CP}}^2$. As before, for a class $\alpha = (a, b_1, b_2, b_3) \in H_2(\mathbb{CP}^2\#3\overline{\mathbb{CP}}^2; \mathbb{Z})$ with $\alpha^2 \geq 0$ there is a diffeomorphism under which the coordinates of the image satisfy $a \geq b_1 + b_2 + b_3$ (and all non-negative). In this step we use the (-2)-sphere representing the homology class $(1, 1, 1, 1) \in H_2(\mathbb{CP}^2\#3\overline{\mathbb{CP}}^2; \mathbb{Z})$. Under the inequality $a \geq b_1 + b_2 + b_3$ the homology class can be represented by a symplectic submanifold: as before, consider three distinct points P_1, P_2, P_3 in $\mathbb{CP}^2$, b_i lines through P_i (and $a - b_1 - b_2 - b_3$ in generic position, avoiding all the P_i) and blow up P_i. After symplectically smoothing the double points of the resulting complex curve, we get the desired symplectic submanifold. This procedure again gives an algorithm, rather than a formula to determine the genus function $g_{\mathbb{CP}^2\#3\overline{\mathbb{CP}}^2}$. Once again, if $\alpha = (a, b_1, b_2, b_3)$ satisfies $b_1 \geq b_2 \geq b_3$ (which can be assumed without loss of generality) and $\alpha^2 = 0$, then after some diffeomorphism $\alpha = (a, a, 0, 0)$, and this class can be represented by a smoothly embedded sphere. Hence

COROLLARY 4.5. *All classes* $\alpha \in H_2(H_2(\mathbb{CP}^2\#3\overline{\mathbb{CP}}^2;\mathbb{Z})$ *with* $\alpha^2 = 0$ *have* $g_{\mathbb{CP}^2\#3\overline{\mathbb{CP}}^2}(\alpha) = 0$.

As in the case of $\mathbb{CP}^2\#2\overline{\mathbb{CP}}^2$, we were not able to determine the genus function on all elements with negative self-intersection in $H_2(\mathbb{CP}^2\#3\overline{\mathbb{CP}}^2;\mathbb{Z})$. The argument for finding a bound earlier for the genera of characteristic elements applies in this case as well, providing some constraint, but in general such results are not sufficient for getting exact values of the genus function.

5. The Akhmedov-Park exotic $\mathbb{CP}^2\#2\overline{\mathbb{CP}}^2$ manifold AP

In [**1**] an infinite family of pairwise non-diffeomorphic smooth four-manifolds has been constructed, each homeomorphic to $\mathbb{CP}^2\#2\overline{\mathbb{CP}}^2$. One of these examples, which will be denoted by AP, was shown to carry a symplectic structure; we recall the definition of these manifolds below. The goal of this section is to partially determine the genus function of these exotic manifolds.

The study of AP (and indeed, all the exotic copies discussed below) is different from $\mathbb{CP}^2\#2\overline{\mathbb{CP}}^2$ in two crucial aspects. First, we do not see the spheres which then provide the diffeomorphisms inducing reflections in homology. (Indeed, we suspect that AP contains no homologically nontrivial sphere at all, cf. Conjecture 5.4.) The other aspect, however, provides more tools to study the genus function: the proof of exoticness shows that the symplectic form $[\omega]$ and its first Chern class $c_1(AP, \omega)$ pair negatively, hence AP has Kodaira dimension $\kappa = 2$ (a 'symplectic four-manifold of general type'). This implies that the adjunction inequality holds true for AP: As this theorem holds for negative self-intersections as well, we find more restrictions in AP for classes of negative square. In a similar manner, the further exotic smooth manifolds admit nontrivial Seiberg-Witten invariants, hence the adjunction inequality also provides bounds on the genera of homology classes (no matter what square they have).

5.1. The construction of the manifold AP. The symplectic exotic example is given as a symplectic normal connected sum of two symplectic manifolds along symplectic genus 2 surfaces, followed by a sequence of appropriately chosen Luttinger surgeries. Indeed, consider $S^3_0(T)$, the result of 0-surgery along the trefoil knot $T \subset S^3$, and let $X_1 = S^3_0(T) \times S^1$. This is a symplectic four-manifold, as it admits a T^2-fibration structure over T^2 (with a section) — resting on the fact that T is a fibered knot of genus 1. Let t_1 be an embedded torus representing the fiber and t_2 representing a section of this bundle. Indeed, the homology class $2[t_1]$ also admits a torus representative (by braiding the generic fiber), which we denote by t_1', and all of these tori can be chosen to be symplectic. As t_1' and t_2 intersect in two points, by blowing up one of them and resolving the other, we get a genus 2 symplectic surface $\Sigma \subset X$ with self-intersection 0 in $X = (S^3_0(T) \times S^1)\#\overline{\mathbb{CP}}^2$. Take the fiber connected sum of (X, Σ) with $(\Sigma_2 \times T^2, \Sigma_2 \times \{p\})$ for the genus 2 surface Σ_2 and for some point p in the 2-torus T^2. The resulting manifold Y will be a symplectic manifold. We can identify a few interesting submanifolds in Y: for example, the symplectic genus 2 surface Σ along which the normal sum was taken (after pushed off) provides a symplectic genus 2 surface in Y with self-intersection 0, which we will denote by A. In a similar manner, two further surfaces B and C can be constructed as follows: take a torus in $\Sigma_2 \times T^2$ intersecting the genus 2 surface we will use in the symplectic normal sum transversely in a single point, and glue to a push-off of either t_1 or t_2. (Notice that t_1 intersects Σ once, so after the gluing we get a genus 2 surface B, while t_2 intersects Σ in two points, so in the construction we need to use two copies of the torus in $\Sigma_2 \times T^2$, and the resulting surface C will have genus 3.) A fourth surface D can be constructed by taking the exceptional divisor of the blow-up in $(S^3_0(T) \times S^1)\#\overline{\mathbb{CP}}^2$. This sphere is punctured twice in the symplectic normal sum procedure, and gluing it to punctured tori on both sides we get a genus 2 surface D. All the above surfaces can be constructed to be symplectic.

The main result of Akhmedov and Park in [**1**] is that by applying Luttinger surgeries in Y we can get a simply connected four-manifold. (As Luttinger surgery does not change signature and Euler characteristic, the result will be necessarily homeomorphic to the two-fold blow-up of $\mathbb{CP}^2$.)

THEOREM 5.1 ([**1**]). *There are four tori in Y (originating from $\Sigma \times T^2$) along which Luttinger surgeries (with appropriately chosen coefficients) provide a simply connected four-manifold, and the symplectic structure of Y extends to the resulting four-manifold AP.*

REMARK 5.2. *In* [**1**] *the coefficients are carefully described, and the fundamental group calculation is given in details.*

The Luttinger surgeries performed in $\Sigma_2 \times T^2$ are disjoint from the surfaces introduced above in Y, hence A, B, C, D give rise to surfaces (also denoted by the same letters) in AP. It is not hard to see that if c_1 denotes the first Chern class of the symplectic structure on AP, then

- $g(A) = g(B) = g(D) = 2$, $g(C) = 3$.
- $c_1(A) = c_1(B) = 2$, $c_1(C) = 4$, $c_1(D) = 3$.
- $A \cdot A = B \cdot B = C \cdot C = 0$, $D \cdot D = -1$.
- $B \cdot D = C \cdot D = 0$
- $A \cdot B = 1$, $A \cdot C = 2$, $B \cdot C = 1$, $A \cdot D = 2$.

Indeed, $H_2(AP; \mathbb{Z})$ is generated by B, C, D, and we have a relation $A = 2B + C - 2D$. As these surfaces are symplectic, Theorem 1.6 implies that the above genera are indeed minimal. This observation then shows

THEOREM 5.3 ([**1**]). *The four-manifold AP is not diffeomorphic to $\mathbb{CP}^2\#2\overline{\mathbb{CP}}^2$.*

PROOF. The genus function is a diffeomorphism invariant, and it vanishes on the subset of $H_2(\mathbb{CP}^2\#2\overline{\mathbb{CP}}^2; \mathbb{Z})$ consisting of elements of zero square, while in AP there are symplectic submanifolds of square zero and nonzero genus (any of A, B, C above will do). □

Indeed, we can determine the function g_{AP} for a large class of elements. Let $\alpha = t_A \cdot A + t_B \cdot B + t_C \cdot C$, where all $t_i \geq 0$ (and at least one of them is nonzero). This class can be represented by a symplectic surface, and its genus can be determined by a simple (although not very enlightening) formula. Indeed, the multiple of a class can be represented by a cover of the 0-section in its (trivial) normal bundle, which can be chosen to be symplectic. The sum of the terms then can be represented by the union of these potential covers, and the positive intersections can be symplectically resolved.

The lower bound provided by Equation (1.1) of the adjunction inequality cannot provide sharp results in general — indeed, when the self-intersection of a homology class is negative, then for sufficiently large multiples of the class the inequality of Equation (1.1) does not provide any information. For this reason, at the moment we do not have effective tools to estimate the genus function for homology classes in general, and this implies that g_X cannot be determined in general.

Notice that the generic homology class in $H_2(\mathbb{CP}^2\#2\overline{\mathbb{CP}}^2;\mathbb{Z})$ can be given as $t_BB + t_CC + t_DD$, but since D has self-intersection -1, even if all coefficients are non-negative, the direct argument given above does not necessarily provide a symplectic submanifold. The same phenomenon can happen if the symplectic surface representing a given homology class happens to be disconnected: in this case smoothly we can tube the components together, but this cannot necessarily be done symplectically. In the following we show some cases where a better understanding of the construction of AP and a surface representing a particular homology class provides results close to optimal (and sometimes even optimal).

Some examples:

- Consider $\alpha = B+C+D$. As B, C, D all can be represented by a symplectic surface, we have a disconnected representative of α (with components of genus 5 and 2); tubing them we get a surface of genus 7 representing α. As $c_1(\alpha) = 9$, the adjunction formula of Theorem 1.7 for any smooth representative gives $g \geq 6$ as a lower bound.
- For $\alpha = B + C - D$ we have $c_1(\alpha) = 3$ and $\alpha^2 = 1$, so adjunction implies $g \geq 3$. A surface representing this class can be given by taking the disconnected symplectic representative of $B + C + D$, reverse the orientation on the component corresponding to D and then tube. The result will be a surface of genus 7. A better result can be achieved by the following simplification: when we construct the individual surfaces B and D, in the $\Sigma_2 \times T^2$-part we consider two punctured genus-1 surfaces (once with coefficient 1 in B and once with coefficient -1 in D). Substituting these subsurfaces with an annulus we get a genus 5 representative of the class α.
- Let $\alpha = B+2C+D$; then $\alpha^2 = 3$ and $c_1(\alpha) = 13$. The lower bound for the genus provided by the adjunction formula is 9. A genus 10 surface can be easily constructed by taking the disconnected symplectic representative (with a genus-8 and a genus 2 components) and tube them.
- For the class $\alpha = 2B + 2C - 2D$ we have $\alpha^2 = 4$ and $c_1(\alpha) = 6$, so the adjunction inequality provides 6 as the minimal genus. Noticing that $A = 2B + C - 2D$ is a symplectic surface of genus 2 intersecting C in two points, after resolving the intersection points, we find a genus 6 symplectic surface representing the homology class at hand.
- Let us take $\alpha = B + D$. In this case $\alpha^2 = -1$ and $c_1(\alpha) = 5$, so the adjunction inequality provides 3 as a lower bound. The disconnected representative (after tubing) provides a genus 4 surface. Now if we use (instead of the two punctured tori, each appearing in B and D on the

$\Sigma_2 \times T^2$-side) a twice-braided torus, which now intersects the normal connect sum surface, we get a representative of genus 3.

After an extensive search we did not find a nontrivial homology class in $H_2(AP; \mathbb{Z})$ which we could represent by a sphere. For classes of self-intersection at least -1 there is no such class, as can be deduced from Seiberg-Witten theory, but for more negative squares this is far from obvious. (Compare with the examples of Lemma 4.2 for $\mathbb{CP}^2\#2\overline{\mathbb{CP}}^2$.) Indeed, the search for embedded tori (representing non-trivial homology classes) did not provide any result either. This led us to the following

CONJECTURE 5.4. *The four-manifold AP does not contain homologically essential embedded spheres or tori.*

As it happened in the case of the homology $S^2 \times S^2$ examples, the last Luttinger surgery in the construction can be replaced by a torus surgery with coefficient n, providing an infinite family Z_n of four-manifolds, distinguished by the value of the Seiberg-Witten function on the two basic classes. (See the computation in Proposition 2.2; above $Z_1 = AP$.) As the above constructions all happened in the complement of all surgeries, the constructions are still valid, and whenever the minimality is verified by the adjunction inequality, the statement holds true for the entire family. (Notice that these manifolds — with the exception of AP — are not symplectic.) In conclusion, whenever we were able to identify the value of the genus function we get $g_{Z_n}(\alpha) = g_{AP}(\alpha)$, leading us to

CONJECTURE 5.5. *The genus functions of the exotic four-manifolds Z_n are all equivalent to g_{AP}.*

REMARK 5.6. *The Seiberg-Witten functions $SW_{Z_n}\colon \mathrm{Spin}^c(Z_n) \to \mathbb{Z}$ all have the same support (the two basic classes of Z_n), but different values on their support. As the adjunction formula (and any genus-related formula) is insensitive for the Seiberg-Witten* ***value****, there is reason to believe that the genus functions for Z_n are all the same. At the same time, this observation raises the question how the Seiberg-Witten value influences the geometry of the underlying four-manifold.*

Conjecture 5.4 naturally extends to the four-manifolds Z_n: we expect that these four-manifolds do not contain homologically essential embedded spheres or tori.

6. The Baldridge-Kirk exotic $\mathbb{CP}^2\#3\overline{\mathbb{CP}}^2$ manifold BK

A similar construction as in the previous section provides exotic structures on $\mathbb{CP}^2\#3\overline{\mathbb{CP}}^2$, as it was demonstrated in **[4]** by Baldridge and Kirk. Indeed, the building blocks are very similar to the ones used earlier: consider the four-manifold $S^3_0(T) \times S^1$ for the trefoil knot T, which admits a T^2-fibration over T^2 with a section. Resolve the union of a section and a fiber and blow up the resulting genus 2 surface twice, getting a genus 2 surface Σ with trivial normal bundle in $(S^3_0(T) \times S^1)\#2\overline{\mathbb{CP}}^2$. Consider the symplectic normal sum of the resulting symplectic surface in the symplectic four-manifold with $\Sigma_2 \times T^2$ along $\Sigma_2 \times \{p\}$ for some $p \in T^2$.

THEOREM 6.1 (**[4]**). *There are four tori in the above normal connected sum (originating from $\Sigma \times T^2$) along which Luttinger surgeries (with appropriately chosen coefficients) provide a simply connected, symplectic four-manifold, which we call BK.*

Using the pieces, and considering surfaces before the Luttinger surgeries, one can find several symplectic submanifolds in BK. First, there is the genus 2 surface used in the symplectic normal sum, which provides a symplectic submanifold with self-intersection 0; the corresponding homology class will be denoted by A. There is a punctured symplectic torus in $\Sigma_2 \times (T^2 \setminus \{p\})$, and as it can be chosen to be disjoint from the Lagrangian tori of the Luttinger surgeries, it descends to BK. From the other side, the section and the fiber of the T^2-fibration on $S^3_0(T) \times S^1$ give punctured tori in the complement of the genus 2 surface; when gluing them with the punctured torus in $\Sigma_2 \times (T^2 \setminus \{p\})$, we get two surfaces B and C of genus 2, self-intersection 0, intersecting each other and A once. The exceptional divisors of the blow-ups on $S^3_0(T) \times S^1$ provide two punctured disks with relative framing (-1), hence after gluing them to the punctured tori in $\Sigma_2 \times (T^2 \setminus \{p\})$ we get two disjoint tori of self-intersection (-1), and call them D_1 and D_2.

A basis of $H_2(BK; \mathbb{Z}) \cong \mathbb{Z}^4$ can be given by B, C, D_1, D_2; the genera and intersection patterns are described above, and the first Chern class c_1 of the symplectic four-manifold BK evaluates as $c_1(B) = c_1(C) = 2, c_1(D_1) = c_1(D_2) = 1$, while $c_1(A)$ is also 2.

Obviously, the genus function g_{BK} is nonzero on the subset of $H_2(BK; \mathbb{Z})$ formed by elements of self-intersection 0 (as g_{BK} takes the value 2 on A, B, C), verifying again that the diffeomorphism invariant g_X distinguishes $\mathbb{CP}^2 \# 3\overline{\mathbb{CP}}^2$ from BK, the straightforward extension of Theorem 5.3 for the manifold BK. (Notice that in the partial determination of g_{BK} we rely on the Symplectic Thom Conjecture Theorem 1.6.)

Once again, homology classes of the form $t_A \cdot A + t_B \cdot B + t_C \cdot C$ with $t_A, t_B, t_C \geq 0$ can be represented by symplectic submanifolds, and hence the genus function on those classes can be algorithmically determined, while in general the adjunction inequality of Theorem 1.7 provides a lower bound on the genus (in terms of the value of c_1 on the homology class and its self-intersection), while specific constructions give upper bounds on g_{BK}. The two bounds, however, do not meet in general.

As before, the last Luttinger surgery in the construction of BK can be replaced by an appropriate sequence of torus surgeries, providing a sequence of four-manifolds V_n (with $V_1 = BK$), which then show the existence of infinitely many distinct smooth structures on $\mathbb{CP}^2 \# 3\overline{\mathbb{CP}}^2$ (see [**4**]). As in the previous section, these manifolds are distinguished by the Seiberg-Witten values on the basic classes. We cannot compare their genus functions, as the values of these functions on most homology classes of negative self-intersection are unknown at the moment. We do expect, however

CONJECTURE 6.2. *The exotic four-manifolds V_n have equivalent genus functions.*

In contrary to AP, in the four-manifold BK it is easy to find tori and spheres. Indeed, the homology classes D_1, D_2 admit toric representatives. In addition, when applying the two blow-ups in $S^3_0(T) \times S^1$, we can arrange the blow-ups so that the resulting four-manifold has a (-2)-sphere disjoint from the genus 2 surface along which the normal sum is taken. Therefore this sphere embeds into BK (and indeed to all V_n's).

Acknowledgments

The first author acknowledges the hospitality of MSRI/SLMath in Berkeley, where part of this project was carried out. We would like to thank the referee for their suggestions which significantly improved the presentation of the results in the paper.

References

[1] Anar Akhmedov and B. Doug Park, *Exotic smooth structures on small 4-manifolds with odd signatures*, Invent. Math. **181** (2010), no. 3, 577–603, DOI 10.1007/s00222-010-0254-y. MR2660453

[2] A. Akhmedov and D. Park, *Exotic smooth structures on* $S^2 \times S^2$, arXiv:1005.3346, 2010

[3] D. Auroux, S. K. Donaldson, and L. Katzarkov, *Luttinger surgery along Lagrangian tori and non-isotopy for singular symplectic plane curves*, Math. Ann. **326** (2003), no. 1, 185–203, DOI 10.1007/s00208-003-0418-9. MR1981618

[4] Scott Baldridge and Paul Kirk, *A symplectic manifold homeomorphic but not diffeomorphic to* $\mathbb{CP}^2 \# 3\overline{\mathbb{CP}}^2$, Geom. Topol. **12** (2008), no. 2, 919–940, DOI 10.2140/gt.2008.12.919. MR2403801

[5] S. K. Donaldson and P. B. Kronheimer, *The geometry of four-manifolds*, Oxford Mathematical Monographs, The Clarendon Press, Oxford University Press, New York, 1990. Oxford Science Publications. MR1079726

[6] Josef G. Dorfmeister and Tian-Jun Li, *The minimal genus problem—a quarter century of progress*, Acta Math. Sci. Ser. B (Engl. Ed.) **42** (2022), no. 6, 2257–2278, DOI 10.1007/s10473-022-0604-z. MR4493676

[7] Ronald Fintushel, B. Doug Park, and Ronald J. Stern, *Reverse engineering small 4-manifolds*, Algebr. Geom. Topol. **7** (2007), 2103–2116, DOI 10.2140/agt.2007.7.2103. MR2366188

[8] Ronald Fintushel and Ronald J. Stern, *Knots, links, and 4-manifolds*, Invent. Math. **134** (1998), no. 2, 363–400, DOI 10.1007/s002220050268. MR1650308

[9] M. Furuta, *Monopole equation and the* $\frac{11}{8}$*-conjecture*, Math. Res. Lett. **8** (2001), no. 3, 279–291, DOI 10.4310/MRL.2001.v8.n3.a5. MR1839478

[10] Hong Zhu Gao, *Representing homology classes of almost definite 4-manifolds*, Topology Appl. **52** (1993), no. 2, 109–120, DOI 10.1016/0166-8641(93)90030-H. MR1241187

[11] Robert E. Gompf and András I. Stipsicz, *4-manifolds and Kirby calculus*, Graduate Studies in Mathematics, vol. 20, American Mathematical Society, Providence, RI, 1999, DOI 10.1090/gsm/020. MR1707327

[12] Robion C. Kirby, *The topology of 4-manifolds*, Lecture Notes in Mathematics, vol. 1374, Springer-Verlag, Berlin, 1989, DOI 10.1007/BFb0089031. MR1001966

[13] P. B. Kronheimer and T. S. Mrowka, *The genus of embedded surfaces in the projective plane*, Math. Res. Lett. **1** (1994), no. 6, 797–808, DOI 10.4310/MRL.1994.v1.n6.a14. MR1306022

[14] Karl Murad Luttinger, *Lagrangian tori in* $\mathbf{R}^4$, J. Differential Geom. **42** (1995), no. 2, 220–228. MR1366546

[15] John W. Morgan, *The Seiberg-Witten equations and applications to the topology of smooth four-manifolds*, Mathematical Notes, vol. 44, Princeton University Press, Princeton, NJ, 1996. MR1367507

[16] John W. Morgan, Tomasz S. Mrowka, and Zoltán Szabó, *Product formulas along* T^3 *for Seiberg-Witten invariants*, Math. Res. Lett. **4** (1997), no. 6, 915–929, DOI 10.4310/MRL.1997.v4.n6.a11. MR1492130

[17] Terry Lawson, *The minimal genus problem*, Exposition. Math. **15** (1997), no. 5, 385–431. MR1486407

[18] Bang-He Li and Tian-Jun Li, *Minimal genus smooth embeddings in* $S^2 \times S^2$ *and* $\mathbf{C}\mathrm{P}^2 \# n\overline{\mathbf{C}\mathrm{P}}^2$ *with* $n \leq 8$, Topology **37** (1998), no. 3, 575–594, DOI 10.1016/S0040-9383(97)00042-6. MR1604887

[19] Peter Ozsváth and Zoltán Szabó, *The symplectic Thom conjecture*, Ann. of Math. (2) **151** (2000), no. 1, 93–124, DOI 10.2307/121113. MR1745017

[20] Daniel Ruberman, *The minimal genus of an embedded surface of non-negative square in a rational surface*, Turkish J. Math. **20** (1996), no. 1, 129–133. MR1392668

[21] Clifford Henry Taubes, *The Seiberg-Witten invariants and symplectic forms*, Math. Res. Lett. **1** (1994), no. 6, 809–822, DOI 10.4310/MRL.1994.v1.n6.a15. MR1306023

[22] Clifford Henry Taubes, *The Seiberg-Witten invariants and 4-manifolds with essential tori*, Geom. Topol. **5** (2001), 441–519, DOI 10.2140/gt.2001.5.441. MR1833751

[23] C. T. C. Wall, *On the orthogonal groups of unimodular quadratic forms. II*, J. Reine Angew. Math. **213** (1963/64), 122–136, DOI 10.1515/crll.1964.213.122. MR155798

[24] Edward Witten, *Monopoles and four-manifolds*, Math. Res. Lett. **1** (1994), no. 6, 769–796, DOI 10.4310/MRL.1994.v1.n6.a13. MR1306021

Rényi Institute of Mathematics, H-1053 Budapest, Reáltanoda utca 13–15, Hungary
Email address: stipsicz.andras@renyi.hu

Department of Mathematics, Princeton University, Princeton, New Jersey, 08544
Email address: szabo@math.princeton.edu

Proceedings of Symposia in Pure Mathematics
Volume **109**, 2024
https://doi.org/10.1090/pspum/109/01998

The Gysin sequence and the $\mathfrak{sl}(N)$ homology of $T(2,m)$

Joshua Wang

ABSTRACT. The $\mathfrak{sl}(N)$ homology of the torus knot or link $T(2,m)$ may be calculated explicitly. By direct comparison, the result is isomorphic to the cohomology of a naturally associated space of $\mathrm{SU}(N)$ representations of the knot group. In honor of Tom Mrowka's 60th birthday, we explain how the Gysin exact sequence may be used to show that these groups are isomorphic without explicitly calculating them.

1. Introduction

For $N \geq 2$, the $\mathfrak{sl}(N)$ homology $\mathrm{KR}_N(L)$ of an oriented link $L \subset \mathbf{R}^3$ is a bigraded abelian group whose graded Euler characteristic is the $\mathfrak{sl}(N)$ link polynomial [**6, 13**]. The invariant is the homology of a chain complex defined combinatorially from a link diagram using a cube of resolutions. With the bigrading suppressed, the invariants of the first few torus links $T(2,m)$ are given in the third column of Table 1.

TABLE 1. The $\mathfrak{sl}(N)$ homology and $\mathrm{SU}(N)$ representation space of the torus link $T(2,m)$ for $m = 1,2,3,4,5$. For all m, it turns out that $\mathrm{KR}_N(T(2,m))$ is isomorphic to $H^*(\mathscr{R}_N(T(2,m)))$.

m	name of $T(2,m)$	$\mathrm{KR}_N(T(2,m))$	$\mathscr{R}_N(T(2,m))$
1	unknot	$\mathbf{Z}^N$	$\mathbf{CP}^{N-1}$
2	Hopf link	$\mathbf{Z}^{N^2}$	$\mathbf{CP}^{N-1} \sqcup \mathbf{F}(1,1;N)$
3	trefoil	$\mathbf{Z}^{3N-2} \oplus \mathbf{Z}/N$	$\mathbf{CP}^{N-1} \sqcup UT\mathbf{CP}^{N-1}$
4	Solomon's knot	$\mathbf{Z}^{N^2+2N-2} \oplus \mathbf{Z}/N$	$\mathbf{CP}^{N-1} \sqcup UT\mathbf{CP}^{N-1} \sqcup \mathbf{F}(1,1;N)$
5	cinquefoil	$\mathbf{Z}^{5N-4} \oplus (\mathbf{Z}/N)^2$	$\mathbf{CP}^{N-1} \sqcup UT\mathbf{CP}^{N-1} \sqcup UT\mathbf{CP}^{N-1}$

These groups turn out to arise in another way, entirely unrelated to the definition of $\mathfrak{sl}(N)$ homology: they are isomorphic to the ordinary cohomology of spaces naturally associated to these links. Let Φ be the diagonal matrix

$$\Phi := e^{\pi i/N}\mathrm{diag}(-1,1,\dots,1) \in \mathrm{SU}(N),$$

2020 *Mathematics Subject Classification.* Primary 57K18.

This material is based upon work supported by the NSF GRFP through grant DGE-1745303.

and for any oriented link $L \subset \mathbf{R}^3$, define the SU(N) representation space $\mathscr{R}_N(L)$ to be the space of homomorphisms

$$\mathscr{R}_N(L) := \{\rho\colon \pi_1(\mathbf{R}^3 \setminus L) \to \mathrm{SU}(N) \mid \rho(\mu) \text{ is conjugate to } \Phi \text{ for each meridian } \mu\}.$$

The space $\mathscr{R}_N(T(2,m))$ is straightforward to determine explicitly from a presentation of $\pi_1(\mathbf{R}^3 \setminus T(2,m))$ having two meridional generators. The spaces associated to the first few torus links $T(2,m)$ are displayed in the fourth column of Table 1 where $UT\mathbf{CP}^{N-1}$ is the unit tangent bundle of the projective space $\mathbf{CP}^{N-1}$, and $\mathbf{F}(1,1;N)$ is the partial flag manifold of pairs of orthogonal one-dimensional linear subspaces of $\mathbf{C}^N$. Using the isomorphisms

$$H^*(\mathbf{CP}^{N-1}) \cong \mathbf{Z}^N \quad H^*(UT\mathbf{CP}^{N-1}) \cong \mathbf{Z}^{2N-2} \oplus \mathbf{Z}/N \quad H^*(\mathbf{F}(1,1;N)) \cong \mathbf{Z}^{N^2-N}$$

we see that the $\mathfrak{sl}(N)$ homology of $T(2,m)$ matches the cohomology of $\mathscr{R}_N(T(2,m))$ for the listed values of m. This observation holds for all m.

OBSERVATION 1.1. $\mathrm{KR}_N(T(2,m))$ is isomorphic to $H^*(\mathscr{R}_N(T(2,m)))$ as an abelian group for any m.

In this note, we explain how to prove Observation 1.1 without directly observing it. More specifically, we use the Gysin sequence to show that $\mathrm{KR}_N(T(2,m))$ and $H^*(\mathscr{R}_N(T(2,m)))$ are isomorphic without needing to explicitly compute either group. Explicitly calculating $H^*(\mathscr{R}_N(T(2,m)))$ is simpler than explicitly calculating $\mathrm{KR}_N(T(2,m))$, so Observation 1.1 may also be thought of as a way to compute $\mathrm{KR}_N(T(2,m))$.

REMARK 1.2. Kronheimer and Mrowka's definition of singular instanton homology was motivated by the observation that the Khovanov homology of $T(2,m)$ coincides with the cohomology of $\mathscr{R}_2(T(2,m))$ [**8**]. Observation 1.1 is the natural extension of this observation to $\mathfrak{sl}(N)$ homology. Also see [**5, 10**].

REMARK 1.3. The original definition of $\mathfrak{sl}(N)$ homology given by Khovanov and Rozansky [**6**] requires $\mathbf{Q}$ coefficients. The $\mathfrak{sl}(N)$ homology over $\mathbf{Q}$ of $T(2,m)$ for m odd was first computed by Rasmussen [**14**]. The author is unaware of a reference in the literature for a computation of $\mathrm{KR}_N(T(2,m))$ over $\mathbf{Z}$, though the result is known to some experts.

REMARK 1.4. A variety of the additional structure on $\mathfrak{sl}(N)$ homology is reflected in the SU(N) representation space. Our proof of Observation 1.1 gives the complete bigrading on $\mathrm{KR}_N(T(2,m))$. For more, see [**18**].

We explain the role of the Gysin sequence in the proof. The key geometric observation is that $U := UT\mathbf{CP}^{N-1}$ is the total space of a circle bundle $S^1 \to U \to \mathbf{F}(1,1;N)$ with base space $\mathbf{F}(1,1;N)$. The Gysin sequence of this circle bundle is the long exact sequence

$$\cdots \longrightarrow H^{i+1}(U) \longrightarrow H^i(\mathbf{F}(1,1;N)) \xrightarrow{\cup e} H^{i+2}(\mathbf{F}(1,1;N)) \longrightarrow H^{i+2}(U) \longrightarrow \cdots$$

where $e \in H^2(\mathbf{F}(1,1;N))$ is the Euler class of the bundle. This class may be determined explicitly. Because $H^i(\mathbf{F}(1,1;N))$ vanishes when i is odd, the kernel of the map

$$H^*(\mathbf{F}(1,1;N)) \xrightarrow{\cup e} H^*(\mathbf{F}(1,1;N))$$

is isomorphic to $H^{\mathrm{odd}}(U)$ while its cokernel is isomorphic to $H^{\mathrm{even}}(U)$. Phrased differently, $H^*(U)$ is isomorphic to the homology of this two-term chain complex. The $\mathfrak{sl}(N)$ complex associated to the trefoil $T(2,3)$ turns out to be homotopy equivalent to the direct sum of this particular two-term complex and the one-term complex associated to the planar unknot diagram. As a result, we are able to deduce that the $\mathfrak{sl}(N)$ homology of $T(2,3)$ is isomorphic to $H^*(\mathbf{CP}^{N-1}) \oplus H^*(U) = H^*(\mathscr{R}_N(T(2,3)))$, without having explicitly calculated the homology of the given two-term complex.

Note that a similar argument using the Gysin sequence of the sphere bundle $S^{2N-3} \to U \to \mathbf{CP}^{N-1}$ identifies $H^*(U)$ with the homology of the complex

$$H^*(\mathbf{CP}^{N-1}) \xrightarrow{\cup\chi(\mathbf{CP}^{N-1})\cdot[\mathbf{CP}^{N-1}]} H^*(\mathbf{CP}^{N-1})$$

where $[\mathbf{CP}^{N-1}] \in H^{2N-2}(\mathbf{CP}^{N-1})$ is the orientation class and χ is the Euler characteristic. Under the standard identification $H^*(\mathbf{CP}^{N-1}) = \mathbf{Z}[X]/X^N$, the above complex is

$$\frac{\mathbf{Z}[X]}{X^N} \xrightarrow{\quad NX^{N-1} \quad} \frac{\mathbf{Z}[X]}{X^N}$$

whose homology is easily seen to be $\mathbf{Z}^{2N-2} \oplus \mathbf{Z}/N$.

In section 2, we compute the $\mathrm{SU}(N)$ representation spaces of $T(2,m)$ from the definition. We then show that $UT\mathbf{CP}^{N-1}$ is a circle bundle over $\mathbf{F}(1,1;N)$ and compute its Euler class. In section 3, we describe the construction of $\mathfrak{sl}(N)$ homology, emphasizing connections to the spaces $\mathbf{CP}^{N-1}$ and $\mathbf{F}(1,1;N)$. Finally, we prove Observation 1.1 in section 4.

2. SU(N) representation spaces

We first compute the $\mathrm{SU}(N)$ representation spaces associated to the torus links $T(2,m)$. We then explain how $UT\mathbf{CP}^{N-1}$ is a circle bundle over $\mathbf{F}(1,1;N)$ and compute its Euler class.

LEMMA 2.1. *The conjugacy class of $\Phi \in \mathrm{SU}(N)$ may be identified with $\mathbf{CP}^{N-1}$ by sending each matrix to its eigenspace with eigenvalue $-e^{i\pi/N}$.*

PROOF. If $e_1, \ldots, e_N$ is the standard basis of $\mathbf{C}^N$, then $\mathbf{C}\cdot e_1$ is the $(-e^{i\pi/N})$-eigenspace of Φ. Let $U \in \mathrm{SU}(N)$, and note that the $(-e^{i\pi/N})$-eigenspace of $U\Phi U^{-1}$ is $U(\mathbf{C}\cdot e_1)$, which is the image of the line $\mathbf{C}\cdot e_1$ under the linear transformation U. Because $\mathrm{SU}(N)$ acts transitively on $\mathbf{CP}^{N-1}$, the map from the conjugacy class of Φ to $\mathbf{CP}^{N-1}$ described in the lemma is surjective. To see that it is also injective, first note that the $(e^{i\pi/N})$-eigenspace of $U\Phi U^{-1}$ is the image of the span of $e_2, \ldots, e_N$ under the map U. This image is precisely the orthogonal complement of $U(\mathbf{C}\cdot e_1)$ because U is unitary. Hence, the $(-e^{i\pi/N})$-eigenspace of $U\Phi U^{-1}$ determines its $(e^{i\pi/N})$-eigenspace and therefore the matrix $U\Phi U^{-1}$ itself. □

LEMMA 2.2. *The fundamental group of the complement of $T(2,m)$ for $m \geq 1$ is given by the presentation*

$$\pi_1(\mathbf{R}^3 \setminus T(2,m)) = \begin{cases} \langle\, a, b \mid (ab)^k a = (ba)^k b \,\rangle & m = 2k+1 \text{ is odd} \\ \langle\, a, b \mid (ab)^k = (ba)^k \,\rangle & m = 2k \text{ is even} \end{cases}$$

where the generators a, b are meridians.

PROOF. This follows from the Wirtinger presentation associated to the diagram of $T(2,m)$ obtained as the braid closure of the two-stranded braid $\sigma_1^m \in B_2 = \mathbf{Z}\langle\sigma_1\rangle$. $\square$

PROPOSITION 2.3. *The* $\mathrm{SU}(N)$ *representation space of* $T(2,m)$ *for* $m \geq 1$ *is*

$$\mathscr{R}_N(T(2,m)) = \begin{cases} \mathbf{CP}^{N-1} \sqcup \bigsqcup^k UT\mathbf{CP}^{N-1} & m = 2k+1 \text{ is odd} \\ \mathbf{CP}^{N-1} \sqcup \bigsqcup^{k-1} UT\mathbf{CP}^{N-1} \sqcup \mathbf{F}(1,1;N) & m = 2k \text{ is even.} \end{cases}$$

PROOF. By Lemma 2.2, a representation $\rho \in \mathscr{R}_N(T(2,m))$ is determined by the values $\rho(a)$ and $\rho(b)$, which lie in the conjugacy class of Φ. By Lemma 2.1, we may identify the conjugacy class of Φ with $\mathbf{CP}^{N-1}$. Hence, there is an embedding

$$\Psi\colon \mathscr{R}_N(T(2,m)) \hookrightarrow \mathbf{CP}^{N-1} \times \mathbf{CP}^{N-1}$$

given by sending ρ to the pair of $(-e^{i\pi/N})$-eigenspaces of $\rho(a)$ and $\rho(b)$. We let $\Lambda_a(\rho)$ and $\Lambda_b(\rho)$ denote these eigenspaces so that $\Psi(\rho) = (\Lambda_a(\rho), \Lambda_b(\rho))$. Next, note that if $U \in \mathrm{SU}(N)$, then the representation $U\rho U^{-1}$ given by $\gamma \mapsto U\rho(\gamma)U^{-1}$ also lies in $\mathscr{R}_N(T(2,m))$. This defines a conjugation action of $\mathrm{SU}(N)$ on $\mathscr{R}_N(T(2,m))$. Now observe that

$$\Lambda_a(U\rho U^{-1}) = U(\Lambda_a(\rho))$$

where $U(\Lambda_a(\rho))$ is the image of the line $\Lambda_a(\rho) \subset \mathbf{C}^N$ under the linear transformation U. Similarly, we have $\Lambda_b(U\rho U^{-1}) = U(\Lambda_b(\rho))$. Hence, the embedding Ψ is $\mathrm{SU}(N)$-equivariant, where $\mathrm{SU}(N)$ acts on $\mathbf{CP}^{N-1} \times \mathbf{CP}^{N-1}$ diagonally.

The orbit space of the diagonal $\mathrm{SU}(N)$ action on $\mathbf{CP}^{N-1} \times \mathbf{CP}^{N-1}$ may be identified with the closed interval $[0, \pi/2]$. The identification is given by sending a pair of lines $(\mathbf{C}\cdot v, \mathbf{C}\cdot w)$ to the Hermitian angle $\theta \in [0,\pi/2]$ between them, which is defined by the equation

$$\|v\|\,\|w\| \cos\theta = |v \cdot w|.$$

We prove this statement in Lemma 2.4 below. Since Ψ is $\mathrm{SU}(N)$-equivariant, it follows that $\Psi(\mathscr{R}_N(T(2,m))) \subset \mathbf{CP}^{N-1} \times \mathbf{CP}^{N-1}$ is a union of orbits.

We now show that $\Psi(\mathscr{R}_N(T(2,m)))$ is the union of the orbits corresponding to the values of $\theta \in [0,\pi/2]$ satisfying $m\theta \in \mathbf{Z}\pi$. Suppose the orbit corresponding to $\theta \in [0,\pi/2]$ lies in $\Psi(\mathscr{R}_N(T(2,m)))$. The lines $\mathbf{C}\cdot e_1$ and $\mathbf{C}\cdot((\cos\theta)e_1 + (\sin\theta)e_2)$ have Hermitian angle θ so there must be a representation $\rho\colon \pi_1(\mathbf{R}^3 \setminus T(2,m)) \to \mathrm{SU}(N)$ for which $\Lambda_a(\rho)$ and $\Lambda_b(\rho)$ are precisely the two given lines. Then $\rho(a) = \Phi$ because the $(-e^{i\pi/N})$-eigenspace of Φ is $\mathbf{C}\cdot e_1$. Since

$$U_\theta = \begin{pmatrix} \cos\theta & -\sin\theta & 0 \\ \sin\theta & \cos\theta & 0 \\ 0 & 0 & \mathrm{Id}_{N-2} \end{pmatrix} \in \mathrm{SU}(N)$$

sends e_1 to $(\cos\theta)e_1 + (\sin\theta)e_2$, it follows that $\rho(b) = U_\theta \Phi U_{-\theta}$, using the identity $(U_\theta)^{-1} = U_{-\theta}$. Now note that

$$\Phi^{-1} = e^{-\pi i/N}\mathrm{diag}(-1,1,\ldots,1) = e^{-2\pi i/N}\Phi$$

from which it follows that $\rho(a)^{-1} = e^{-2\pi i/N}\rho(a)$ and $\rho(b)^{-1} = e^{-2\pi i/N}\rho(b)$. We may therefore express the relation of Lemma 2.2 as

$$\begin{aligned}
&(e^{-2\pi i/N}\rho(b)\rho(a))^m = \mathrm{Id} \\
\iff \quad &\begin{cases} (\rho(a)\rho(b))^k\rho(a) = (\rho(b)\rho(a))^k\rho(b) & m = 2k+1 \text{ is odd} \\ (\rho(a)\rho(b))^k = (\rho(b)\rho(a))^k & m = 2k \text{ is even.} \end{cases}
\end{aligned}$$

Note that $(e^{-i\pi/N}\Phi)U_{-\theta}(e^{-i\pi/N}\Phi) = U_\theta$ because $(e^{-i\pi/N}\Phi) = \mathrm{diag}(-1,1,\dots,1)$ is reflection in the first coordinate while $U_{-\theta}$ is rotation by $-\theta$ in the first two coordinates. Thus

$$e^{-2\pi i/N}\rho(b)\rho(a) = e^{-2i\pi/N}U_\theta\Phi U_{-\theta}\Phi = (U_\theta)^2 = U_{2\theta}$$

so the relation can be further simplified to $U_{2m\theta} = \mathrm{Id}$. Since $U_{2m\theta}$ is rotation by $2m\theta$ in the first two coordinates, we see that $2m\theta \in 2\mathbf{Z}\pi$ which is equivalent to $m\theta \in \mathbf{Z}\pi$.

We have proven that if the orbit corresponding to θ lies in $\Psi(\mathscr{R}_N(T(2,m)))$, then $m\theta \in \mathbf{Z}\pi$. The work we have done also verifies the converse. Given $\theta \in [0,\pi/2]$ satisfying $m\theta \in \mathbf{Z}\pi$, we define the representation $\rho\colon \pi_1(\mathbf{R}^3 \setminus T(2,m)) \to \mathrm{SU}(N)$ by sending $\rho(a) = \Phi$ and $\rho(b) = U_\theta\Phi U_{-\theta}$. To finish the proof, we must verify that the orbit in $\mathbf{CP}^{N-1} \times \mathbf{CP}^{N-1}$ corresponding to $\theta \in [0,\pi/2]$ is

$$\text{Orbit corresponding to } \theta = \begin{cases} \mathbf{CP}^{N-1} & \theta = 0 \\ UT\mathbf{CP}^{N-1} & 0 < \theta < \pi/2 \\ \mathbf{F}(1,1;N) & \theta = \pi/2 \end{cases}$$

which we prove in Lemma 2.5 below. See Figure 1. □

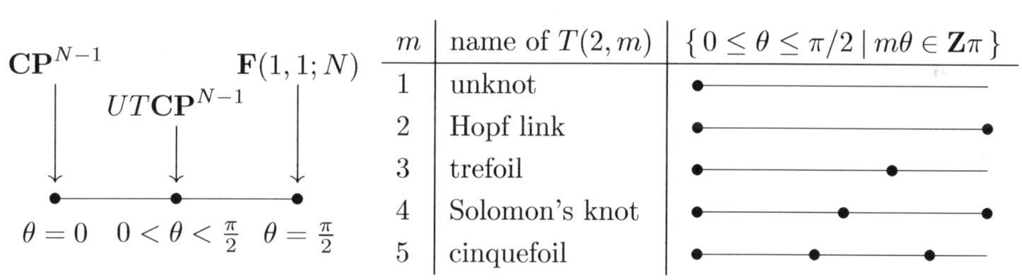

FIGURE 1. On the left are the orbits of the diagonal $\mathrm{SU}(N)$ action on $\mathbf{CP}^{N-1} \times \mathbf{CP}^{N-1}$. On the right are the orbits whose union is $\mathscr{R}_N(T(2,m))$ for $m = 1,2,3,4,5$.

LEMMA 2.4. *The map from $\mathbf{CP}^{N-1} \times \mathbf{CP}^{N-1}$ to $[0,\pi/2]$ sending a pair of lines to the Hermitian angle between them gives an identification between the orbit space of the diagonal $\mathrm{SU}(N)$ action on $\mathbf{CP}^{N-1} \times \mathbf{CP}^{N-1}$ and $[0,\pi/2]$.*

PROOF. The Hermitian angle is invariant under the diagonal action because

$$\frac{|Uv \cdot Uw|}{\|Uv\|\,\|Uw\|} = \frac{|v \cdot w|}{\|v\|\,\|w\|}$$

for any $U \in \mathrm{SU}(N)$. It suffices to show that $\mathrm{SU}(N)$ acts transitively on pairs of lines having a fixed Hermitian angle between them. If $e_1,\dots,e_N$ is the standard basis for $\mathbf{C}^N$, then any pair of lines may be mapped into the span of e_1 and e_2 by an element of $\mathrm{SU}(N)$. This follows from the fact that $\mathrm{SU}(N)$ acts transitively on the Grassmannian $\mathbf{G}(2, N)$ of 2-planes in $\mathbf{C}^N$. Using the inclusion $\mathrm{SU}(2) \times \mathrm{Id}_{N-2} \subset \mathrm{SU}(N)$ by block diagonal matrices, we have reduced to the case that $N = 2$.

First, we describe the geometric picture. Given two points in $\mathbf{CP}^1$, the Hermitian angle between them is precisely the distance between them in the Fubini–Study metric, which is the round metric on S^2 with radius $1/2$. Furthermore, $\mathrm{SU}(2)$ acts on $\mathbf{CP}^1$ through isometries, realizing $\mathrm{SU}(2)$ as the double cover of $\mathrm{SO}(3)$. The quotient of $S^2 \times S^2$ by the diagonal $\mathrm{SO}(3)$ action is clearly a closed interval given by the distance between the points, as claimed.

With the picture in place, it is easy to verify the result in coordinates and formulas. We think of $[0 : 1]$ as a basepoint, and consider the set of points with a fixed Hermitian angle θ from $[0 : 1]$. If $\theta = 0$, then the set is the singleton $[0 : 1]$, while if $\theta = \pi/2$, then the set is the singleton $[1 : 0]$. Any other point may be described in homogeneous coordinates as $[\zeta : 1]$ with $\zeta \neq 0$. The Hermitian angle θ between $[\zeta : 1]$ and $[0 : 1]$ is given by

$$\cos \theta = \frac{1}{\sqrt{1 + |\zeta|^2}}$$

so the points with a fixed Hermitian angle $0 < \theta < \pi/2$ from $[0 : 1]$ form a circle. They are the points $[\zeta : 1]$ with a fixed value of $|\zeta| > 0$. It suffices to show that the matrices in $\mathrm{SU}(2)$ that fix $[0 : 1]$ act transitively on each of these circles. The matrices fixing $[0 : 1]$ are precisely the diagonal matrices $\mathrm{diag}(\omega, \omega^{-1})$ with $\omega \in \mathrm{U}(1)$. The matrix $\mathrm{diag}(\omega, \omega^{-1})$ sends $[\zeta : 1]$ to $[\omega^2 \zeta : 1]$, which is transitive as required. □

LEMMA 2.5. *The orbit of the diagonal* $\mathrm{SU}(N)$ *action on* $\mathbf{CP}^{N-1} \times \mathbf{CP}^{N-1}$ *consisting of pairs of lines with a fixed Hermitian angle* $\theta \in [0, \pi/2]$ *between them is given by*

$$\textit{Orbit corresponding to } \theta = \begin{cases} \mathbf{CP}^{N-1} & \theta = 0 \\ UT\mathbf{CP}^{N-1} & 0 < \theta < \pi/2 \\ \mathbf{F}(1,1;N) & \theta = \pi/2. \end{cases}$$

PROOF. If the Hermitian angle between lines Λ_A and Λ_B is zero, then $\Lambda_A = \Lambda_B$, and the orbit is clearly a copy of $\mathbf{CP}^{N-1}$. The Hermitian angle between Λ_A and Λ_B is $\pi/2$ if and only if they are orthogonal, so their orbit is $\mathbf{F}(1,1;N)$ by definition of $\mathbf{F}(1,1;N)$.

To describe the orbits corresponding to $0 < \theta < \pi/2$, we again first describe the geometric picture. Given two points in $\mathbf{CP}^{N-1}$, the Hermitian angle between them is precisely the distance between them in the Fubini–Study metric. The exponential map at a point $\Lambda_A \in \mathbf{CP}^{N-1}$ is a diffeomorphism when restricted to the open ball of radius $\pi/2$, and it is surjective on the closed ball of radius $\pi/2$. The image of the sphere of radius $0 < \theta \leq \pi/2$ under the exponential is precisely the set of points $\Lambda_B \in \mathbf{CP}^{N-1}$ whose Hermitian angle from Λ_A is θ. Thus the exponential map, precomposed with rescaling, allows us to identify $UT\mathbf{CP}^{N-1}$ with the orbit corresponding to $0 < \theta < \pi/2$.

We verify the picture in coordinates and formulas. Think of $[0 : \cdots : 0 : 1]$ as a basepoint, and note that the points $[z_1 : \cdots : z_N]$ with $z_N = 0$ all have Hermitian angle $\pi/2$ from $[0 : \cdots : 0 : 1]$. The Hermitian angle θ between $[0 : \cdots : 0 : 1]$ and $[\zeta_1 : \cdots : \zeta_{N-1} : 1]$ is given by

$$\cos\theta = \frac{1}{\sqrt{1 + |\zeta_1|^2 + \cdots + |\zeta_{N-1}|^2}}$$

so the points with a fixed Hermitian angle $0 < \theta < \pi/2$ from $[0 : \cdots : 0 : 1]$ form a $(2N-3)$-sphere consisting of those points $[\zeta_1 : \cdots : \zeta_{N-1} : 1]$ with a fixed value of $|\zeta_1|^2 + \cdots + |\zeta_{N-1}|^2$. Thus the orbit corresponding to θ is a $(2N-3)$-sphere bundle over $\mathbf{CP}^{N-1}$, and it clear from our description that it is the unit tangent bundle. □

We now explain how $UT\mathbf{CP}^{N-1}$ is a circle bundle over $\mathbf{F}(1,1;N)$, and compute its Euler class. First, note that there are two tautological line bundles $\mathscr{A}, \mathscr{B}$ over $\mathbf{F}(1,1;N)$. The fiber of $\mathscr{A}$ over the point $(\Lambda_A, \Lambda_B) \in \mathbf{F}(1,1;N)$ is the 1-dimensional vector space Λ_A, while the fiber of $\mathscr{B}$ over (Λ_A, Λ_B) is Λ_B.

Proposition 2.6. *The space $UT\mathbf{CP}^{N-1}$ is a circle bundle over $\mathbf{F}(1,1;N)$ whose Euler class $e \in H^2(\mathbf{F}(1,1;N))$ is*

$$e = c_1(\mathscr{A}) - c_1(\mathscr{B}).$$

Proof. View $UT\mathbf{CP}^{N-1}$ as the space of pairs (Λ, v) where $\Lambda \in \mathbf{CP}^{N-1}$ and v is a tangent vector at Λ of length $\pi/2$ in the Fubini–Study metric. Then the exponential $\exp_\Lambda(v)$ is a point in $\mathbf{CP}^{N-1}$ whose Hermitian angle from Λ is $\pi/2$. The bundle map $UT\mathbf{CP}^{N-1} \to \mathbf{F}(1,1;N)$ is given by $(\Lambda, v) \mapsto (\Lambda, \exp_\Lambda(v))$. Both $UT\mathbf{CP}^{N-1}$ and $\mathbf{F}(1,1;N)$ are fiber bundles over $\mathbf{CP}^{N-1}$, with fibers S^{2N-3} and $\mathbf{CP}^{N-2}$, respectively. The map $UT\mathbf{CP}^{N-1} \to \mathbf{F}(1,1;N)$ sends fibers to fibers, and the map $S^{2N-3} \to \mathbf{CP}^{N-2}$ is just the standard quotient map with circle fiber generalizing the Hopf fibration. Alternatively, we may view $UT\mathbf{CP}^{N-1}$ as the orbit in $\mathbf{CP}^{N-1} \times \mathbf{CP}^{N-1}$ corresponding to a fixed Hermitian angle $0 < \theta < \pi/2$. Given a point (Λ_A, Λ_B) in this orbit, consider the unique projective line ℓ in $\mathbf{CP}^{N-1}$ through these points. The map $UT\mathbf{CP}^{N-1} \to \mathbf{F}(1,1;N)$ can also be defined by sending (Λ_A, Λ_B) to (Λ_A, Λ_C) where Λ_C is the antipodal point of Λ_A in ℓ.

We compute the Euler class of this circle bundle. First, consider the map $\pi\colon \mathbf{F}(1,1;N) \to \mathbf{G}(2,N)$ sending a pair of orthogonal lines to their 2-dimensional span in $\mathbf{C}^N$. This map is a fiber bundle with fiber $\mathbf{CP}^1$. Let V be the vertical tangent bundle of $\mathbf{F}(1,1;N)$ with respect to π so that

$$T\mathbf{F}(1,1;N) = V \oplus \pi^*(T\mathbf{G}(2,N)).$$

The bundle V is a complex line bundle over $\mathbf{F}(1,1;N)$ whose unit circle bundle is $UT\mathbf{CP}^{N-1}$. To see this, fix $0 < \theta < \pi/2$ and consider the space of quadruples $(\Lambda_A, \Lambda_B, \Lambda_C, \Lambda_D)$ of points in $\mathbf{CP}^{N-1}$ where

- the Hermitian angle between Λ_A and Λ_B is θ,
- Λ_C is the antipodal point of Λ_A on the unique projective line ℓ through Λ_A and Λ_B,
- Λ_D is the antipodal point of Λ_B on ℓ.

Clearly, the pair (Λ_A, Λ_B) determines the quadruple so this space can be identified with $UT\mathbf{CP}^{N-1}$. On the other hand, we can view this quadruple as a pair $((\Lambda_A, \Lambda_C), (\Lambda_B, \Lambda_D))$ of points in $\mathbf{F}(1,1;N)$. The point (Λ_B, Λ_D) can be identified with the exponential of a vector in the fiber of V over (Λ_A, Λ_C) of length θ under a natural generalization of the Fubini–Study metric to $\mathbf{F}(1,1;N)$.

For cosmetic reasons, we orient the circle bundle $UT\mathbf{CP}^{N-1}$ to be opposite that of the natural orientation on the unit circle bundle of V. The Euler class of $UT\mathbf{CP}^{N-1}$ is therefore the negative of the first Chern class of V, so

$$-e = c_1(V) = c_1(T\mathbf{F}(1,1;N)) - \pi^*(c_1(T\mathbf{G}(2,N))).$$

Let $\mathscr{C} \to \mathbf{F}(1,1;N)$ be the rank $N-2$ vector bundle whose fiber over (Λ_A, Λ_B) is the orthogonal complement of $\Lambda_A \oplus \Lambda_B$, and note that $\mathscr{A} \oplus \mathscr{B} \oplus \mathscr{C}$ is the trivial rank N bundle. It follows that $c_1(\mathscr{C}) = -c_1(\mathscr{A}) - c_1(\mathscr{B})$. Let S, Q be the tautological vector bundles over $\mathbf{G}(2,N)$ so that their fibers over the point $\Lambda \in \mathbf{G}(2,N)$ are Λ and $\Lambda^\perp$ respectively. Note that $\pi^*(S) = \mathscr{A} \oplus \mathscr{B}$ and $\pi^*(Q) = \mathscr{C}$. Recalling that $T(\mathbf{G}(2,N)) \cong \mathrm{Hom}(S,Q) \cong S^* \otimes Q$ and $c_1(V \otimes W) = \mathrm{rk}(V)c_1(W) + \mathrm{rk}(W)c_1(V)$, we have

$$\begin{aligned}\pi^*(c_1(T\mathbf{G}(2,N))) &= \pi^*(-\mathrm{rk}(Q)c_1(S) + \mathrm{rk}(S)c_1(Q)) \\ &= -(N-2)(c_1(\mathscr{A}) + c_1(\mathscr{B})) + 2(-c_1(\mathscr{A}) - c_1(\mathscr{B})) \\ &= -N(c_1(\mathscr{A}) + c_1(\mathscr{B})).\end{aligned}$$

Since $T\mathbf{F}(1,1;N) \cong \mathrm{Hom}(\mathscr{A},\mathscr{B}) \oplus \mathrm{Hom}(\mathscr{A},\mathscr{C}) \oplus \mathrm{Hom}(\mathscr{B},\mathscr{C})$ [**9**], we find that

$$\begin{aligned}c_1(T\mathbf{F}(1,1;N)) &= (-c_1(\mathscr{A}) + c_1(\mathscr{B})) + (-(N-2)c_1(\mathscr{A}) - c_1(\mathscr{A}) - c_1(\mathscr{B})) \\ &\quad + (-(N-2)c_1(\mathscr{B}) - c_1(\mathscr{A}) - c_1(\mathscr{B})) \\ &= (-N-1)c_1(\mathscr{A}) + (-N+1)c_1(\mathscr{B}).\end{aligned}$$

Thus, we have

$$-e = (-N-1)c_1(\mathscr{A}) + (-N+1)c_1(\mathscr{B}) + N(c_1(\mathscr{A}) + c_1(\mathscr{B})) = c_1(\mathscr{B}) - c_1(\mathscr{A})$$

as claimed. □

3. $\mathfrak{sl}(N)$ homology

We describe the construction of $\mathfrak{sl}(N)$ homology, drawing from the exposition in [**17**, section 2].

DEFINITION 3.1. A *web W (with labels* 1 *and* 2*)* is a trivalent graph embedded in the plane where multiple edges and circles without vertices are permitted. Each edge is oriented and labeled 1 or 2, and the orientations and labels of the three edges incident to each vertex must match one of the two models given in Figure 2.

FIGURE 2. Orientations and labels of the edges incident to a vertex in a web.

Associated to a web W and integer $N \geq 2$ is a polynomial $P_N(W) \in \mathbf{Z}[q, q^{-1}]$ with nonnegative coefficients called its MOY polynomial [**12**]. This polynomial is sometimes denoted $\langle W \rangle$, though it does depend on N. There are local relations that the polynomial satisfies which also determine it, which we summarize in the following result (see for example [**11**, Figure 3]). The MOY polynomial is multiplicative under disjoint union, so in particular, the MOY polynomial of the empty web is 1.

MOY CALCULUS. *The MOY polynomial of a web is determined by the local relations in Figure* 3.

Figure 3 uses the following notation. For integers $k, N \geq 0$, the polynomials $[N], \left[{N \atop k}\right] \in \mathbf{Z}[q, q^{-1}]$ are

$$[N] = \frac{q^N - q^{-N}}{q - q^{-1}} = q^{N-1} + q^{N-3} + \cdots + q^{-N+1}$$

$$\begin{bmatrix} N \\ k \end{bmatrix} = \begin{cases} \dfrac{[N]!}{[k]![N-k]!} & 0 \leq k \leq N \\ 0 & \text{else} \end{cases}$$

where $[k]! = [k] \cdots [2][1]$. These polynomials are just the Poincaré polynomials of $\mathbf{CP}^{k-1}$ and the complex Grassmannian $\mathbf{G}(k, n)$, shifted downwards to be invariant under $q \mapsto q^{-1}$. After a similar shift, the Poincaré polynomial of $\mathbf{F}(1, 1; N)$ is $[N][N-1]$. This can be seen from the fact that the Serre spectral sequence of the bundle $\mathbf{CP}^{N-2} \to \mathbf{F}(1, 1; N) \to \mathbf{CP}^{N-1}$ collapses on the second page for lacunary reasons.

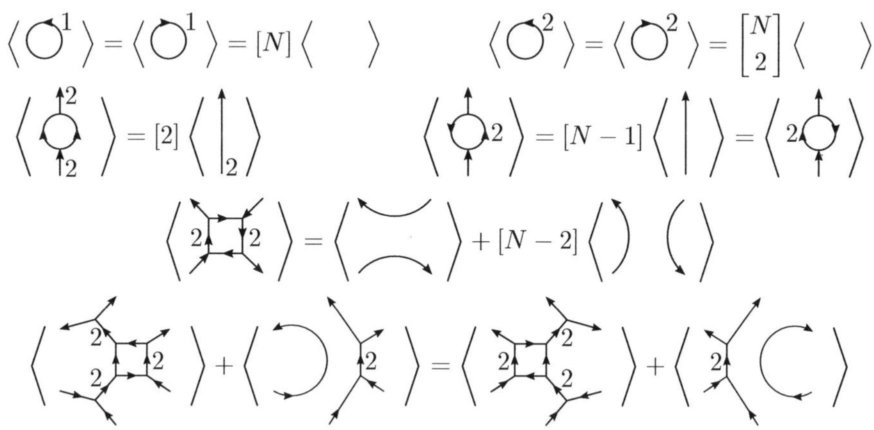

FIGURE 3. Local relations in MOY calculus. Edges without explicit labels are labeled 1.

Given a link diagram, the $\mathfrak{sl}(N)$ polynomial can be computed as a weighted sum of MOY polynomials obtained by taking complete resolutions of the diagram, where one of the two possible resolutions introduces trivalent vertices and an edge labeled 2. See Figure 4.

EXAMPLE 3.2. Consider applying the first relation of Figure 4 to the positive crossing in the diagram of unknot described by $T(2, 1)$. The first resolution consists of two planar loops which can be thought of as $T(2, 0) = O \sqcup O$, while the second resolution is called the θ-*web*:

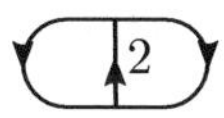

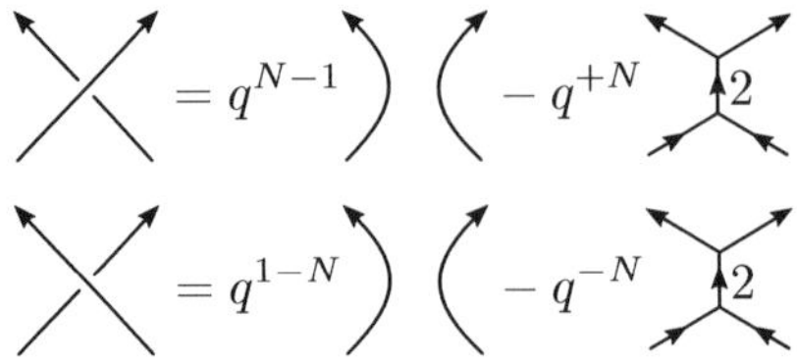

FIGURE 4. The $\mathfrak{sl}(N)$ polynomial from the MOY polynomial. Edges without explicit labels are labeled 1.

The relation states that $P_N(T(2,1)) = q^{N-1}P_N(O \sqcup O) - q^N P_N(\text{CD})$. Using MOY calculus (Figure 3), we find that $P_N(O \sqcup O) = [N]^2$ and $P_N(\text{CD}) = [N][N-1]$ from which it follows that

$$\begin{aligned} P_N(T(2,1)) &= [N](q^{N-1}[N] - q^N[N-1]) \\ &= [N]\left(\frac{q^{N-1}(q^N - q^{-N}) - q^N(q^{N-1} - q^{-N+1})}{q - q^{-1}}\right) = [N] \end{aligned}$$

thereby illustrating one instance of invariance of P_N under the first Reidemeister move.

The construction of $\mathfrak{sl}(N)$ homology, which categorifies the $\mathfrak{sl}(N)$ polynomial, begins by categorifying the MOY polynomial of an $\mathfrak{sl}(N)$ web. Associated to each web W is a *state space* $C_N(W)$, which takes the form of a finitely-generated free abelian group equipped with a $\mathbf{Z}$-grading called the q-grading. The Poincaré polynomial of $C_N(W)$ is $P_N(W)$. Of course, the polynomial $P_N(W)$ therefore determines the graded isomorphism type of $C_N(W)$, so as a group by itself, the state space $C_N(W)$ is contains nothing new. What makes $C_N(W)$ a categorification of $P_N(W)$ is *functoriality.* There is a notion of a cobordism between webs called a *foam*, and a foam $F\colon W_0 \to W_1$ induces a map $C_N(F)\colon C_N(W_0) \to C_N(W_1)$ on state spaces which is functorial under gluing. Each foam F has a *degree* $d(F)$, and the induced map is homogeneous of degree $d(F)$.

Just as a web is a 1-complex in $\mathbf{R}^2$ with trivalent singularities and labels on its edges, a foam is a 2-complex in $[0,1] \times \mathbf{R}^2$ with constrained singularities and labels on its 2-dimensional facets. For example, the *identity foam* Id_W of a web W is just the product $[0,1] \times W \subset [0,1] \times \mathbf{R}^2$ whose additional pieces of data, such as orientations and labels, are inherited from those of W. One novel feature of foams is that facets may carry finitely many marked points called *dots*, which are directly analogous to the dots that cobordisms may carry in Khovanov homology. For a precise definition of a foam, see for example [**15**, **18**].

EXAMPLE 3.3. Let O denote an oriented planar loop labeled 1 thought of as a web. Since $P_N(O) = [N]$, we know that $C_N(O)$ is graded-isomorphic to $q^{1-N}H^*(\mathbf{CP}^{N-1})$ where q^{1-N} denotes the grading shift required so that its Poincaré polynomial is $[N]$. This grading shift is just the complex dimension of $\mathbf{CP}^{N-1}$, which is half its real dimension. It turns out that there is a particular isomorphism, which we will denote by an equality $C_N(O) = q^{1-N}H^*(\mathbf{CP}^{N-1})$, that is particularly nice in the following way. Consider the identity foam of O with a single dot added to its interior. This foam induces a map $C_N(O) \to C_N(0)$ of degree 2 called the *dot map*. Under the identification $C_N(O) = q^{1-N}H^*(\mathbf{CP}^{N-1})$, the dot map is cup product with the positive generator of $q^{1-N}H^2(\mathbf{CP}^{N-1})$. In particular, it is cup product with the first Chern class of the hyperplane line bundle $\mathscr{O}(1)$ over $\mathbf{CP}^{N-1}$, which is dual to the tautological line bundle $\mathscr{O}(-1)$. The map may also be viewed as multiplication by X under the standard identification $H^*(\mathbf{CP}^{N-1}) = \mathbf{Z}[X]/X^N$.

REMARK 3.4. Let W be a web, and consider a foam $F\colon \emptyset \to W$ where $\emptyset$ denotes the empty web. The state space of the empty web is $\mathbf{Z}$, and the foam F induces a map $\mathbf{Z} \to C_N(W)$. We may think of the foam F itself as an element of the state space $C_N(W)$ by identifying it with the image of $1 \in \mathbf{Z}$ under its induced map. It turns out that the state space of any web is generated by foams. As an example, consider the "cup" foam $D\colon \emptyset \to O$ consisting of just a single disc. The map induced by the cup sends $1 \in \mathbf{Z}$ to $1 \in q^{1-N}H^0(\mathbf{CP}^{N-1})$. The foams obtained by adding i dots to D for $i = 0, 1, \ldots, N-1$ are therefore a basis for the state space of O by Example 3.3.

Given two foams $F, G\colon \emptyset \to W$, there is a *pairing* $\langle F, G\rangle \in \mathbf{Z}$ defined in the following way. The *mirror* of G is a foam $\overline{G}\colon W \to \emptyset$ obtained by reflecting G through a plane. The foam obtained by gluing F to $\overline{G}$ along their common boundary W is a foam $F \cup_W \overline{G}\colon \emptyset \to \emptyset$. This foam induces a map $\mathbf{Z} \to \mathbf{Z}$, which is given by multiplication by an integer. This integer is precisely the pairing $\langle F, G\rangle \in \mathbf{Z}$. This pairing is defined on all of $C_N(W)$ and turns out to be nondegenerate. The pairing on $C_N(O) = q^{1-N}H^*(\mathbf{CP}^{N-1})$ is precisely the Poincaré pairing with respect to the negative orientation on $\mathbf{CP}^{N-1}$. In particular, the map $\mathbf{Z} \to \mathbf{Z}$ induced by a sphere with $N-1$ dots is multiplication by -1.

The mirror of a foam $H\colon W_0 \to W_1$ is a foam $\overline{H}\colon W_1 \to W_0$, and the induced maps $C_N(H)\colon C_N(W_0) \to C_N(W_1)$ and $C_N(\overline{H})\colon C_N(W_1) \to C_N(W_0)$ are transposes of each other with respect to the described nondegenerate pairings.

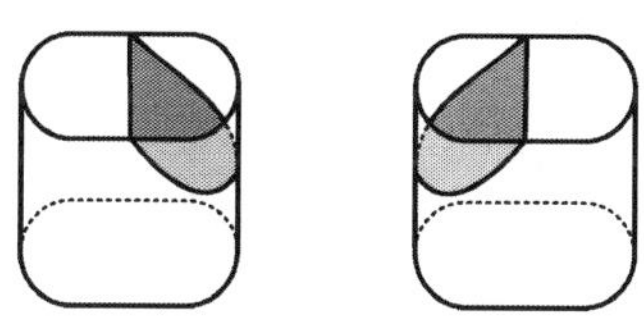

FIGURE 5. Two foams from O, on the bottom, to the θ-web, on the top. The maps induced on state spaces are the maps induced by the two projections $\pi_A, \pi_B\colon \mathbf{F}(1,1;N) \to \mathbf{CP}^{N-1}$.

EXAMPLE 3.5. Consider the θ-web defined in Example 3.2 whose MOY polynomial is $[N][N-1]$. It follows that C_N(⦶) and $q^{3-2N}H^*(\mathbf{F}(1,1;N))$ are graded-isomorphic. We note that the complex dimension of $\mathbf{F}(1,1;N)$ is $2N-3$. Again, there turns out to be a particularly nice isomorphism between them that we denote by an equality C_N(⦶) $= q^{3-2N}H^*(\mathbf{F}(1,1;N))$. The foams in Figure 5 induce maps from $C_N(O) = q^{3-2N}H^*(\mathbf{CP}^{N-1})$ to C_N(⦶) $= q^{3-2N}H^*(\mathbf{F}(1,1;N))$. These two maps are precisely the maps on cohomology induced by the two projections $\pi_A, \pi_B\colon \mathbf{F}(1,1;N) \to \mathbf{CP}^{N-1}$ given by sending (Λ_A, Λ_B) to Λ_A or Λ_B, respectively. Our convention is that the foam on the left induces $(\pi_A)^*$ while the one on the right induces $(\pi_B)^*$.

Just as in section 2, let $\mathscr{A}$ and $\mathscr{B}$ denote the tautological line bundles over $\mathbf{F}(1,1;N)$ whose fibers over the point (Λ_A, Λ_B) are the vector spaces Λ_A and Λ_B, respectively. Note that $\mathscr{A}$, $\mathscr{B}$ may be obtained by pulling back the tautological line bundle $\mathscr{O}(-1)$ over $\mathbf{CP}^{N-1}$ by the projections π_A, π_B. Starting with the identity foam of the θ-web, consider adding a dot to either the leftmost facet or to the rightmost facet. These two foams are denoted

These foams induce endomorphisms of degree 2 on the state space of the θ-web. Under the identification $C_N(\Theta) = q^{3-2N}H^*(\mathbf{F}(1,1;N))$, these endomorphisms are precisely cup product with the negatives of the first Chern classes of $\mathscr{A}$ and $\mathscr{B}$, respectively. Our convention is that the foam on the left induces $-c_1(\mathscr{A})$ while the foam on the right induces $-c_1(\mathscr{B})$.

Once the MOY polynomial of an $\mathfrak{sl}(N)$ web has been categorified, $\mathfrak{sl}(N)$ homology for links is obtained by categorifying the relations in Figure 4. The state space of a web is thought of as a chain complex supported in homological grading zero. The complex associated to a diagram with ordered crossings is inductively defined using Figure 6. In particular, the complex associated to the diagram on the left is defined to be the mapping cone of the complexes associated to the diagrams on the right. In other words, the chain complex assigned to a diagram is obtained from the cube of resolutions construction where the edge maps are given by the local foams given in Figure 6. The bigraded chain homotopy type of this complex is invariant under Reidemeister moves, and its bigraded homology group is the $\mathfrak{sl}(N)$ homology of the oriented link.

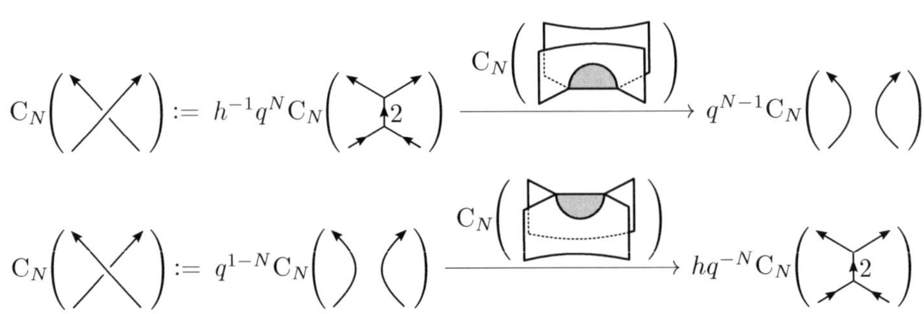

FIGURE 6. Edge maps in the cube of resolutions defining $\mathfrak{sl}(N)$ homology. The notation $h^i q^j$ denotes a shift in the homological grading by i and a shift in the q-grading by j. Our convention for signs in a mapping cone is that an odd shift in homological grading of a complex negates the differential.

EXAMPLE 3.6. Consider the diagram of the unknot with a single negative crossing, which we can think of as $T(2,-1)$. Then $C_N(T(2,-1))$ is supported in homological gradings 0 and 1. In homological grading 0, the complex is $q^{1-N}C_N(T(2,0)) = q^{1-N}(q^{1-N}H^*(\mathbf{CP}^{N-1}) \otimes q^{1-N}H^*(\mathbf{CP}^{N-1}))$ while in homological grading 1, the complex is $q^{-N}C_N(\Theta) = q^{-N}q^{3-2N}H^*(\mathbf{F}(1,1;N))$. The differential is precisely the map on cohomology induced by $(\pi_A, \pi_B)\colon \mathbf{F}(1,1;N) \to \mathbf{CP}^{N-1} \times \mathbf{CP}^{N-1}$. This can be seen by comparing the foam defining the differential with the foams in Figure 5 together with the facts asserted in Remark 3.4 and Example 3.5.

This map is surjective because the first Chern classes of the line bundles $\mathscr{A}$ and $\mathscr{B}$ generate $H^*(\mathbf{F}(1,1;N))$. It follows that the homology of this complex is supported in homological grading zero, and it is free abelian with graded rank $[N]$ thereby illustrating one instance of invariance of $\mathfrak{sl}(N)$ homology under the first Reidemeister move.

Now consider the complex associated to the diagram $T(2,1)$. Then $C_N(T(2,1))$ is supported in homological gradings -1 and 0. By Remark 3.4, the differential is just the transpose of the differential in the complex $C_N(T(2,-1))$ so the map is injective and its cokernel is free abelian of graded rank $[N]$.

Before turning to the proof of Observation 1.1, we record two facts that we will use. The first is a special case of the *dot migration relation* [**13**, Equation (3.9)], and the second is an isomorphism that categorifies the first equality in the second row of Figure 3, and is an analogue of delooping in Khovanov homology [**2**]. The second follows from [**13**, Equations 3.10 and 3.11].

LEMMA 3.7. *Generalizing the notation used in Example 3.5, we let a diagram of a web W with a dot on an edge denote the identity foam of W with a dot on the facet corresponding to the edge. We have the equality of maps:*

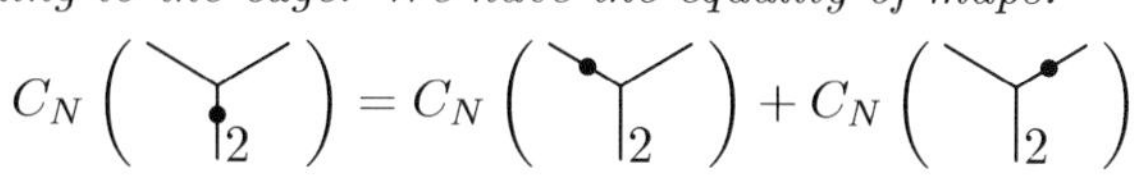

The relation is valid if all edges are oriented upwards or if all edges are oriented downwards.

LEMMA 3.8. *The maps*

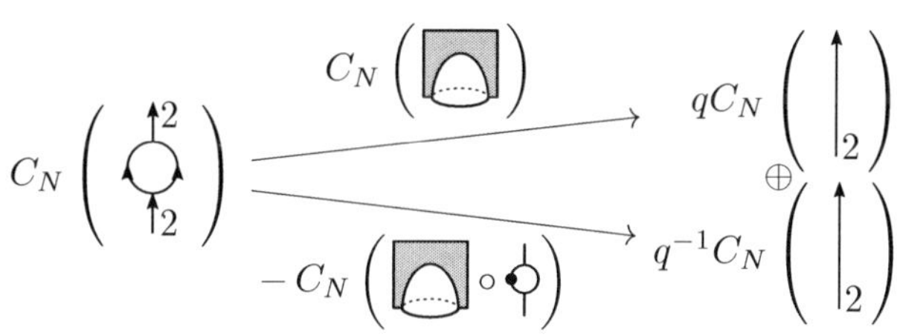

and

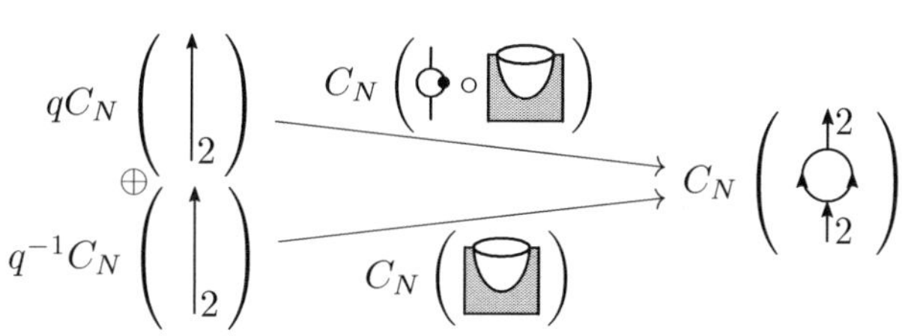

are inverse isomorphisms. In particular, we have the identities

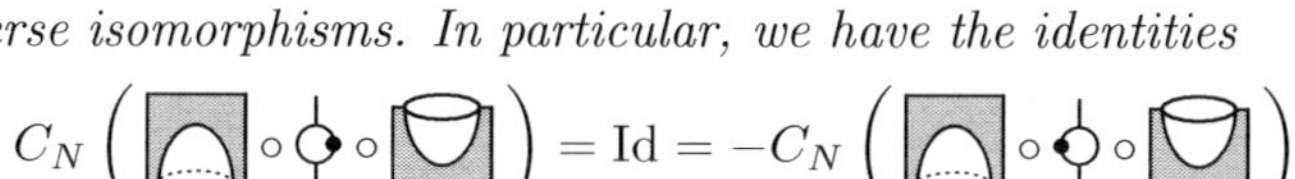

and

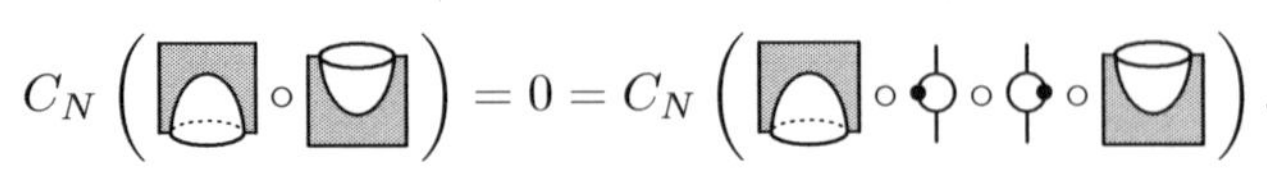

4. Proof of Observation 1.1

The following proposition, which is well-known to experts, is the key result on the $\mathfrak{sl}(N)$ homology side needed to prove Observation 1.1. See for example [**7**, **16**, **19**].

PROPOSITION 4.1. *For each $m \geq 1$, there is a homotopy equivalence*

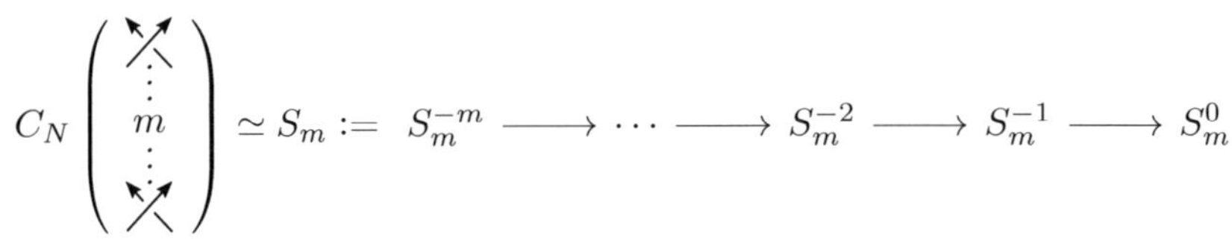

where

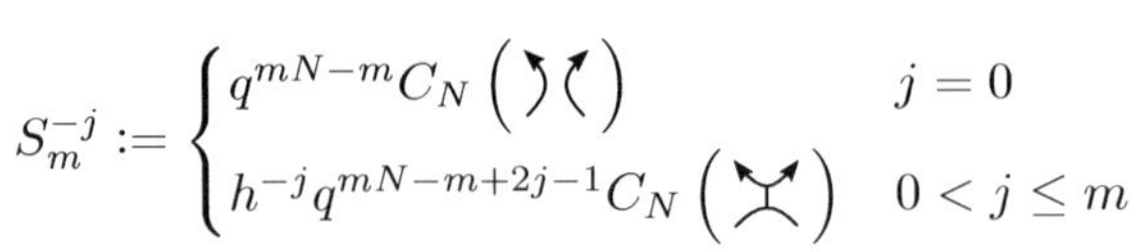

and

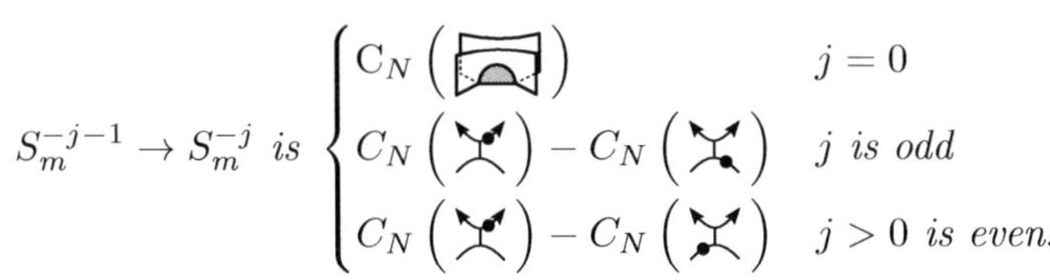

We view these pictures as local modifications of a fixed diagram, but we note that there is also a theory for tangle diagrams (see for example [**4**]) analogous to [**1**] for which Proposition 4.1 is valid.

REMARK 4.2. Omitting the symbols $C_N(-)$ as a matter of notation, the complexes $q^{i-iN}S_i$ for $i = 1, 2, 3, 4$ are

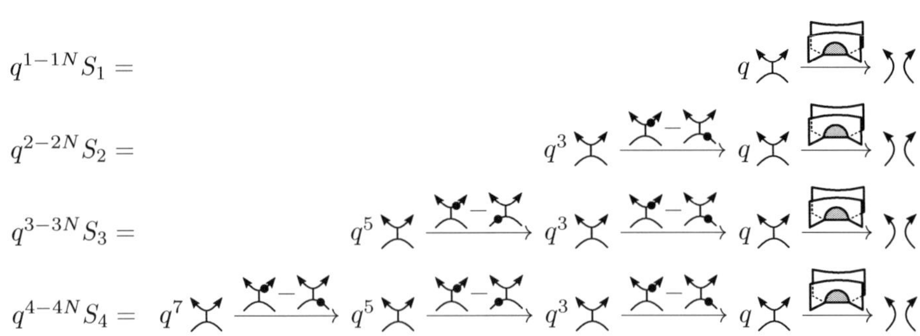

where the rightmost term in each complex is in homological degree 0.

PROOF OF PROPOSITION 4.1. We omit the symbols $C_N(-)$ for notational simplicity as in Remark 4.2. We prove the result by induction on $m \geq 1$. The base case $m = 1$ is true by definition (Figure 6). For the inductive case, we must prove that

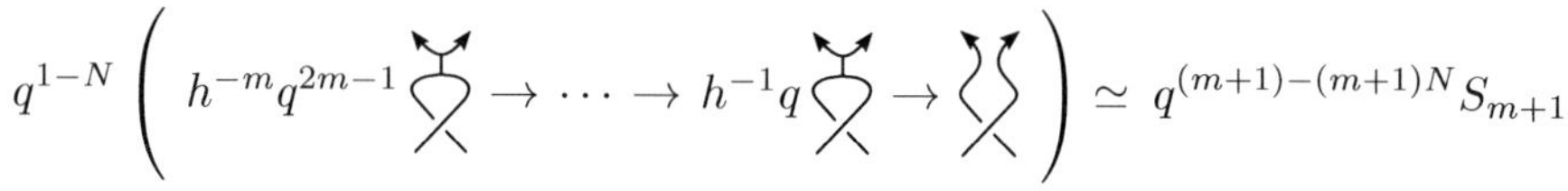

where the arrows in the complex on the lefthand side are induced by those in S_m. The complex expands to

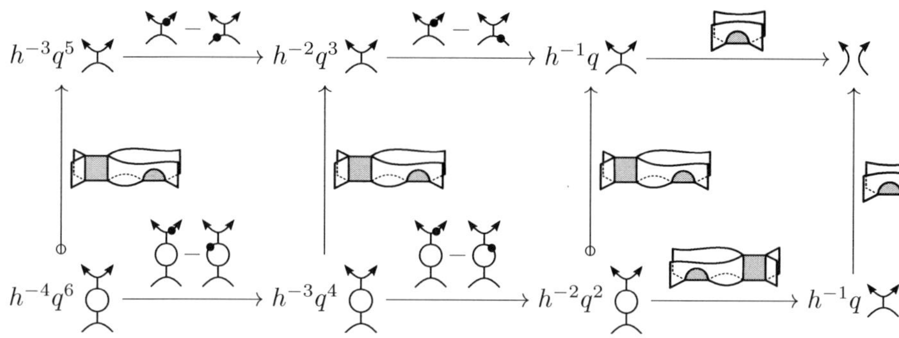

where we have drawn the $m = 3$ case for clarity. The small circles drawn on the vertical arrows in every other column indicate negation (see the caption of Figure 6). By Lemma 3.8, we have inverse isomorphisms

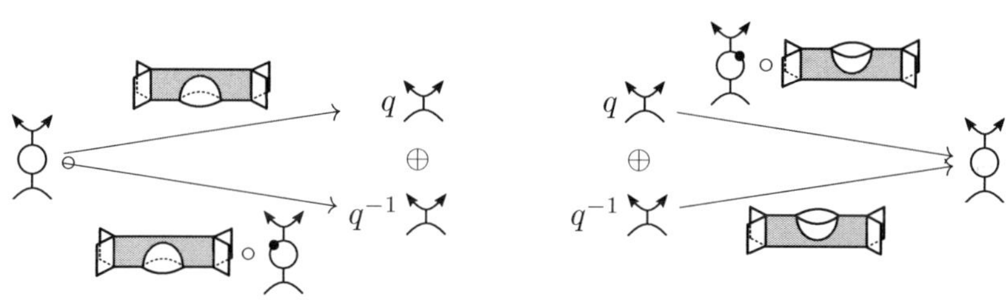

Using this isomorphism, we may rewrite the complex as

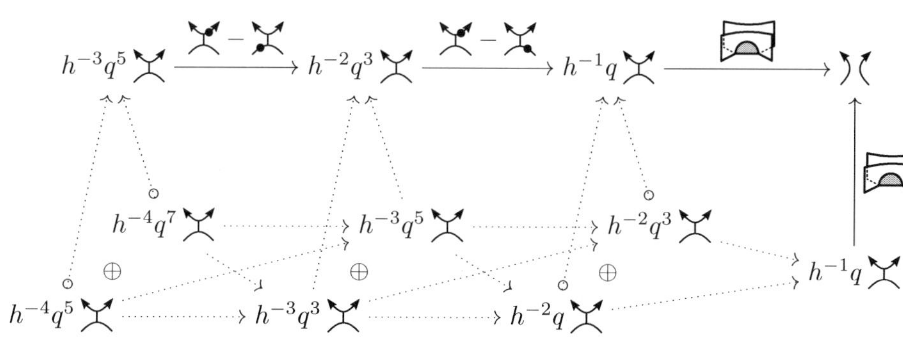

where the dotted arrows are composites, which we now consider. The vertical arrows can be determined using the observation that

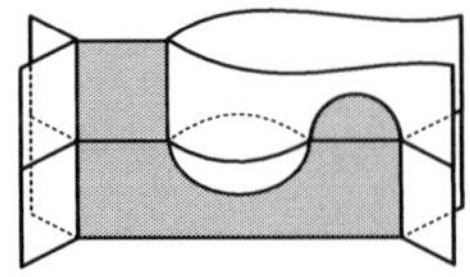

is isotopic to the identity foam. A similar observation allows us to also determine the rightmost horizontal arrows. We therefore have

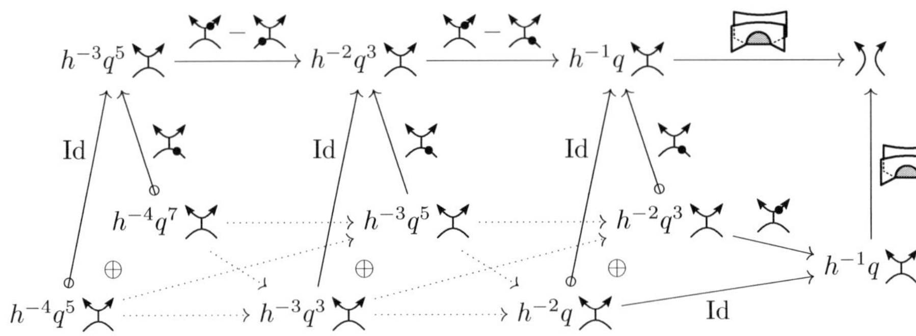

We may therefore simplify the complex by Gaussian elimination along the $m(= 3)$ vertical identity maps (for Gaussian elimination, see for example [**2**, Lemma 4.2] or [**3**, Lemma 3.1]). Only half of the remaining dotted maps need to be computed explicitly to determine the result of Gaussian elimination. They are

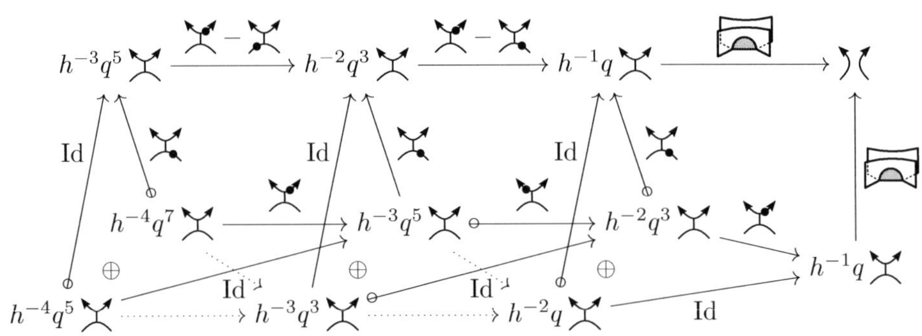

To be explicit, we are asserting that

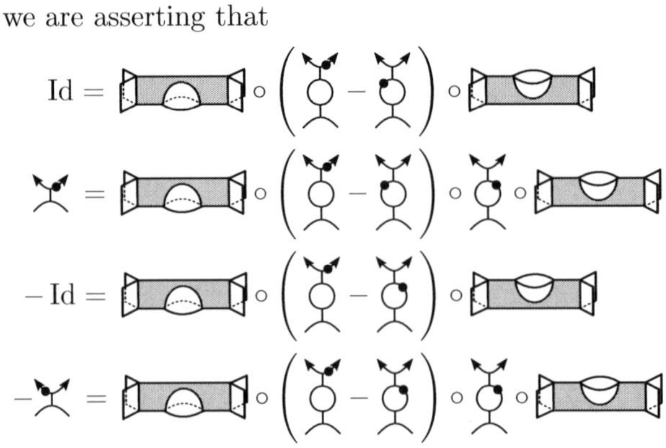

The first three computations follow from Lemma 3.8, while the last computation uses the identity

which follows from Lemma 3.7. Finally, we perform the stated Gaussian elimination to obtain

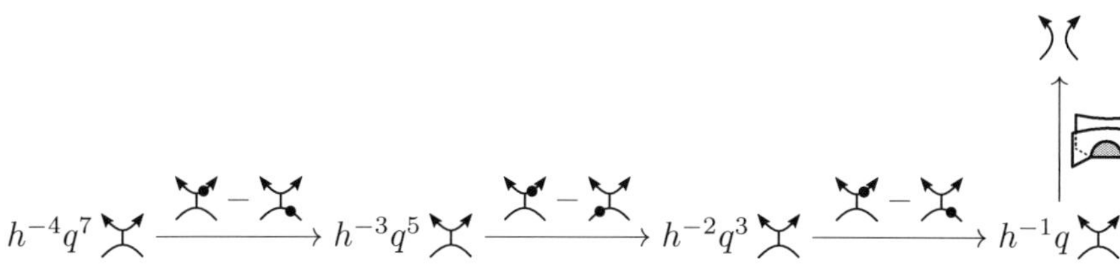

which is precisely $q^{(m+1)-(m+1)N}S_{m+1}$ as required. □

PROOF OF OBSERVATION 1.1. Let O denote the crossingless diagram of the unknot, and note that when $m = 0$, the torus link $T(2,0)$ is just the two-component unlink $O \sqcup O$. Since $\mathscr{R}_N(O \sqcup O) = \mathbf{CP}^{N-1} \times \mathbf{CP}^{N-1}$, the $m = 0$ case of the observation follows from $\mathrm{KR}_N(O) \cong q^{1-N}H^*(\mathbf{CP}^{N-1})$ using $\mathrm{KR}_N(O \sqcup O) \cong \mathrm{KR}_N(O) \otimes \mathrm{KR}_N(O)$ and $H^*(\mathbf{CP}^{N-1} \times \mathbf{CP}^{N-1}) \cong H^*(\mathbf{CP}^{N-1}) \otimes H^*(\mathbf{CP}^{N-1})$. We may assume that $m \geq 1$, since $\mathscr{R}_N(T(2,-m))$ and $\mathscr{R}_N(T(2,m))$ are the same, and $\mathrm{KR}_N(T(2,-m))$ and $\mathrm{KR}_N(T(2,m))$ are related by the universal coefficient theorem, which can be shown using Remark 3.4.

Proposition 4.1 gives a homotopy equivalence between $C_N(T(2,m))$ and a much simpler complex. For the first few values of m, for example, we form the braid closures of the local diagrams in Remark 4.2 to obtain

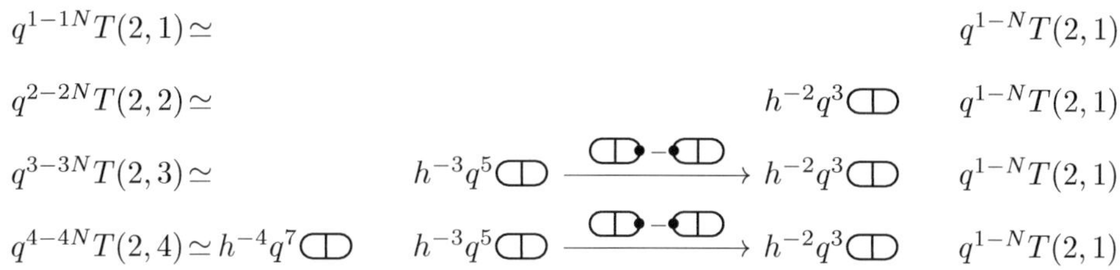

where we have again surpressed the symbols $C_N(-)$. Note that the maps from even homological degree to odd homological degree are all zero. Furthermore, since $T(2,1)$ represents the unknot, $C_N(T(2,1))$ is homotopy equivalent to $C_N(O)$. Thus, if we set

then $q^{m-mN}C_N(T(2,m))$ is homotopy equivalent to

$$\begin{cases} \left(\bigoplus_{i=1}^{k} h^{-2i}q^{4i}A\right) \oplus \left(q^{1-N}C_N(O)\right) & m = 2k+1 \text{ is odd} \\ \left(h^{-2k}q^{4k-1}C_N(\bigcirc\!\!\bigcirc)\right) \oplus \left(\bigoplus_{i=1}^{k-1} h^{-2i}q^{4i}A\right) \oplus \left(q^{1-N}C_N(O)\right) & m = 2k \text{ is even.} \end{cases}$$

Proposition 2.3 gives an explicit calculation of $\mathscr{R}_N(T(2,m))$: it consists of a single copy of $\mathbf{CP}^{N-1}$, a single copy of $\mathbf{F}(1,1;N)$ when m is even, and $\lfloor (m-1)/2 \rfloor$ copies of

$UT\mathbf{CP}^{N-1}$. We recall from Examples 3.3 and 3.5 that $C_N(O) = q^{1-N}H^*(\mathbf{CP}^{N-1})$ and $C_N(\mathbb{O}) = q^{3-2N}H^*(\mathbf{F}(1,1;N))$. Thus, in order to show that $\mathrm{KR}_N(T(2,m))$ and $H^*(\mathscr{R}_N(T(2,m)))$ are isomorphic as abelian groups, it suffices to show that the homology of the complex A is isomorphic to $H^*(UT\mathbf{CP}^{N-1})$.

By Proposition 2.6, we know that $U := UT\mathbf{CP}^{N-1}$ is a circle bundle over $\mathbf{F}(1,1;N)$ with Euler class $c_1(\mathscr{A}) - c_1(\mathscr{B})$, where $\mathscr{A}$ and $\mathscr{B}$ are the two tautological line bundles over $\mathbf{F}(1,1;N)$. Since the cohomology of $\mathbf{F}(1,1;N)$ is supported in even degrees, the Gysin sequence for this bundle is the exact sequence

$$0 \longrightarrow H^{2i+1}(U) \longrightarrow H^{2i}(\mathbf{F}(1,1;N)) \xrightarrow{\cup e} H^{2i+2}(\mathbf{F}(1,1;N)) \longrightarrow H^{2i+2}(U) \longrightarrow 0$$

Under the identification $C_N(\mathbb{O}) = q^{3-2N}H^*(\mathbf{F}(1,1;N))$ given in Example 3.5, the complex A is precisely

$$h^{-1}q^{4-2N}H^*(\mathbf{F}(1,1;N)) \xrightarrow{\cup e} q^{2-2N}H^*(\mathbf{F}(1,1;N))$$

from which it immediately follows that the homology of A is isomorphic to the cohomology of $U = UT\mathbf{CP}^{N-1}$, without needing to actually compute either group.

Notice that we may actually describe the homology of A as a bigraded group in terms of the cohomology of $UT\mathbf{CP}^{N-1}$, which thereby gives the complete bigrading information on $\mathrm{KR}_N(T(2,m))$. The homology of A is supported in homological degrees -1 and 0, and in even q-gradings. In homological degree 0, the q-graded homology of A is simply the direct sum of the even-degree cohomology groups of $UT\mathbf{CP}^{N-1}$, shifted in grading by $2-2N$. In homological degree -1, the q-graded homology of A is the direct sum of the odd-degree cohomology groups of $UT\mathbf{CP}^{N-1}$, shifted in grading by $3-2N$.

We can explicitly determine the cohomology of $UT\mathbf{CP}^{N-1}$ using the Gysin sequence for the $(2N-3)$-sphere bundle $UT\mathbf{CP}^{N-1} \to \mathbf{CP}^{N-1}$ whose Euler class is $N = \chi(\mathbf{CP}^{N-1})$ times a generator of $H^{2N-2}(\mathbf{CP}^{N-1})$. We find that

$$H^i(UT\mathbf{CP}^{N-1}) = \begin{cases} \mathbf{Z} & i \text{ is even and } 0 \le i \le 2N-4 \\ \mathbf{Z}/N & i = 2N-2 \\ \mathbf{Z} & i \text{ is odd and } 2N-1 \le i \le 4N-5 \\ 0 & \text{else.} \end{cases}$$

Thus, in homological degree 0, the homology of A consists of a copy of $\mathbf{Z}$ in even q-gradings i satisfying $2-2N \le i \le -2$ and a copy of $\mathbf{Z}/N$ in q-grading 0. In homological degree -1, the homology of A consists of a copy of $\mathbf{Z}$ in even q-gradings i satisfying $2 \le i \le 2N-2$. □

Acknowledgment

Many thanks to my advisor Peter Kronheimer for his continued guidance, support, and encouragement.

References

[1] Dror Bar-Natan, *Khovanov's homology for tangles and cobordisms*, Geom. Topol. **9** (2005), 1443–1499, DOI 10.2140/gt.2005.9.1443. MR2174270

[2] Dror Bar-Natan, *Fast Khovanov homology computations*, J. Knot Theory Ramifications **16** (2007), no. 3, 243–255, DOI 10.1142/S0218216507005294. MR2320156

[3] Joshua Batson and Cotton Seed, *A link-splitting spectral sequence in Khovanov homology*, Duke Math. J. **164** (2015), no. 5, 801–841, DOI 10.1215/00127094-2881374. MR3332892

[4] Michael Ehrig, Daniel Tubbenhauer, and Paul Wedrich, *Functoriality of colored link homologies*, Proc. Lond. Math. Soc. (3) **117** (2018), no. 5, 996–1040, DOI 10.1112/plms.12154. MR3877770

[5] Jonathan Grant, *The moduli problem of Lobb and Zentner and the colored* $\mathfrak{sl}(N)$ *graph invariant*, J. Knot Theory Ramifications **22** (2013), no. 10, 1350060, 16, DOI 10.1142/S0218216513500600. MR3125899

[6] Mikhail Khovanov and Lev Rozansky, *Matrix factorizations and link homology*, Fund. Math. **199** (2008), no. 1, 1–91, DOI 10.4064/fm199-1-1. MR2391017

[7] Daniel Krasner, *A computation in Khovanov-Rozansky homology*, Fund. Math. **203** (2009), no. 1, 75–95, DOI 10.4064/fm203-1-7. MR2491784

[8] P. B. Kronheimer and T. S. Mrowka, *Knot homology groups from instantons*, J. Topol. **4** (2011), no. 4, 835–918, DOI 10.1112/jtopol/jtr024. MR2860345

[9] Kee Yuen Lam, *A formula for the tangent bundle of flag manifolds and related manifolds*, Trans. Amer. Math. Soc. **213** (1975), 305–314, DOI 10.2307/1998048. MR431194

[10] Andrew Lobb and Raphael Zentner, *The quantum* sl(N) *graph invariant and a moduli space*, Int. Math. Res. Not. IMRN **7** (2014), 1956–1972, DOI 10.1093/imrn/rns275. MR3190356

[11] Marco Mackaay, Marko Stošić, and Pedro Vaz, $\mathfrak{sl}(N)$*-link homology* $(N \geq 4)$ *using foams and the Kapustin-Li formula*, Geom. Topol. **13** (2009), no. 2, 1075–1128, DOI 10.2140/gt.2009.13.1075. MR2491657

[12] Hitoshi Murakami, Tomotada Ohtsuki, and Shuji Yamada, *Homfly polynomial via an invariant of colored plane graphs*, Enseign. Math. (2) **44** (1998), no. 3-4, 325–360. MR1659228

[13] Hoel Queffelec and David E. V. Rose, *The* $\mathfrak{sl}_n$ *foam 2-category: a combinatorial formulation of Khovanov-Rozansky homology via categorical skew Howe duality*, Adv. Math. **302** (2016), 1251–1339, DOI 10.1016/j.aim.2016.07.027. MR3545951

[14] Jacob Rasmussen, *Khovanov-Rozansky homology of two-bridge knots and links*, Duke Math. J. **136** (2007), no. 3, 551–583, DOI 10.1215/S0012-7094-07-13635-4. MR2309174

[15] Louis-Hadrien Robert and Emmanuel Wagner, *A closed formula for the evaluation of foams*, Quantum Topol. **11** (2020), no. 3, 411–487, DOI 10.4171/qt/139. MR4164001

[16] Benjamin Thompson, *Khovanov complexes of rational tangles*, `arXiv:1701.07525`, 2017.

[17] Joshua Wang, *On* $\mathfrak{sl}(N)$ *link homology with mod N coefficients*, Quantum Topol. **15** (2024), no. 1, 87–121, DOI 10.4171/qt/194. MR4706035

[18] Joshua Wang, *Colored sl(N) homology and SU(N) representations of the trefoil*, `arXiv:2211.08409`, 2022.

[19] Hao Wu, *Simplified Khovanov-Rozansky chain complexes of open 2-braids*, Topology Appl. **159** (2012), no. 14, 3190–3203, DOI 10.1016/j.topol.2012.06.008. MR2948277

Department of Mathematics, Harvard University, Cambridge, Massachusetts 02138

Proceedings of Symposia in Pure Mathematics
Volume **109**, 2024
https://doi.org/10.1090/pspum/109/01999

Heegaard Floer multicurves of double tangles

Claudius Zibrowius

ABSTRACT. We describe a simple formula for computing the Heegaard Floer multicurve invariant of double tangles from the Heegaard Floer multicurve invariant of knot complements. A comparison with a similar multicurve invariant for Conway tangles in the setting of Khovanov homology confirms that knot Floer homology and Khovanov homology behave very differently under satellite operations, echoing recent observations from [**LZ22**]. We also obtain a new characterisation of L-space knots in terms of Heegaard Floer A-link satellites. Along the way, we find the first example of a satellite knot whose knot Floer homology is thin.

1. Introduction

FIGURE 1. The double tangle (right) of the right-handed trefoil knot (left).

The double tangle T_K associated with an oriented knot $K \subset S^3$ is the Conway tangle obtained by cutting the knot K open to a tangle with two ends and adding another tangle strand parallel to the first, as illustrated in Figure 1. This family of Conway tangles plays a key role in understanding satellites of wrapping number 2: Any such satellite knot $P(K)$ can be written as

$$P(K) = T_K \cup T_P,$$

the union of the double tangle T_K associated with the companion knot K and a second Conway tangle T_P corresponding to the pattern P. This perspective allows

The author was funded through the Emmy Noether Programme of the DFG, project number 412851057, and an individual research grant of the DFG, project number 505125645.

us to investigate satellite formulae for any knot invariants that admit glueable generalisations to Conway tangles. For example, in joint work with Lukas Lewark, I recently applied this strategy to prove such satellite formulae for Rasmussen invariants [**LZ22**], using the multicurve tangle invariant generalising Bar-Natan homology [**KWZ19**].

The focus of this paper lies on a similar multicurve invariant: the tangle invariant $\gamma_T := \mathrm{HFT}(T)$ from Heegaard Floer theory [**Zib20, Zib19b**]. The main goal is to show that γ_{T_K} can be read off easily from yet another multicurve invariant, namely the Hanselman-Rasmussen-Watson multicurve $\gamma_K := \widehat{\mathrm{HF}}(S^3 \smallsetminus \mathring{\nu}(K))$ associated with the knot exterior [**HRW16, HRW22**]. Before explaining how this works, we briefly recall those features of the two multicurve invariants that are relevant to this story. The reader familiar with both invariants may skip straight to the statement of the Main Theorem, which is illustrated in Figure 2.

1.1. A brief review of the multicurve invariant γ_K. The knot invariant γ_K is a multicurve, a finite set of free homotopy types of closed curves[1], on the one-punctured torus. This torus is naturally identified with the boundary of the knot exterior $S^3 \smallsetminus \mathrm{int}(\nu(K))$, and the puncture can be thought of as some basepoint z on said boundary.

To describe key properties of γ_K, we introduce some notation: Let μ_K denote a meridian of K and λ_K a longitude of K, both of which go through the basepoint z. We may assume without loss of generality that γ_K intersects the meridian μ_K and longitude λ_K transversally and minimally. Then by [**HRW16**, Example 6.8], the intersection points of γ_K with μ_K generate the knot Floer homology of K:

$$\widehat{\mathrm{HFK}}(K) \cong \mathbb{F}_2\langle \gamma_K \pitchfork \mu_K \rangle.$$

(We will work over the field $\mathbb{F}_2$ of two elements throughout this paper.) These intersection points cut the multicurve γ_K into *curve segments*, which can be characterized by how their ends approach μ_K and by how they intersect λ_K. To make this more precise, let $\mathbb{R}^2$ be the universal cover of the torus such that the integer lattice $\mathbb{Z}^2 \subset \mathbb{R}^2$ agrees with the preimage of the basepoint z. This is illustrated on the left of Figure 2. The preimage of μ_K under this covering map is equal to the union of the vertical lines $\tilde{\mu}_K := \mathbb{Z} \times \mathbb{R}$. Given a curve segment c of γ_K, consider a lift $\tilde{c}$ to $\mathbb{R}^2$. It starts and ends on some connected components $\tilde{\mu}_K^+$ and $\tilde{\mu}_K^-$ of $\tilde{\mu}_K$, respectively. We now say the curve segment c is of type

$$\begin{cases} \mathbf{u}_\ell & \text{if } \tilde{\mu}_K^+ = \tilde{\mu}_K^- \text{ and } \tilde{c} \text{ lies to the right of } \tilde{\mu}_K^\pm; \\ \mathbf{v}_\ell & \text{if } \tilde{\mu}_K^+ = \tilde{\mu}_K^- \text{ and } \tilde{c} \text{ lies to the left of } \tilde{\mu}_K^\pm; \\ \mathbf{d}_\ell & \text{if } \tilde{\mu}_K^+ \neq \tilde{\mu}_K^- \text{ (in which case } \tilde{c} \text{ lies between } \tilde{\mu}_K^+ \text{ and } \tilde{\mu}_K^-). \end{cases}$$

In the first two cases, we call the subscript $\ell \in \mathbb{Z}$ the *length* of c; in the third case, we call it the *slope* of c. In all three cases, $|\ell|$ equals the number of intersection points of c with λ_K. In the first two cases, this already determines the curve segments up to homotopy (relative to their endpoints), so we choose $\ell \geq 0$. In fact, $\ell > 0$ for any such curve segment, since we assume that γ_K intersects μ_K minimally. In the third case, we choose the sign of ℓ to be positive if and only if the slope of $\tilde{c}$ is positive (after some homotopy relative endpoints). Clearly, the type of a curve

[1] As part of the data of a multicurve, each curve is equipped with a local system and some grading data. Here, in the introduction, we will ignore any of these additional curve decorations.

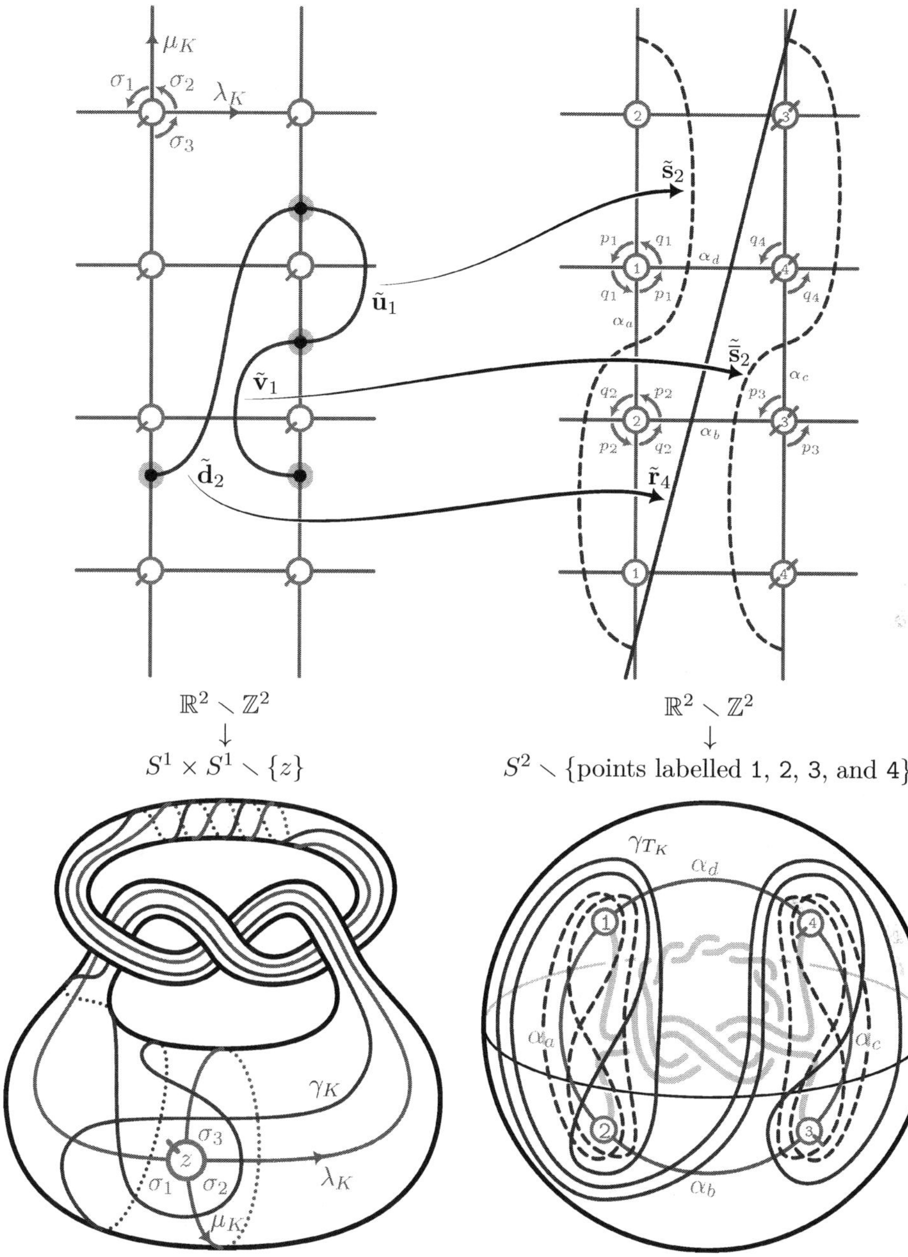

FIGURE 2. Executive summary of the Main Theorem: To construct γ_{T_K}, first cut γ_K along the generators $\bullet$ of $\widehat{\mathrm{HFK}}(K)$ into curve segments. This always produces a single curve segment $\mathbf{d}_{2\tau(K)}$ of slope $2\tau(K)$ as well as some number of curve segments $\mathbf{u}_\ell$ and $\mathbf{v}_\ell$ that are right and left caps, respectively, of some length $\ell \in \mathbb{Z}^{>0}$. Then γ_{T_K} contains a rational component $\mathbf{r}_{4\tau(K)} := \mathbf{r}(4\tau(K))$ as well as a special component $\mathbf{s}_{2\ell} := \mathbf{s}_{2\ell}(\infty; 1, 2)$ and $\bar{\mathbf{s}}_{2\ell} := \mathbf{s}_{2\ell}(\infty; 3, 4)$ for every $\mathbf{u}_\ell$ and $\mathbf{v}_\ell$ in γ_K, respectively. Note: In all pictures except the bottom left, we are looking at the manifolds from the outside.

segment c is independent of the choice of lift $\tilde{c}$ and invariant under homotopy of c relative endpoints.

REMARK 1.1. The multicurve invariant γ_K is equivalent to a version of knot Floer homology that takes the form of a chain complex over the polynomial ring $\mathbb{F}_2[U,V]/(UV=0)$; see for example [**KWZ20**, Theorem 2]. The curve segments $\mathbf{u}_\ell$ and $\mathbf{v}_\ell$ respectively correspond to components U^ℓ and V^ℓ of the differential in this chain complex.

The first part of the following structure theorem follows from [**LOT18**, Theorem 11.26]. The second part follows from the usual conjugation symmetry on knot Floer homology; see also [**HRW22**, Theorem 7].

THEOREM 1.2. *For any knot $K \subset S^3$, the multicurve γ_K contains exactly one curve segment $\mathbf{d}_{2\tau(K)}$, where $\tau(K)$ is the Ozsváth-Szabó concordance invariant, and every other curve segment is of type $\mathbf{u}_\ell$ or $\mathbf{v}_\ell$ for some $\ell \in \mathbb{Z}^{>0}$. Moreover, for any $\ell \in \mathbb{Z}^{>0}$,*

$$\#\{\textit{curve segments in } \gamma_K \textit{ of type } \mathbf{u}_\ell\} = \#\{\textit{curve segments in } \gamma_K \textit{ of type } \mathbf{v}_\ell\}.$$

1.2. A brief review of the multicurve invariant γ_T. The tangle invariant γ_T is also a multicurve; in this case, the multicurve lives on the four-punctured sphere which is naturally identified with the boundary of the three-dimensional ball containing the tangle minus the four tangle ends. Again, it is useful to consider the curves in a certain planar cover $\mathbb{R}^2 \smallsetminus \mathbb{Z}^2$ of the four-punctured sphere, which is illustrated on the top right of Figure 2. The integer lattice corresponds to the four punctures of the sphere, labelled $\texttt{1}$, $\texttt{2}$, $\texttt{3}$, and $\texttt{4}$, the vertical lines $\mathbb{Z}\times\mathbb{R}$ are the preimage of the arcs connecting tangle ends ($\texttt{1}$ and $\texttt{2}$) and ($\texttt{3}$ and $\texttt{4}$) and similarly, the horizontal lines $\mathbb{R}\times\mathbb{Z}$ correspond to the arcs connecting ($\texttt{1}$ and $\texttt{4}$) and ($\texttt{2}$ and $\texttt{3}$).

The components of the multicurve γ_T are completely classified [**Zib19b**, Theorem 0.5]. We do not need the full classification here. We only describe those curves that show up as components of γ_{T_K}. First, an embedded simple closed curve in the four-punctured sphere is called a *rational curve of slope* $k \in \mathbb{Z}$, denoted by $\mathbf{r}(k)$ if it lifts to a straight line of slope k in the cover $\mathbb{R}^2 \smallsetminus \mathbb{Z}^2$. Second, let $(\texttt{i},\texttt{j}) = (\texttt{1},\texttt{2})$ or $(\texttt{3},\texttt{4})$. Consider an infinite straight line in $\mathbb{R}^2 \smallsetminus \mathbb{Z}^2$ of slope ∞ going through some lattice points corresponding to the punctures $\texttt{i}$ and $\texttt{j}$. The lattice points divide this line into intervals of equal length. For a fixed integer $\ell \in 2\mathbb{Z}^{>0}$, let us mark every ℓth interval of the line. Then consider a small equivariant push-off of this line such that it intersects only the marked intervals and each of them exactly once. Finally, let $\mathbf{s}_\ell(\infty;\texttt{i},\texttt{j})$ be the immersed, primitive curve in the four-punctured sphere that lifts to this push-off. We call $\mathbf{s}_\ell(\infty;\texttt{i},\texttt{j})$ the *special curve of slope* ∞ *and length* ℓ through the punctures $\texttt{i}$ and $\texttt{j}$. In this paper, we will use the following simplified notation:

$$\mathbf{r}_k := \mathbf{r}(k), \quad \mathbf{s}_\ell := \mathbf{s}_\ell(\infty;\texttt{1},\texttt{2}), \quad \text{and} \quad \bar{\mathbf{s}}_\ell := \mathbf{s}_\ell(\infty;\texttt{3},\texttt{4}).$$

The following structure theorem for γ_{T_K} is a special case of a general result for cap-trivial tangles, analogous in both statement and proof to [**KWZ22a**, Theorem 3.1].

THEOREM 1.3. *For any knot $K \subset S^3$, the multicurve γ_{T_K} contains exactly one curve segment $\mathbf{r}_k$, for some $k \in 2\mathbb{Z}$, and every other curve segment is of type $\mathbf{s}_\ell$ or $\bar{\mathbf{s}}_\ell$ for some $\ell \in 2\mathbb{Z}^{>0}$. Moreover, for any $\ell \in 2\mathbb{Z}^{>0}$,*

$$\#\{\textit{components } \mathbf{s}_\ell \textit{ of } \gamma_{T_K}\} = \#\{\textit{components } \bar{\mathbf{s}}_\ell \textit{ of } \gamma_{T_K}\}.$$

1.3. The main theorem. The similarity between the two structure theorems for γ_K and γ_{T_K} suggests a close relationship between the two invariants. And indeed, it is as simple as it can possibly be!

MAIN THEOREM. *The components of γ_{T_K} are in one-to-one correspondence with the curve segments obtained by splitting the curve γ_K along its intersection points with the meridian μ_K. More specifically, we have the following correspondence:*

$$\text{in } \gamma_K \quad \begin{bmatrix} \mathbf{d}_{2\tau(K)} & \longleftrightarrow & \mathbf{r}_{4\tau(K)} \\ \mathbf{u}_\ell & \longleftrightarrow & \mathbf{s}_{2\ell} \\ \mathbf{v}_\ell & \longleftrightarrow & \overline{\mathbf{s}}_{2\ell} \end{bmatrix} \quad \text{in } \gamma_{T_K}.$$

For a relatively bigraded version of this result, see Theorem 4.1.

EXAMPLE 1.4. Figure 2 illustrates the Main Theorem for the right-handed trefoil knot $K = T_{2,3}$. The curve γ_K decomposes into three curve segments, namely $\mathbf{d}_2$, $\mathbf{u}_1$, and $\mathbf{v}_1$. Hence, γ_{T_K} consists of three components: $\mathbf{r}_4$, $\mathbf{s}_2$, and $\overline{\mathbf{s}}_2$. Interestingly, as an ungraded curve, this agrees with the invariant of the quotient tangle $T_{K,h}$ of K under the unique strong inversion h on K [**KWZ22a**, Figure 4]. (The tangle $T_{K,h}$ is equal to the $(2,-3)$-pretzel tangle, up to some number of twists, and its invariant γ_T was computed in [**Zib20**, Example 2.26].) However, as graded invariants, γ_{T_K} and $\gamma_{T_{K,h}}$ are distinct; see Example 4.2.

REMARK 1.5. The multicurve invariants γ_K are very computable thanks to Szabó's computer program [**OS17**] and Hanselman's implementation of the arrow sliding algorithm [**Han19**]. Hanselman has compiled a list of the invariants for all knots up to 15 crossings, which is available on github along with the source code of his program. While there also exists software [**Zib18**] for computing the tangle invariant γ_T, it is neither as efficient nor as practical. Consequently, γ_T had so far only been known for very few tangles T other than the family of two-stranded pretzel tangles considered in [**Zib20**, Theorem 6.9]. The Main Theorem now provides an infinite family of new tangles for which the invariants are very easy to compute. The python script [**Zib23**] automates this computation, taking as input the multicurves γ_K in the format used in Hanselman's database [**Han19**].

1.4. A reformulation of the Main Theorem. Recall the following well-known structure theorem for $\mathrm{HFK}^-(K)$, a version of knot Floer homology taking the form of a bigraded module over $\mathbb{F}_2[U]$; see for example [**OSS15**, Chapter 7].

THEOREM 1.6. *For any knot $K \subset S^3$, there exist integers $n \in \mathbb{Z}^{\geq 0}$ and $\ell_1, \dots, \ell_n \in \mathbb{Z}^{>0}$ such that*

$$\mathrm{HFK}^-(K) \cong \mathbb{F}_2[U] \oplus \bigoplus_{i=1}^{n} \mathbb{F}_2[U]/U^{\ell_i}.$$

The integers $\ell_1, \dots, \ell_n$ are unique up to permutation.

The module $\mathrm{HFK}^-(K)$ is generally not sufficient to determine γ_K. However, the curve segments of γ_K can be determined from $\mathrm{HFK}^-(K)$. First, the Alexander grading of the generator of $\mathbb{F}_2[U]$ is equal to $-\tau(K)$, by definition. Second, using the conjugation symmetry of knot Floer homology, one can show that for every U-torsion summand $\mathbb{F}_2[U]/U^{\ell_i}$ in $\mathrm{HFK}^-(K)$, γ_K contains a pair of curve segments $\mathbf{u}_{\ell_i}$ and $\mathbf{v}_{\ell_i}$. We therefore obtain the following reformulation of the Main Theorem.

THEOREM 1.7. *Given a knot $K \subset S^3$, let $n \in \mathbb{Z}^{\geq 0}$ and $\ell_1, \dots, \ell_n \in \mathbb{Z}^{>0}$ be the integers determined by Theorem* 1.6. *Then*

$$\gamma_{T_K} = \mathbf{r}_{4\tau(K)} \cup \bigcup_{i=1}^{n} \mathbf{s}_{2\ell_i} \cup \bar{\mathbf{s}}_{2\ell_i}. \qquad \square$$

1.5. Consequences and questions: satellites, thinness, and A-links. In 2021, Steve Boyer, Cameron Gordon, and Ying Hu showed that the Khovanov homology of any satellite knot is not thin, i.e. it is supported in more than one δ-grading [**BGH21**]. Up to now, the corresponding result for knot Floer homology was only known to be true for some specific patterns such as cables and the Mazur pattern [**Dey19, PW21**]. In fact, it is false in general:

THEOREM 1.8. *There exists a satellite knot $K \subset S^3$ for which $\widehat{\mathrm{HFK}}(K)$ is thin.*

The knot in question is shown in Figure 3. At the time of writing, I do not know if this is the only such example.

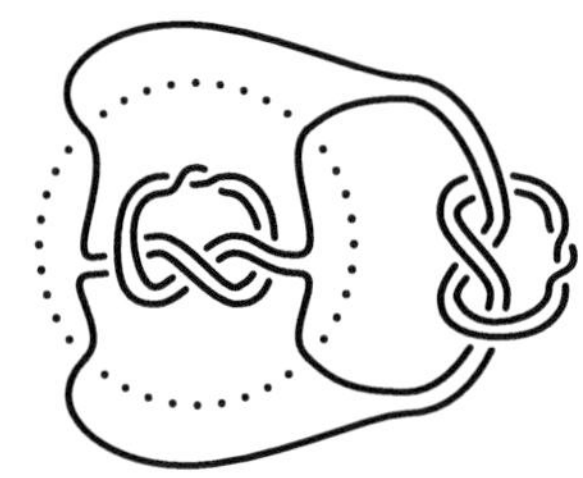

FIGURE 3. A satellite knot with thin knot Floer homology. The dotted circle indicates a Conway sphere splitting the knot into two tangles that agree with the double tangles of the left- and right-handed trefoil knots up to some twisting.

QUESTION 1.9. Is this the only satellite knot with thin knot Floer homology?

By applying thinness criteria for unions of Conway tangles from joint work with Artem Kotelskiy and Liam Watson [**KWZ21**] to the graded version of the Main Theorem (Theorem 4.1), we can give a partial answer to Question 1.9:

PROPOSITION 1.10. *Let $K \subset S^3$ be a knot and P a pattern of wrapping number 2. Suppose K is neither the unknot nor a trefoil knot. Then $\widehat{\mathrm{HFK}}(P(K))$ is not thin.*

Following [**KWZ21**], we call links whose knot Floer homology is supported in δ-gradings of the same parity *Heegaard Floer A-links*, or simply *A-links* for short. The set of A-links contains all non-split alternating links, and more generally the set of thin links as a proper subset. This definition is inspired by the definition of L-spaces, which include all lens spaces [**OS05a**]. Recall that an L-space knot is a knot that admits any non-trivial L-space surgery. Similar to Proposition 1.10, we show:

PROPOSITION 1.11. *Let $K \subset S^3$ be a knot and P a pattern of wrapping number 2. Suppose K is not an L-space knot. Then the satellite knot $P(K)$ is not an A-link.*

Given $p/q \in \mathbb{Q}\mathrm{P}^1$, let $Q_{p/q}$ denote the rational tangle of slope p/q. We define the *p/q-rational filling* $T(p/q)$ of a Conway tangle T as the union $Q_{-p/q} \cup T$. If a rational filling of a tangle is an A-link, we call it a *rational A-link filling* [**KWZ21**]. Note that for any knot K, $T_K(\infty)$ is the unknot, so T_K always admits at least one rational A-link filling. This is analogous to the fact that there is at least one L-space surgery for any knot $K \subset S^3$, namely the ∞-surgery. This analogy between rational A-link fillings and L-space surgeries goes much further:

THEOREM 1.12. *A knot $K \subset S^3$ admits a non-trivial L-space surgery if and only if its double tangle T_K admits a non-trivial rational A-link filling.*

We compare the relevant filling slopes in Theorem 5.3.

1.6. Consequences and questions: growth under cabling. An open question about the Heegaard Floer tangle invariant γ_T (in fact, any of the multicurve invariants now existing in low-dimensional topology) is how to interpret the number of their components. Thanks to the Main Theorem, double tangles $T = T_K$ are the first family of tangles for which we can give a fairly satisfying answer:

COROLLARY 1.13. *For any knot $K \subset S^3$,*

$$\#\{\textit{connected components in } \gamma_{T_K}\} = \dim \widehat{\mathrm{HFK}}(K). \qquad \square$$

This has the following consequence for the growth of knot Floer homology of cable knots $C_{2,2t+1}(K)$ of winding number 2.

PROPOSITION 1.14. *Given a knot $K \subset S^3$, let $d = \dim \widehat{\mathrm{HFK}}(K)$ and let $\overline{\ell}$ be the average torsion order of* $\mathrm{HFK}^-(K)$, *i.e. the arithmetic mean of the integers ℓ_i in Theorems 1.6 and 1.7. Then for any $t \in \mathbb{Z}$,*

$$\dim \widehat{\mathrm{HFK}}(C_{2,2t+1}(K)) = 2(d-1)\overline{\ell} + |2t+1-4\tau(K)|.$$

In particular,

$$2(d-1)\ell_{max} + |2t+1-4\tau(K)| \geq \dim \widehat{\mathrm{HFK}}(C_{2,2t+1}(K)) \geq 2d-1,$$

where ℓ_{max} is the maximum torsion order of $\mathrm{HFK}^-(K)$.

The average torsion order $\overline{\ell}$ may be arbitrarily large, as demonstrated by the fact that for any positive integer n, the value of $\overline{\ell}$ for the torus knot $K = T_{n,n+1}$ is equal to $\frac{n}{2}$. Usually, however, $\overline{\ell}$ is very close to 1. In fact, among all the knots up to 15 crossings, the average value of $\overline{\ell}$ is about 1.00077 [**Han19**, **Zib23**]. Its maximum value is 2 among those knots (realized by $13n_{4639} = C_{2,5}(T_{2,3})$ and $15n_{41185} = T_{4,5}$). So one might interpret Proposition 1.14 as saying that $\dim \widehat{\mathrm{HFK}}(C_{2,2t+1}(K))$ depends approximately linearly on $d = \dim \widehat{\mathrm{HFK}}(K)$. This is in stark contrast to the non-linear behaviour of reduced Khovanov homology:

PROPOSITION 1.15. *Given a knot $K \subset S^3$, let $d = \dim \widetilde{\mathrm{Kh}}(K)$. Then for any $t \in \mathbb{Z}$,*

$$\dim \widetilde{\mathrm{Kh}}(C_{2,2t+1}(K)) \geq 2d^2 - 2 + |2t+1-2\vartheta_2(K)|,$$

where $\vartheta_2(K)$ is the concordance homomorphism introduced in [**LZ22**].

This result is basically an application of the link splitting spectral sequence in Khovanov homology of Batson and Seed [**BS15**]. Computations suggest that this behaviour extends to the number of components in the reduced Khovanov multicurve invariant $\widetilde{\mathrm{Kh}}(T_K)$ of double tangles [**KWZ19**]. We will explore the following conjecture in upcoming work.

CONJECTURE 1.16. *For any knot $K \subset S^3$, the number of special components in $\widetilde{\mathrm{Kh}}(T_K)$ is bigger or equal to $\frac{1}{2}(d^2-1)$, where d is the dimension of the reduced Khovanov homology of K. For two-bridge knots, this bound is sharp.*

On a perhaps more speculative note, the decomposition of the multicurve γ_K into curve segments might also be relevant to the Khovanov multicurve invariant $\widetilde{\mathrm{Kh}}(T)$, specifically in view of the spectral sequence [**OS05b**].

OBSERVATION 1.17. For all strongly invertible knots (K, h) with up to 9 crossings, the number of special components of $\widetilde{\mathrm{Kh}}(T_{K,h})$ is bounded below by the number of curve segments $\mathbf{u}_\ell$ in γ_K. In fact we have equality for all such knots, except 9_{46}, the only such knot with two strong inversions h_1 and h_2 for which the (ungraded) Khovanov multicurve invariants do not agree [**KWZ22a**].

It would be interesting to compute the Heegaard Floer multicurve invariant γ_T of quotient tangles of strongly invertible knots to see if the identity of the tangle invariants in Example 1.4 is part of a general pattern.

Outline of the paper. In Section 2, we recall the construction of the multicurve invariants γ_K and γ_{T_K} and we discuss gradings in both settings. In Section 3, we compute the bordered sutured type AD bimodule that turns the type D structure corresponding to the knot complement into the type D structure for the corresponding double tangle. In Section 4, we interpret the action of this bimodule in terms of multicurves, proving the Main Theorem. In Sections 5 and 6, we prove the results from Subsections 1.5 and 1.6, respectively. Throughout this paper, we will assume familiarity with bordered and bordered sutured Heegaard Floer homology [**LOT18, Zar11**] as well as basic techniques for modifying type D structures within their homotopy class; see for instance [**Zib20**, Section 1]. The terms 'chain complex' and 'type D structure' will be used synonymously.

2. Review and conventions

2.1. A more detailed review of the multicurve invariant γ_K. The invariant γ_K is a topological interpretation of a certain bordered Heegaard Floer invariant. We start with a description of the latter and then explain how it is related to γ_K.

The bordered structure on the knot exterior. The exterior $X_K := S^3 \smallsetminus \mathrm{int}(\nu(K))$ of an oriented knot $K \subset S^3$ is a three-manifold with torus boundary. On this torus ∂X_T, we fix two simple closed curves, namely a meridian μ_K and a longitude λ_K of K, that intersect in a single point z, which we may regard as a basepoint on ∂X_K. The longitude is a push-off of K, inheriting its orientation. The orientation of the meridian is determined by the right-hand rule. This decoration of ∂X_K can be turned into a bordered structure as follows: Around z, we add a small circle that intersects μ_K and λ_K minimally and which one may think of as a suture. The restriction of μ_K to the complement of the interior of the disc bounded by the suture is an oriented arc, which we denote by $\mathbf{a}_\bullet$. Similarly, λ_K gives rise to an oriented arc

$\mathbf{a}_\circ$ with ends on the suture. We now add a basepoint to the suture; by convention, it sits immediately between the endpoints of $\mathbf{a}_\bullet$ and $\mathbf{a}_\circ$. This is illustrated at the bottom left of Figure 2. The arc diagram $\mathcal{Z}_\alpha$ consisting of the suture together with this basepoint and the arcs $\mathbf{a}_\bullet$ and $\mathbf{a}_\circ$ embedded on ∂X_K is a bordered structure for X_K. Now the algebraic invariant corresponding to the multicurve γ_K is the type D structure $\widehat{\mathrm{CFD}}(X_K)^{\mathcal{A}}$ associated with X_K equipped with this bordered structure. For the construction of $\widehat{\mathrm{CFD}}(X_K)^{\mathcal{A}}$, we refer the reader to [**LOT18**]; we only recall its formal properties, following [**LOT18**, Section 11].

The torus algebra. We define $\widehat{\mathrm{CFD}}(X_K)^{\mathcal{A}}$ as a graded right type D structure over the bordered algebra $\mathcal{A}(\mathcal{Z}_\alpha, 1)$ with one moving strand, which we identify with the quiver algebra

$$\mathcal{A} := \mathbb{F}_2$$ 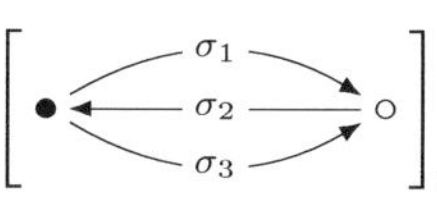 $$\Big/ \Big(\sigma_1\sigma_2 = 0 = \sigma_2\sigma_3\Big).$$

We denote the idempotents in $\mathcal{A}$ by $\iota_\bullet$ and $\iota_\circ$, respectively. We also write $\sigma_{12} := \sigma_2\sigma_1$, $\sigma_{23} := \sigma_3\sigma_2$, and $\sigma_{123} := \sigma_3\sigma_2\sigma_1$. The algebra $\mathcal{A}$ is graded by some non-commutative groups, which we will discuss later. For the moment, we will also ignore the grading on $\widehat{\mathrm{CFD}}(X_K)$, which takes values in the set of certain cosets of these groups.

REMARK 2.1. We have renamed the elements of the algebra $\mathcal{A}$, usually denoted by the letter ρ instead of σ, to avoid confusion with the elements p_i of the algebra $\mathcal{B}$ that we will introduce below. More importantly, the invariant of X_K was originally defined as a *left* type D structure ${}^{\mathcal{A}^{op}}\widehat{\mathrm{CFD}}(X_K)$ over the opposite algebra $\mathcal{A}^{op}$ of $\mathcal{A}$ [**LOT18, HRW16**]. Our conventions have the advantage that the multiplication in the torus algebra $\mathcal{A}$ is read from right to left, like morphisms in a category. In fact, we usually consider $\mathcal{A}$ as a category with two objects and we regard type D structures simply as chain complexes over this category; see [**Zib20**, Section 1].

EXAMPLE 2.2. For the unknot U and the right-handed trefoil knot $T_{2,3}$, we have

$$\widehat{\mathrm{CFD}}(X_U)^{\mathcal{A}} \cong \Big[\bullet \circlearrowleft \sigma_{12}\Big] \quad \text{and} \quad \widehat{\mathrm{CFD}}(X_{T_{2,3}})^{\mathcal{A}} \cong$$ 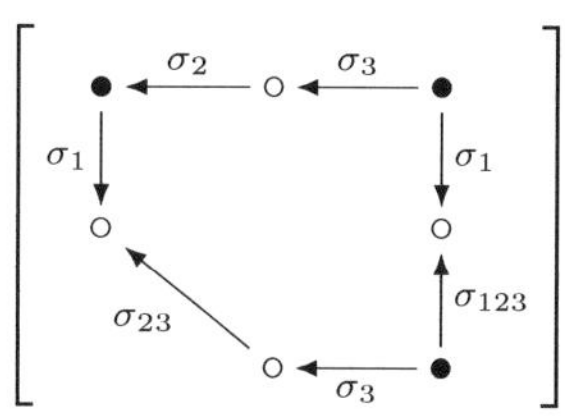 .

These labelled graphs should be understood as follows: Each vertex represents a generator/basis element of the underlying module of the chain complex. Each arrow corresponds to a non-zero entry of the square matrix representing the differential when written with respect to the chosen basis. The entries of this square matrix are the elements of $\mathcal{A}$ specified by the arrow labels.

The complexes in the above example have the remarkable property that each of them arises from an immersed curve on a one-punctured torus via the following general construction.

From curves to chain complexes. Let Σ be an oriented, one-punctured torus $S^1 \times S^1 \smallsetminus \mathrm{int}(D^2)$, decorated by two arcs $\mathbf{a}_\bullet$ and $\mathbf{a}_\circ$, just like ∂X_K above. In fact, if K is given, we identify Σ with ∂X_K minus a disc bounded by the suture.

Let γ be a collection of non-contractible and primitive immersed curves $S^1 \looparrowright \Sigma$ that intersects $\mathbf{a}_\bullet$ and $\mathbf{a}_\circ$ minimally and transversely. For the second complex in the above example, γ is the blue curve γ_K on the bottom left of Figure 2; for the first complex, the curve agrees with the longitude of U. Each point in $\gamma \cap \mathbf{a}_\bullet$ corresponds to a generator in idempotent $\iota_\bullet$, and similarly, each point in $\gamma \cap \mathbf{a}_\circ$ corresponds to a generator in idempotent $\iota_\circ$. The arcs cut γ into curve segments, each of which corresponds to an arrow (= a component of the differential) in the complex, connecting the two generators for the ends of the curve segment. To determine the direction and labels of these arrows, we first need to fix a labelling of some paths on $\partial\Sigma$. Up to reparametrization, there are four elementary paths on $\partial\Sigma$, i.e. paths that start and end on $\partial(\mathbf{a}_\bullet \cup \mathbf{a}_\circ)$, that follow the induced orientation on $\partial\Sigma$, and whose interiors avoid $\partial(\mathbf{a}_\bullet \cup \mathbf{a}_\circ)$. By convention, we label three of these consecutive paths by σ_1, σ_2, and σ_3 as shown in Figure 2. Now, every curve segment of γ is homotopic (relative to $\mathbf{a}_\bullet \cup \mathbf{a}_\circ$) to either σ_1, σ_2, or σ_3, or one of their compositions, and each curve segment inherits its orientation and label from this path on $\partial\Sigma$.

Chain complexes over $\mathcal{A}$ that (up to chain homotopy) arise from this construction and that can be appropriately graded (see below) are called *loop-type* [**HW15**]. However, as we will see in a moment, not every chain complex is loop-type.

Curves with local systems. Let γ be a curve in the torus as above, but consisting of a single component only. Also, let n be some positive integer and let (C, d) be a chain complex that arises from γ using the previous construction. Then $(C \otimes \mathbb{F}_2^n, d \otimes \mathrm{id}_{\mathbb{F}_2^n})$ is also a well-defined chain complex. We now modify its differential as follows: First, we fix an orientation of γ. Next, let x and y be two basis elements for (C, d) such that the differential d has a component a from x to y. In the new chain complex, this component corresponds to a component $a \otimes \mathrm{id}_{\mathbb{F}_2^n}$. We now change this component to $a \otimes X$ or $a \otimes X^{-1}$, depending on certain orientation conventions, where $X \in \mathrm{GL}_n(\mathbb{F}_2)$. One can show that the chain homotopy type of the resulting chain complex is independent of the choice of basis elements x and y (see the remarks at the end of the proof of [**HRW16**, Theorem 1.5]). We can therefore denote this chain complex simply by $C(\gamma, X)$. We call X the local system of the oriented curve γ. For $n > 0$, the local systems $X = \mathrm{id}_{\mathbb{F}_2^n}$ are called trivial.

DEFINITION 2.3. Given a collection $\{(\gamma_i, X_i)\}_i$ of curves with local systems, we define the associated chain complex as the direct sum of the chain complexes $C(\gamma_i, X_i)$. Two collections $\{(\gamma_i, X_i)\}_i$ and $\{(\gamma'_j, X'_j)\}_j$ of curves with local systems are considered equal (and can be shown to induce the same chain complexes up to homotopy) if for any curve γ, the matrices

$$\bigoplus_{i:\gamma_i=\gamma} X_i \oplus \bigoplus_{i:\gamma_i=-\gamma} X_i^{-1} \quad \text{and} \quad \bigoplus_{j:\gamma'_j=\gamma} X'_j \oplus \bigoplus_{j:\gamma'_j=-\gamma} X'^{-1}_j$$

are similar. Here, $-\gamma$ denotes the curve γ with the opposite orientation and two oriented immersed curves are considered equal if they are homotopic.

Classification result. Chain complexes arising from curves with local systems are *extendable*, which is a certain algebraic property that is immaterial to our discussion. We only need to know the following result of Hanselman-Rasmussen-Watson [**HRW16**, Theorem 1.4], which is essentially due to Lipshitz-Ozsváth-Thurston [**LOT18**, Chapter 11].

THEOREM 2.4. *For every knot $K \subset S^3$, $\widehat{\mathrm{CFD}}(X_K)^{\mathcal{A}}$ is extendable.*

This is relevant in the context of the following classification result of Hanselman-Rasmussen-Watson [**HRW16**, Theorem 1.5].

THEOREM 2.5. *Every extendable graded chain complex over $\mathcal{A}$ is chain homotopic to a chain complex corresponding to a collection of curves with (possibly trivial) local systems. This collection of curves with local systems is unique up to the equivalence from Definition* 2.3.

Combining these two results, we can now finally define:

DEFINITION 2.6. For any knot $K \subset S^3$, let γ_K denote the collection of immersed curves with local systems on ∂X_K corresponding to $\widehat{\mathrm{CFD}}(X_K)^{\mathcal{A}}$.

DEFINITION 2.7. Given a collection $\gamma = \{(\gamma_i, X_i)\}_i$ of curves with local systems, we define the curve segments $S_{\text{✂}}(\gamma)$ as the set of (graded) type D structures obtained as follows: Let D_i be the type D structure corresponding to the curve γ_i. Split D_i along each generator in idempotent $\bullet$ like so:

$$(\cdots \longleftrightarrow \bullet \longleftrightarrow \cdots) \quad \text{becomes} \quad (\cdots \longleftrightarrow \bullet) \quad (\bullet \longleftrightarrow \cdots).$$

Let $\{D_{i,j}\}_j$ be the connected components of the resulting type D structure. We now define $S_{\text{✂}}(\gamma)$ as the set of type D structures that for each pair (i,j) contains $\dim X_i$ copies of $D_{i,j}$.

Clearly, the elements of $S_{\text{✂}}(\gamma)$ correspond to the curve segments $\mathbf{u}_\ell$, $\mathbf{v}_\ell$, and $\mathbf{d}_\ell$ from the introduction.

2.2. A more detailed review of the multicurve invariant γ_T. Like the multicurve γ_K, the tangle invariant γ_{T_K} is a topological interpretation of a certain bordered (in fact: bordered sutured) Heegaard Floer invariant. But before describing any of this, we fix some conventions regarding Conway tangles, largely following [**Zib20**, Section 2.1] and [**LZ22**, Section 2.1].

Conway tangles. An *oriented Conway tangle* T is a proper embedding of two intervals into a three-dimensional ball B^3, considered up to ambient isotopy. A *framing* of B^3 is a circle Φ on $S^2 = \partial B^3$ together with a choice of four points on this circle that are consecutively labelled 1, 2, 3, and 4. We consecutively label the four arcs $\Phi \smallsetminus \{1,2,3,4\}$ by α_a, α_b, α_c, and α_d, starting with the arc between 1 and 2. A *framed* oriented Conway tangle is a Conway tangle together with a choice of framing of B^3 which identifies the endpoints of the two intervals with the four points 1, 2, 3, and 4. We consider framed tangles up to ambient isotopy which identifies the framings. In the following, we will refer to framed, oriented Conway tangles simply as *Conway tangles*.

The *double tangle* T_K of an oriented knot K is the Conway tangle that is defined as follows: Let L be the oriented link obtained as the union of K with one of its zero-framed longitudes, regarded as a small push-off of K. Then L bounds an embedded (unoriented) annulus in S^3, namely the image of a suitably chosen homotopy between K and the longitude. Now choose a small open three-ball b which intersects this annulus in a trivially embedded band. Then T_K is defined as the intersection of L with the complement B^3 of b. We choose the framing of T_K that is (up to homotopy and up to interchanging the two components of T_K) uniquely characterized by the following properties:

(1) the intersection of ∂B^3 with the band is equal to the arcs α_a and α_c,

(2) in the closure of the three-ball b, the restriction of L and the arcs α_b and α_d cobound two discs (ensuring that the linking number of the two strands in T_K vanishes), and

(3) the orientation of the tangle points out of B^3 at the tangle ends 1 and 2.

In particular, if follows that

(4) the tangle connects the tangle end 1 with 4 and 2 with 3.

A Conway tangle satisfying (4) is said to have *horizontal connectivity*. We implicitly assume from now that all Conway tangles that we consider satisfy (3) and (4).

The bordered sutured structure on the tangle exterior. The framing Φ of a Conway tangle T enables us to distinguish between the two components of $\partial B^3 \smallsetminus \Phi$: The *front* is the component where the labelling of the points 1, 2, 3, 4 increases in counter-clockwise direction when viewed from outside B^3; we call the other component the *back* of T. On both front and back, we now choose a small disc whose boundary we may consider as a suture. We connect the two discs by four arcs β_a, β_b, β_c, and β_d that each intersects the circle Φ only once, namely within the arcs α_a, α_b, α_c, and α_d, respectively. As in the case of the suture on the knot complement X_K in the previous subsection, we label the elementary paths on the front (resp. back) suture by p_i (resp. q_i) if they are adjacent to the tangle end labelled i. We now place basepoints on the elementary paths labelled p_4 and q_3. Finally, we remove a sufficiently small open tubular neighbourhood of the tangle T that avoids the arcs β_a, β_b, β_c, and β_d and the sutures. We denote the result by $X_T := B^3 \smallsetminus \operatorname{int}(\nu(T))$, the exterior of the tangle T. Since T is a tangle of horizontal connectivity, the basepoints lie in different components of the complement of the sutures and arcs on ∂X_T. Hence the arc diagram $\mathcal{Z}_\beta$ consisting of the sutures with basepoints and the β-arcs β_a, β_b, β_c, and β_d embedded on ∂X_T gives rise to a well-defined bordered sutured structure on X_T. This is illustrated in Figure 6b. Now the algebraic invariant corresponding to the multicurve γ_T is the associated type D structure $\widehat{\mathrm{BSD}}(X_T)^{\mathcal{B}}$. For the construction of $\widehat{\mathrm{BSD}}(X_T)^{\mathcal{B}}$, we refer the reader to [**Zar11**]; we only recall its formal properties, following [**Zib20, Zib19b**] and specializing to the case that the tangle T satisfies (3) and (4).

The peculiar algebra. $\widehat{\mathrm{BSD}}(X_T)^{\mathcal{B}}$ is a graded chain complex over the bordered sutured algebra $\mathcal{A}(\mathcal{Z}_\beta, 1)$ with one moving strand, which can be identified with the quiver algebra

$$\mathcal{B} := \mathcal{A}^{\partial}_{43} = \mathbb{F}_2$$

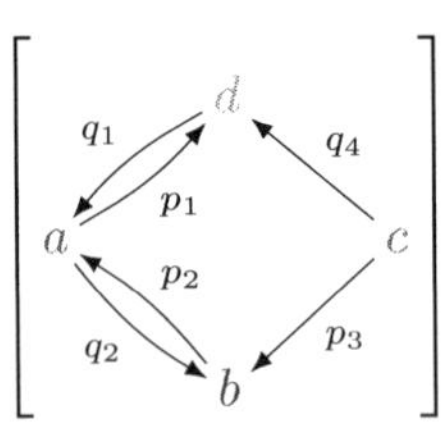

$$\Big/ \left(p_i q_i = 0 = q_i p_i\right)_{i=1,2}.$$

We denote the idempotents in $\mathcal{B}$ by ι_a, ι_b, ι_c, and ι_d, respectively. We also write $p_{12} := p_1p_2$, $p_{23} := p_2p_3$, $p_{123} := p_1p_2p_3$, $q_{21} := q_2q_1$, $q_{32} := q_3q_2$, and $q_{321} := q_3q_2q_1$. This algebra carries two gradings, a $\frac{1}{2}\mathbb{Z}$-grading δ determined by

$$\delta(p_1) = \delta(p_2) = \delta(p_3) = \tfrac{1}{2} = \delta(q_1) = \delta(q_2) = \delta(q_4)$$

and a $\mathbb{Z}^2$-grading, called the Alexander grading, determined by

$$-A(p_1) = -A(q_1) = (1,0) = A(q_4) \qquad -A(p_2) = -A(q_2) = (0,1) = A(p_3).$$

The chain complex $\widehat{\mathrm{BSD}}(X_T)^{\mathcal{B}}$ is equipped with a δ-grading, which increases along the differential by 1, and an Alexander grading, which is preserved along the differential. We sometimes denote elements in this bigrading by $[r; a_1, a_2]$, where $r \in \frac{1}{2}\mathbb{Z}$ is the δ-grading and $(a_1, a_2) \in \mathbb{Z}^2$ is the Alexander grading. We sometimes indicate the bigrading of a generator x using super- and subscripts: ${}^{r}x^{a_1}_{a_2}$. Given a bigraded chain complex C over $\mathcal{B}$, let $\delta^r t_1^{a_1} t_2^{a_2} C$ denote the chain complex obtained from C by increasing the Alexander and δ-gradings of all generators by (a_1, a_2) and r, respectively.

Classification result. To describe the classification of graded chain complexes over $\mathcal{B}$, we introduce a decoration of ∂X_T which is in some sense "dual" to the bordered sutured structure described above. We consider the tangle exterior with a meridional suture around each tangle end. The sutures decompose ∂X_T into a four-punctured sphere S^2_4 and two annuli. The four arcs α_a, α_b, α_c, and α_d of the framing Φ restrict to four arcs on ∂X_T, which we denote by the same combination of letters. We label the elementary path on the suture corresponding to the tangle end i along the front (resp. back) by p_i (resp. q_i). Finally, we place basepoints on the elementary paths labelled p_4 and q_3. This is illustrated on the right of Figure 2. (Note that with minor modifications, this can be turned into an actual bordered sutured structure: First, one adds basepoints to the sutures at the tangle ends 1 and 2; second, one also adds two arcs around those basepoints such that the corresponding gluing surface remains a four-punctured sphere; for details, see the proof of [**Zib20**, Theorem 3.7]. The simpler bordered sutured structure on X_T that we use in this paper was first introduced in [**Zib19b**, Section 5].)

Similar to the relationship between curves on the one-punctured torus and complexes over the algebra $\mathcal{A}$, every non-contractible and primitive immersed curve (with or without local system) on the four-punctured sphere S^2_4 gives rise to a chain complex over $\mathcal{B}$. For example, the solid curve on the four-punctured sphere in Figure 2 corresponds to the chain complex

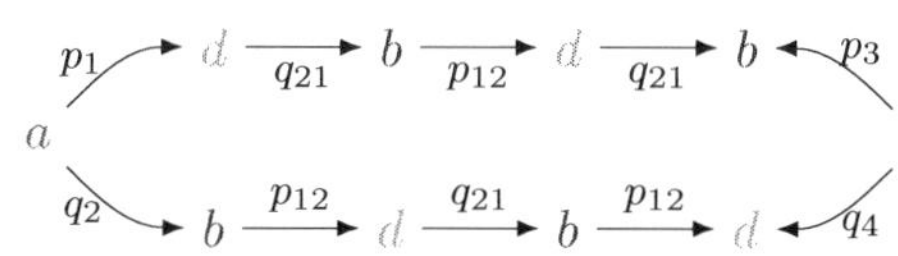

where each generator $x \in \{a, b, c, d\}$ sits in idempotent ι_x. Likewise, the dashed curves in this picture correspond to

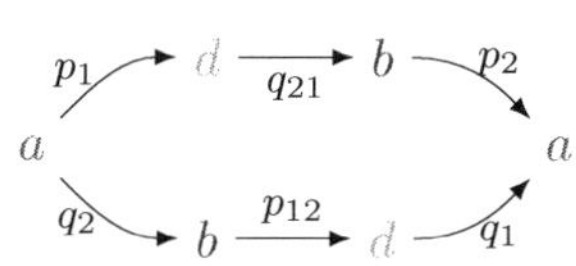

and

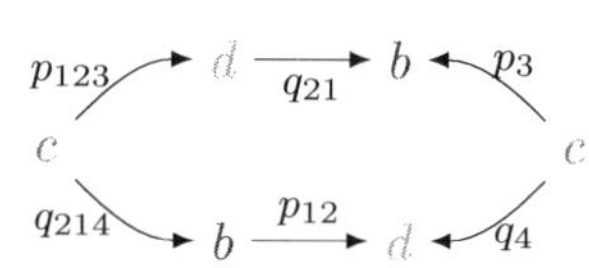

A bigrading of a curve is simply a choice of bigrading of the corresponding chain complex over $\mathcal{B}$. We will therefore use the same notation for grading shifts. Note that to specify a bigrading for a single curve, it suffices to specify a bigrading of a single generator of the corresponding chain complex.

Chain complexes arising from curves with local systems are again *extendable* and we have the following result, which is apparent from [**Zib19b**, Theorem 5.3].

THEOREM 2.8. *$\widehat{\mathrm{BSD}}(X_T)^{\mathcal{B}}$ is extendable for every Conway tangle T satisfying* (3) *and* (4).

The following result is [**Zib20**, Theorem 0.4].

THEOREM 2.9. *Every extendable graded chain complex over $\mathcal{B}$ is chain homotopic to a chain complex corresponding to a collection of curves on S^2_4 with (possibly trivial) local systems. This collection of curves is unique up to the equivalence from Definition* 2.3.

Combining these two results, we now define:

DEFINITION 2.10. For any Conway tangle T satisfying (3) and (4), let γ_T denote the collection of immersed curves with local systems on ∂X_T corresponding to $\widehat{\mathrm{BSD}}(X_T)^{\mathcal{B}}$.

PROOF OF THEOREM 1.3. By the gluing formula [**Zib20**, Theorem 5.9], the Lagrangian Floer homology of γ_{T_K} with a rational curve of slope ∞ (a simple closed curve lifting to a straight line of slope ∞ in the planar cover) computes $\mathbb{F}_2^2 \otimes \widehat{\mathrm{HFK}}(U) \cong \mathbb{F}_2^2$. Combining this with the classification of components [**Zib19b**, Theorem 0.5] and detection of tangle connectivity [**Zib20**, Observation 6.1], we see that γ_{T_K} consists of exactly one rational component of even integer slope with the trivial one-dimensional local system and a (potentially vanishing) number of special components of slope ∞ with trivial local systems. The second part follows from [**Zib19b**, Theorem 0.10]. □

2.3. Gradings. Recall from [**Zar11**] that in bordered (sutured) Heegaard Floer theory, the unreduced grading gr (aka unrefined grading gr′ in the terminology of [**LOT18**]) of the algebra $\mathcal{A}(\mathcal{Z})$ associated with an arc diagram $\mathcal{Z} = (\mathbf{Z}, \mathbf{a}, M)$ takes values in the non-commutative group $\mathrm{Gr}(\mathcal{Z})$, which is defined as a certain semi-direct product $\frac{1}{2}\mathbb{Z} \rtimes H_1(\mathbf{Z}, \mathbf{a})$. For convenience, we enlarge the grading set by working with rational coefficients, i.e. we replace $\mathrm{Gr}(\mathcal{Z})$ by $\mathrm{Gr}(\mathcal{Z}) \otimes \mathbb{Q} = \mathbb{Q} \rtimes H_1(\mathbf{Z}, \mathbf{a}; \mathbb{Q})$. There is also a reduced (refined) grading $\underline{\mathrm{gr}}$ of $\mathcal{A}(\mathcal{Z})$, which takes values in $\underline{\mathrm{Gr}}(\mathcal{Z})$, which we define as $\mathbb{Q} \rtimes H_1(\mathcal{F}(\mathcal{Z}); \mathbb{Q})$, where $\mathcal{F}(\mathcal{Z})$ is the surface obtained by thickening the arc diagram $\mathcal{Z}$. By default, we work with the gradings associated with type A sides.

Identification with the usual grading on the knot invariant. From [**LOT18**, Section 11.1], recall the reduced grading

$$\underline{\mathrm{Gr}}(\mathcal{Z}_\alpha) := \mathbb{Q} \rtimes H_1(\mathcal{F}(\mathcal{Z}_\alpha); \mathbb{Q}) \cong \mathbb{Q} \rtimes \mathbb{Q}^2$$

with the group law defined by

$$(j_1; p_1, q_1) \cdot (j_2; p_2, q_2) = \left(j_1 + j_2 - \det \begin{pmatrix} p_1 & q_1 \\ p_2 & q_2 \end{pmatrix}; p_1 + p_2, q_1 + q_2 \right)$$

for any $(j_i; p_i, q_i) \in \underline{\mathrm{Gr}}(\mathcal{Z}_\alpha)$, $i = 1, 2$. (When comparing this formula with [**LOT18**, Section 11.1], note that the minus sign in front of the determinant comes from the fact that our algebra $\mathcal{A}$ is the opposite algebra of the usual torus algebra; see Remark 2.1.) The grading on $\mathcal{A}$ is defined by

$$\underline{\mathrm{gr}}(\sigma_1) = (-\tfrac{1}{2}; \tfrac{1}{2}, -\tfrac{1}{2}) \qquad \underline{\mathrm{gr}}(\sigma_2) = (-\tfrac{1}{2}; \tfrac{1}{2}, \tfrac{1}{2}) \qquad \underline{\mathrm{gr}}(\sigma_3) = (-\tfrac{1}{2}; -\tfrac{1}{2}, \tfrac{1}{2})$$

The grading on $\widehat{\mathrm{CFD}}(X_K)^{\mathcal{A}}$ takes values in the coset $\mathcal{P}_1 \backslash \underline{\mathrm{Gr}}(\mathcal{Z}_\alpha)$, where $\mathcal{P}_1$ is the subgroup generated by $(-\frac{1}{2}; -1, 0)$. Moreover, it suffices to specify the grading of the generators in idempotent $\iota_\bullet$, which are in bijection with the ($\mathbb{Q}^2$-graded) generators of $\widehat{\mathrm{HFK}}(K)$. (These two facts are explained in [**LOT18**, Theorem 11.26

and Remark 11.28], noting that by Theorem 2.5, the twisting parameter n can be chosen to be 0.) The first $\mathbb{Q}$-grading on $\widehat{\mathrm{HFK}}(K)$ is called the δ-grading and the second $\mathbb{Q}$-grading is called the Alexander grading A. We will write $[m;n]$ for an element $(m,n) \in \mathbb{Q}^2$ of this bigrading. (Often, the Maslov grading M is considered instead of the δ-grading; one can be determined from the other using the identity $M + \delta = A$.) Then if $\mathbf{x}$ is a (homogeneous) generator of $\widehat{\mathrm{HFK}}(K)$ corresponding to a generator $\mathbf{x}_0$ of $\widehat{\mathrm{CFD}}(X_K).\iota_\bullet$, then

$$\begin{aligned}\underline{\mathrm{gr}}(\mathbf{x}_0) &= (M(\mathbf{x}) - \tfrac{3}{2}A(\mathbf{x}); 0, -A(\mathbf{x})) \\ &= (-\delta(\mathbf{x}) - \tfrac{1}{2}A(\mathbf{x}); 0, -A(\mathbf{x})) \in \mathcal{P}_1 \backslash \underline{\mathrm{Gr}}(\mathcal{Z}_\alpha).\end{aligned}$$

Given a bigraded chain complex C over $\mathcal{A}$ and $r, a \in \mathbb{Q}$, let $\delta^r t^a C$ denote the chain complex obtained from C by increasing the δ- and Alexander gradings of all generators by r and a, respectively. A grading on a curve or a curve segment on the one-punctured torus is simply a grading of the corresponding chain complex; therefore, we will use the same notation for grading shifts.

Identification with the usual grading on the tangle invariant. Let T be a Conway tangle satisfying (3) and (4). The arc diagram $\mathcal{Z}_\beta = (\mathbf{Z}_\beta, \mathbf{a}_\beta, M_\beta)$ on the tangle exterior X_T is shown in Figure 4. The surface $\mathcal{F}(\mathcal{Z}_\beta)$ is a four-punctured sphere; in particular, it is planar. Hence, the intersection pairing on $H_1(\mathcal{F}(\mathcal{Z}_\beta))$ is trivial and $\underline{\mathrm{Gr}}(\mathcal{Z}_\beta)$ is actually identical to $\mathbb{Q} \times H_1(\mathcal{F}(\mathcal{Z}_\beta); \mathbb{Q})$ as a group. To determine the reduced grading on the algebra $\mathcal{B} = \mathcal{A}(\mathcal{Z}_\beta)$, we choose the paths $\{p_1, p_2, p_3, q_2, q_1, q_4\}$ as a basis for $H_1(\mathbf{Z}_\beta, \mathbf{a}_\beta; \mathbb{Q})$ so that the unreduced gradings of the algebra generators are as follows:

$$\begin{aligned}
\mathrm{gr}(p_1) &= (-\tfrac{1}{2}; 1,0,0; 0,0,0), & \mathrm{gr}(q_2) &= (-\tfrac{1}{2}; 0,0,0; 1,0,0),\\
\mathrm{gr}(p_2) &= (-\tfrac{1}{2}; 0,1,0; 0,0,0), & \mathrm{gr}(q_1) &= (-\tfrac{1}{2}; 0,0,0; 0,1,0),\\
\mathrm{gr}(p_3) &= (-\tfrac{1}{2}; 0,0,1; 0,0,0), & \mathrm{gr}(q_4) &= (-\tfrac{1}{2}; 0,0,0; 0,0,1).
\end{aligned}$$

Recall from [**Zib19b**, Definition 1.7] the usual grading group $\mathfrak{A}$ for $\mathcal{B}$, which we will replace by $\mathfrak{A} \otimes \mathbb{Q}$:

$$\mathfrak{A} := \mathbb{Q}^4 / \operatorname{im}(\mathbb{Q} \to \mathbb{Q}^4, n \mapsto (n,n,n,n)).$$

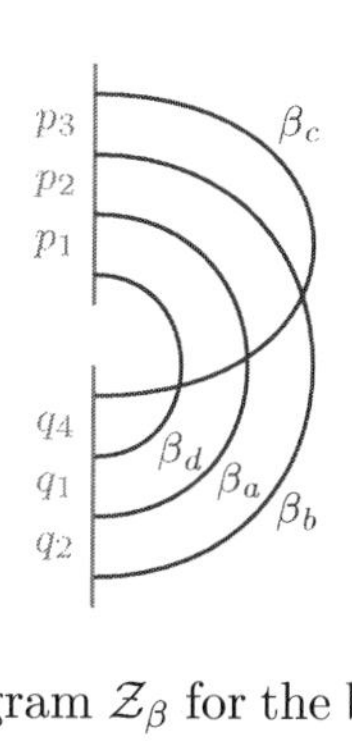

FIGURE 4. The arc diagram $\mathcal{Z}_\beta$ for the bordered sutured structure on X_T

We identify $H_1(\mathcal{F}(\mathcal{Z}_\beta);\mathbb{Q})$ with $\mathfrak{A}$ via the homomorphism from $\mathfrak{A}$ to $H_1(\mathcal{F}(\mathcal{Z}_\beta);\mathbb{Q})$ defined by

$$(1,0,0,0)\mapsto(\tfrac{1}{2},0,0;0,\tfrac{1}{2},0),\qquad (0,1,0,0)\mapsto(0,\tfrac{1}{2},0;\tfrac{1}{2},0,0),$$
$$(0,0,1,0)\mapsto(0,0,\tfrac{1}{2};-\tfrac{1}{2},-\tfrac{1}{2},-\tfrac{1}{2}),\qquad (0,0,0,1)\mapsto(-\tfrac{1}{2},-\tfrac{1}{2},-\tfrac{1}{2};0,0,\tfrac{1}{2}).$$

We now choose a grading reduction r with ι_a as the base idempotent (so $r(\iota_a)=0$) and

$$r(\iota_b)=(0;0,-\tfrac{1}{2},0;\tfrac{1}{2},0,0),$$
$$r(\iota_c)=(-\tfrac{1}{2};0,-\tfrac{1}{2},-\tfrac{1}{2};0,-\tfrac{1}{2},-\tfrac{1}{2}),$$
$$r(\iota_d)=(0;\tfrac{1}{2},0,0;0,-\tfrac{1}{2},0).$$

Given a basic algebra element $\xi=\iota_t.\xi.\iota_s\in\mathcal{B}$ for some $t,s\in\{a,b,c,d\}$, the reduced grading is computed using the formula

$$\underline{\mathrm{gr}}(\xi)=(r(\iota_t))^{-1}\,\mathrm{gr}(\xi)\,r(\iota_s). \tag{2.1}$$

An elementary calculation shows that the induced reduced grading is given by

$$\underline{\mathrm{gr}}(p_1)=\underline{\mathrm{gr}}(q_1)=(-\tfrac{1}{2};1,0,0,0),\qquad \underline{\mathrm{gr}}(p_3)=(-\tfrac{1}{2};0,0,1,0),$$
$$\underline{\mathrm{gr}}(p_2)=\underline{\mathrm{gr}}(q_2)=(-\tfrac{1}{2};0,1,0,0),\qquad \underline{\mathrm{gr}}(q_4)=(-\tfrac{1}{2};0,0,0,1).$$

(To determine $\underline{\mathrm{gr}}(p_3)$ and $\underline{\mathrm{gr}}(q_4)$, it might be easier to compute $\underline{\mathrm{gr}}(p_{23})$ and $\underline{\mathrm{gr}}(q_{14})$ first.) Finally, since the tangle T satisfies (3) and (4), we simplify the grading further via the homomorphism $\varphi\colon\mathfrak{A}\to\mathbb{Q}^2$ sending

$$(1,0,0,0)\mapsto(-1,0),\qquad (0,0,0,1)\mapsto(1,0),$$
$$(0,1,0,0)\mapsto(0,-1),\qquad (0,0,1,0)\mapsto(0,1).$$

Then the gradings (δ,A) and $((-\mathrm{id}_{\mathbb{Q}})\times\varphi)\circ\underline{\mathrm{gr}}$ on the algebra $\mathcal{B}$ agree.

The grading on $\widehat{\mathrm{BSD}}(X_T)$ takes values in $\mathcal{P}_2\backslash\underline{\mathrm{Gr}}(\mathcal{Z}_\beta)$, where $\mathcal{P}_2$ is the subgroup $\underline{\mathrm{gr}}(\pi_2(\mathbf{x},\mathbf{x};\mathbb{Q}))$ defined as follows: First we fix a Heegaard diagram for X_T and we pick some base generator $\mathbf{x}$ of $\widehat{\mathrm{BSD}}(X_T).\iota_a$. (This is always possible.) We now consider $\pi_2(\mathbf{x},\mathbf{x};\mathbb{Q})$, the group of periodic domains on a Heegaard diagram for X_T (with coefficients in $\mathbb{Q}$). Each domain B has a certain unreduced grading $\mathrm{gr}(B)$. Let $R(\mathrm{gr}(B))$ be the element of $\mathrm{Gr}(\mathcal{Z}_\beta)$ obtained from $\mathrm{gr}(B)$ by multiplying all components except the first by -1. Then we define $\underline{\mathrm{gr}}(B)$ as the element of $\underline{\mathrm{Gr}}(\mathcal{Z}_\beta)$ corresponding to $R(\mathrm{gr}(B))$ using the above grading reduction r. $\mathcal{P}_2$ is now the subgroup generated by the elements $\underline{\mathrm{gr}}(B)$ for all periodic domains $B\in\pi_2(\mathbf{x},\mathbf{x};\mathbb{Q})$. To understand this subgroup better, we choose a Heegaard diagram with a single α-circle parallel to the front suture. (This is always possible.) The periodic domains of such a Heegaard diagram are generated by a single domain ψ. An elementary computation (similar to the proof of [**Zib19a**, Lemma 5.12]) shows that the reduced grading of ψ is equal to $\pm(0;1,0,0,1)$. Hence, we can use the above identification of the gradings on the algebra $\mathcal{B}$ to also identify the gradings on $\widehat{\mathrm{BSD}}(X_T)$. (Alternatively, in the case that T is a double tangle T_K for some knot K, this identification will also emerge from the proof of Theorem 3.1).

3. Computations in bordered sutured Heegaard Floer theory

The goal of this section is to prove the following result.

THEOREM 3.1. *Let X_K be the complement of an open tubular neighbourhood of a knot K in S^3. We equip ∂X_K with the standard bordered parametrization by meridian and longitude. Then*

$$\widehat{\mathrm{BSD}}(X_{T_K})^{\mathcal{B}} \cong \widehat{\mathrm{CFD}}(X_K)^{\mathcal{A}} \boxtimes {}_{\mathcal{A}}\mathcal{Y}^{\mathcal{B}},$$

where ${}_{\mathcal{A}}\mathcal{Y}^{\mathcal{B}}$ is the type AD bimodule shown in Figure 5. The reduced grading of ${}_{\mathcal{A}}\mathcal{Y}^{\mathcal{B}}$ takes values in $(\underline{\mathrm{Gr}}(\mathcal{Z}_\alpha) \times \mathfrak{A})/\mathcal{P}_3$ where $\mathcal{P}_3$ is the subgroup generated by

$$(0;0,0;0,1,1,0) \quad \textit{and} \quad (-\tfrac{1}{2};0,1;0,0,2,2).$$

Moreover, if $\underline{\mathrm{gr}}(a) = \mathcal{P}_3$ then $\underline{\mathrm{gr}}(c) = (-\tfrac{1}{2};-1,0;0,0,0,0)\cdot\mathcal{P}_3$.

PROOF. Let us equip the knot complement X_K and the tangle complement X_{T_K} with the bordered/bordered sutured structures described in Section 2. They are illustrated in Figures 6a and 6b, respectively. Consider the bordered sutured manifold Y represented by the bordered sutured Heegaard diagram in Figure 6c. Its underlying three-manifold with boundary is obtained by thickening a one-punctured genus two surface Σ_2 and attaching a two-handle along a non-separating curve. In Figure 6c, Σ_2 should be thought of as the Heegaard surface with a disc attached to the boundary component/suture on the left. The attaching curve of the two-handle is α_1. There are two sutures on ∂Y. One of them is equal to $\partial\Sigma_2 \subset Y$ and it decomposes ∂Y into a component $\partial_1 Y$ of genus one and a component $\partial_2 Y$ of genus two. The boundary component $\partial_1 Y$ is parametrized by the α-arcs and can be identified with $-\partial X_K$. The other boundary component $\partial_2 Y$ contains the second suture, which bounds a disc in $\partial_2 Y$, and it is parametrized as ∂X_{T_K} by the β-arcs. So we see that gluing Y to the knot complement X_K produces the tangle

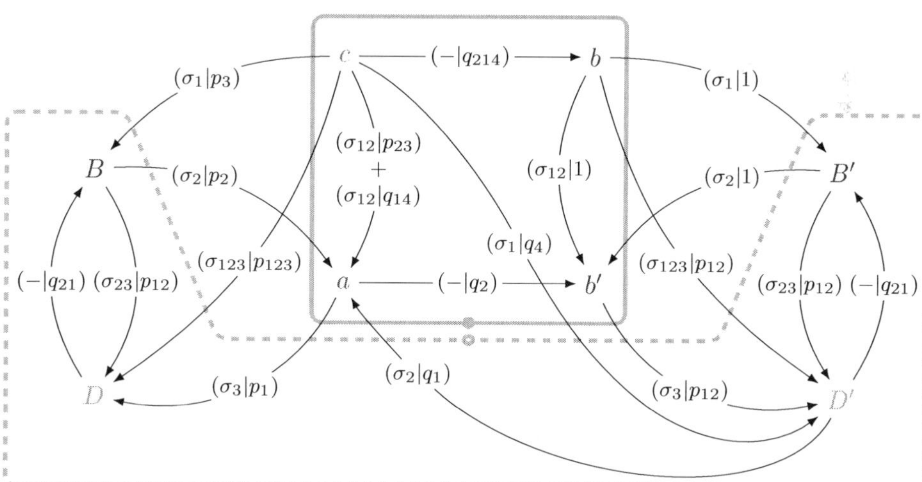

FIGURE 5. The type AD bimodule ${}_{\mathcal{A}}\mathcal{Y}^{\mathcal{B}}$. The right idempotents of the generators are determined by the letter of the generator names; for example, the right idempotent of the generator named B' is ι_b. Generators named by a lower-case letter sit in the left idempotent $\iota_\bullet$, those named by an upper-case letter belong to the left idempotent $\iota_\circ$.

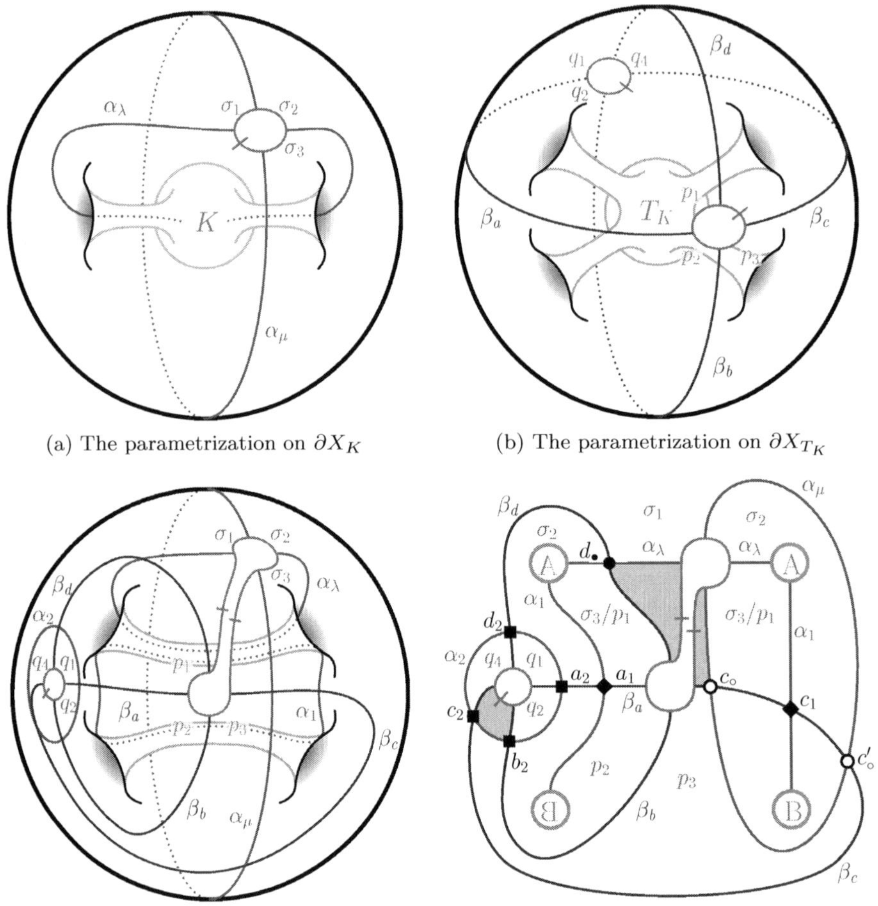

(a) The parametrization on ∂X_K

(b) The parametrization on ∂X_{T_K}

(c) A Heegaard decomposition for Y

(d) The Heegaard diagram $\mathcal{H}_Y$ for Y

FIGURE 6. The tangle complement X_{T_K} (b) is obtained from the knot complement X_K (a) by gluing the bordered sutured manifold Y (c) to it. Figure (d) shows a Heegaard diagram from which we compute the type AD bimodule ${}_{\mathcal{A}}\mathcal{Y}^{\mathcal{B}}$.

complement X_{T_K}. Therefore, by the pairing theorem [**Zar11**, Theorem 12.3.2],

$$\widehat{\mathrm{BSD}}(X_{T_K})^{\mathcal{B}} = \widehat{\mathrm{CFD}}(X_K)^{\mathcal{A}} \boxtimes {}_{\mathcal{A}}\widehat{\mathrm{BSAD}}(Y)^{\mathcal{B}}.$$

(Here, ${}_{\mathcal{A}}\widehat{\mathrm{BSAD}}(Y)^{\mathcal{B}}$ denotes the direct summand of the type AD structure $\widehat{\mathrm{BSAD}}(Y)$ whose generators occupy exactly one of α_μ and α_λ.) So it suffices to show that ${}_{\mathcal{A}}\mathcal{Y}^{\mathcal{B}}$ and ${}_{\mathcal{A}}\widehat{\mathrm{BSAD}}(Y)^{\mathcal{B}}$ are graded chain homotopic. In fact, we will show that

$${}_{\mathcal{A}}\mathcal{Y}^{\mathcal{B}} = {}_{\mathcal{A}}\widehat{\mathrm{BSAD}}(\mathcal{H}_Y)^{\mathcal{B}}$$

for the Heegaard diagram $\mathcal{H}_Y$ of Y shown in Figure 6d.

The type AD structure ${}_{\mathcal{A}}\widehat{\mathrm{BSAD}}(\mathcal{H}_Y)^{\mathcal{B}}$ consists of eight generators in total. We name the generators by the concatenation of the intersection points they are composed of. They are shown as the vertices of the graph in Figure 7. The idempotents

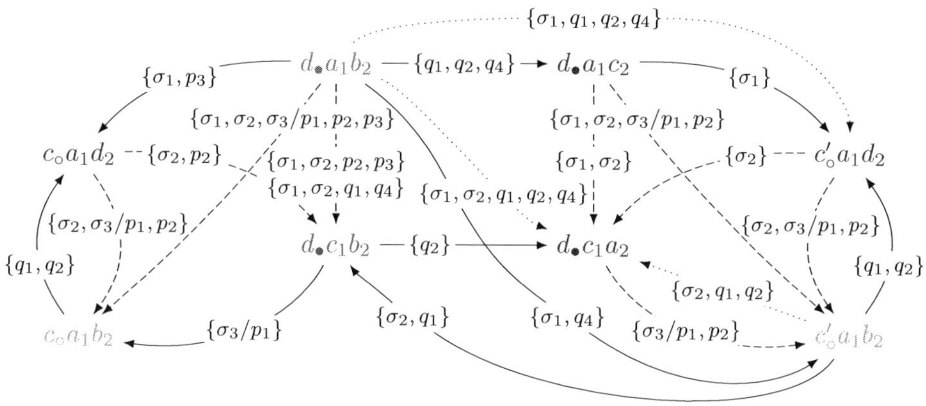

FIGURE 7. Domains connecting generators

of $\mathcal{A}$ acting on the generators on the left can be read off the subscript of the first letter: The four generators in middle two columns lie in the meridional idempotent $\iota_\bullet$, the other four generators belong to the longitudinal idempotent $\iota_\circ$. The basic idempotents of $\mathcal{B}$ acting on the generators on the right can also be deduced from the name: A generator belongs to idempotent ι_x for $x \in \{a, b, c, d\}$ if the letter x does not appear in the name. (This is also indicated by the colours.)

Next, we consider the regions in the Heegaard diagram. As for the generators, we use suggestive notation and label them by algebra elements of $\mathcal{B}$ and $\mathcal{A}$, i.e. p_i, q_j, and σ_ℓ, or combinations thereof. In the following, let us write domains D as formal differences $D_+ - D_-$ of unordered multisets of regions D_+ and D_- with $D_+ \cap D_- = \varnothing$ such that

$$D = \sum_{r \in D_+} r - \sum_{r \in D_-} r.$$

Let us calculate some connecting domains between the generators. The solid arrows in Figure 7 indicate some n-gons with boundary punctures, which contribute to the type AD structure. Furthermore, elementary combinatorial arguments show that the group of periodic domains of $\mathcal{H}_Y$ is freely generated by the domains

$$B_1 := \{q_1, q_4\} - \{p_2, p_3\}, \quad \text{and} \quad B_2 := \{\sigma_2, \sigma_3/p_1, p_2, q_1, q_2\}.$$

Figure 7 shows all connecting domains satisfying the following conditions: the multiplicity of each region is non-negative; the multiplicity of each region labelled by p_i and q_j is at most 1; the multiplicity 1 regions do not include *both* those labelled by p_i and q_j. (This follows from routine calculations, most of which can be outsourced to a computer using the python script [**Zib22**].) These are the only domains that can possibly contribute to the differential in the type AD structure ${}_{\mathcal{A}}\widehat{\mathrm{BSAD}}(\mathcal{H}_Y)^{\mathcal{B}}$, but we do not know (yet) whether they do. However, it turns out, this is all we need to compute from the Heegaard diagram.

The domains on the three dotted arrows do not contribute to the differential ${}_{\mathcal{A}}\widehat{\mathrm{BSAD}}(\mathcal{H}_Y)^{\mathcal{B}}$, because the expected dimension of the corresponding moduli spaces is not 1; see Figure 8. Comparing Figures 5 and 7, it remains to see that all dashed arrows contribute to the differential. We check this indirectly by pairing ${}_{\mathcal{A}}\widehat{\mathrm{BSAD}}(\mathcal{H}_Y)^{\mathcal{B}}$ with certain test type D structures $\mathcal{T}^{\mathcal{A}}$ for which we already know the result

$$\mathcal{S}^{\mathcal{B}} := \mathcal{T}^{\mathcal{A}} \boxtimes {}_{\mathcal{A}}\widehat{\mathrm{BSAD}}(\mathcal{H}_Y)^{\mathcal{B}}$$

up to homotopy. To construct such $\mathcal{T}^{\mathcal{A}}$, let $X_K(i)$ be the knot complement equipped with the bordered structure given by the meridian μ and the longitude $\lambda+i\mu$, where λ is the homological longitude. Then, setting $K = U$ and $\mathcal{T} = \mathcal{T}_i = \widehat{\mathrm{CFD}}(X_U(i))$, we know that $\mathcal{S}^{\mathcal{B}}$ is chain homotopic to $\widehat{\mathrm{BSD}}(X_{Q_{-2i}})^{\mathcal{B}}$, where Q_{-2i} is the rational tangle of slope $-2i$. Figure 9 compares the expected and the computed result for $\mathcal{S}^{\mathcal{B}}$ for some values of i. We conclude that all dashed arrows need to contribute, which concludes the identification of ${}_{\mathcal{A}}\widehat{\mathrm{BSAD}}(\mathcal{H}_Y)^{\mathcal{B}}$ with ${}_{\mathcal{A}}\mathcal{Y}^{\mathcal{B}}$, up to gradings.

domain B	$e(B)$	$n_{\mathbf{x}}(B)$	$n_{\mathbf{y}}(B)$	$\#\vec{\boldsymbol{\rho}}_1$	$\#\vec{\boldsymbol{\rho}}_2$	$\iota(\boldsymbol{\rho}_1)$	$\iota(\boldsymbol{\rho}_2)$	$\mathrm{ind}(B, \vec{\boldsymbol{\rho}}_1, \vec{\boldsymbol{\rho}}_2)$
$\{\sigma_1, q_1, q_2, q_4\}$	$-\frac{1}{2}$	$\frac{1}{2}$	1	1	1	$-\frac{1}{2}$	$-\frac{1}{2}$	**2**
$\{\sigma_2, q_1, q_2\}$	-1	$\frac{3}{4}$	$\frac{5}{4}$	1	1	$-\frac{1}{2}$	$-\frac{1}{2}$	**2**
$\{\sigma_1, \sigma_2, q_1, q_2, q_4\}$	$-\frac{3}{2}$	1	$\frac{3}{2}$	1	≥ 1	$-\frac{1}{2}$	$-\frac{1}{2}$	$\geq \mathbf{2}$

FIGURE 8. Computation of the embedded indices of the domains on the dotted arrows in Figure 7 following [**Zar11**, Definition 8.4.2]. Here, $B \in \pi_2(\mathbf{x}, \mathbf{y})$ denotes a domain connecting two generators $\mathbf{x}$ and $\mathbf{y}$. The symbols $\vec{\boldsymbol{\rho}}_1$ and $\vec{\boldsymbol{\rho}}_2$ denote sets of Reeb chords on the type D side and type A side, respectively.

So it remains to consider gradings. The unreduced grading of ${}_{\mathcal{A}}\widehat{\mathrm{BSAD}}(\mathcal{H}_Y)^{\mathcal{B}}$ takes values in $\mathbb{Q} \rtimes H_1(\mathbf{Z}_\alpha, \mathbf{a}_\alpha; \mathbb{Q}) \times H_1(\mathbf{Z}_\beta, \mathbf{a}_\beta; \mathbb{Q})$. The unreduced grading gr of the periodic domains $B_1, B_2 \in \pi_2(d_\bullet c_1 b_2, d_\bullet c_1 b_2)$ is equal to

$$\mathrm{gr}(B_1) = (0; 0, 0, 0; 0, -1, -1; 0, 1, 1) \quad \text{and} \quad \mathrm{gr}(B_2) = (-\tfrac{1}{2}; 0, 1, 1; 1, 1, 0; 1, 1, 0)$$

and the unreduced grading of the domain $B_3 = \{\sigma_1, \sigma_2, p_2, p_3\} \in \pi_2(d_\bullet a_1 b_2, d_\bullet c_1 b_2)$ is equal to

$$\mathrm{gr}(B_3) = (0; 1, 1, 0; 0, 1, 1; 0, 0, 0).$$

Using the same grading reductions as in Section 2 (after multiplying the components of $H_1(\mathbf{Z}_\beta, \mathbf{a}_\beta; \mathbb{Q})$ by -1), the reduced gradings of these three domains is

$$\underline{\mathrm{gr}}(B_1) = (0; 0, 0; 0, 2, 2, 0) \quad \text{and} \quad \underline{\mathrm{gr}}(B_2) = (-\tfrac{1}{2}; 0, 1; 0, 0, 2, 2)$$

and

$$\begin{aligned}\underline{\mathrm{gr}}(B_3) &= (r(\iota_c))^{-1} R(\mathrm{gr}(B_3)) \\ &= (\tfrac{1}{2}; 1, 1, 0; 0, -\tfrac{1}{2}, -\tfrac{1}{2}; 0, \tfrac{1}{2}, \tfrac{1}{2}) = (\tfrac{1}{2}; 1, 0; 0, -1, -1, 0);\end{aligned}$$

compare [**LOT18**, Formulas 3.44 and 10.36] and (2.1). Let $\mathcal{P}_3 = \langle \underline{\mathrm{gr}}(B_1), \underline{\mathrm{gr}}(B_2) \rangle$ and choose a as the base generator, i.e. $\underline{\mathrm{gr}}(a) = P_3$. Then, since $-B_3 \in \pi_2(c, a)$,

$$\begin{aligned}\underline{\mathrm{gr}}(c) &= \underline{\mathrm{gr}}(-B_3) \cdot \mathcal{P}_3 \\ &= (-\tfrac{1}{2}; -1, 0; 0, 1, 1, 0) \cdot \mathcal{P}_3 = (-\tfrac{1}{2}; -1, 0; 0, 0, 0, 0) \cdot \mathcal{P}_3. \qquad \square\end{aligned}$$

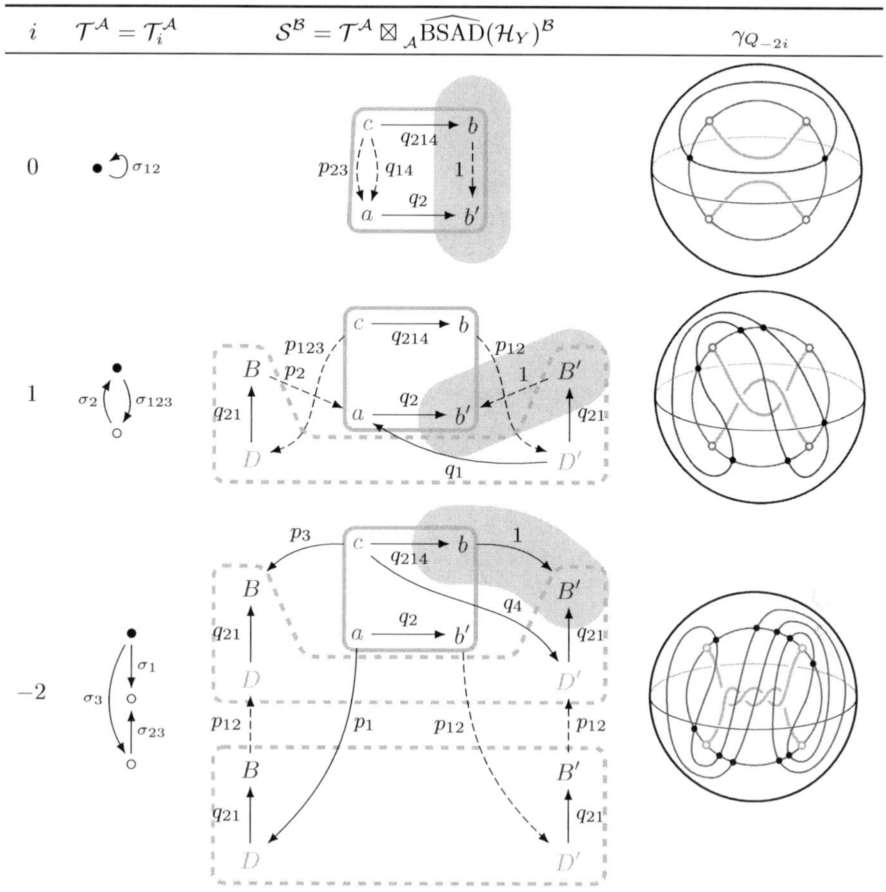

FIGURE 9. Test pairings to determine which pseudoholomorphic curves contribute to the differential in ${}_{\mathcal{A}}\widehat{\mathrm{BSAD}}(\mathcal{H}_Y)^{\mathcal{B}}$. Specifically, the dashed arrows originate from the problematic differentials in ${}_{\mathcal{A}}\widehat{\mathrm{BSAD}}(\mathcal{H}_Y)^{\mathcal{B}}$. Likewise, the solid arrows correspond to the unproblematic differentials. The highlighted arrows labelled 1 can be cancelled using [**Zib20**, Lemma 1.22]. A comparison of the last two columns demonstrates that indeed, all dashed arrows have to be part of the differential of $\mathcal{S}^{\mathcal{B}}$.

4. Proof of the Main Theorem

We now prove the following graded version of the Main Theorem.

THEOREM 4.1. *The components of γ_{T_K} are in one-to-one correspondence with the curve segments in $S_{\times}(\gamma_K)$ (Definition 2.7). More specifically, for any $r, a \in \frac{1}{2}\mathbb{Z}$ and $\ell \in \mathbb{Z}^{>0}$, we have the following correspondence:*

$$\text{in } \gamma_K \begin{bmatrix} \mathbf{d}_{2\tau(K)} & \longleftrightarrow & \mathbf{r}_{4\tau(K)} \\ \delta^r t^a\, \mathbf{u}_\ell & \longleftrightarrow & \delta^{r+a} t_1^{2a} t_2^{2a}\, \mathbf{s}_{2\ell} \\ \delta^r t^{-a}\, \mathbf{v}_\ell & \longleftrightarrow & \delta^{r+a} t_1^{-2a} t_2^{-2a}\, \overline{\mathbf{s}}_{2\ell} \end{bmatrix} \text{ in } \gamma_{T_K}$$

where the gradings on the curves/curve segments is as in Figures 10 *and* 11.

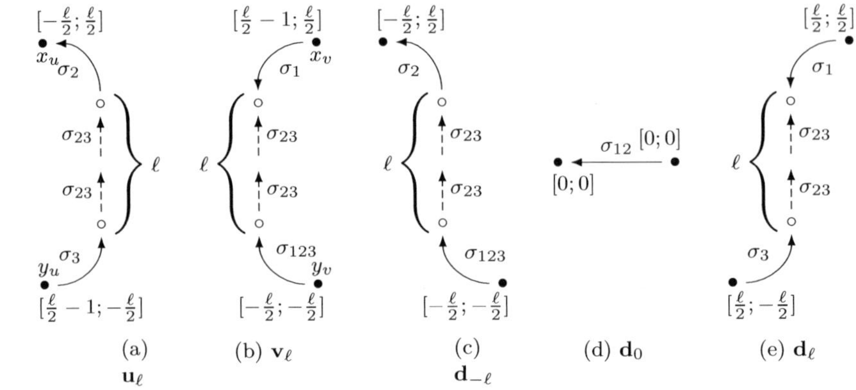

FIGURE 10. The graded chain complexes corresponding to the curve segments that can occur in $S_{\times}(\gamma_K)$. Note $\ell \in \mathbb{Z}^{>0}$.

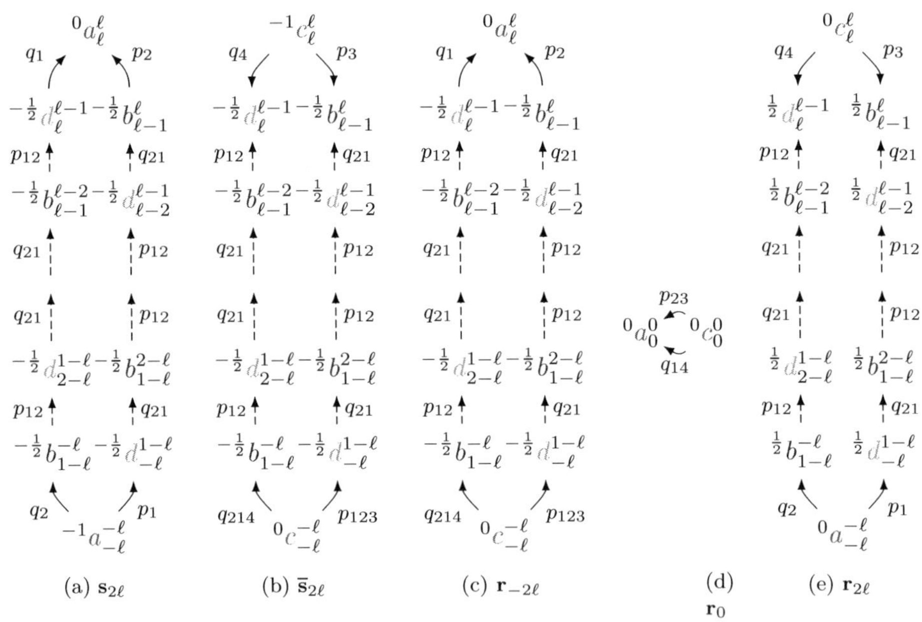

FIGURE 11. The graded chain complexes corresponding to curves that can appear as components of γ_{T_K}. Note $\ell \in \mathbb{Z}^{>0}$.

EXAMPLE 4.2. For the right-handed trefoil knot $K = T_{2,3} = 3_1$,

$$S_{\times}(\gamma_K) = \{\mathbf{d}_2, \delta^{\frac{3}{2}} t^{+\frac{1}{2}}\, \mathbf{u}_1, \delta^{\frac{3}{2}} t^{-\frac{1}{2}}\, \mathbf{v}_1\};$$

see for example [**Han19**]. Hence

$$\gamma_{T_K} = \mathbf{r}_4 \cup \delta^2 t_1^{+1} t_2^{+1} \mathbf{s}_2 \cup \delta^2 t_1^{-1} t_2^{-1} \overline{\mathbf{s}}_2.$$

As one might expect [**KWZ21**, Example 7.15], T_K is an example of a Heegaard Floer exceptional tangle, since $\delta(\mathbf{s}, \mathbf{r}_4) = 2$ for both special components $\mathbf{s}$ of γ_{T_K}.

EXAMPLE 4.3. For the figure-eight knot $K = 4_1$,

$$\begin{aligned} S_{\ast}(\gamma_K) &= \{\mathbf{d}_0, \delta^{\frac{1}{2}} t^{+\frac{1}{2}} \mathbf{u}_1, \delta^{\frac{1}{2}} t^{-\frac{1}{2}} \mathbf{u}_1, \delta^{\frac{1}{2}} t^{-\frac{1}{2}} \mathbf{v}_1, \delta^{\frac{1}{2}} t^{+\frac{1}{2}} \mathbf{v}_1\}, \\ \gamma_{T_K} &= \mathbf{r}_0 \cup \delta^1 t_1^{+1} t_2^{+1} \mathbf{s}_2 \cup \delta^0 t_1^{-1} t_2^{-1} \mathbf{s}_2 \\ &\quad \cup \delta^1 t_1^{-1} t_2^{-1} \overline{\mathbf{s}}_2 \cup \delta^0 t_1^{+1} t_2^{+1} \overline{\mathbf{s}}_2. \end{aligned}$$

EXAMPLE 4.4. For the $(3,4)$-torus knot $K = T_{3,4} = 8_{19}$,

$$\begin{aligned} S_{\ast}(\gamma_K) &= \{\mathbf{d}_6, \delta^{\frac{7}{2}} t^{+\frac{5}{2}} \mathbf{u}_1, \delta^3 t^{-1} \mathbf{u}_2, \delta^{\frac{7}{2}} t^{-\frac{5}{2}} \mathbf{v}_1, \delta^3 t^{+1} \mathbf{v}_2\}, \\ \gamma_{T_K} &= \mathbf{r}_{12} \cup \delta^6 t_1^{+5} t_2^{+5} \mathbf{s}_2 \cup \delta^2 t_1^{-2} t_2^{-2} \mathbf{s}_4 \\ &\quad \cup \delta^6 t_1^{-5} t_2^{-5} \overline{\mathbf{s}}_2 \cup \delta^2 t_1^{+2} t_2^{+2} \overline{\mathbf{s}}_4. \end{aligned}$$

REMARK 4.5. The curve segments $\mathbf{d}_k$ in Figure 10 are graded such that for any knot K, $\mathbf{d}_{2\tau(K)}$ is equal to an element of $S_{\ast}(\gamma_K)$ without any additional grading shift. This is ensured by the symmetry with respect to the Alexander grading and the convention that the Maslov grading of the right generators $\bullet$ in Sections 4 to 4 vanishes. (These are the generators that are mapped to the generators of $\widehat{\mathrm{HF}}(S^3)$ under the spectral sequence corresponding to setting $U = 0$ and $V = 1$.)

REMARK 4.6. The graded version of conjugation symmetry for the multicurve γ_K (Theorem 1.2) corresponds precisely to the graded version of conjugation symmetry for the multicurve γ_{T_K} (Theorem 1.3). Indeed, the former says that for any $\ell \in \mathbb{Z}^{>0}$ and $r, a \in \frac{1}{2}\mathbb{Z}$,

$$\#\{\text{curve segments } \delta^r t^a \mathbf{u}_\ell \text{ in } S_{\ast}(\gamma_K)\} = \#\{\text{curve segments } \delta^r t^{-a} \mathbf{v}_\ell \text{ in } S_{\ast}(\gamma_K)\}$$

and the latter that for any $\ell \in \mathbb{Z}^{>0}$ and $r, a_1, a_2 \in \frac{1}{2}\mathbb{Z}$,

$$\#\{\text{components } \delta^r t_1^{a_1} t_2^{a_2} \mathbf{s}_{2\ell} \text{ of } \gamma_{T_K}\} = \#\{\text{components } \delta^r t_1^{-a_1} t_2^{-a_2} \overline{\mathbf{s}}_{2\ell} \text{ of } \gamma_{T_K}\}.$$

PROOF OF THEOREM 4.1. Let us assume first that all local systems on γ_K are one-dimensional and thus trivial. With a view towards applying Theorem 3.1, we first compute the effect of applying $- \boxtimes {}_{\mathcal{A}}\mathcal{Y}^{\mathcal{B}}$ to the chain complexes over $\mathcal{A}$ corresponding to curve segments, ignoring gradings for a moment. We carry out these computations in Figure 12. Clearly, $\mathbf{u}_\ell \boxtimes {}_{\mathcal{A}}\mathcal{Y}^{\mathcal{B}}$ contains a direct summand that is chain homotopic to (the chain complex over $\mathcal{B}$ corresponding to) $\mathbf{s}_{2\ell}$, and the same is true for $\mathbf{v}_\ell \boxtimes {}_{\mathcal{A}}\mathcal{Y}^{\mathcal{B}}$ and $\overline{\mathbf{s}}_{2\ell}$, as well as for $\mathbf{d}_k \boxtimes {}_{\mathcal{A}}\mathcal{Y}^{\mathcal{B}}$ and $\mathbf{r}_{2k}$ for all $\ell \in \mathbb{Z}^{>0}$ and $k \in \mathbb{Z}$. However, in each case, four additional generators remain, two on each end, so to speak. This is where we make use of Hanselman and Watson's puzzle piece notation [**HW15**, Figure 1]. This notation is chosen such that two puzzle pieces fit together if and only if the corresponding curve segments can be joined at the respective ends to a valid curve on the torus. The following observation is crucial: At each male end of a puzzle piece, the remaining two generators are the same, namely c and b. Similarly, at the female ends, the generators a and b' remain. This justifies a posteriori that the computation of $\widehat{\mathrm{CFD}}(X_K)^{\mathcal{A}} \boxtimes {}_{\mathcal{A}}\mathcal{Y}^{\mathcal{B}}$ can indeed be done locally on each curve segment. In other words, $\widehat{\mathrm{CFD}}(X_K)^{\mathcal{A}} \boxtimes {}_{\mathcal{A}}\mathcal{Y}^{\mathcal{B}}$ is indeed a direct sum of chain complexes over $\mathcal{B}$, namely one for each curve segment, as stated in the correspondence. We now apply Theorem 3.1 to conclude.

It remains to verify the gradings and to consider curves with non-trivial local systems. We start with the latter. Let us suppose that γ_K contains some component γ with an n-dimensional local system for some $n \geq 2$. Let us choose some orientation of γ and represent this local system by the matrix X. We may put this matrix on any arrow of the complexes in Figure 12 corresponding to the curve segments of γ. For convenience, we may put it always on the first arrow; see the discussion before Definition 2.3. Suppose it lies on the first arrow of $\mathbf{u}_\ell$. Then the corresponding chain complex over $\mathcal{B}$ looks like the one for $\mathbf{u}_\ell$, except that first, it should be tensored with $\mathbb{F}^n$ and second, the labels $p_1 \otimes \mathrm{id}_{\mathbb{F}^n}$ and $p_{12} \otimes \mathrm{id}_{\mathbb{F}^n}$ on the arrows leaving $a \otimes \mathbb{F}^n$ and $b' \otimes \mathbb{F}^n$ should be replaced by $p_1 \otimes X^{\pm 1}$ and $p_{12} \otimes X^{\pm 1}$, respectively. Whether it is X^{-1} or X^{+1} depends on the chosen orientation of γ; crucially, it is the same for both labels. So after a change of basis, we recognize the local system on the resulting curve as $XX^{-1} = \mathrm{id}_{\mathbb{F}^n}$. By inspecting the other complexes in Figure 12 in a similar fashion, we see that any local system has the same effect on the result as a trivial local system of the same dimension.

To check the gradings, it clearly suffices to compute the grading of the generators in idempotents a and c. Note that any element the double coset $\mathcal{P}_1 \backslash (\underline{\mathrm{Gr}}(\mathcal{Z}_\alpha) \times \mathfrak{A})/\mathcal{P}_3$ can be uniquely written as $\mathcal{P}_1 \cdot (x; 0, 0; 0, 0, y, z) \cdot \mathcal{P}_3$ for some $x, y, z \in \mathbb{Q}$. Identifying this element with $[-x; y, z]$, this recovers the usual bigrading on the tangle invariant. So, given a generator of $\widehat{\mathrm{CFD}}(X_K)^{\mathcal{A}}.\iota_\bullet$ in grading $[m; n]$, the corresponding generator of $\widehat{\mathrm{CFD}}(X_K)^{\mathcal{A}} \boxtimes {}_{\mathcal{A}}\mathcal{Y}^{\mathcal{B}}$ in idempotent a has grading

$$\begin{aligned}
&\mathcal{P}_1 \cdot (-m - \tfrac{n}{2}; 0, -n; 0, 0, 0, 0) \cdot \mathcal{P}_3 \\
&= \mathcal{P}_1 \cdot (-m - n; 0, 0; 0, 0, 2n, 2n)(\tfrac{1}{2}; 0, -1; 0, 0, -2, -2)^n \cdot \mathcal{P}_3 \\
&= \mathcal{P}_1 \cdot (-m - n; 0, 0; 0, 0, 2n, 2n) \cdot \mathcal{P}_3 \\
&= [m + n; 2n, 2n]
\end{aligned}$$

and the corresponding generator in idempotent c has grading

$$\begin{aligned}
&\mathcal{P}_1 \cdot (-m - \tfrac{n}{2}; 0, -n; 0, 0, 0, 0)(-\tfrac{1}{2}; -1, 0; 0, 0, 0, 0) \cdot \mathcal{P}_3 \\
&= \mathcal{P}_1 \cdot (-m - \tfrac{n}{2} - \tfrac{1}{2} + n; -1, -n; 0, 0, 0, 0) \cdot \mathcal{P}_3 \\
&= \mathcal{P}_1 \cdot (-\tfrac{1}{2}; -1, 0; 0, 0, 0, 0)(-m + n + \tfrac{n}{2}; 0, -n; 0, 0, 0, 0) \cdot \mathcal{P}_3 \\
&= \mathcal{P}_1 \cdot (-m + n; 0, 0; 0, 0, 2n, 2n)(\tfrac{1}{2}; 0, -1; 0, 0, -2, -2)^n \cdot \mathcal{P}_3 \\
&= \mathcal{P}_1 \cdot (-m + n; 0, 0; 0, 0, 2n, 2n) \cdot \mathcal{P}_3 \\
&= [m - n; 2n, 2n].
\end{aligned}$$

So it suffices to check the identities above for $r = 0 = a$. This is elementary to check; we simply compare the gradings in Figures 10 and 11. □

5. Satellites, thinness, and A-links

We now prove the results from Subsection 1.5 of the introduction.

PROOF OF THEOREM 1.8. Zoltan Szabó's program [**OS17**] computes the following knot Floer homology of the knot K shown in Figure 3:

$$\widehat{\mathrm{HFK}}(K) \cong \delta^0 t^1 \mathbb{F}^4 \oplus \delta^0 t^0 \mathbb{F}^9 \oplus \delta^0 t^{-1} \mathbb{F}^4.$$

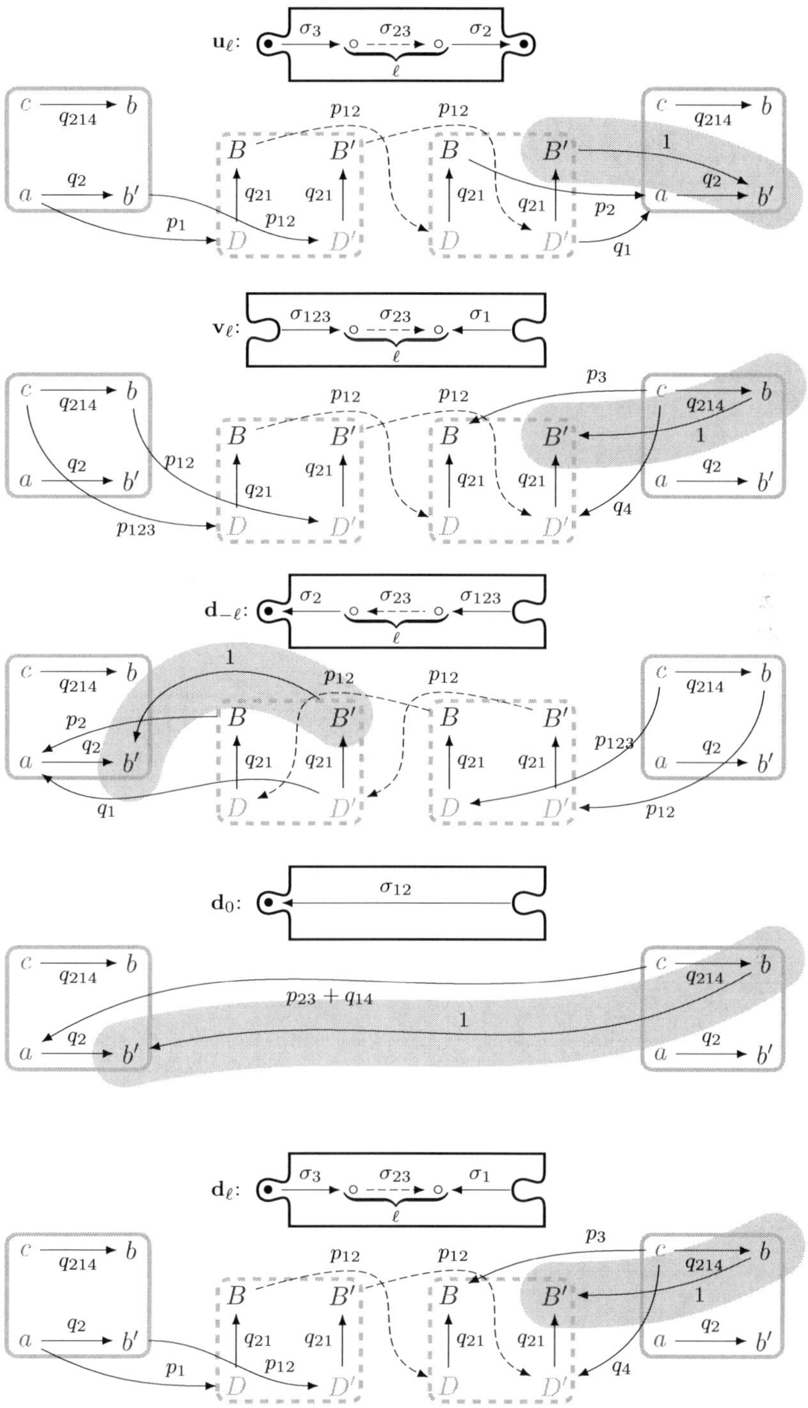

FIGURE 12. The main computation for the proof of Theorem 4.1, not including gradings. Each puzzle piece represents a curve segment in $S_{\times}(\gamma_K)$, which, when tensored with the type AD structure ${}_{\mathcal{A}}\mathcal{Y}^{\mathcal{B}}$, becomes the chain complex over $\mathcal{B}$ shown below the puzzle piece. Note $\ell \in \mathbb{Z}^{>0}$.

A conceptually more interesting proof, which also explains how this example was found, uses the multicurve invariant of the double tangle of the trefoil knot computed in Example 4.2. Note that the Conway tangle $T_\star$ highlighted on the left of Figure 3 agrees with said double tangle up to two full twists. Using the naturality of the multicurve invariant under twisting [**Zib19b**, Theorem 3.1], we see that $\gamma_{T_\star}$ consists of a single rational component $\mathbf{r}_0$ of slope 0 and a pair of conjugate special components of slope ∞, denoted here by $\mathbf{S}_\infty$. The complementary tangle $T'_\star$ on the right of Figure 3 is obtained from $T_\star$ by rotation in the plane about 90 degrees and reversing all crossings. Thus, the corresponding multicurve invariant consists of a single rational component $\mathbf{r}_\infty$ of slope ∞ and a pair $\mathbf{S}_0$ of conjugate special components of slope 0. The Lagrangian Floer homology between a rational and a special component of the same slope is zero [**KWZ21**, Lemma 4.20]. Hence, by the gluing theorem [**Zib20**, Theorem 5.9],

$$\widehat{\mathrm{HFK}}(T'_\star \cup T_\star) \otimes \mathbb{F}^2 \cong \mathrm{HF}(-\gamma_{T'_\star}, \gamma_{T_\star}) \cong \mathrm{HF}(-\mathbf{r}_\infty, \mathbf{r}_0) \oplus \mathrm{HF}(-\mathbf{S}_0, \mathbf{S}_\infty).$$

These two summands are thin and supported in the same δ-grading. This can be seen either by direct computation or as an application of [**KWZ21**, Lemma 4.17]. □

PROPOSITION 5.1. *For any knot $K \subset S^3$ except the unknot and the trefoil knots, the multicurve γ_{T_K} contains two special components in distinct δ-gradings.*

PROOF. We first introduce some notation and make some preliminary computations. Given a curve segment $\mathbf{u}_\ell$ or $\mathbf{v}_\ell$ in $S_{\times}(\gamma_K)$, consider the two generators of $\widehat{\mathrm{HFK}}(K)$ at the ends of this curve segment. If the segment is of type $\mathbf{u}_\ell$, we denote these two generators by x_u and y_u as shown in Section 4; if it is of type $\mathbf{v}_\ell$, we denote them by x_v and y_v as shown in Section 4. The second column of the table in Figure 13 shows the graded curve segments whose generators $x \in \{x_u, y_u, x_v, y_v\}$, determined in the first column, sit in bigrading $[r; a]$. The third column shows the corresponding components of γ_{T_K} computed using Theorem 4.1. Note that the δ-grading of the components of γ_{T_K} are independent of the components' length ℓ.

generator x	component of $S_{\times}(\gamma_K)$	component of γ_{T_K}
x_u	$\delta^{r+\frac{\ell}{2}} t^{a-\frac{\ell}{2}} \mathbf{u}_\ell$	$\delta^{r+a} t_1^{2a-\ell} t_2^{2a-\ell} \mathbf{s}_{2\ell}$
y_u	$\delta^{r+1-\frac{\ell}{2}} t^{a+\frac{\ell}{2}} \mathbf{u}_\ell$	$\delta^{r+1+a} t_1^{2a+\ell} t_2^{2a+\ell} \mathbf{s}_{2\ell}$
x_v	$\delta^{r+1-\frac{\ell}{2}} t^{a-\frac{\ell}{2}} \mathbf{v}_\ell$	$\delta^{r+1-a} t_1^{2a-\ell} t_2^{2a-\ell} \overline{\mathbf{s}}_{2\ell}$
y_v	$\delta^{r+\frac{\ell}{2}} t^{a+\frac{\ell}{2}} \mathbf{v}_\ell$	$\delta^{r-a} t_1^{2a+\ell} t_2^{2a+\ell} \overline{\mathbf{s}}_{2\ell}$

FIGURE 13. Computations for the proof of Proposition 5.1.

We now study how two such curve segments can be joined up along a generator x of $\widehat{\mathrm{HFK}}(K)$ to form part of the curve γ_K. We distinguish two cases:

(1) Suppose $x = x_u = x_v$ or $x = y_u = y_v$. We compare the first and third, respectively second and fourth entry of the last column in the table in Figure 3. We see that in this case, γ_{T_K} contains two special components in distinct δ-gradings, since a, being the Alexander grading of a generator of $\widehat{\mathrm{HFK}}(K)$, is an integer.

(2) Suppose $x = x_u = y_v$ or $x = y_u = x_v$. In this case, γ_{T_K} also contains two special components in distinct δ-gradings, unless $a = 0$.

Let $K \subset S^3$ be a knot and suppose γ_{T_K} does not contain two special components in distinct δ-gradings. Then γ_K consists of a single connected component. Indeed, if there were two components, then in particular there would be a component of γ_K not containing the curve segment $\mathbf{d}_{2\tau(K)}$; a generator of maximal Alexander grading in that component would be a generator of type (1). Similarly, one can see that γ_K has the structure of the knot Floer homology of an L-space knot: All generators sit in distinct Alexander gradings and every curve segment in $S_{✂}(\gamma_K)$ except $\mathbf{d}_{2\tau(K)}$ connects generators in consecutive Alexander gradings. (This is sometimes known as the "staircase structure".) If $\dim \widehat{\mathrm{HFK}}(K) > 3$ this implies that there is more than one generator of type (2), and only one of them can be in Alexander grading 0. So we deduce that $\dim \widehat{\mathrm{HFK}}(K) \leq 3$. But the only knots with such small knot Floer homology are the unknot and the two trefoil knots [**OS04**, **Ghi08**, **HW18**]. □

Proof of Proposition 1.10. Let K be as in the statement of Proposition 1.10 and suppose for contradiction that $P(K)$ is a satellite knot whose knot Floer homology is thin for some pattern P with wrapping number 2. Then $P(K) = T_P \cup T_K$ for some pattern tangle T_P and we may write

$$\widehat{\mathrm{HFK}}(P(K)) \otimes \mathbb{F}^2 \cong \mathrm{HF}(-\gamma_{T_P}, \gamma_{T_K}).$$

By Proposition 5.1, T_K contains two special components in distinct δ-grading. Thus, γ_{T_P} only contains component of slope ∞, because otherwise, $\mathrm{HF}(-\gamma_{T_P}, \gamma_{T_K})$ would not be thin by [**KWZ21**, Lemma 4.17]. Furthermore, by the same reasoning, using [**KWZ21**, Lemma 4.20], no component of γ_{T_P} can be special. So γ_{T_P} consists of rational components of slope ∞ only. By [**LMZ22**, Theorem 4.1], the tangle T_P is vertically split and hence the wrapping number of the pattern P is 0, contrary to our assumptions. □

Lemma 5.2. *A knot $K \subset S^3$ is an L-space knot if and only if all special components in γ_{T_K} sit in δ-gradings of the same parity.*

Proof. This follows from the computations in the proof of Proposition 5.1, observing the fact that generators of type (1) cannot occur. □

Proof of Proposition 1.11. Let P and K be as in the statement of Proposition 1.11 and suppose for contradiction that $P(K)$ is an A-link. We now argue as in the proof of Proposition 1.10, using the fact that by Lemma 5.2, γ_{T_K} contains two special components in δ-gradings of different parity. □

Proof of Theorem 1.12. By Lemma 5.2, it remains to see that there exists a non-trivial rational A-link filling of T_K if and only if all special components in γ_{T_K} sit in δ-gradings of the same parity. But this follows from [**KWZ21**, Lemma 4.17]. □

It is interesting to compare the following spaces [**RR17**, **HRW16**, **KWZ21**]:

$$\mathcal{L}(K) := \{p/q \in \mathbb{Q}\mathrm{P}^1 \mid S^3_{p/q}(K) \text{ is an L-space}\},$$
$$\mathrm{A}(T_K) := \{p/q \in \mathbb{Q}\mathrm{P}^1 \mid T_K(p/q) \text{ is an A-link}\}.$$

Theorem 5.3. *For any knot $K \subset S^3$, exactly one of the following holds:*

(1) $\mathcal{L}(K) = \{\infty\} = \mathrm{A}(T_K)$;
(2) $\mathcal{L}(K) = \mathbb{Q}\mathrm{P}^1 \smallsetminus \{0\} = \mathrm{A}(T_K)$; *in this case, K is the unknot;*
(3) $\mathcal{L}(K) = [2\tau(K)-1, \infty]$ *and* $\mathrm{A}(T_K) = (4\tau(K), \infty]$; *in this case,* $\tau(K) > 0$;
(4) $\mathcal{L}(K) = [\infty, 2\tau(K)+1]$ *and* $\mathrm{A}(T_K) = [\infty, 4\tau(K))$; *in this case,* $\tau(K) < 0$.

PROOF. Let $K \subset S^3$ be a knot. By Theorem 1.12, $\mathcal{L}(K) = \{\infty\}$ if and only if $\mathrm{A}(T_K) = \{\infty\}$. So it suffices to see that if K is an L-space knot, then exactly one of (2)–(4) holds. We claim that the three cases correspond precisely to the cases $\tau(K) = 0$, $\tau(K) > 0$, and $\tau(K) < 0$, respectively. Note that the unknot is the only L-space knot with $\tau(K) = 0$.

The values of $\mathcal{L}(K)$ in these three cases can be easily determined from the multicurve γ_K; this is explained in [**HRW16**, Section 7.5]. So it remains to determine the value of $\mathrm{A}(T_K)$ in these cases.

For case (2), observe that T_K is the rational tangle of slope 0, and so any rational filling of T_K is a 2-bridge link. All two-bridge links are non-split alternating, except the unlink, which corresponds to the 0-rational filling and whose knot Floer homology is supported in two consecutive δ-gradings. So indeed $\mathrm{A}(T_K) = \mathbb{Q}\mathrm{P}^1 \smallsetminus \{0\}$.

For cases (3) and (4), we first note that $\mathrm{A}(T_K)$ contains some $p/q \neq \infty$ by Theorem 1.12. So $\mathrm{A}(T_K)$ must be an interval [**KWZ21**, Theorem 1.8]. In fact, as explained in [**KWZ21**, Sections 2.1 and 4.2], the boundary points of the interval must be equal to ∞ and $4\tau(K)$, since γ_{T_K} is supported in these slopes. Moreover, as the curve of slope $4\tau(K)$ is rational, this slope is not contained in the interval and we already know that $\infty \in \mathrm{A}(T_K)$. So $\mathrm{A}(T_K)$ is either equal to $(4\tau(K), \infty]$ or $[\infty, 4\tau(K))$. We determine which one it is by observing $0 \notin \mathrm{A}(T_K)$. This is because the determinant of the link $T_K(0) = C_{2,0}(K)$ is zero. □

COROLLARY 5.4. *For any knot $K \subset S^3$ and $n \gg 0$, $S^3_n(K)$ is an L-space if and only if $C_{2,n}(K)$ is an A-link.*

PROOF. For any integer n, $T_K(n) = C_{2,n}(K)$. Now choose $n > 4|\tau(K)|$ and apply Theorem 5.3. □

It is interesting to compare this to Hanselman and Watson's cabling formula for γ_K [**HW19**].

6. Growth of knot Floer and Khovanov homology under cabling

We now compare the Heegaard Floer multicurve invariant γ_T and the Khovanov multicurve invariant $\widetilde{\mathrm{Kh}}(T;\mathbb{F}_2)$ [**KWZ19**] for double tangles $T = T_K$. To emphasize the similarity between the two invariants, we will in this section denote the former by γ_T^{HF} and the latter by γ_T^{Kh}. Like γ_T^{HF}, the multicurve invariant γ_T^{Kh} satisfies a gluing theorem that allows one to compute the Khovanov homology of the union of two tangles in terms of Lagrangian Floer homology [**KWZ19**, Theorem 1.9]. Even more remarkably, the components of γ_T^{Kh} are also subject to very similar geography restrictions [**KWZ22b**, Theorem 1.2]. We do not need the full statement of this classification; it suffices to say that there is a class of *rational* curves $\mathbf{r}^{\mathrm{Kh}}(k)$, parametrized by their *slope* $k \in \mathbb{Z}$, and a class of *special* curves $\mathbf{s}_\ell^{\mathrm{Kh}}(\infty)$, parametrized by their *length* $\ell \in 2\mathbb{Z}$. To simplify notation, we will write

$$\mathbf{s}_\ell^{\mathrm{Kh}} := \mathbf{s}_\ell^{\mathrm{Kh}}(\infty) \quad \text{and} \quad \mathbf{r}_k^{\mathrm{Kh}} := \mathbf{r}^{\mathrm{Kh}}(k).$$

For clarity, we will in this section write $\mathbf{s}_\ell^{\mathrm{HF}}$, $\overline{\mathbf{s}}_\ell^{\mathrm{HF}}$, and $\mathbf{r}_k^{\mathrm{HF}}$ for the curves previously denoted by $\mathbf{s}_\ell$, $\overline{\mathbf{s}}_\ell$, and $\mathbf{r}_k$, respectively, for all $\ell \in \mathbb{Z}^{>0}$ and $k \in \mathbb{Z}$.

We have the following structure theorem for $\gamma_{T_K}^{\mathrm{Kh}}$, which is implicit in [**LZ22**].

THEOREM 6.1. *For any knot $K \subset S^3$, the multicurve γ_{T_K} contains a single rational component $\mathbf{r}_{2\vartheta_2(K)}^{\mathrm{Kh}}$, where $\vartheta_2(K) \in \mathbb{Z}$ is the concordance homomorphism from* [**LZ22**]*, and every other curve segment is of type $\mathbf{s}_{\ell_i}^{\mathrm{Kh}}$ for some $\ell_i > 0$.*

PROOF. The tangle T_K is the quotient tangle of a strongly invertible knot, namely $K\#r(K)$, where $r(K)$ is the reverse of K. So by [**KWZ22a**, Theorem 3.1], every component except one is of type $\mathbf{s}_{\ell_i}^{\mathrm{Kh}}$ for some $\ell_i \in 2\mathbb{Z}^{>0}$, and the remaining one is of type $\mathbf{r}_k^{\mathrm{Kh}}$ for some slope $k \in \mathbb{Z}$. The slope k is divisible by 2 because rational components (of odd length) detect how tangle ends are connected [**KWZ22b**, Theorem 5.6]. (This implies that also in [**KWZ22a**, Theorem 3.1], the slope k is always divisible by 2.)

It remains to relate the slope k to the concordance invariant $\vartheta_2(K)$. By [**LZ22**, Corollary 5.14], $2\vartheta_2(K)$ is the slope of the multicurve invariant $\widetilde{\mathrm{BN}}(T_K)$ near the bottom right tangle end. By naturality of the mapping class group action on $\widetilde{\mathrm{BN}}(T_K)$ and $\widetilde{\mathrm{Kh}}(T_K)$ [**KWZ19**, Theorem 1.13], it suffices to show that the rational component of the twisted curve $\widetilde{\mathrm{Kh}}(T_K+Q_{-2\vartheta_2(K)})$ is 0. This follows from a simple computation carried out in [**LZ22**, Proof of Lemma 5.11]. □

REMARK 6.2. Combining Theorems 4.1 and 6.1, we see that the slopes of the rational components in $\gamma_{T_K}^{\mathrm{HF}}$ and $\gamma_{T_K}^{\mathrm{Kh}}$ are $4\tau(K)$ and $2\vartheta_2(K)$, respectively. So, as already observed in [**LZ22**, Section 1.3], $2\tau(K)$ plays the same role in knot Floer homology as the invariant $\vartheta_2(K)$ in Khovanov homology.

For the proofs of Propositions 1.14 and 1.15, we need the following basic result.

LEMMA 6.3. *Let $k, k' \in \mathbb{Z}$ with $k \neq k'$ and $\ell \in 2\mathbb{Z}^{>0}$. Then*

$$\dim \mathrm{HF}(\gamma,\gamma') = \begin{cases} 2\ell & \text{if } (\gamma,\gamma') \in \{(\mathbf{r}_k^{\mathrm{Kh}},\mathbf{s}_\ell^{\mathrm{Kh}}),(\mathbf{r}_k^{\mathrm{HF}},\mathbf{s}_\ell^{\mathrm{HF}}),(\mathbf{r}_k^{\mathrm{HF}},\overline{\mathbf{s}}_\ell^{\mathrm{HF}})\} \\ 2|k'-k| & \text{if } (\gamma,\gamma') \in \{(\mathbf{r}_k^{\mathrm{Kh}},\mathbf{r}_{k'}^{\mathrm{Kh}}),(\mathbf{r}_k^{\mathrm{HF}},\mathbf{r}_{k'}^{\mathrm{HF}})\} \\ 4 & \text{if } \gamma=\gamma'=\mathbf{r}_k^{\mathrm{Kh}} \\ 2 & \text{if } \gamma=\gamma'=\mathbf{r}_k^{\mathrm{HF}} \end{cases}$$

PROOF. These are straightforward computations. □

PROOF OF PROPOSITION 1.14. By the gluing theorem, combined with the Main Theorem,

$$\begin{aligned} 2\dim \widehat{\mathrm{HFK}}(C_{2,2t+1}(K)) &= \dim \mathrm{HF}(\mathbf{r}_{2t+1}^{\mathrm{HF}},\gamma_{T_K}^{\mathrm{HF}}) \\ &= \dim \mathrm{HF}(\mathbf{r}_{2t+1}^{\mathrm{HF}},\mathbf{r}_{4\tau(K)}^{\mathrm{HF}}) + \sum_{\mathbf{s}} \dim \mathrm{HF}(\mathbf{r}_{2t+1}^{\mathrm{HF}},\mathbf{s}) \end{aligned}$$

where the sum is over all special connected components $\mathbf{s}$ in $\gamma_{T_K}^{\mathrm{HF}}$. The first summand is equal to $2|2t+1-4\tau(K)|$ by Lemma 6.3. Each term in the second summand is equal to twice the length of the special curve $\mathbf{s}$. The number of terms is $(d-1)$ by the Main Theorem, so the second term is equal to $4(d-1)\overline{\ell}$. The desired identity follows.

The upper bound for $\dim \widehat{\mathrm{HFK}}(C_{2,2t+1}(K))$ is clear. The lower bound follows from the observation that $\overline{\ell} \geq 1$ and $|2t+1-4\tau(K)|$ is odd. □

Alternative proof of Proposition 1.14. We briefly outline an alternative argument that uses Hanselman and Watson's cabling formula for the multicurve invariant γ_K [**HW19**]. As noted in the introduction, for any knot $K \subset S^3$, the dimension of $\widehat{\mathrm{HFK}}(K)$ is equal to the minimal number of intersection points of γ_K with the meridian μ_K (counted with multiplicity of the dimension of any local system). Equivalently, $\dim \widehat{\mathrm{HFK}}(K)$ is equal to the number of elements in $S_{\text{✂}}(\gamma_K)$. The same is true, of course, if we replace K by $K' := C_{2,2t+1}(K)$. Using Hanselman and Watson's cabling formula, we can easily compute the multicurve $\gamma_{K'}$ from γ_K; in particular, we determine $S_{\text{✂}}(\gamma_{K'})$ from $S_{\text{✂}}(\gamma_K)$ as follows: Every component $\mathbf{u}_\ell$ contributes a single component $\mathbf{u}_{\ell'}$ (for some $\ell' \in \mathbb{Z}^{>0}$), ℓ components $\mathbf{u}_1$ and $(\ell-1)$ components $\mathbf{v}_1$ in $S_{\text{✂}}(\gamma_{K'})$. Similarly, $\mathbf{v}_\ell$ contributes a total of 2ℓ components in $S_{\text{✂}}(\gamma_{K'})$. From this, we get the first summand in Proposition 1.14. The second summand comes from the contribution of $\mathbf{d}_{2\tau(K)}$. This curve contributes a single component $\mathbf{d}_{2\tau(K')}$ to $S_{\text{✂}}(\gamma_{K'})$ as well as $\lfloor |\frac{2t+1}{2} - 2\tau(K)| \rfloor$ pairs of components $\mathbf{u}_1$ and $\mathbf{v}_1$. □

Proof of Proposition 1.15. We start by observing that each of the two components of the link $C_{2,2t}(K)$ is equal to K. So by [**BS15**, Corollary 1.6],

$$\dim \mathrm{Kh}(C_{2,2t}(K)) \geq (\dim \mathrm{Kh}(K))^2,$$

where $\mathrm{Kh}(\cdot)$ denotes unreduced Khovanov homology over $\mathbb{F}_2$. Thus, for reduced Khovanov homology $\widetilde{\mathrm{Kh}}(\cdot)$, we obtain

$$\dim \widetilde{\mathrm{Kh}}(C_{2,2t}(K)) \geq 2(\dim \widetilde{\mathrm{Kh}}(K))^2 = 2d^2, \tag{6.1}$$

since $\dim \mathrm{Kh}(J) = 2 \dim \widetilde{\mathrm{Kh}}(J)$ for any link J [**Shu04**, Corollary 3.2.C]. By the gluing theorem, combined with Theorem 6.1,

$$\begin{aligned} 2 \dim \widetilde{\mathrm{Kh}}(C_{2,k}(K)) &= \dim \mathrm{HF}(\mathbf{r}_k^{\mathrm{Kh}}, \gamma_{T_K}^{\mathrm{Kh}}) \\ &= \dim \mathrm{HF}(\mathbf{r}_k^{\mathrm{Kh}}, \mathbf{r}_{2\vartheta_2(K)}^{\mathrm{Kh}}) + \sum_{\mathbf{s}} \dim \mathrm{HF}(\mathbf{r}_k^{\mathrm{Kh}}, \mathbf{s}) \end{aligned}$$

for any integer k, where the sum is over all special components $\mathbf{s}$ in $\gamma_{T_K}^{\mathrm{Kh}}$. The first summand is equal to $2|k - 2\vartheta_2(K)|$, unless $k = 2\vartheta_2(K)$, in which case it is equal to 4. The second summand is independent of k. Setting $k = 2t + 1$ and $k = 2\vartheta_2(K)$, we obtain

$$\dim \widetilde{\mathrm{Kh}}(C_{2,2t+1}(K)) = \dim \widetilde{\mathrm{Kh}}(C_{2,2\vartheta_2(K)}(K)) - 2 + |2t + 1 - 2\vartheta_2(K)|$$

We now combine this identity with (6.1) and obtain the desired inequality. □

I close with a few remarks about Conjecture 1.16. I have verified this conjecture for a selection of about 30 knots with up to 11 crossings using the computer program [**Zib21**]. The conjecture is also consistent with Proposition 1.15—at least, if we assume the following conjecture, which generalises [**KWZ22a**, Conjecture 3.10] and for which there is now overwhelming computational evidence.

Conjecture 6.4. *For any Conway tangle T, the length of every special component of $\gamma_T^{\mathrm{Kh}} = \widetilde{\mathrm{Kh}}(T; \mathbb{F}_2)$ is divisible by 4.*

Note this conjecture is known to be false if $\widetilde{\mathrm{Kh}}(T; \mathbb{F}_2)$ is computed over a field of different characteristic. Now, assuming Conjectures 1.16 and 6.4, Proposition 1.15

follows by the same argument as Proposition 1.14. By the gluing theorem,

$$\begin{aligned}2\dim\widetilde{\mathrm{Kh}}(C_{2,2t+1}(K)) &= \dim\mathrm{HF}(\mathbf{r}^{\mathrm{Kh}}_{2t+1},\gamma^{\mathrm{Kh}}_{T_K})\\ &= \dim\mathrm{HF}(\mathbf{r}^{\mathrm{Kh}}_{2t+1},\mathbf{r}^{\mathrm{Kh}}_{2\vartheta_2(K)})+\sum_{\mathbf{s}}\dim\mathrm{HF}(\mathbf{r}^{\mathrm{Kh}}_{2t+1},\mathbf{s}),\end{aligned}$$

where the sum is over all special connected components $\mathbf{s}$ in $\gamma^{\mathrm{Kh}}_{T_K}$. By assumption, there are at least $\frac{1}{2}(d^2-1)$ such components (where $d=\dim\widetilde{\mathrm{Kh}}(K)$) and each of them contributes at least 8 to the expression above. So the overall contribution of special components of $\gamma^{\mathrm{Kh}}_{T_K}$ to $\dim\widetilde{\mathrm{Kh}}(C_{2,2t+1}(K))$ is $2(d^2-1)$, i.e. the first two summands in Proposition 1.15. The remaining summand is equal to the contribution of the rational component of $\gamma^{\mathrm{Kh}}_{T_K}$.

Acknowledgments

This paper was motivated by joint work with Lukas Lewark [**LZ22**]. I am very grateful to him for his generous support over the past two years and for giving me the freedom to pursue my own research as his postdoc. I would also like to thank Jake Rasmussen, Liam Watson, and Paul Wedrich for helpful conversations about the behaviour of knot Floer and Khovanov homology under cabling. I thank Jen Hom and Biji Wong for their helpful questions and comments after a talk I gave about this work at the low-dimensional topology workshop in Budapest in March 2023, which motivated me to explore some further applications of the Main Theorem and which allowed me to improve the exposition. Finally, I would like to thank the anonymous referee for their detailed comments.

References

[BGH21] Steven Boyer, Cameron McA Gordon, and Ying Hu, *Slope detection and toroidal 3-manifolds*, 2021, `arXiv:2106.14378`.

[BS15] Joshua Batson and Cotton Seed, *A link-splitting spectral sequence in Khovanov homology*, Duke Math. J. **164** (2015), no. 5, 801–841, DOI 10.1215/00127094-2881374. MR3332892

[Dey19] Subhankar Dey, *Cable knots are not thin*, 2019, `arXiv:1904.11591`.

[Ghi08] Paolo Ghiggini, *Knot Floer homology detects genus-one fibred knots*, Amer. J. Math. **130** (2008), no. 5, 1151–1169, DOI 10.1353/ajm.0.0016. MR2450204

[Han19] Jonathan Hanselman, *github repository for immersed curve invariants of knot complements*, 2019, `https://github.com/hanselman/CFK-immersed-curves/`.

[HRW16] Jonathan Hanselman, Jacob Rasmussen, and Liam Watson, *Bordered Floer homology for manifolds with torus boundary via immersed curves*, 2016, `arXiv:1604.03466`.

[HRW22] Jonathan Hanselman, Jacob Rasmussen, and Liam Watson, *Heegaard Floer homology for manifolds with torus boundary: properties and examples*, Proc. Lond. Math. Soc. (3) **125** (2022), no. 4, 879–967, DOI 10.1112/plms.12473. MR4500201

[HW15] Jonathan Hanselman and Liam Watson, *A calculus for bordered Floer homology*, Geom. Topol. **27** (2023), no. 3, 823–924, DOI 10.2140/gt.2023.27.823. MR4599308

[HW18] Matthew Hedden and Liam Watson, *On the geography and botany of knot Floer homology*, Selecta Math. (N.S.) **24** (2018), no. 2, 997–1037, DOI 10.1007/s00029-017-0351-5. MR3782416

[HW19] Jonathan Hanselman and Liam Watson, *Cabling in terms of immersed curves*, Geom. Topol. **27** (2023), no. 3, 925–952, DOI 10.2140/gt.2023.27.925. MR4599309

[KWZ19] Artem Kotelskiy, Liam Watson, and Claudius Zibrowius, *Immersed curves in Khovanov homology*, 2019, `arXiv:1910.14584`.

[KWZ20] Artem Kotelskiy, Liam Watson, and Claudius Zibrowius, *A mnemonic for the Lipshitz-Ozsváth-Thurston correspondence*, 2020, `arXiv:2005.02792`, to appear in Algebr. Geom. Topol.

[KWZ21] Artem Kotelskiy, Liam Watson, and Claudius Zibrowius, *Thin links and Conway spheres*, 2021, arXiv:2105.06308.

[KWZ22a] Artem Kotelskiy, Liam Watson, and Claudius Zibrowius, *Khovanov homology and strong inversions*, Gauge theory and low-dimensional topology—progress and interaction, Open Book Ser., vol. 5, Math. Sci. Publ., Berkeley, CA, 2022, pp. 223–244. MR4600725

[KWZ22b] Artem Kotelskiy, Liam Watson, and Claudius Zibrowius, *Khovanov multicurves are linear*, 2022, arXiv:2202.01460.

[LMZ22] Tye Lidman, Allison H. Moore, and Claudius Zibrowius, *L-space knots have no essential Conway spheres*, Geom. Topol. **26** (2022), no. 5, 2065–2102, DOI 10.2140/gt.2022.26.2065. MR4520302

[LOT18] Robert Lipshitz, Peter S. Ozsvath, and Dylan P. Thurston, *Bordered Heegaard Floer homology*, Mem. Amer. Math. Soc. **254** (2018), no. 1216, viii+279, DOI 10.1090/memo/1216. MR3827056

[LZ22] Lukas Lewark and Claudius Zibrowius, *Rasmussen invariants of Whitehead doubles and other satellites*, 2022, arXiv:2208.13612.

[OS04] Peter Ozsváth and Zoltán Szabó, *Holomorphic disks and genus bounds*, Geom. Topol. **8** (2004), 311–334, DOI 10.2140/gt.2004.8.311. MR2023281

[OS05a] Peter Ozsváth and Zoltán Szabó, *On knot Floer homology and lens space surgeries*, Topology **44** (2005), no. 6, 1281–1300, DOI 10.1016/j.top.2005.05.001. MR2168576

[OS05b] Peter Ozsváth and Zoltán Szabó, *On the Heegaard Floer homology of branched double-covers*, Adv. Math. **194** (2005), no. 1, 1–33, DOI 10.1016/j.aim.2004.05.008. MR2141852

[OS17] Peter Ozsváth and Zoltán Szabó, *Knot Floer homology calculator*, https://web.math.princeton.edu/~szabo/HFKcalc.html, 2017.

[OSS15] Peter S. Ozsváth, András I. Stipsicz, and Zoltán Szabó, *Grid homology for knots and links*, Mathematical Surveys and Monographs, vol. 208, American Mathematical Society, Providence, RI, 2015, DOI 10.1090/surv/208. MR3381987

[PW21] Ina Petkova and Biji Wong, *Twisted Mazur pattern satellite knots & bordered Floer theory*, Michigan Math. J. **73** (2023), no. 2, 255–304, DOI 10.1307/mmj/20205927. MR4584863

[RR17] Jacob Rasmussen and Sarah Dean Rasmussen, *Floer simple manifolds and L-space intervals*, Adv. Math. **322** (2017), 738–805, DOI 10.1016/j.aim.2017.10.014. MR3720808

[Shu04] Alexander N. Shumakovitch, *Torsion of Khovanov homology*, Fund. Math. **225** (2014), no. 1, 343–364, DOI 10.4064/fm225-1-16. MR3205577

[Zar11] Rumen Zarev, *Bordered Sutured Floer Homology*, ProQuest LLC, Ann Arbor, MI, 2011. Thesis (Ph.D.)–Columbia University. MR2941830

[Zib18] Claudius Zibrowius, *PQM.m*, 2018, Mathematica package for computing the Heegaard Floer multicurve invariant for Conway tangles, available from the author's website.

[Zib19a] Claudius Bodo Zibrowius, *Kauffman states and Heegaard diagrams for tangles*, Algebr. Geom. Topol. **19** (2019), no. 5, 2233–2282, DOI 10.2140/agt.2019.19.2233. MR4023317

[Zib19b] Claudius Zibrowius, *On symmetries of peculiar modules; or, δ-graded link Floer homology is mutation invariant*, 2019, arXiv:1909.04267v2, to appear in J. Eur. Math. Soc.

[Zib20] Claudius Zibrowius, *Peculiar modules for 4-ended tangles*, J. Topol. **13** (2020), no. 1, 77–158, DOI 10.1112/topo.12120. MR3999673

[Zib21] Claudius Zibrowius, *kht++, a program for computing Khovanov invariants for links and tangles*, https://cbz20.raspberryip.com/code/khtpp/docs/, 2021.

[Zib22] Claudius Zibrowius, *python script domains.py, ancillary file to this article*, 2022, available at https://cbz20.raspberryip.com/code/ancillary/HF2HFT/domains.py.

[Zib23] Claudius Zibrowius, *python script hf2hft.py, ancillary file to this article*, 2023, available at https://cbz20.raspberryip.com/code/ancillary/HF2HFT/hf2hft.py.

Department of Mathematical Sciences, Durham University, United Kingdom
Email address: cbz20@posteo.net
URL: https://cbz20.raspberryip.com/

Selected Published Titles in This Series

109 **Paul M. N. Feehan, Lenhard L. Ng, and Peter S. Ozsváth, Editors,** Frontiers in Geometry and Topology, 2024

107 **Ron Donagi, Adrian Langer, Piotr Sułkowski, and Katrin Wendland, Editors,** String-Math 2022, 2024

106 **Gilles Godefroy, Mohammad Sal Moslehian, and Juan Benigno Seoane-Sepúlveda, Editors,** Geometry of Banach Spaces and Related Fields, 2024

105 **A. Connes, C. Consani, B. I. Dundas, M. Khalkhali, and H. Moscovici, Editors,** Cyclic Cohomology at 40: Achievements and Future Prospects, 2023

104 **A. Kechris, N. Makarov, D. Ramakrishnan, and X. Zhu, Editors,** Nine Mathematical Challenges, 2021

103 **Sergey Novikov, Igor Krichever, Oleg Ogievetsky, and Senya Shlosman, Editors,** Integrability, Quantization, and Geometry, 2021

102 **David T. Gay and Weiwei Wu, Editors,** Breadth in Contemporary Topology, 2019

101 **Avraham Aizenbud, Dmitry Gourevitch, David Kazhdan, and Erez M. Lapid, Editors,** Representations of Reductive Groups, 2019

100 **Chiu-Chu Melissa Liu and Motohico Mulase, Editors,** Topological Recursion and its Influence in Analysis, Geometry, and Topology, 2018

99 **Vicente Muñoz, Ivan Smith, and Richard P. Thomas, Editors,** Modern Geometry, 2018

98 **Amir-Kian Kashani-Poor, Ruben Minasian, Nikita Nekrasov, and Boris Pioline, Editors,** String-Math 2016, 2018

97 **Tommaso de Fernex, Brendan Hassett, Mircea Mustaţă, Martin Olsson, Mihnea Popa, and Richard Thomas, Editors,** Algebraic Geometry: Salt Lake City 2015 (Parts 1 and 2), 2018

96 **Si Li, Bong H. Lian, Wei Song, and Shing-Tung Yau, Editors,** String-Math 2015, 2017

95 **Izzet Coskun, Tommaso de Fernex, and Angela Gibney, Editors,** Surveys on Recent Developments in Algebraic Geometry, 2017

94 **Mahir Bilen Can, Editor,** Algebraic Groups: Structure and Actions, 2017

93 **Vincent Bouchard, Charles Doran, Stefan Méndez-Diez, and Callum Quigley, Editors,** String-Math 2014, 2016

92 **Kailash C. Misra, Daniel K. Nakano, and Brian J. Parshall, Editors,** Lie Algebras, Lie Superalgebras, Vertex Algebras and Related Topics, 2016

91 **V. Sidoravicius and S. Smirnov, Editors,** Probability and Statistical Physics in St. Petersburg, 2016

90 **Ron Donagi, Sheldon Katz, Albrecht Klemm, and David R. Morrison, Editors,** String-Math 2012, 2015

89 **D. Dolgopyat, Y. Pesin, M. Pollicott, and L. Stoyanov, Editors,** Hyperbolic Dynamics, Fluctuations and Large Deviations, 2015

88 **Ron Donagi, Michael R. Douglas, Ljudmila Kamenova, and Martin Rocek, Editors,** String-Math 2013, 2014

87 **Helge Holden, Barry Simon, and Gerald Teschl, Editors,** Spectral Analysis, Differential Equations and Mathematical Physics: A Festschrift in Honor of Fritz Gesztesy's 60th Birthday, 2013

86 **Kailash C. Misra, Daniel K. Nakano, and Brian J. Parshall, Editors,** Recent Developments in Lie Algebras, Groups and Representation Theory, 2012

85 **Jonathan Block, Jacques Distler, Ron Donagi, and Eric Sharpe, Editors,** String-Math 2011, 2012

For a complete list of titles in this series, visit the
AMS Bookstore at **www.ams.org/bookstore/pspumseries/**.